INTERNATIONAL SERIES OF MONOGRAPHS IN
NATURAL PHILOSOPHY
GENERAL EDITOR: D. TER HAAR

VOLUME 61

PROBLEMS IN ELECTRONICS

PROBLEMS IN ELECTRONICS

(INCLUDING LUMPED CONSTANTS, TRANSMISSION LINES AND HIGH FREQUENCIES)

J. AUVRAY and M. FOURRIER

University of Paris

Translated by A. J. POINTON

Preface by

J. UEBERSFELD

Professor of Electronics
in the Faculty of Science at Paris

PERGAMON PRESS

OXFORD · NEW YORK · TORONTO
SYDNEY · BRAUNSCHWEIG

Pergamon Press Ltd., Headington Hill Hall, Oxford
Pergamon Press Inc., Maxwell House, Fairview Park, Elmsford, New York 10523
Pergamon of Canada Ltd., 207 Queen's Quay West, Toronto 1
Pergamon Press (Aust.) Pty. Ltd., 19a Boundary Street, Rushcutters Bay, N.S.W. 2011, Australia
Vieweg & Sohn GmbH, Burgplatz 1, Braunschweig

French editions 1967

First English edition 1973

Library of Congress Cataloging in Publication Data

Auvray, Jean.
Problems in electronics.

(International series of monographs in natural philosophy, v. 61)
An updated translation of the ed. published in 1967 under title: Problèmes d'électronique.
1. Electronics--Problems, exercises, etc.
I. Fourrier, Michel, joint author. II. Title.
TK7862.A8813 1973 621.3815'076 73-7617
ISBN 0-08-016982-1

Printed in Hungary

Contents

Preface

IT IS a great pleasure to present this excellent collection of problems written by J. Auvray and M. Fourrier who are at present lecturers in the Faculty of Science at Paris, one in charge of practical work and the other in charge of the work for the Certificate in Electronics.

This book is the result of a close co-operation between the authors. It is evidence for the value of the collaboration which is necessary, in all fields, between those responsible for practical work and those responsible for theoretical instruction and this is particularly the case in electronics. It has given me great pleasure to be present at their discussions and so to help establish this close relationship between the two areas of instruction.

It is well known that electronics is not learnt in the lecture theatre; that only the basic ideas, the mathematics and physics of electronics may be learnt from a lecture course. The practical details and the building of actual circuits can only be undertaken in the laboratory. The student must be shown how the basic ideas may be used in actual calculations and how these calculations lead to practical circuits. Thus, exercises and problems play a double role; to aid the understanding of the basic theory and to lead on to circuit design and construction. The choice of problems to satisfy these conditions is often difficult. I think I can say that, in this book, Messrs Auvray and Fourrier, because of their experience in both electronics and teaching, have overcome these difficulties.

The universality of electronics and the variety of the possibilities open to electronic techniques provide another obstacle for the authors of a collection of problems. The aim must be, not so much to illustrate exhaustively all the immense variety of applications, but to bring out the unity of the theoretical methods as applied to different systems. Here also the authors have succeeded. The exercises and problems bear directly on the essentials and foundations of electronics. Among these I would only mention those applied to high frequencies where the need for such exercises has made itself particularly felt.

This book of Auvray and Fourrier will be useful to a large number of students, because it has been prepared especially for the Certificate in Electronics which forms a part of the new masters course. It will also be useful, however, to all those research workers and engineers who work with electronics either directly or as an instrument of research.

I wish this splendid work the success which it deserves.

J. Uebersfeld

Professor of Electronics in the
Faculty of Science at Paris

Introduction

The French edition of this book has appeared in the series "exercises et problèmes de maîtrise" published by Dunod. It reflected the philosophy behind this series in giving, in the first part, exercises designed to illustrate the course material and, in the second, problems on the examination subjects.

A problem in electronics is the theoretical analysis of an actual system or circuit designed to fulfil a number of practical functions. We have tried to illustrate the study of these functions and the means by which they can be realized with passive and active components which modern solid state physics has made available. The electronic valve has tended to disappear except at the extremes of power and frequency and they have therefore been left out except for one or two cases which have particular intrinsic interest.

The breadth of the subject meant a drastic limitation of choice if the book was not to be unreasonably long. We have tried to illustrate general methods which are applicable in the largest number of cases while remaining as down to earth as possible, taking in particular those circuits which can be found easily in the laboratory.

In the parts dealing with lumped elements, the evolution of electronic techniques since the first appearance of this text, and in particular the growth in importance of integrated circuits as their price has decreased, has required a fundamental revision of certain chapters. In addition, the methods of synthesizing passive and active circuits from active networks such as the gyrator or the negative-impedance-converter (NIC) could not be left out, nor could the present applications of operational amplifiers. Consequently a number of more classical examples have had to be sacrificed. This introduction to the more modern electronic systems as well as the importance which we have placed on the high frequency systems should ensure that this text will not overlap with the excellent collections of exercises which already exist.

In the second part, which is reserved for high frequencies, the concept of the reflection coefficient is dealt with at length, as is the practical use of the Smith Chart. It has also seemed imperative to consider in detail the problems of power transfer and to devote a chapter to the study of resonant cavities, of which the importance in physics and electronics is well known.

The final section consists of the detailed solutions of examination questions. Our experience has shown that, in the absence of explanations and intermediate steps, students often find it impossible to follow a solution. On the other hand it cannot be bad to show how a solution may be obtained. The knowledge of the final result is not always sufficient to allow an isolated student to piece together the complete solution, and it is with such students well in mind that we have prepared this text.

We are aware that, in its present state, this text may contain many errors and omissions and we would thank, in advance, all those who might send us their observations and corrections. Our thanks are also due to all those who have helped in the preparation of this book. First of all to Professor Uebersfeld who was originally responsible for the work being entrusted to the authors and who has been good enough to write the preface. The numerous discussions which we have had with him have been most useful in deciding both content and presentation and we wish to thank him warmly. We would also like to thank Professors Mezencev and Motchane who authorized the use of examination papers set in their departments and Miss Defaix who has accomplished the task of typing the manuscript in a charming and efficient manner. Finally we would express our pleasure at the agreeable manner in which we have been able to work with our publisher, Mr Dunod.

The new parts which have been introduced in the present edition have been taken from the numerous examples set to the students of the University of Paris over the past three years.

We are grateful for the honour which Pergamon Press have done us in publishing a translation of our book. Dr A. J. Pointon has undertaken the task of translation of the old and new material with care and clarity and we would wish to thank him here.

PART ONE

Exercises

Notation

Small letters i and v are used to represent variations of current and voltage about a mean working (or bias) position. The large letters I and V then represent the total current and voltage including the steady components. Thus,

$$i = \Delta I$$
$$v = \Delta V$$

i: in general the alternating component input current of a four-terminal network

v: in general the alternating component of the input voltage of a four-terminal network

i_g, v_g: grid current and voltage, or current and voltage supplied by a generator

v_c: a.c. collector voltage of a transistor

V_c: d.c. collector voltage of a transistor

Z: general impedance

Y: general admittance

R: pure resistance

G: pure conductance, $G = 1/R$. (Sometimes, when there is no ambiguity, the relation $Y = 1/R$ has been used.)

ϱ: internal resistance

ϱ_g: internal resistance of a generator

The hybrid parameters of a four-terminal network are h_{11}, h_{21}, h_{12}, h_{22}. In the case of a transistor we have: $h_{21i} = \beta$, $h_{22i} = 1/r_c$, $|h_{21B}| = \alpha$.

g: the conductance of a valve or of a field effect transistor.

μ: the amplification factor of a triode ($\mu = R_a g$)

G_i, G_v: the current and voltage gain respectively

φ: a phase angle

n: a pure number, often the number of turns on a transformer

s: the symbol for the operator $s = j\omega$

(In all which follows the term "frequency" is often used, incorrectly but as in common usage, for the pulsatance $\omega = 2\pi f$.)

θ: a temperature

P: a power

$R_1 // R_2$: $\dfrac{R_1 R_2}{R_1 + R_2}$ being the equivalent resistance of R_1 and R_2 in parallel

Chapter 1

Passive Circuits

Fourier and Laplace Transforms

EXERCISE No. 1

Bridged-T Network

1. Calculate the transfer matrix T given by

$$\begin{pmatrix} v_1 \\ i_1 \end{pmatrix} = (T) \begin{pmatrix} v_2 \\ -i_2 \end{pmatrix}$$

for a T network as shown in Fig. 1 starting from the transfer matrices of the

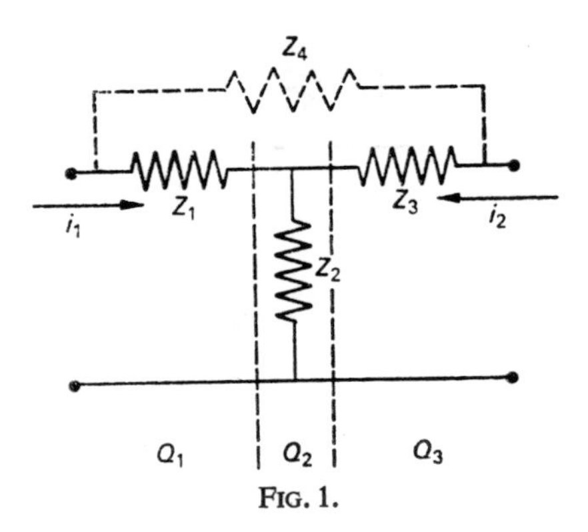

FIG. 1.

elementary networks Q_1, Q_2, and Q_3. Hence deduce the admittance matrix Y. (Put: $S = Z_1Z_2+Z_2Z_3+Z_3Z_1$.)

2. An impedance Z_4 is placed in parallel so as to form a bridged-T network. Calculate the matrix Y for Z_4.

3. The four-terminal network so formed is supplied by a source of voltage u_0 and feeds current into a load Z_L. Calculate the input impedance v_1/i_1 and the gain

$$v_2/v_1 = v_2/u_0 = G_v.$$

What relation must exist between the elements of the network if $G_v = 0$ independent of Z_L (i.e. such that the bridge is balanced)?

Solution. 1. For the network Q_1:

$$i_1 = -i_2$$
$$v_1 = Z_1 i_1 + v_2 = -Z_1 i_2 + v_2$$

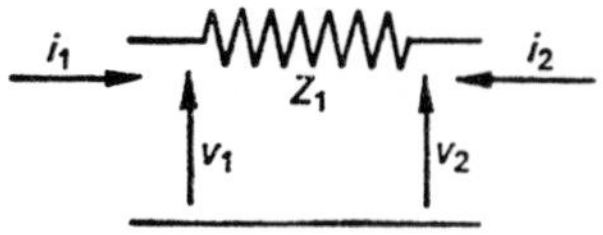

FIG. 2.

and hence

$$(T_1) = \begin{pmatrix} 1 & +Z_1 \\ 0 & +1 \end{pmatrix}.$$

Similarly

$$(T_3) = \begin{pmatrix} 1 & +Z_3 \\ 0 & +1 \end{pmatrix}$$

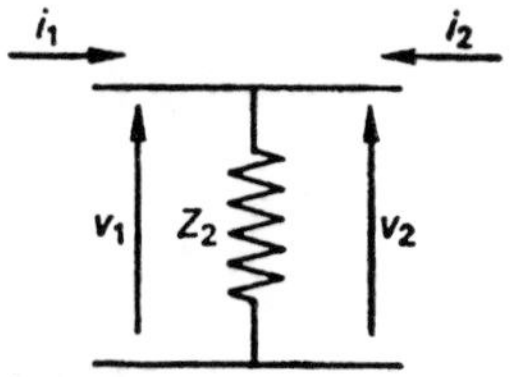

FIG. 3.

For Q_2:

$$\begin{cases} v_1 = v_2 \\ i_1 = \dfrac{v_1}{Z_2} - i_2 = \dfrac{v_2}{Z_2} - i_2, \end{cases}$$

so that

$$(T_2) = \begin{pmatrix} 1 & 0 \\ \dfrac{1}{Z_2} & +1 \end{pmatrix}.$$

The total matrix (T) is given by the product

$$(T) = \begin{pmatrix} 1 & Z_1 \\ 0 & 1 \end{pmatrix} \begin{pmatrix} 1 & 0 \\ \dfrac{1}{Z_2} & 1 \end{pmatrix} \begin{pmatrix} 1 & Z_3 \\ 0 & 1 \end{pmatrix} = \begin{pmatrix} 1 + \dfrac{Z_1}{Z_2} & Z_3 + Z_1 \left(1 + \dfrac{Z_3}{Z_2}\right) \\ \dfrac{1}{Z_2} & 1 + \dfrac{Z_3}{Z_2} \end{pmatrix}.$$

We now consider the matrix Y where

$$\begin{cases} v_1 = T_{11} v_2 - T_{12} i_2 \\ i_1 = T_{21} v_2 - T_{22} i_2 \end{cases} \tag{1}$$

$$\begin{cases} i_1 = Y_{11} v_1 + Y_{12} v_2 \\ i_2 = Y_{21} v_1 + Y_{22} v_2 \end{cases} \tag{2}$$

with $Y_{11} = (i_1/v_1)_{v_2=0}$. On putting $v_2 = 0$ in the equations (1) we find

$$Y_{11} = \frac{T_{22}}{T_{12}}$$

and similarly for the other terms. For example:

$$Y_{12} = -\frac{T_{11}T_{22}-T_{12}T_{21}}{T_{12}}.$$

For a passive network $|T| = 1$ and

$$Y_{12} = -\frac{1}{T_{12}}, \quad Y_{21} = -\frac{1}{T_{12}}, \quad Y_{22} = \frac{T_{11}}{T_{12}}.$$

From these equations we find for the passive network

$$(Y_1) = \frac{1}{T_{12}}\begin{pmatrix} T_{22} & -1 \\ -1 & T_{11} \end{pmatrix} = \frac{1}{S}\begin{pmatrix} Z_2+Z_3 & -Z_2 \\ -Z_2 & Z_1+Z_2 \end{pmatrix}.$$

2. The matrix Y for the network formed by Z_4 alone is

$$(Y_4) = \frac{1}{Z_4}\begin{pmatrix} 1 & -1 \\ -1 & 1 \end{pmatrix}$$

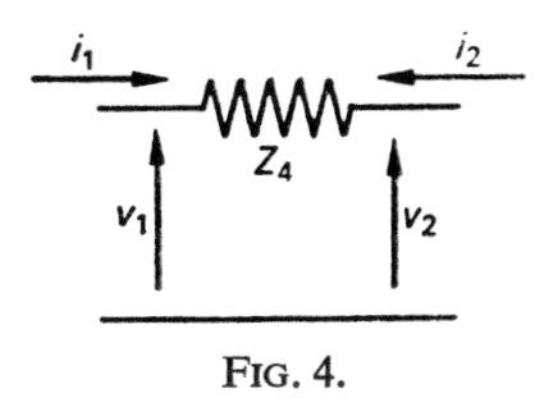

FIG. 4.

so that, for the T network formed with Z in parallel with the original network,

$$(Y) = (Y_1)+(Y_4) = \begin{pmatrix} \frac{1}{Z_4}+\frac{Z_2+Z_3}{S} & -\left(\frac{1}{Z_4}+\frac{Z_2}{S}\right) \\ -\left(\frac{1}{Z_4}+\frac{Z_2}{S}\right) & \frac{1}{Z_4}+\frac{Z_1+Z_2}{S} \end{pmatrix}.$$

3. The classical expressions for gain and input impedance give

$$G_v = \frac{v_2}{v_1} = \frac{-Y_{21}}{Y_{22}+Y_L} = \frac{S+Z_2Z_4}{S+Z_4Z_1+Z_4Z_2+\frac{SZ_4}{Z_L}},$$

and

$$Y_i = \frac{1}{Z_i} = Y_{11} - \frac{Y_{12}Y_{21}}{Y_{22}+Y_L}.$$

It is clear that G_v will be zero if $S+Z_2Z_4 = 0$ or

$$Z_1Z_2+Z_2Z_3+Z_3Z_1+Z_2Z_4 = 0.$$

EXERCISE No. 2

What is the total equivalent inductance of two coils placed in series with individual self-inductance L_1 and L_2 and mutual inductance M?

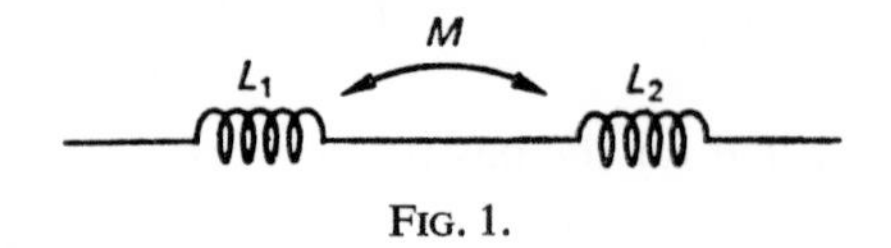

FIG. 1.

Solution. The voltage v_1 across L_1 consists of two terms: $j\omega L_1 i$ due to the current i in L_1 and $j\omega Mi$ due to the current in L_2. Thus

$$v_1 = j\omega(L_1+M)i$$

and, with a similar expression for v_2 across L_2,

$$v = v_1+v_2 = j\omega i(L_1+L_2+2M) = j\omega Li$$

so that

$$L = (L_1+L_2+2M).$$

EXERCISE No. 3

Impedance Matching by Autotransformer

What is the input admittance of the circuit shown in Fig. 1 if the magnetic coupling can be assumed to be perfect?

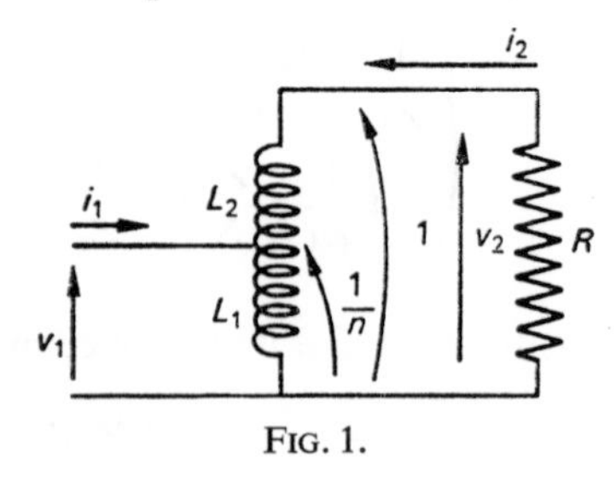

FIG. 1.

(Express the admittance in terms of $1/R$ and the transformer ratio n.)

Solution. The voltage across each section of the transformer winding is the sum of two terms, one due to the self inductance of the section and the other due to its mutual inductance with the remaining section. Thus

$$v_1 = j\omega L_1(i_1+i_2)+j\omega M i_2,$$

while

$$v_2 = j\omega L_2 i_2+j\omega M(i_1+i_2)+v_1,$$

so that

$$v_2 = j\omega L_2 i_2+j\omega M(i_1+i_2)+j\omega L_1(i_1+i_2)+j\omega M i_2.$$

Also

$$v_2 = -Ri_2.$$

Eliminating v_2 and i_2 between these equations then gives the input admittance as

$$\frac{i_1}{v_1} = \frac{1}{j\omega L_1}+\frac{L_1+L_2+2M}{L_1}\cdot\frac{1}{R}.$$

Now the inductance of a given section of winding is proportional to the square of the number of turns on that section, so that

$$L_1 = a(1/n)^2$$

$$L_2 = a\left(1-\frac{1}{n}\right)^2$$

$$M = \sqrt{L_1 L_2} = a\frac{1}{n}\left(1-\frac{1}{n}\right)$$

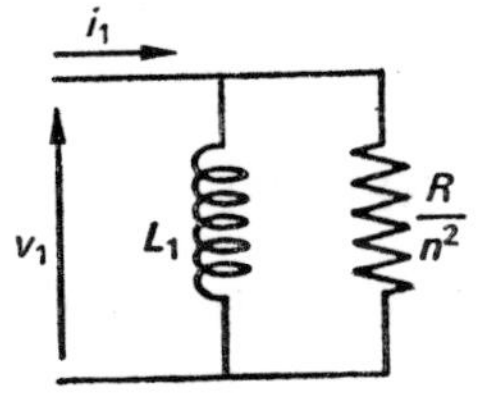

FIG. 2.

where a is the coefficient of proportionality. Then

$$\frac{i_1}{v_1} = \frac{1}{j\omega L_1}+\frac{n^2}{R}$$

which is the admittance of a self-inductance L_1 in parallel with a resistance R/n^2.

EXERCISE No. 4

Calculate the y parameters of the four terminal network formed by a transformer with coupling coefficient k, primary inductance L_1 and secondary inductance L_2.

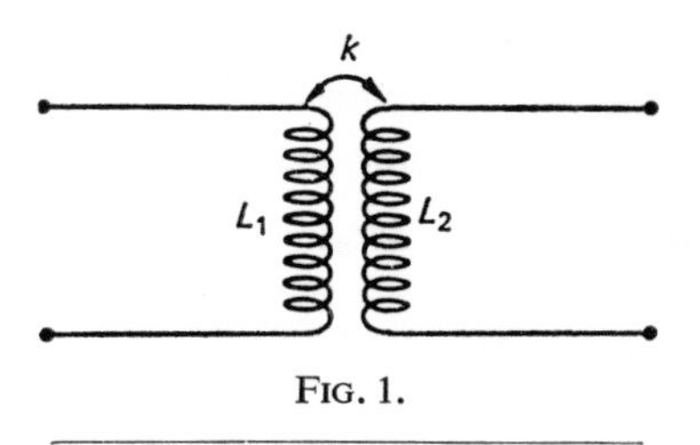

FIG. 1.

Solution. If the output is short circuited, the two equations of the transformer give immediately

$$y_{11} = \left(\frac{i_1}{v_1}\right)_{v_2=0} \quad \text{and} \quad y_{21} = \left(\frac{i_2}{v_1}\right)_{v_2=0}$$

from which

$$y_{11} = \frac{1}{j\omega L_1(1-k^2)}$$

$$y_{21} = y_{12} = \frac{-jk}{\omega\sqrt{L_1L_2}(1-k^2)}$$

and by symmetry,

$$y_{22} = \frac{1}{j\omega L_2(1-k^2)}.$$

EXERCISE No. 5

Using the results of the previous exercise and the formulae for associated four terminal networks, calculate the y parameters for the network shown in Fig. 1, if the coupling coefficient is k.

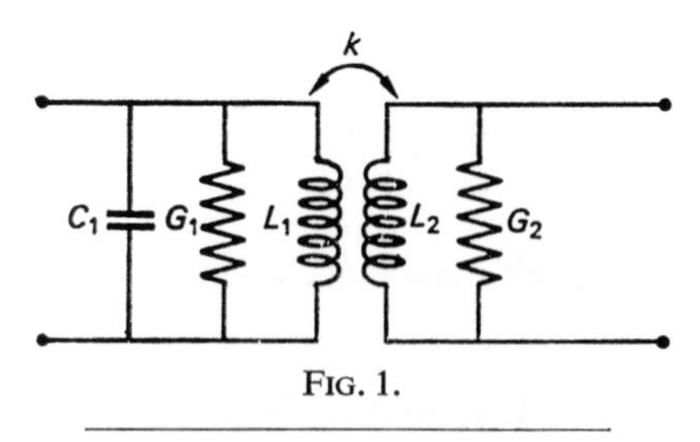

FIG. 1.

Solution. If admittances Y_1 and Y_2 are placed in parallel with the input and output respectively of a four terminal network, the y parameters of the composite network will be

$$\begin{cases} Y_{11} = y_{11}+Y_1 \\ Y_{12} = y_{12} \\ Y_{21} = y_{21} \\ Y_{22} = y_{22}+Y_2. \end{cases}$$

where y_{11} etc. are the y-parameters of the original network.

Here we have

$$Y_1 = G_1+j\omega C_1$$
$$Y_2 = G_2,$$

so that

$$Y_{11} = \frac{1}{j\omega L_1(1-k^2)}+G_1+j\omega C_1,$$

$$Y_{22} = \frac{1}{j\omega L_2(1-k^2)}+G_2,$$

$$Y_{21} = Y_{12} = \frac{jk}{\omega\sqrt{L_1L_2}(1-k^2)}.$$

EXERCISE No. 6

Consider that, in the transformer system shown in Fig. 1 below, the inductance between the points B and E is L, that the transformer is perfect and that any damping is negligible. Hence calculate the pass band of the system at the 3 dB points and the gain v_2/u_0 of the system at its matching frequency.

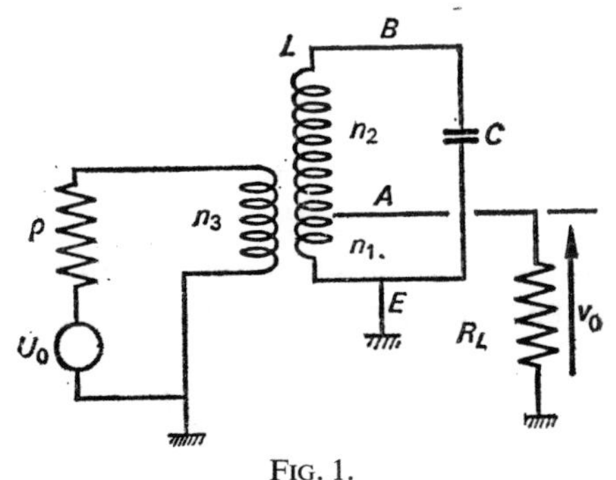

FIG. 1.

Take n_1 as the number of turns between A and E and n_2 between B and E.

Solution. If all the impedances are considered to be across the terminals of C, the equivalent circuit will be as in Fig. 2.

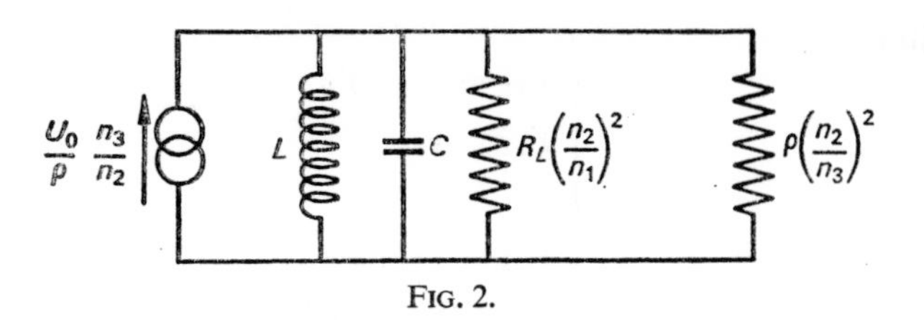

FIG. 2.

From this circuit the bandwidth is seen to be

$$B = \frac{f_0}{Q} = f_0 G L \omega_0 = \frac{G}{2\pi C} = \frac{1}{2\pi C}\left[\frac{1}{R_L}\left(\frac{n_1}{n_2}\right)^2 + \frac{1}{\varrho}\left(\frac{n_3}{n_2}\right)^2\right]$$

where G is the total parallel conductance.

The useful gain is

$$\frac{v_2}{u_0} = \left(\frac{n_3}{n_2}\cdot\frac{1}{\varrho}\right)\cdot\left(\frac{1}{\frac{1}{R_L}\left(\frac{n_1}{n_2}\right)^2 + \frac{1}{\varrho}\left(\frac{n_3}{n_2}\right)^2}\right)\frac{n_1}{n_2}.$$

EXERCISE No. 7

Given an impedance

$$Z(j\omega) = \frac{R_0}{\omega_0}\cdot\frac{11\omega_0^2 + 11 j\omega_0\omega - 6\omega^2}{\omega_0 + 2j\omega}$$

express this impedance in a normalized form $z(s)$ by putting

$$x = \frac{\omega}{\omega_0}, \quad s = jx \quad \text{and} \quad z(s) = \frac{Z(s)}{R_0}.$$

Using only passive elements, construct a two terminal network having this impedance.

Solution. With the chosen normalization, the unit of resistance is R_0, the unit of inductance will thus be R_0/ω_0 and that of capacity $1/R_0\omega_0$. Now, in terms of s and R_0

$$z(s) = \frac{6s^2 + 11s + 11}{2s + 1},$$

which can be written in the form $As + B + C/(2s+1)$ so that

$$z(s) = 3s + 4 + \frac{1}{\frac{2s}{7} + \frac{1}{7}}.$$

These three terms correspond to a series connection of (i) a reduced inductance of value 3 (i.e. $L = 3R_0/\omega_0$); (ii) a reduced resistance of value 4 (i.e. $R_1 = 4R_0$) and (iii) a parallel RC combination with $R_2 = 7R_0$ and $C = 2/7R_0\omega_0$. The circuit is thus that shown in Fig. 1.

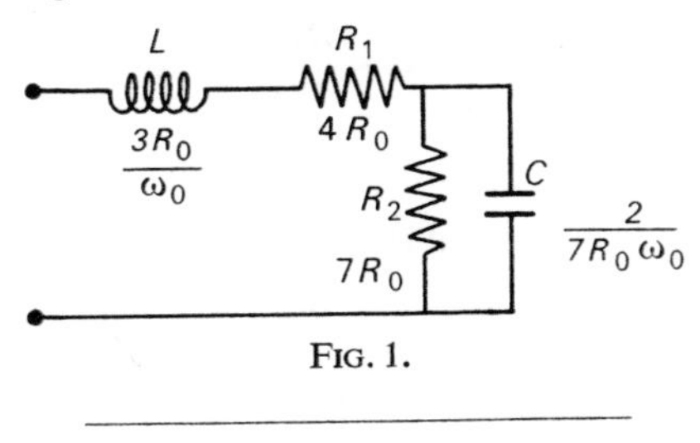

FIG. 1.

EXERCISE No. 8

Descending Cauer Synthesis

Show how the normalized impedance of a two terminal network

$$z(s) = \frac{10s^3+47s^2+14s+15}{10s^2+17s+8}$$

could be obtained by means of a passive ladder circuit.

Solution. It is required to form a ladder circuit as indicated in Fig. 1 below, for which the input impedance is

$$z = z_1 + \cfrac{1}{y_1 + \cfrac{1}{z_2 + \cfrac{1}{y_2 + \cfrac{1}{z_3 + \ldots}}}}$$

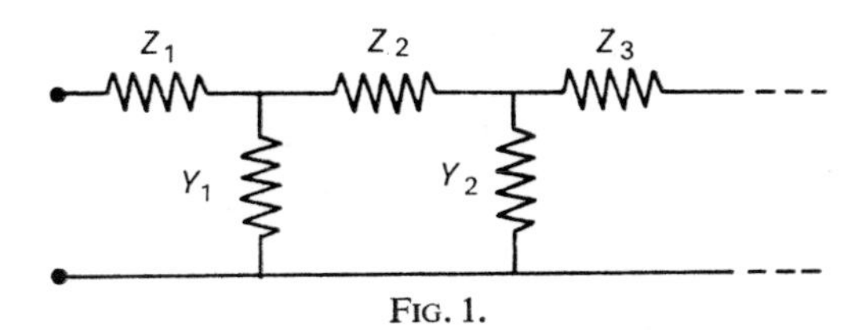

FIG. 1.

Now we can write

$$z(s) = N(s)/D(s),$$

with

$$N(s) = Q_1(s)\, D(s) + R_1(s),$$

$Q_1(s)$ being the quotient and $R_1(s)$ the remainder.

Then

$$z(s) = Q_1(s) + \frac{R_1(s)}{D(s)},$$

which can be written as

$$z(s) = Q_1(s) + \frac{1}{\dfrac{D(s)}{R_1(s)}}.$$

Dividing $D(s)$ by $R(s)$ in a similar procedure gives

$$D(s) = Q_2(s)\, R_1(s) + R_2(s)$$

and

$$z(s) = Q_1(s) + \cfrac{1}{Q_2(s) + \cfrac{1}{\dfrac{R_1(s)}{R_2(s)}}}.$$

The successive quotients $Q_1(s)$, $Q_2(s)$ etc. can be identified as z_1, y_1 etc. The two polynomials for $z(s)$ can then be separated by replacing, after each step of the division, the dividend or numerator by the preceding divisor and the divisor by the preceding remainder. Then, the first division gives:

$$\begin{array}{r|l} 10s^3+47s^2+14s+15 & 10s^2+17s+8 \\ \hline R_1(s) \to 30s^2+\ 6s+15 & s \to Q_1(s) = z_1(s) \end{array}.$$

Thus $z_1(s) = s$ corresponds to an inductance $L = 1$.

Second division:

$$\begin{array}{r|l} 10s^2+17s+8 & 30s^2+6s+15 \quad \to R_1(s) \\ \hline R_2(s) \to 15s+3 & \frac{1}{3} \qquad\qquad\quad \to Q_2(s) \end{array}$$

$y_1(s) = \frac{1}{3}$ corresponding to a resistance $R = 3$.

Third division:

$$\begin{array}{r|l} 30s^2+6s+15 & 15s+3 \\ \hline 15 & 2s \end{array}$$

$z_2(s) = 2s$, an inductance $L = 2$.

Fourth division:

$$\begin{array}{r|l} 15s+3 & 15 \\ \hline 3 & s \end{array}$$

$y_2(s) = s$ corresponding to a capacity $C = 1$.

Fifth division

$$\begin{array}{c|c} 0 & 3 \\ \hline 15 & 5 \end{array}$$

$z_3(s) = 5$, a resistance $R = 5$
and, finally,

$$\begin{array}{c|c} 3 & 0 \\ \hline & \infty \end{array}$$

$y_3(s) = \infty$.

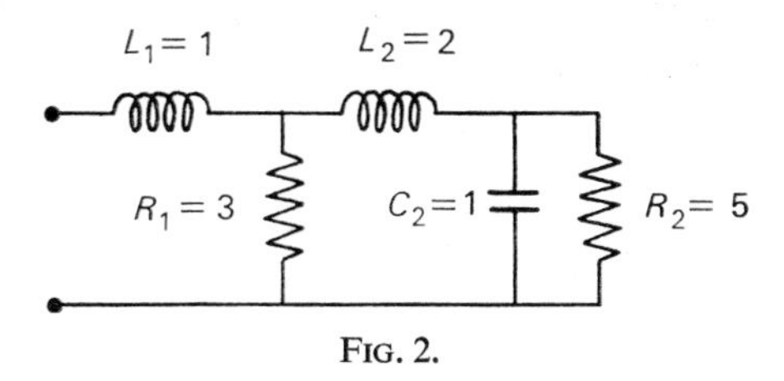

FIG. 2.

The complete ladder circuit is thus as shown in Fig. 2.

EXERCISE No. 9

By combining the methods of the ascending and descending Cauer syntheses, calculate the elements of a passive two terminal network, which has a normalized transfer function,

$$z(s) = \frac{20s^3 + 12s^2 + 31s + 7}{10s^3 + 21s^2 + 7s}.$$

Solution. The division of the one polynomial by the other when they are arranged in descending powers of s leads to a remainder having a negative term which cannot be synthesized by means of passive elements. The division is therefore begun with the polynomials arranged in ascending powers of s.

$$\begin{array}{r|l} 7 + 31s + 12s^2 + 20s^3 & 7s + 21s^2 + 10s^3 \\ \hline & \dfrac{1}{s} \\ 10s + 2s^2 + 20s^3 & \end{array} \qquad Q_1(s) = z_1(s)$$

$z_1(s)$ thus corresponds to a capacity $C = 1$.

The next step of the division with ascending powers again leads to a negative remainder. We therefore change to descending powers to give

$$\begin{array}{r|l} 10s^3 + 21s^2 + 7s & 20s^3 + 2s^2 + 10s \\ \hline 20s^2 + 2s & \frac{1}{2} \end{array}$$

$y_1 = \frac{1}{2}$ corresponding to a resistance $R = 2$.

$$\begin{array}{r|l} 20s^3+2s^2+10s & \underline{20s^2+2s} \\ 10s & s \end{array}$$

$z_2(s) = s$: $L = 1$.

$$\begin{array}{r|l} 20s^2+2s & \underline{10s} \\ 0 & 2s+\frac{1}{5} \end{array}$$

$y_2(s) \equiv$ capacity $C = 2$ in parallel with a resistance $R = 5$. Thus we have the complete circuit shown in Fig. 1

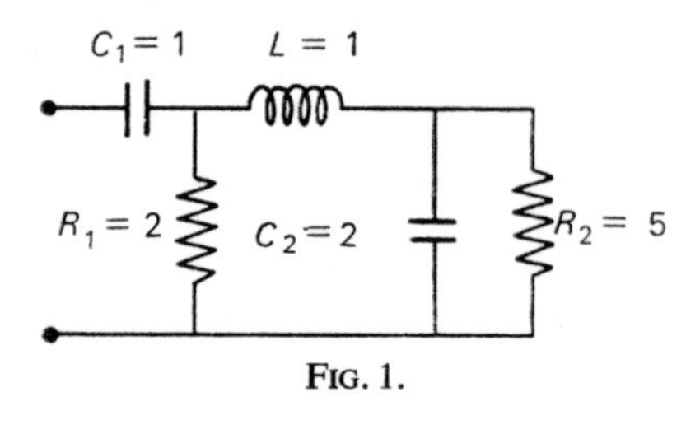

FIG. 1.

EXERCISE No. 10

The transfer function of a given filter is

$$G(\omega) = \frac{v_2(\omega)}{v_1(\omega)} = G_0 \frac{\omega_2}{\omega_1} \cdot \frac{\omega_1 + j\omega}{\omega_2 + j\omega} = |G(\omega)|\, e^{j\varphi(\omega)}.$$

Indicate the position of the poles and zeros of this function in the complex plane.

Trace the asymptotic curve for the gain in the form

$$\log |G(\omega)| = f(\log \omega)$$

in both cases

$$\omega_1 \ll \omega_2 \quad \text{and} \quad \omega_2 \ll \omega_1.$$

Using geometric reasoning find the frequency for which the phase angle $|\varphi(\omega)|$ is a maximum.

Solution. The function

$$G(s) = G_0 \frac{\omega_2}{\omega_1} \frac{s+\omega_1}{s+\omega_2}$$

has a zero at $s = -\omega_1$ and a pole at $s = -\omega_2$.

First case; $\omega_1 \ll \omega_2$

If $\omega \ll \omega_1$, then $\omega \ll \omega_2$ and $G(\omega)$ reduces to G_0, a constant.

If $\omega \gg \omega_1$ and $\omega \gg \omega_2$, then $(j\omega+\omega_1)/(j\omega+\omega_2) \to 1$ and G reduces to $G_0\ \omega_2/\omega^2$.

If $\omega_1 \ll \omega \ll \omega_2$

$$G(\omega) \simeq G_0 \frac{\omega_2}{\omega_1} \cdot \frac{j\omega}{\omega_2} = j\frac{G_0\omega}{\omega_1}.$$

The pole and zero are thus as shown at Z and P in Fig. 1 while the asymptotic

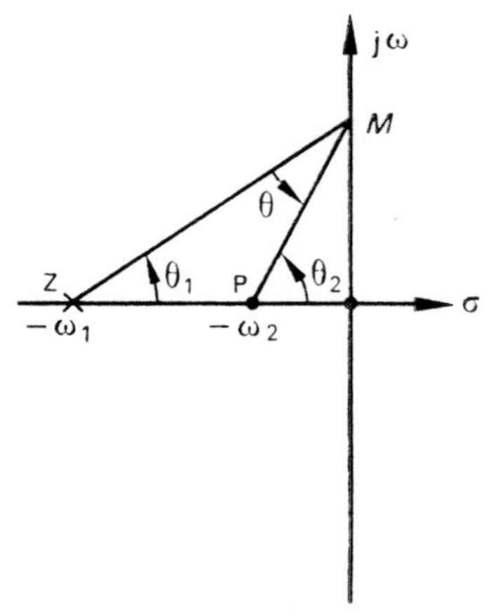

FIG. 1.

curve for the gain between ω_1 and ω_2 has a slope 1 (6 dB/oct) since

$$\log |G(\omega)| = \log \frac{G_0}{\omega_1} + \log \omega$$

the full line being shown in Fig. 2.

The phase angle is zero for $\omega \ll \omega_1$ or $\omega \gg \omega_2$ and approaches 90° for $\omega_1 \ll \omega \ll \omega_2$.

Second case; $\omega_2 \ll \omega_1$.

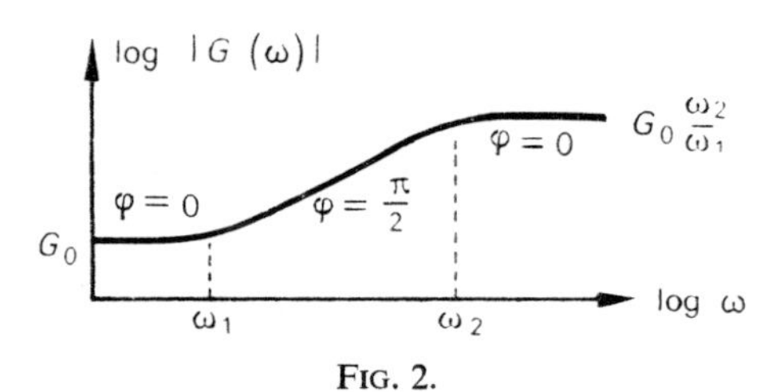

FIG. 2.

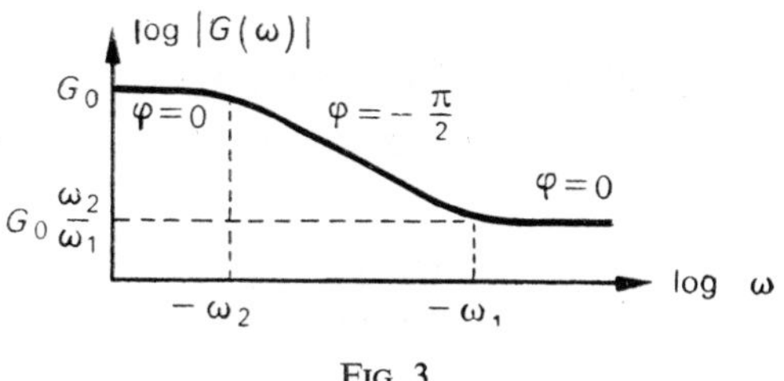

FIG. 3.

The curve is the inverse of that of Fig. 2 as is also the phase angle, the diagram being shown in Fig. 3.

If M is a point $j\omega$ on the imaginary axis, then $(j\omega+\omega_1)$ is a complex number represented by the vector $\boldsymbol{ZM}$ and $(j\omega+\omega_2)$ by the vector $\boldsymbol{PM}$, as seen in Fig. 1. The quotient of these two quantities has for argument the angle between the vectors

$$\theta = \theta_1 - \theta_2 = \widehat{\boldsymbol{PMZ}}.$$

We now look for the point on the imaginary axis at which the segment ZP subtends the maximum angle. If a circle Γ is drawn through Z and P it will, in general, cut the axis $j\omega$ in two points M_1 and M_2 as shown in Fig. 4.

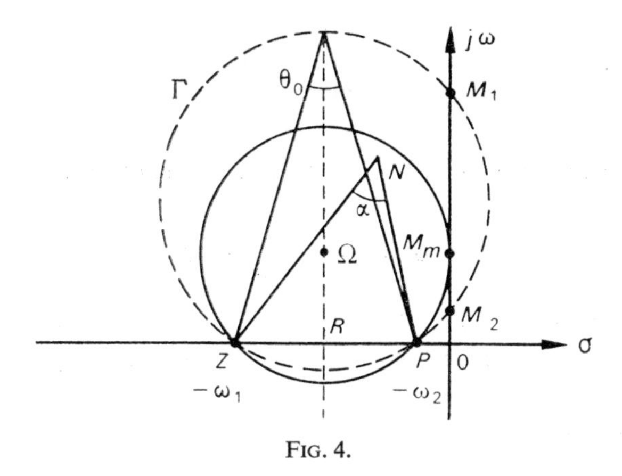

FIG. 4.

For both M_1 and M_2 the angle subtended is θ_0, and the angle subtended would be bigger for all points inside the circle Γ such as the point N for which $\alpha > \theta_0$. The limit to θ_0 is then clearly when the circle is tangential to $j\omega$ at M_m when $\theta_0 \to \theta_m$.

The centre of this circle is at Ω on the perpendicular bisector of ZP such that the circle radius $\Omega P = RO$.

EXERCISE No. 11

Determine the poles and zeros for the transfer function $G(s) = v_2(s)/v_1(s)$ for a circuit having an asymptotic curve for the gain as shown in Fig. 1.

Give an explicit form for this transfer function.

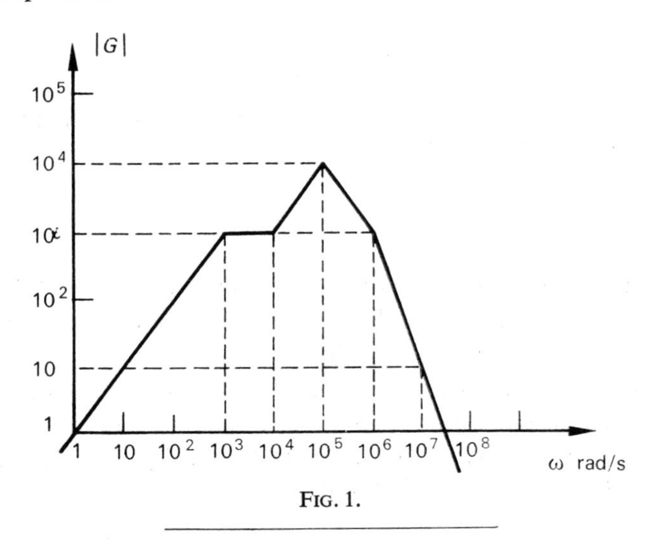

FIG. 1.

Solution. The slopes of the gain curve are, successively, +6 dB/oct, 0, +6 dB/oct, −6 dB/oct, −12 dB/oct. (In terms of decades these would be +20 dB/D, 0, 20 dB/D, −20 dB/D and −40 dB/D.) Now, as the frequency increases, the slope increases by 6 dB/oct each time the frequency passes a zero and decreases by the same amount each time that it passes a pole (see, for example, the preceding exercise). The asymptotic curve shown would thus be expected if there were:

a zero at the origin to give the initial slope
a pole at $\omega = 10^3$ which would compensate for the original zero and so render the gain constant
a zero at $\omega = 10^4$
a *double* pole at $\omega = 10^5$
a pole at $\omega = 10^6$.

The transfer function would therefore be of the form

$$G(j\omega) = K\frac{\omega(j\omega+10^4)}{(j\omega+10^3)(j\omega+10^5)^2(j\omega+10^6)}.$$

The coefficient K can be determined from the condition

$$|G(j\omega)| = 1 \quad \text{for} \quad \omega = 1$$

or, with the quantity $j\omega = j$ neglected in the brackets compared to the terms 10^3, 10^4 etc. to give $K = 10^{15}$, the transfer function may be written

$$G(s) = 10^{15}\frac{s(s+10^4)}{(s+10^3)(s^2+2\cdot 10^5 s+10^{10})(s+10^6)}.$$

EXERCISE No. 12

It is required to construct a circuit producing the minimum phase change but for which the gain varies with frequency as

$$|G(j\omega)|^2 = K\frac{\omega^2+1}{1-\omega^2+\omega^4}.$$

Mark on the complex plane the poles and zeros of $F(s^2) = |G(s)|^2$. Explain why, in the most general case, these poles and zeros can be grouped in fours at the corners of rectangles centred on the origin.

Suppose that $A(s)$ and $B(s)$ are two functions which have the poles and zeros of $F(s^2)$ which are situated respectively on the left and right of the imaginary axis. Write down a simple relation between $A(s)$ and $B(s)$, $A(j\omega)$ and $B(j\omega)$.

Calculate $|G(j\omega)|^2$ in terms of $|A(j\omega)|^2$ and hence deduce $G(j\omega)$.

Solution.

$$F(s^2) = K\cdot\frac{1-s^2}{1+s^2+s^4},$$

where $j\omega$ is replaced by s. This function has two zeros; $s = +1$ and $s = -1$. It has four poles, the roots of $(1+s^2+s^4) = 0$. These are:

$$P_1 = -\frac{1}{2}+j\frac{\sqrt{3}}{2}$$

$$P_2 = -\frac{1}{2}-j\frac{\sqrt{3}}{2}$$

$$P_3 = +\frac{1}{2}+j\frac{\sqrt{3}}{2}$$

$$P_4 = +\frac{1}{2}-j\frac{\sqrt{3}}{2}$$

which are the corners of a rectangle as seen in Fig. 1.

In the general case, if $(a+jb)$ is a pole (or zero) of $F(s^2)$, then its complex conjugate $(a-jb)$ will also be a pole (or zero), the coefficients of $F(s^2)$ necessarily

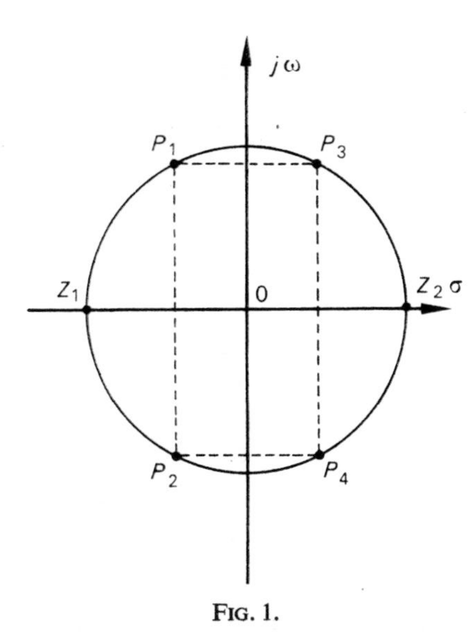

FIG. 1.

being real if this quantity is the square of an amplitude. Because of the parity of the function, that is the fact that $+s$ can be replaced by $-s$, the quantities $-(a+jb)$ and $-(a-jb)$ must equally be poles (or zeros). The co-ordinates $\pm(a\pm jb)$ then define the corners of a rectangle centred on the origin.

If, in the two functions $A(s)$ and $B(s)$, s is changed for $-s$ the poles and the zeros of $A(s)$ will evidently be interchanged for those of $B(s)$ and vice versa. Thus

$$B(s) = A(-s) \quad \text{or} \quad B(j\omega) = A(-j\omega).$$

However, for a physical transfer function,

$$A(-j\omega) = A^*(j\omega)$$

so that

$$B(j\omega) = A^*(j\omega).$$

Now we see that

$$|G(j\omega)|^2 = F(-\omega^2) = A(j\omega)\cdot B(j\omega)$$
$$= A(j\omega)A^*(j\omega) = |A(j\omega)|^2$$

and so we may write

$$G(j\omega) \equiv A(j\omega) \quad \text{or} \quad G(s) \equiv A(s).$$

In the present case, with zeros and poles Z_1, P_1 and P_2

$$A(s) = \frac{s+1}{\left(s+\frac{1}{2}-j\frac{\sqrt{3}}{2}\right)\left(s+\frac{1}{2}+j\frac{\sqrt{3}}{2}\right)} = \frac{s+1}{s^2+s+1}$$

EXERCISE No. 13

Calculate the transfer function for the circuit of Fig. 1 below and hence deduce the quality factor for a circuit for which the transfer function is

$$\frac{v_2(s)}{v_1(s)} = \frac{Ks}{\alpha s^2+\beta s+\gamma}.$$

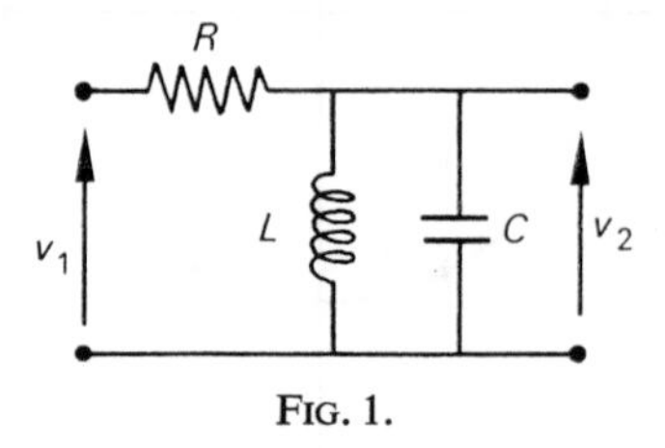

FIG. 1.

Solution. Writing the ratio of the voltages as the ratio of the impedances we see

$$\frac{v_2}{v_1} = \frac{s}{RCs^2+s+\dfrac{R}{L}} = \frac{1}{RC}\cdot\frac{s}{s^2+\dfrac{s}{RC}+\dfrac{1}{LC}}.$$

Putting $\omega_0^2 = 1/LC$ and $Q = R/\omega_0 L$ for the quality factor of the circuit (considered as a transmission system) this becomes

$$\frac{v_2}{v_1} = \frac{\omega_0}{Q}\cdot\frac{s}{s^2+\dfrac{\omega_0}{Q}\cdot s+\omega_0^2},$$

which is the required transfer function.

The constants can be identified by writing

$$\frac{v_2}{v_1} = \frac{K}{\alpha}\cdot\frac{s}{s^2+\dfrac{\beta}{\alpha}\cdot s+\dfrac{\gamma}{\alpha}}$$

$$= \frac{K}{\beta}\cdot\frac{\beta}{\alpha}\cdot\frac{s}{s^2+\dfrac{\beta}{\alpha}\cdot s+\dfrac{\gamma}{\alpha}}$$

so that

$$\frac{\beta}{\alpha} = \frac{\omega_0}{Q}, \qquad \frac{\gamma}{\alpha} = \omega_0^2$$

and

$$\omega_0 = \sqrt{\frac{\gamma}{\alpha}}, \qquad Q = \frac{\sqrt{\alpha\gamma}}{\beta}.$$

In this general case we see that there is, in effect, an amplification factor K/β which does not, however, change the form of the response curve.

EXERCISE No. 14

Any periodic signal with pulsatance ω can be decomposed into a spectrum with components ω, 2ω, 3ω etc. The total power is then the sum of the partial powers of the different harmonics.

A real amplifier will have a frequency-dependent gain such that, generally, the high harmonics of a signal will not be amplified. One may then assume that there will be little distortion of the amplified signal provided that some 98% of the incident energy is transmitted through the lower harmonics. With this condition, what must be the minimum pass-band of an amplifier required to receive at its input a periodic square wave voltage of the form

$$v(t) = 0 \quad \text{for} \quad 0 < t < T/2$$
$$= A \quad \text{for} \quad T/2 < t < T.$$

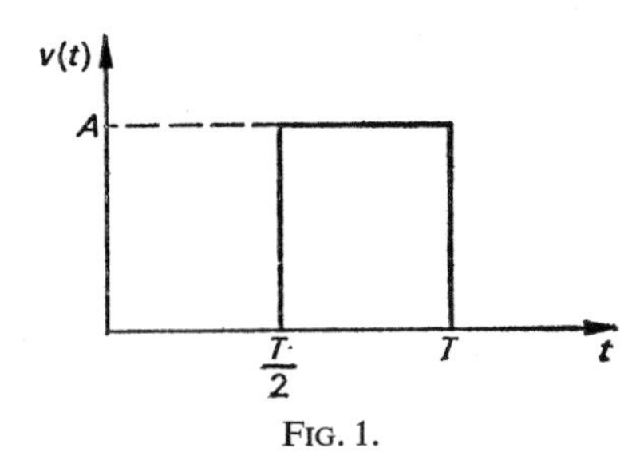

FIG. 1.

Solution. The signal can be expressed as a Fourier series as

$$v(t) = \frac{A}{2} - \frac{2A}{\pi}\left(\sin \omega t + \frac{1}{3}\sin 3\omega t + \frac{1}{5}\sin 5\omega t + \ldots\right).$$

The incident power is proportional to

$$P_0 = \frac{1}{T}\int_0^T v^2(t)\,dt = \frac{1}{T}\int_{T/2}^T A^2\,dt = \frac{A^2}{2}.$$

The general harmonic of the form

$$v_n = A_n \sin n\omega t,$$

carries an energy proportional to $A_n^2/2$ since A_n is the peak amplitude. The total incident power can be written in terms of the Fourier series as

$$P_0 = A^2\left(\frac{1}{4}+\frac{2}{\pi^2}\left(1+\frac{1}{9}+\frac{1}{25}+\ldots\right)\right)$$

If we put $S_m = \left(1+\frac{1}{3^2}+\frac{1}{5^2}+\ldots+\frac{1}{m^2}\right)$ then $P_m = A^2\left(\frac{1}{4}+\frac{2}{\pi^2}S_m\right)$ is the energy carried by the first m harmonics. We look for m such that, say,

$$P_m > 0{\cdot}98P_0$$

so that

$$\frac{1}{4}+\frac{2}{\pi^2}S_m > 0{\cdot}98\times\frac{1}{2},$$

or

$$S_m > 0{\cdot}24\cdot\frac{\pi^2}{2} = 1{\cdot}18308.$$

Now

$$S_1 = 1,$$

$$S_3 = 1+\frac{1}{3^2} = 1{\cdot}1111,$$

$$S_5 = 1+\frac{1}{3^2}+\frac{1}{5^2} = 1{\cdot}1511,$$

$$S_7 = 1{\cdot}1715,$$

$$S_9 = 1{\cdot}1838.$$

The amplifier must accept the 9th harmonic, so that its pass-band must be at least 9ω. (Generally the band is taken as $10\ \omega$.)

EXERCISE No. 15

A circuit has a response function

$$\frac{v_2}{v_1}(\omega) = \begin{cases} K\omega & \text{for} \quad \omega < \omega_0, \quad K \text{ real} \\ 0 & \text{for} \quad \omega > \omega_0 \end{cases}.$$

What will be the form of the output signal $v_2(t)$ if this circuit receives a rectangular input pulse of amplitude A and width τ centred on the origin?

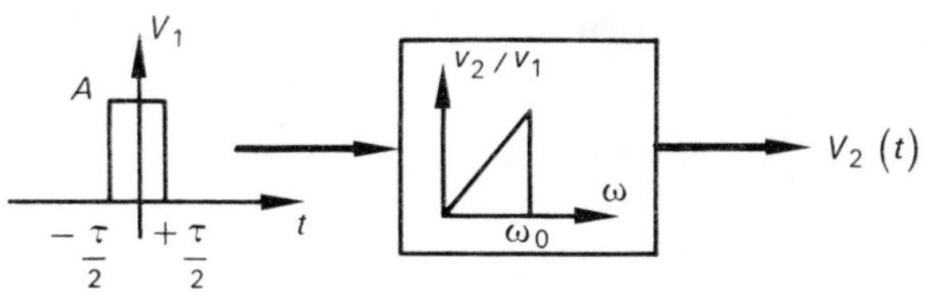

FIG. 1.

Solution. The process may be envisaged as in Fig. 1.

The Fourier transform $R(j\omega)$ of the response of a circuit is equal to the product of the Fourier transform of the input signal $F(j\omega)$ and the transfer function of the circuit $H(j\omega)$,

$$R(j\omega) = H(j\omega)\cdot F(j\omega).$$

For $F(j\omega)$ we have

$$F(j\omega) = \int_{-\infty}^{\infty} v_1(t)\, e^{-j\omega t}\, dt$$

which simplifies to

$$F(j\omega) = 2\int_{0}^{\infty} v_1(t) \cos \omega t\, dt,$$

since $v_1(t)$ is an even function of the time such that $v_1(t) = v_1(-t)$. Then

$$F(j\omega) = 2\int_{0}^{\tau/2} A \cos \omega t\, dt = \frac{2A}{\omega} \sin \frac{\omega\tau}{2}.$$

We have

$$\begin{aligned} H(j\omega) &= 0 \quad \text{for} \quad |\omega| > \omega_0 \\ &= K\omega \quad \text{for} \quad 0 < \omega < \omega_0 \\ &= -K\omega \quad \text{for} \quad -\omega_0 < \omega < 0, \end{aligned}$$

since a real transfer function is always symmetric about $\omega = 0$. Then

$$\begin{aligned} R(j\omega) &= 0 \quad \text{for} \quad |\omega| > \omega_0 \\ &= 2AK \sin \frac{\omega\tau}{2} \quad \text{for} \quad 0 > \omega > \omega_0 \\ &= -2AK \sin \frac{\omega\tau}{2} \quad \text{for} \quad -\omega_0 < \omega < 0. \end{aligned}$$

Taking the inverse Fourier transfer to return to the time domain gives

$$v_2(t) = \frac{1}{2\pi}\int_{-\infty}^{\infty} R(j\omega)\, e^{j\omega t}\, d\omega$$

$$= \frac{AK}{\pi}\left(\int_{-\omega_0}^{0} -\sin\frac{\omega\tau}{2}\, e^{j\omega t}\, d\omega + \int_{0}^{\omega_0} \sin\frac{\omega\tau}{2}\, e^{j\omega t}\, d\omega\right)$$

$$= \frac{2AK}{\pi}\int_{0}^{\omega_0} \sin\frac{\omega\tau}{2}\cos\omega t\, d\omega,$$

or

$$v_2(t) = \frac{AK}{\pi}\left(\frac{1}{t-\frac{\tau}{2}}\cdot\left[1-\cos\left(\frac{\tau}{2}-t\right)\omega_0\right] - \frac{1}{t+\frac{\tau}{2}}\left[1-\cos\left(\frac{\tau}{2}+t\right)\omega_0\right]\right).$$

EXERCISE No. 16

An input signal is applied to an amplifier in the form $f(t)$ such that the Fourier transform of $f(t)$ is zero for $\omega > \omega_c$. What conditions must the amplifier transfer function $H(j\omega)$ satisfy if the output signal $r(t)$ is to be the same form as $f(t)$?

What distortion will be introduced if the transfer function is

$$H(j\omega) = \left(1+a\cos\frac{\omega\tau}{2}\right)e^{-j\omega t_0} \qquad \omega < \omega_c$$

$$= 0 \qquad \omega > \omega_c$$

where $a \leqslant 1$.

Solution. If the output signal is to be of the same form as the input signal, the amplifier can only produce a change of amplitude (gain A) and a delay in the signal. Thus

$$r(t) = Af(t-\tau).$$

Going over to the Fourier transforms and using the theorem of retarded signals gives

$$R(j\omega) = AF(j\omega)e^{-j\omega\tau}.$$

Now we know that

$$R(j\omega) = H(j\omega)\cdot F(j\omega)$$

and so, for frequencies less than ω_c

$$H(j\omega) = Ae^{-j\omega\tau},$$

corresponding to a constant amplitude but with a phase angle proportional to frequency. This is a classical result. (For $\omega > \omega_c$ the form of the transfer function is not important for the particular function $f(t)$.)

The transfer function suggested in the problem has an amplitude which is a function of frequency as well as having a phase angle proportional to frequency. Replacing $\cos(\omega\tau/2)$ by the complex exponential gives

$$H(j\omega) = e^{-j\omega t_0} + \tfrac{1}{2}a\, e^{-j\omega(t_0-\tau/2)} + \tfrac{1}{2}a\, e^{-j\omega(t_0+\tau/2)}.$$

The corresponding response of the amplifier is

$$r(t) = \frac{1}{2\pi}\int_{-\infty}^{\infty} F(j\omega)\, H(j\omega)\, e^{j\omega t}\, d\omega$$

$$= \frac{1}{2\pi}\int_{-\infty}^{\infty} F(j\omega)\left(e^{-j\omega t_0} + \frac{1}{2}\, a\, e^{-j\omega(t_0-\tau/2)} + \frac{1}{2}\, a\, e^{-j\omega(t_0+\tau/2)}\right) e^{j\omega t}\, d\omega.$$

This is just the sum of three terms having retardation times t_0, $t_0-\tau/2$ and $t_0+\tau/2$, the former having an amplification factor unity, the latter two amplification factors $a/2$. Thus

$$r(t) = f(t-t_0) + \frac{a}{2} f\left(t - t_0 + \frac{\tau}{2}\right) + \frac{a}{2} f\left(t - t_0 - \frac{\tau}{2}\right).$$

Thus the retarded signal is accompanied by two "echoes" as shown in Fig. 1.

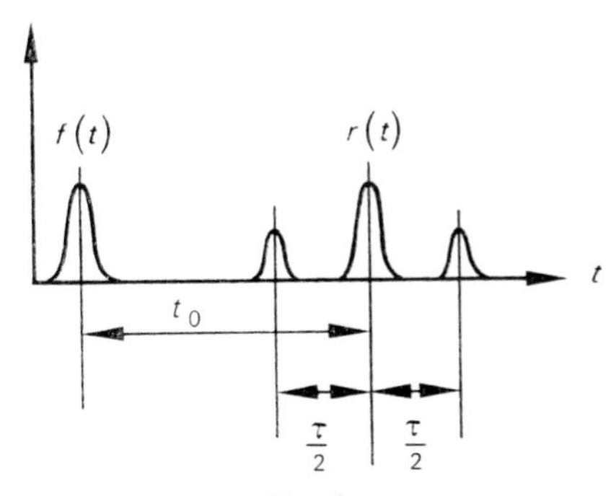

FIG. 1.

Chapter 2

Amplifiers at High and Audio Frequencies

Biassing. Active Filters

EXERCISE No. 1

Find the bias resistance R_B for the circuit of Fig. 1 such that the working point of the transistor is at the centre of the load line.

The characteristics of the transistor are plotted out in Fig. 2.

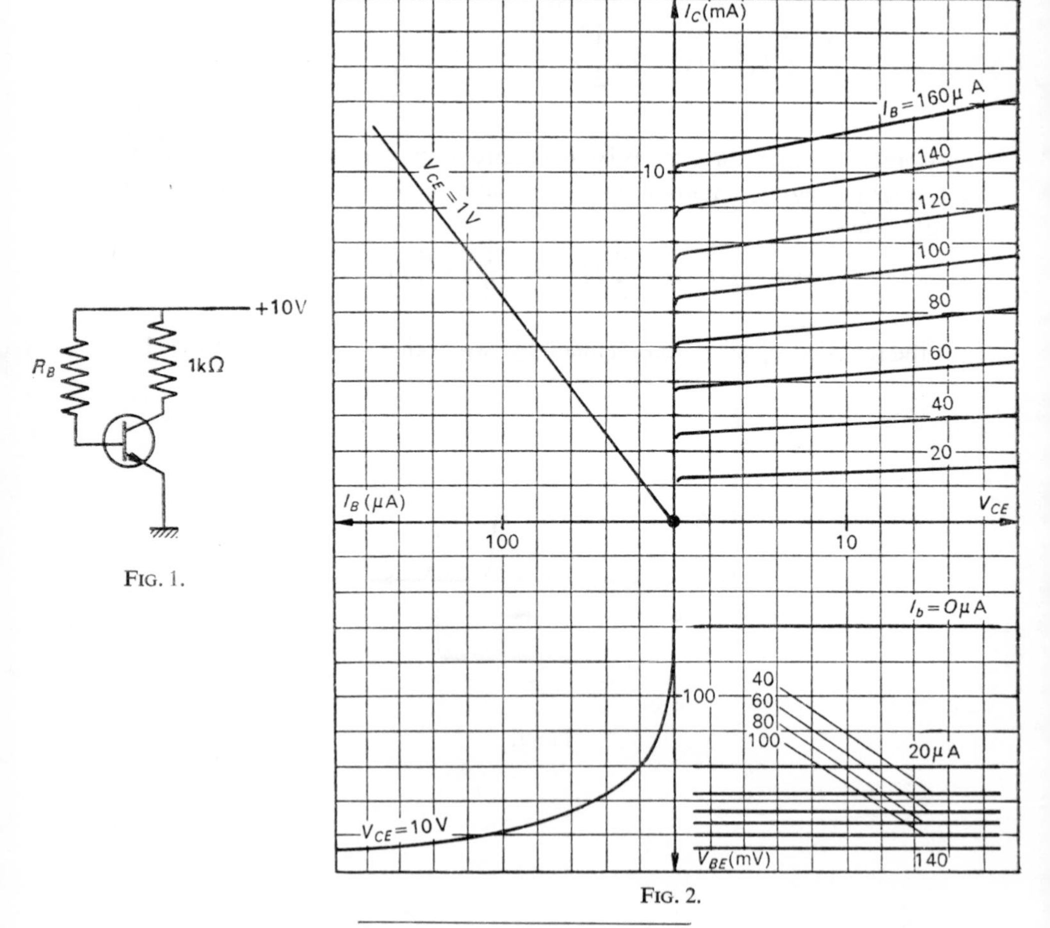

FIG. 1.

FIG. 2.

Solution. The load line has the equation $V_C = 10-1000\,I$ and cuts the current axis at $I = 10$ mA. The central point (M in Fig. 3) has co-ordinates (5V, 5 mA).

For $V_{CE} = 5$ V a line parallel to the V_{CE} axis from M will intercept the $I_C = f(I_b)$ characteristic at N. Reference to Fig. 2 shows this to give $I_B = 76$ μA.

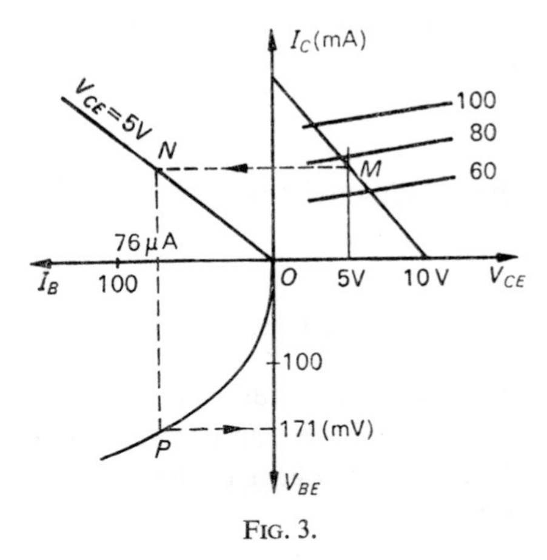

FIG. 3.

Because the internal h parameter h_{12} is always small the characteristics of the third quadrant are practically coincident. Thus the current $I_B = 76$ μA, corresponds to the point P or

$$V_{BE} = 171 \text{ mV}.$$

To calculate R_B we write

$$E = R_B I_B + V_{BE}$$

or

$$R_B = \frac{E-V_{BE}}{I_B} \simeq 129\,000\ \Omega.$$

(The approximate value of R_B obtained taking $V_{BE} \ll E$ would give 131 500 Ω!)

EXERCISE No. 2

The resistance R_B in the circuit in Fig. 1 of the previous exercise has a value 200 kΩ. Find the working point of the transistor using the characteristics given in the corresponding Fig. 2.

Solution. In the third quadrant the input load line has the equation

$$V_{BE} = 10-2\cdot 10^5\, I_B.$$

This is practically vertical and cuts the input characteristics at

$$I_B \simeq 50\ \mu\text{A}, \qquad V_{BE} = 160\ \text{mV}.$$

If we take a linear interpolation (which is justified by the form of the I_C, I_B characteristic in the second quadrant) the corresponding current I_C is approximately 3·4 mA. This value occurs on the load line

$$V_{CE} = 10 - 10^3 I_C$$

at $V_{CE} \simeq 6{\cdot}6$ V. Thus the working point is (6·6 V, 3·4 mA).

EXERCISE No. 3

(i) The gate of an *n*-type field effect transistor is connected to earth as shown in Fig. 1. If the characteristics of the transistor are as shown in Fig. 2 below, determine graphically the quiescent voltage of the source with respect to earth for $R_0 = 500\ \Omega$, 1 kΩ, 2 kΩ, 4 kΩ.

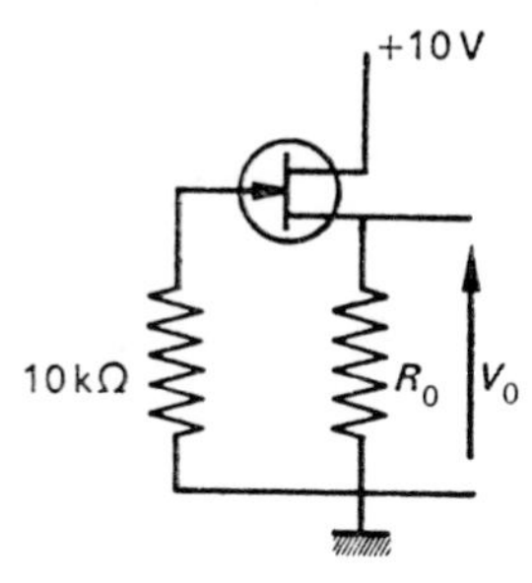

FIG. 1.

(ii) What is the voltage V_0 in the four given cases if the gate to earth voltage is (a) −3 V, (b) +3 V?

Solution. (i) With V_g as the gate voltage and V_a the voltage between the source and the drain, the quiescent point will be such that

$$V_0 + V_g = 0$$

(since there will be no gate current) while

$$V_0 = E - V_a.$$

We look for a point Q on the load-line Δ, indicated in Fig. 3, such that the gate voltage $a = -V_g$ is exactly equal to V_0. The easiest way to find this point Q is by a subsidiary graphical construction as indicated in Fig. 4. We consider the system

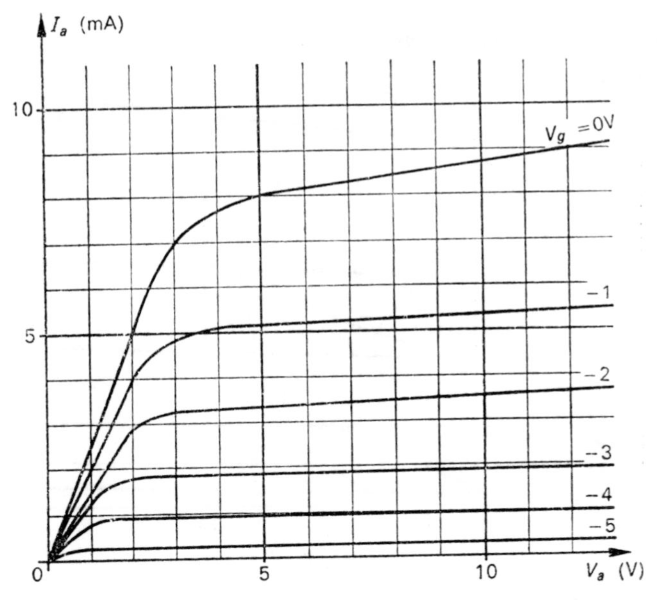

FIG. 2.

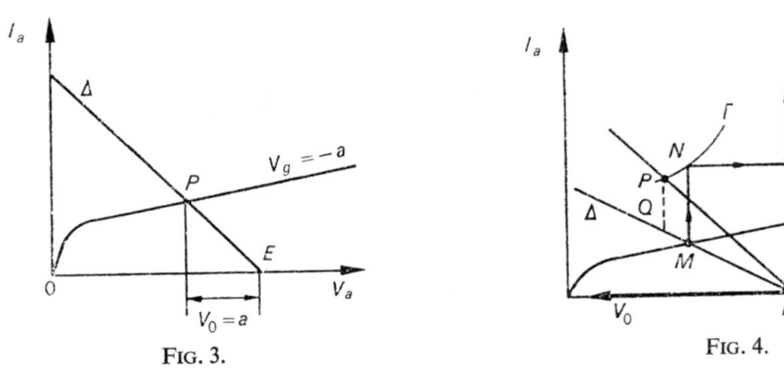

FIG. 3.

FIG. 4.

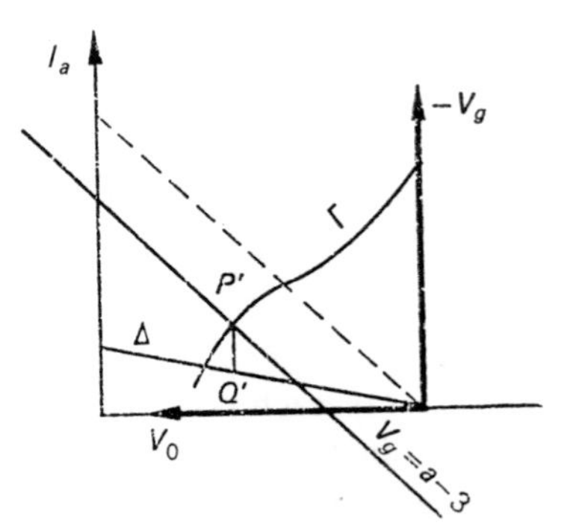

FIG. 5.

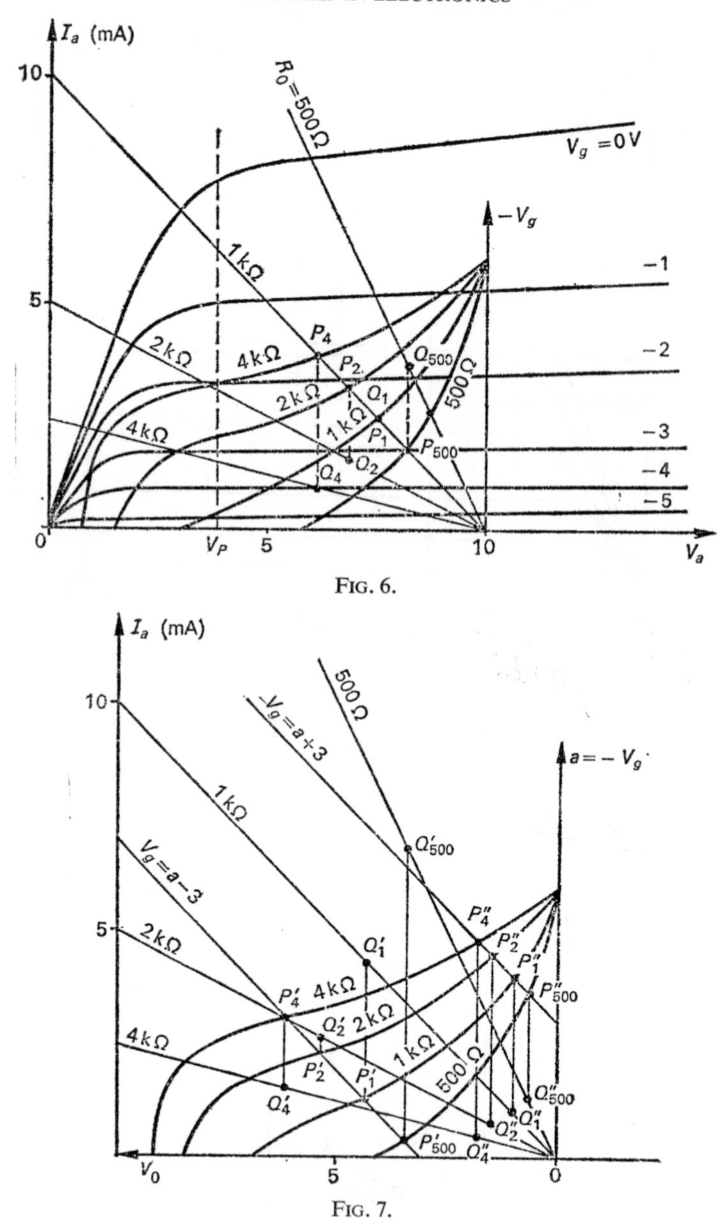

FIG. 6.

FIG. 7.

of co-ordinates with origin at E. For each point M on the load-line a point N is chosen with abscissa V_0 and ordinate $a = -V_g$, the gate potential at M. As M follows the load-line, N will describe a curve Γ which cuts the bisector of the $-V_g$, V_0 axes at the point P which is just the transform of a point Q on Δ which satisfies the stated condition. For each value of R there will be a different load-line and a different curve Γ. These are shown in Fig. 6 from which were obtained the results of the final table.

(ii) The preceding curves can also be used to solve the second question. Here we require that

$$V_0 + V_g = \pm 3\ V$$

and points P' are found by the intercept of the particular Γ by the lines $V_g = -3+a$ and $V_g = 3+a$, the former of which is shown in Fig. 5.

The corresponding point Q' is formed on the relevant load-line as shown in Fig. 7 from which the results are included in the final table.

The results are thus:

R_0	$V_0(V)$		
	$V_g = -3$	$V_g = 0$	$V_g = +3$
500 Ω	0·65	1·9	2·45
1 kΩ	1	2·4	4·4
2 kΩ	1·4	3·2	5·4
4 kΩ	1·75	3·9	6·2

EXERCISE No. 4

An AD140 transistor has a thermal resistance of 1·5 °C/W between the junction and the case. If the maximum junction temperature is 90 °C and the case is maintained at 25 °C, what power can be dissipated by the transistor?

Solution. The temperature difference between the case and the junction can be written

$$\theta_j - \theta_c = R_{th}.P$$

from which the maximum power which can be dissipated is

$$P_{\max} = \frac{90-25}{1{\cdot}5} = 43{\cdot}3\ \text{W}.$$

EXERCISE No. 5

It is wished to use an OC26 transistor in a pure class A push–pull amplifier. The thermal resistance between the junction and the bottom of the case is 1·2 °C/W and a disc of lead and an insulating layer of mica are placed between the transistor case and a blackened aluminium radiator as indicated in Fig. 1.

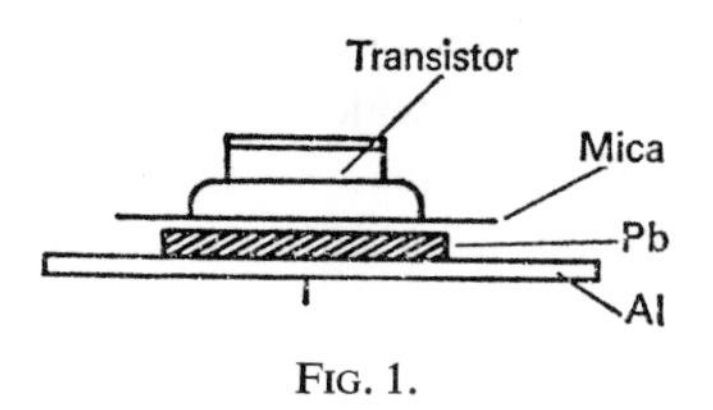

FIG. 1.

If it is given that (a) the thermal resistance between the base of the transistor case and the external face of the lead disc is 0·5 °C/W; (b) the dissipation resistance of the radiator is 1 °C/W; (c) the maximum ambient temperature is 35 °C; (d) the maximum junction temperature is 90 °C and (e) the supply voltage is 12 V, what is the transistor working point which will allow the maximum a.c. output to be obtained? (The resistance of the transformer windings may be neglected.)

Solution. The total thermal resistance between the junction and the surrounding air is

$$R_{\text{th}} = 1{\cdot}2+0{\cdot}5+1 = 2{\cdot}7\ ^\circ\text{C/W}.$$

The maximum power which may be dissipated is thus limited to

$$P_{\max} = \frac{90-35}{2{\cdot}7} = 20{\cdot}4\ \text{W}.$$

The transistor load-line in the case given is vertical and the working point is thus at $V_{CE} = 12$ V. The corresponding current is

$$I_{CE} = \frac{P_{\max}}{V_{CE}} = 1{\cdot}7\ \text{A}.$$

EXERCISE No. 6

The collector of a transistor is connected to an external load R by a large capacitor C as indicated in Fig. 1. If R_C is small compared with $1/h_{22e}$, calculate the value of R_B which allows the largest sinusoidal voltage to be obtained across R in terms of R, R_C and the mean current gain $\beta = I_C/I_B$.

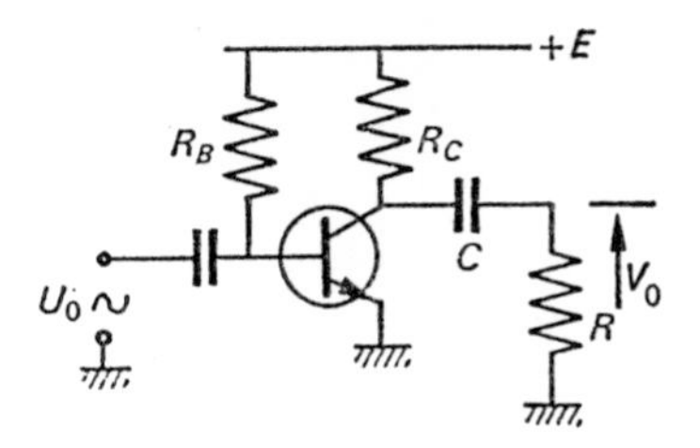

FIG. 1.

Express this voltage as a peak-to-peak value, taking $I_{CE_0} = 0$ and assuming that the characteristics can be extrapolated to $V_{CE} = 0$.

As a numerical example take

$$R = R_C = 1\ \text{k}\Omega$$
$$E = 12\ \text{V}$$
$$\beta = 80.$$

Solution. The maximum output voltage, without distortion due to cut-off, will be obtained when the working point is at the centre of the useful section of the dynamic load-line.

The static load-line has the equation

$$V_C = E - R_C I_C$$

while the dynamic load-line is

$$V'_C = 2\lambda E - \frac{RR_C}{R+R_C} \cdot I_C.$$

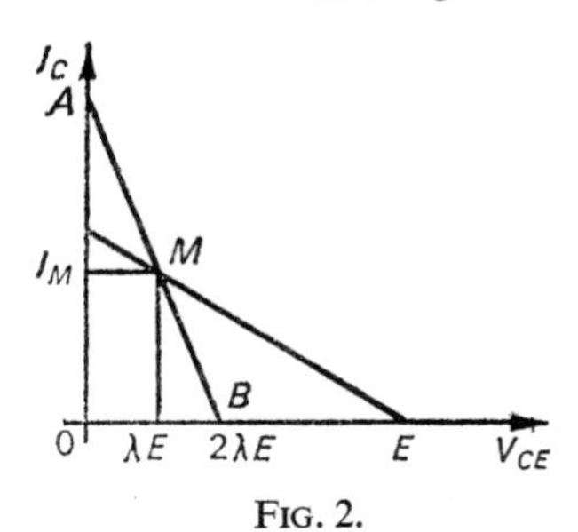

FIG. 2.

These are indicated in Fig. 2. If, now, we note that the intersection of these two lines, which is just the working point, will occur at the mid-point of AB, it is obvious that this point will have the abscissa $V_C = \lambda E$. Then

$$\lambda = \frac{R/R_C}{2\dfrac{R}{R_C}+1}$$

from which the bias current is

$$I_M = \frac{E}{R_C}\left(\frac{\frac{R}{R_C}+1}{\frac{R}{2R_C}+1}\right).$$

Neglecting V_{BE} with respect to E gives R_B as

$$R_B = E\frac{\beta}{I_M} = \beta\left(\frac{\frac{R}{R_C}+1}{2\frac{R}{R_C}+1}\right)R_C.$$

At the most the voltage current point can follow the line AB and this corresponds to a maximum peak-to-peak variation of $2\lambda E$ or

$$V_{pp\,\max} = \frac{2R/R_C}{2R/R_C+1}\cdot E.$$

With the figures given in the numerical example we have

$$\lambda = \tfrac{1}{3}, \quad V_{pp\,\max} = 8\ \text{V} \quad \text{and} \quad R_B = 120\ \text{k}\Omega.$$

EXERCISE No. 7

Consider the classical stabilization system with the potential divider in the base circuit as shown in Fig. 1. Obtain the general expression for the variation of the collector current in terms of the thermal fluctuations of the residual collector

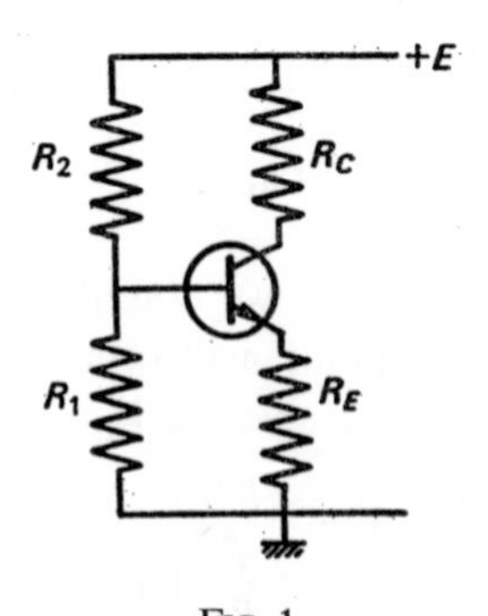

FIG. 1.

current I_{CE_0}, the base-emitter voltage V_{BE} and the current gain β. (Assume that $1/h_{22}$ is big enough for the characteristics to be given by

$$I_C = I_{CE_0}+\beta I_B.$$

Solution. By Thévenin's theorem the divider or bridge R_1, R_2 can be replaced by a single resistance

$$r = \frac{R_1 R_2}{R_1+R_2}$$

connected to a source

$$E' = \frac{ER_1}{R_1+R_2},$$

We can then write

$$\frac{ER_1}{R_1+R_2} = rI_B+V_{BE}+R_E I_E$$

which becomes, on differentiating,

$$0 = r\,\Delta I_B+\Delta V_{BE}+R_E\,\Delta I_E = (r+R_E)\,\Delta I_B+\Delta V_{BE}+R_E\,\Delta I_C\,.$$

Now, from the characteristic equation,

$$\Delta I_C = \Delta I_{CE_0}+\beta\,\Delta I_B+I_B\,\Delta\beta.$$

Eliminating ΔI_B between these equations gives

$$\Delta I_C = \frac{\Delta I_{CE_0}}{1+\dfrac{\beta R_E}{R_E+r}}+\frac{I_B\,\Delta\beta}{1+\dfrac{\beta R_E}{R_E+r}}-\frac{\beta\,\Delta V_{BE}}{r+R_E(\beta+1)}\,.$$

EXERCISE No. 8

(i) The output stage of a transistor amplifier is shown in Fig. 1 below. If it is supposed originally that $R \gg R_E$, what working point must be chosen to give the maximum sinusoidal voltage output across the load R? (It is known that the residual collector current, $I_{CE_0} = 250$ μA at 25 °C and that the saturation resistance of the transistor $R_S = 10\ \Omega$.)

(ii) Supposing that I_{CE_0} increases exponentially with temperature in doubling every 10 °C above 25 °C, calculate this current at 55 °C.

(iii) If the bias voltage had been produced by $R_B = R_2$ alone (i.e. $R_1 = \infty$, $R_E = 0$) what would be the stable condition of the transistor at 55 °C and the

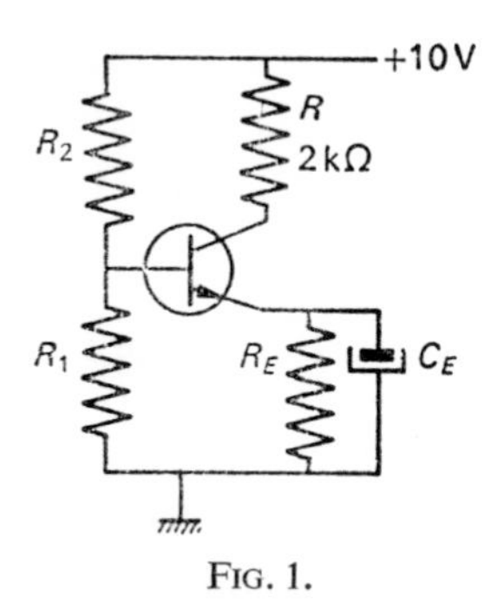

FIG. 1.

maximum output voltage? (Suppose that β and V_{BE} are independent of temperature.)

(iv) What is the increase in the collector current which would be required, allowing for the increase in I_{CE_0}, to give the maximum output voltage at 55 °C? Calculate the corresponding stabilization factor $S = \Delta I_{CE_0}/\Delta I_C$ which would be required.

(v) In order that the approximation in (i) regarding R_E should be valid, let us take $R_E = 200\ \Omega$. Then, taking $\beta = 100$ and $V_{BE} = 0{\cdot}15$ V, calculate the values of R_1 and R_2 in the biassing bridge.

Solution. (i) The useful part of the load-line (which is drawn in Fig. 2 for 2 kΩ since $R \gg R_E$) extends from the point A, the intersection with the characteristic $I_B = 0$ at $V_A = 9{\cdot}5$ V, to the point B where it crosses the saturation characteristic corresponding to

$$V_C = I_C \times 10\ \Omega \qquad (V_B = 50\ \text{mV}).$$

The stable point at the centre of AB then has co-ordinates

$$I_{CM} = 2{\cdot}61\ \text{mA}, \qquad V_{CM} = 4{\cdot}77\ \text{V}.$$

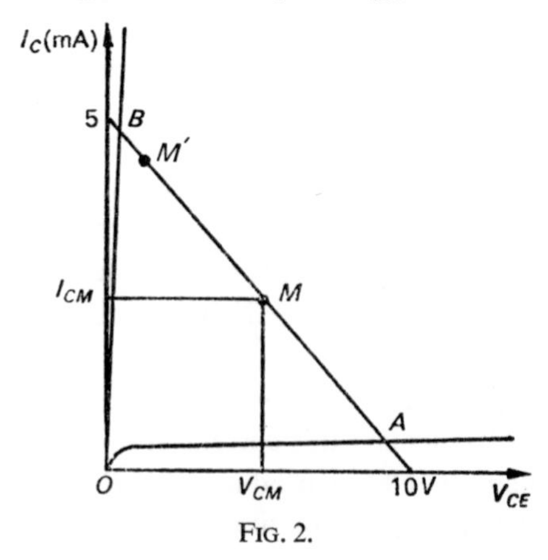

FIG. 2.

(ii) $$I_{CE_0\,(55\,^\circ C)} = I_{CE_0\,(25\,^\circ C)} \cdot 2^{(55-25)/10} = 2 \text{ mA}.$$

(iii) We can write

$$I_C = I_{CE_0} + \beta I_B.$$

At 25 °C, the value of R_B has been chosen to give a collector current of 2·61 mA or

$$\beta I_B = I_C - I_{CE_0} = 2{\cdot}36 \text{ mA}.$$

With this type of bias, I_B is maintained constant so that, at 55 °C,

$$I_C = 2{\cdot}36 + 2 = 4{\cdot}36 \text{ mA}.$$

The stable position will be moved then to M' with

$$I_{CM'} = 4{\cdot}36 \text{ mA}, \qquad V_{CM'} = 1{\cdot}28 \text{ V}.$$

The output voltage is limited by saturation at the point B and the maximum peak-to-peak value is

$$V_{pp\ \max} = 2(V_{CM'} - V_B) = 2{\cdot}46 \text{ V}.$$

(iv) At 55 °C the optimum bias point is found following the method of the previous question (Exercise 7) with $\Delta\beta = \Delta V_{BE} = 0$. We find

$$I_{CM''} = 3{\cdot}95 \text{ mA}, \qquad V_{CM''} = 3{\cdot}025 \text{ V}$$

and

$$V_{pp\ \max} = 5{\cdot}95 \text{ V}.$$

Between 25 and 55 °C

$$\Delta I_C = 1{\cdot}34 \text{ mA}.$$

Then

$$S = \frac{\Delta I_{CE_0}}{\Delta I_C} = \frac{2 - 0{\cdot}250}{1{\cdot}34} = 1{\cdot}3.$$

(v) The first relation is given by the factor S, i.e.

$$S = 1 + \beta \frac{R_E}{R_E + r} = 1{\cdot}3$$

from which, with the value given for β,

$$r = 332\ R_E = 33{\cdot}2 \text{ k}\Omega.$$

Also, in the base circuit,

$$E \frac{R_1}{R_1 + R_2} = rI_B + V_{BE} + R_E I_E.$$

At 25 °C, I_B, V_{BE} and I_E are known and so we can find

$$R_2 = 277\ \Omega, \qquad R_1 = 37{\cdot}5\ \Omega.$$

Note. This exercise shows that, for germanium transistors having a relatively large I_{CE_0}, it is not always desirable to have a stabilization coefficient above a certain optimum value, since the output voltage may be limited at high temperatures by an upper, rather than a lower, cut-off. That is, there is a risk of over-correction.

EXERCISE No. 9

With a given source voltage it is required to design a single-stage transistor amplifier capable of giving the greatest output voltage without saturation. That is, the working point will be situated at the centre of the load-line. Show that, under these conditions, the voltage gain obtained is almost independent of the load resistance as long as $R_C \ll 1/h_{22e}$. Is the same effect observed for the useful gain v_O/v_i where v_O is the output voltage and v_i is the input voltage from a source of internal resistance ϱ?

Solution. With a given supply voltage the working point must necessarily be at $V_C = E/2$. If $R_C \ll 1/h_{22e}$, the voltage gain is given by

$$G_v = \frac{\beta R_C}{h_{11}}.$$

β really only depends on V_C and is almost independent of I_C.
On the other hand h_{11} is inversely proportional to the current

$$h_{11e} = \frac{K}{I_C} = K \cdot \frac{2R_C}{E}.$$

Then

$$G_v = \frac{\beta E}{2K}, \quad \text{which is independent of } R_C.$$

The useful gain takes account of the input mismatch so that

$$\frac{v_O}{v_i} = \frac{h_{11}}{\varrho + h_{11}} \cdot G_v = \frac{\beta R_C}{h_{11} + \varrho} = \beta \frac{R_C}{\frac{2K}{E} R_C + \varrho}.$$

This quantity increases as R_C increases but only slightly if ϱ is small compared with h_{11}.

EXERCISE No. 10

The output stage of an amplifier is as shown in Fig. 1 where the resistances of the primary and secondary windings of the transformer are 50 Ω and, effectively, zero respectively. The maximum allowable dissipation in the transistor is 250 mW. Supposing that the transformer is perfect and that the working point may be anywhere in the (V_C, I_C) quadrant, calculate the maximum power which may be supplied to the load.

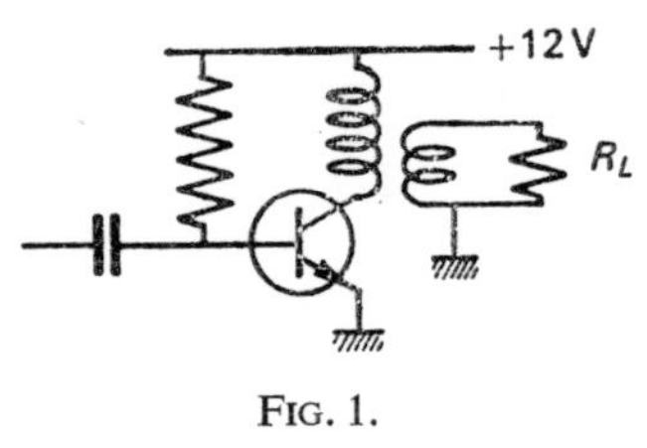

FIG. 1.

Solution. The maximum possible output from a class A amplifier is $\frac{1}{2}$. This corresponds to a maximum modulated power of 125 mW of which part will be dissipated in the resistance of the transformer windings.

To find the optimum working point, we draw the static load-line

$$V_C = 12 - 50\, I_C$$

and the dissipation hyperbola

$$V_C I_C = \tfrac{1}{4}$$

as shown in Fig. 2. There are then two points of intersection M_1 and M_2, of which only that, M_1, corresponding to the smaller current is useful, since, with transformer coupling, the dynamic load is always greater than the static value and the dynamic

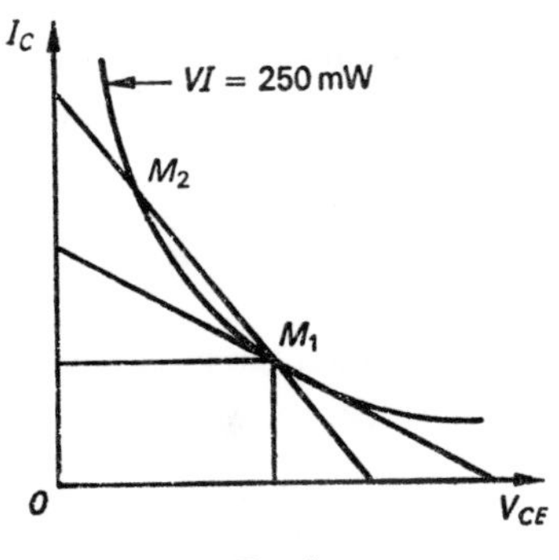

FIG. 2.

load-line is always closer to the horizontal axis. The working point must therefore be at

$$I_{CM_1} = 23 \text{ mA}, \quad V_{CM_1} = 10{\cdot}85 \text{ V}$$

and the dynamic load must be

$$R = \frac{V_{CM_1}}{I_{CM_1}} = 472 \, \Omega.$$

Allowing for the 50 Ω resistance of the primary, the maximum power which can be supplied to an external load is

$$P_0 = 125 \times \frac{472-50}{472} = 112 \text{ mW}.$$

EXERCISE No. 11

The output transformer of a transistor amplifier stage is constructed on a ferrite core which allows complete coupling between the primary and secondary.

We wish to obtain the maximum power in the load within the limits imposed by the power which may be dissipated in the transistor, i.e. $P_{max} = 5$ W. The reduction in the gain for very low frequencies must not exceed 3 dB at 15·9 Hz ($\omega_{min} = 100$). The secondary load is a loud-speaker which has an effective resistance of $R_S = 2{\cdot}5 \, \Omega$, while the windings have negligible resistance.

Calculate the numbers of turns on the primary and secondary of the transformer which will be necessary to give P_{max}, if 10 turns of wire give 1 mH of inductance and the supply voltage is $E = 12$ V. (The d.c. component of the collector current may be neglected.)

Solution. The load of the transistor is made up of the transferred resistance $R = (n_1/n_2)^2 R_S$ in parallel with the primary inductance L, so that

$$Z = \frac{j\omega L_1 R_L}{j\omega L_1 + R_L}.$$

The stage gain is reduced by 3 dB when the modulus of this impedance is divided by $\sqrt{2}$ from its high frequency value, i.e. when

$$\omega = \frac{R_L}{L_1}.$$

The inductance of the primary must thus be at least

$$L_1 = \frac{R_L}{\omega_{\min}}.$$

Since the resistance of the windings are negligible the static load-line will be vertical and cut the dissipation hyperbola at

$$I_{CM} = \frac{P_{\max}}{E} = 0{\cdot}417 \text{ A}.$$

This point will be the centre of the dynamic load-line if

$$R_L = \frac{V_{CM}}{I_{CM}} = \frac{E}{I_{CM}} = 28{\cdot}8\ \Omega.$$

Thus

$$L_1 = 0{\cdot}288 \text{ H}$$

from which

$$n_1 = 10 \times \sqrt{288} \simeq 170 \text{ turns}$$

$$n_2 = n_1 \times \sqrt{\frac{2{\cdot}5}{28{\cdot}8}} \simeq 50 \text{ turns}.$$

EXERCISE No. 12

The emitter circuit of a transistor is composed of a resistance R_E decoupled by a capacitor C_E. If ϱ_g is the internal impedance of the signal source of e.m.f. v_i, at what frequency will the apparent gain $|v_{\bar{O}}/v_i|$ drop by 3 dB from its value at medium frequencies? (This defines the lower limit of the pass band of the stage.) Assume that the load resistance is small.

As a numerical example take

$$\varrho_g = 1 \text{ k}\Omega, \quad h_{11} = 1 \text{ k}\Omega, \quad \beta = 100, \quad R_E = 100\ \Omega \quad \text{and} \quad C_E = 25\ \mu\text{F}.$$

Solution. The input impedance of a transistor having an impedance Z_E in the emitter is, for R_C small,

$$r_i = h_{11} + (\beta+1)Z_E$$

and the corresponding voltage gain is

$$G_v = \frac{\beta R_C}{r_i}.$$

The apparent gain, allowing for the input mismatch, is

$$\frac{\bar{o}}{v_i} = \frac{\beta R_C}{r_i+\varrho_g}.$$

Then, substituting $Z_E^{-1} = j\omega C_E + R_E^{-1}$ in r_i,

$$\frac{v_{\bar{O}}}{v_i} = \frac{\beta R_C}{\varrho_g+h_{11}} \cdot \frac{s+\dfrac{1}{R_E C_E}}{s+\dfrac{1}{R_E C_E}\left(1+\dfrac{(\beta+1)R_E}{\varrho_g+h_{11}}\right)} \quad \text{where} \quad s = j\omega$$

which is of the form

$$\frac{v_{\bar{O}}}{v_i} = \text{A}\cdot\frac{s+\omega_0}{s+\omega_1}.$$

The modulus gives

$$\left|\frac{v_{\bar{O}}}{v_i}\right|^2 = A^2\frac{\omega^2+\omega_0^2}{\omega^2+\omega_1^2}$$

and the frequency for which the gain drops by 3 dB from its value where $\omega^2 \gg \omega_0^2$ and ω_1^2 is given by the condition $|v_{\bar{O}}/v_i|^2 \to A^2/2$, i.e.

$$\omega_{1/2}^2 = \omega_1^2 - 2\omega_0^2.$$

Generally

$$(\beta+1)R_E \gg \varrho_g + h_{11}$$

so that

$$\omega_1^2 \gg \omega_0^2 \quad \text{and}$$

$$\omega_{1/2} \simeq \omega_1 = \frac{1}{R_E C_E}\left[1+\frac{(\beta+1)R_E}{\varrho_g+h_{11}}\right].$$

The numerical example gives, on substitution $\omega = 2420$ rad/s (385 Hz).

EXERCISE No. 13

The Miller Effect

Calculate the input admittance of the circuit shown in Fig. 1 below for the case of low frequencies (i.e. $\omega \ll \omega_\beta$ of the transistor) and where $R_L \ll 1/h_{22e}$ and $h_{21e} = \beta$.

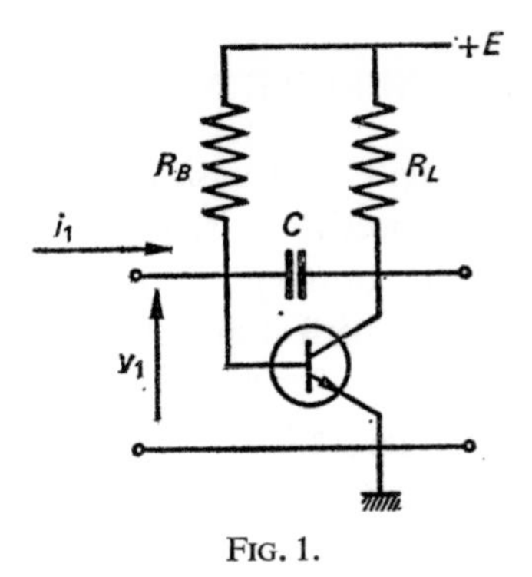

FIG. 1.

Express this admittance as an equivalent circuit of a resistance R in parallel with a series $R'C'$ circuit.

Solution. With $R_L \ll 1/h_{22e}$ the equivalent circuit becomes as shown in Fig. 2.

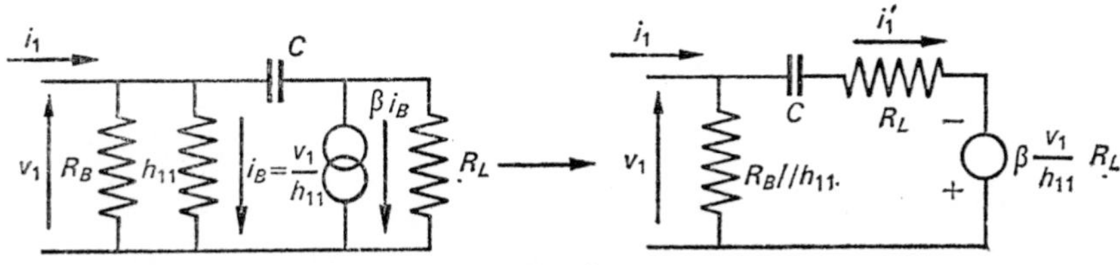

FIG. 2.

Then

$$i_1' = \frac{v_1\left(1+\dfrac{\beta R_L}{h_{11}}\right)}{R_L + \dfrac{1}{j\omega C}} = v_1 \frac{1}{\dfrac{R_L}{1+\dfrac{\beta R_L}{h_{11}}} + \dfrac{1}{j\omega C\left(1+\dfrac{\beta R_L}{h_{11}}\right)}}.$$

This suggests an equivalent circuit as in Fig. 3.

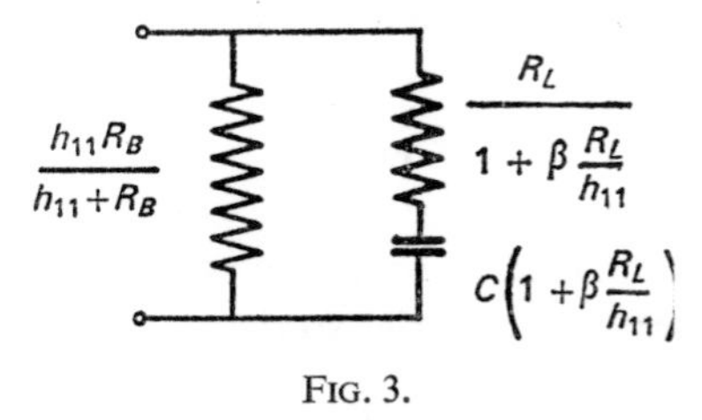

FIG. 3.

Note that $\beta R_L/h_{11}$ is the classic voltage gain.

EXERCISE No. 14

In the push–pull system of Fig. 1 the transistors may be considered as current sources in parallel with their respective resistances r_{C_1} and r'_{C_1}. If the number of turns on each of the half-windings of the primary is n_1 and of the secondary is n_2, the magnetic coupling is assumed perfect and the winding resistance negligible, what is the value of the inductance L of each half of the primary for the attenuation of low frequencies to be 3 dB at $\omega_0/2\pi$?

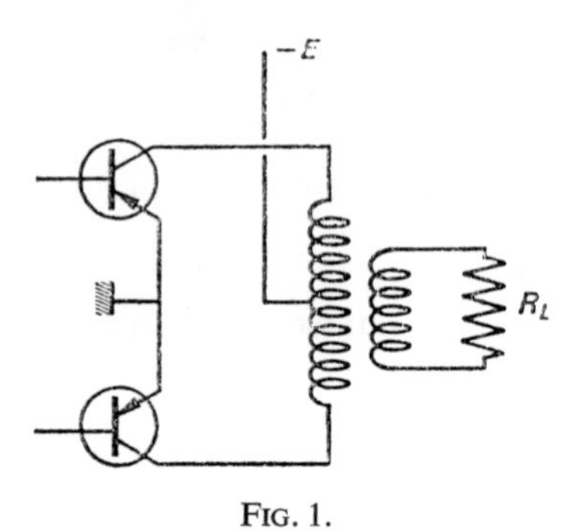

FIG. 1.

As a numerical example, take:

$$R_L = 2{\cdot}5\,\Omega,\quad n_1/n_2 = 10,\quad r_{C_1} = 1000\,\Omega \quad \text{and} \quad \omega_0 = 100.$$

Solution. The equivalent circuit may be taken as in Fig. 2.

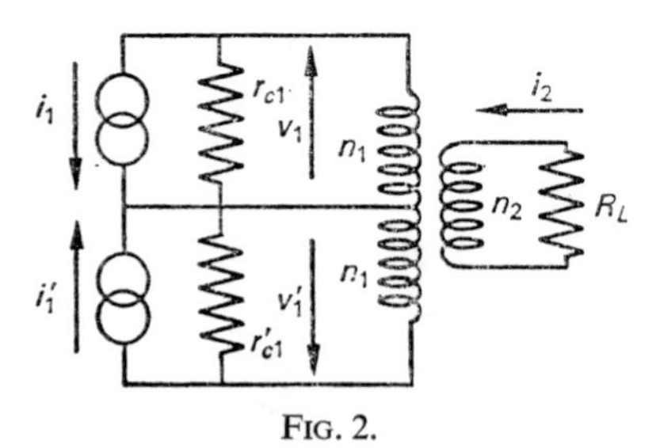

FIG. 2.

To calculate the equivalent load resistance of each transistor we take i_1 and i'_1 as the currents in the transformer primaries so that

$$\left.\begin{aligned} v_1 &= j\omega L_1 i_1 + j\omega M_{11'} i'_1 + j\omega M_{12} i_2 \\ v'_1 &= j\omega L_{1'} i'_1 + j\omega M_{11'} i_1 + j\omega M_{12'} i_2 \\ 0 &= R_L i_2 + j\omega L_2 i_2 + j\omega M_{21} i_1 + j\omega M_{21'} i'_1 \end{aligned}\right\} A$$

From the assumptions made and the sense of the windings $L_1' = L_1$; $M_{11'} = -L_1$ (mutual inductance between the two halves of the primary); $M_{12} = (n_2/n_1).L_1 = -M_{1'2}$; $L_2 = L_1(n_2/n_1)^2$

Since the two transistors are fed in opposition we know that

$$i_1 = -i_1'$$

so that the equations A give

$$\frac{v_1}{i_1} = \frac{2j\omega L_1 R_L}{R_L + j\omega(n_2/n_1)^2 L_1} = \frac{j\omega(2L_1).2R_L(n_1/n_2)^2}{2R_L(n_1/n_2)^2 + j\omega(2L_1)}.$$

The total equivalent circuit for a single transistor is thus as shown in Fig. 3 with r_{C_1}, $2L_1$ and $2R_L(n_1/n_2)^2$ in parallel.

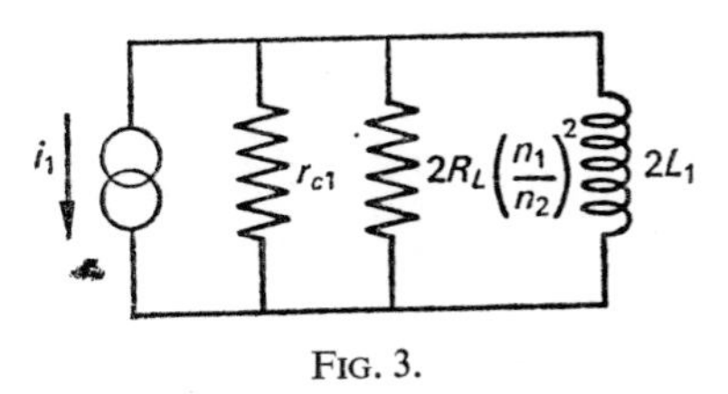

FIG. 3.

The gain will be reduced by 3 dB when

$$2\omega_0 L_1 = \frac{2\,R_L(n_1/n_2)^2 r_c}{2R_L(n_1/n_2)^2 + r_c}.$$

Numerical example:

$$L_1 = \tfrac{5}{3}\ \text{H}.$$

EXERCISE No. 15

A transistor is used at high frequencies with a real load R_L. Show that, if R_L is sufficiently small, i.e. $R_L \ll 1/g_{ce}$ and $1/\omega C_{b'c}$, the Giacoletto circuit may be replaced by a simpler π circuit. (Simplify the Giacoletto circuit by taking account of the relative magnitudes of the impedances and calculating y_1 from the circuit of Fig. 1.)

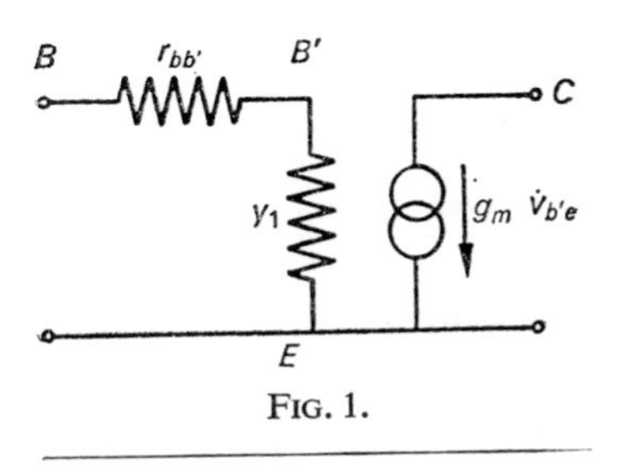

FIG. 1.

Solution. With the assumptions made, the current in g_{ce} and that flowing internally from the collector to the base B' will be negligible compared with i_c. Thus we can take

$$v_2 = -g_m v_{b'e} R_L .$$

If we also neglect $g_{b'e}$ relative to $\omega C_{b'c}$ we get the circuit of Fig. 2.

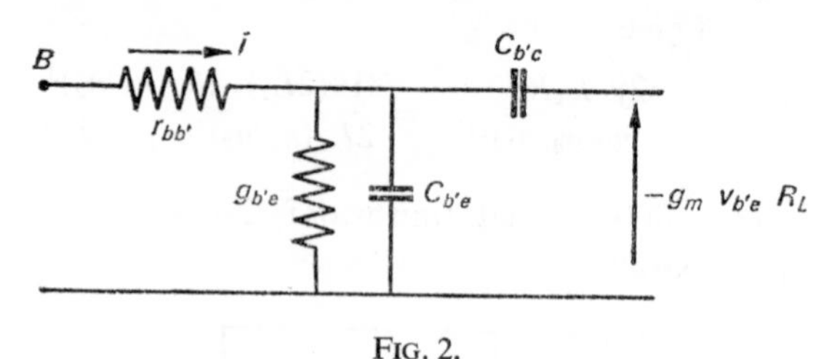

FIG. 2.

At the point corresponding to B' of Fig. 1 the equation is

$$(g_{b'e}+j\omega C_{b'e})v_{b'e} = i-j\omega C_{b'c}(1+g_m R_L)v_{b'e} .$$

Because of the relative magnitudes of $C_{b'e}$ and $C_{b'c}$ this can be written to give

$$i = v_{b'e}[g_{b'e}+j\omega(C_{b'e}+g_m R_L C_{b'c})].$$

The simplified circuit will thus be identical with the original if we take

$$y_1 = g_{b'e}+j\omega(C_{b'e}+g_m R_L C_{b'c}),$$

which is equivalent to placing a supplementary capacitance in parallel with $C_{b'e}$ in the internal impedance as in Fig. 3.

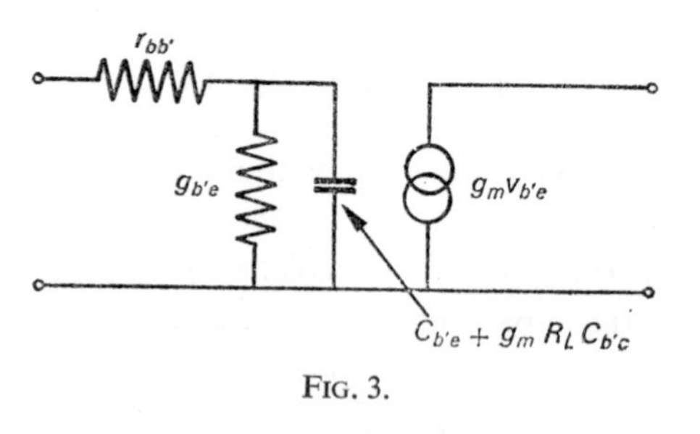

FIG. 3.

EXERCISE No. 16

Assuming that the resistive load is sufficiently small that the internal impedance can be neglected, calculate the upper cut-off frequency of a one-stage transistor amplifier. (This will give the pass-band of the apparent gain u_f/u_0, u_0 being the input e.m.f. from a source of internal resistance ϱ_g.) How does this frequency depend on ϱ_g?

Solution. With the internal impedance neglected, this being justified for a sufficiently small load, the Giacoletto circuit simplifies to Fig. 1.

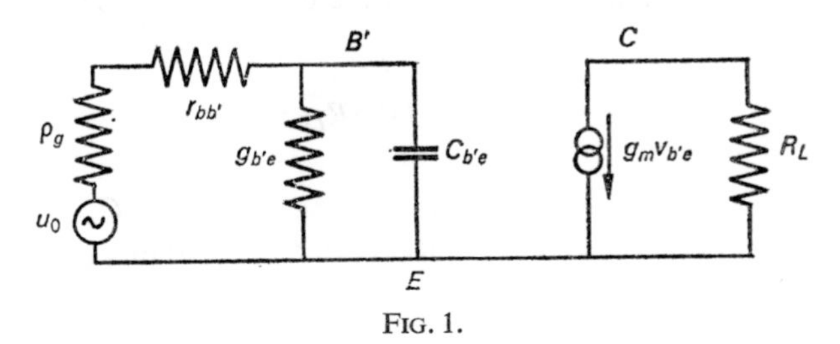

FIG. 1.

Then

$$v_{b'e} = u_0/[1+y_{b'e}(\varrho_g+r_{bb'})] \quad \text{with} \quad y_{b'e} = g_{b'e}+j\omega C_{b'e}$$

$$u_f = -R_L g_m v_{b'e}.$$

The apparent gain is thus

$$G_A = \frac{u_f}{u_0} = \frac{-R_L g_m}{1+(g_{b'e}+j\omega C_{b'e})(\varrho_g+r_{bb'})}$$

which can be written as

$$G_A = \frac{-R_L g_m}{1+g_{b'e}(\varrho_g+r_{bb'})} \cdot \frac{\omega_1}{j\omega+\omega_1}$$

with

$$\omega_1 = \frac{1+g_{b'e}(\varrho_g+r_{bb'})}{C_{b'e}(\varrho_g+r_{bb'})}$$

as the 3 dB cut-off frequency.

If we write the cut-off frequency for the current gain as

$$\omega_\beta = \frac{1}{r_{b'e}C_{b'e}}$$

we have

$$\omega_1 = \omega_\beta\left[1+\frac{r_{b'e}}{\varrho_g+r_{bb'}}\right],$$

This frequency varies from the value ω_β when ϱ_g is large (i.e. the source is effectively a current source) to

$$\omega_\beta\left[1+\frac{r_{b'e}}{r_{bb'}}\right]$$

for ϱ_g small, which is the slope or α cut-off frequency for a voltage source.

EXERCISE No. 17

Show that, because of the internal impedance of a transistor represented by the Giacoletto circuit, the pass-band of an amplifier with a purely resistive load decreases rapidly as this load increases. Put the cut-off frequency in the form

$$f_c = \frac{f_{c0}}{1+(R_L/R_{LC})}$$

where f_{c0} is the cut-off frequency for $R_L = 0$ and we take $(\omega C_{b'c})^{-1} \gg R_L$.

As a numerical example take $C_{b'e} = 1000$ pF, $C_{b'c} = 10$ pF, $g_m = 40$ mA/V.

Solution. The calculation follows that of the preceding exercise with $C_{b'e}$ replaced by

$$C_{b'e}+g_m R_L C_{b'c}$$

by virtue of the results of Exercise 15. Thus

$$\omega_c = \frac{\omega_{c0}}{1+(R_L/R_{LC})} \quad \text{with} \quad R_{LC} = \frac{1}{g_m} \cdot \frac{C_{b'e}}{C_{b'c}} = 2{,}500\ \Omega.$$

EXERCISE No. 18

Consider a high frequency amplifier stage of the form given in Fig. 1, the supply connection being taken to the exact centre of the primary winding as shown.

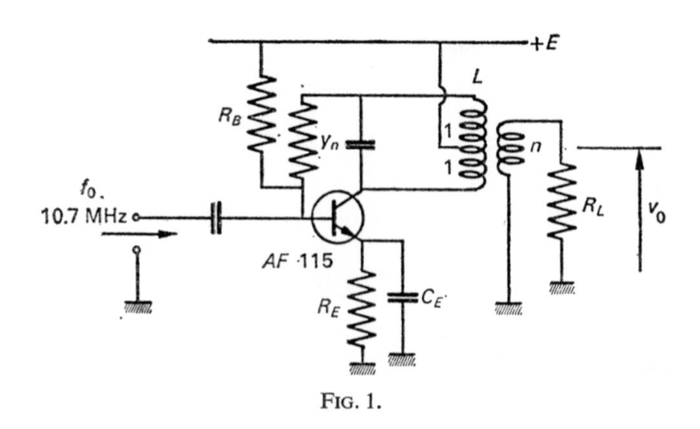

FIG. 1.

For a working frequency of 10·7 MHz and a collector current of 1 mA, the manufacturer gives the y parameters of the transistor AF 115 as

$$y_{11} = g_{11}+j\omega C_{11} = 10^{-3}+j\omega\ 70\ \text{pF}$$
$$y_{22} = 25\times 10^{-6}+j\omega\ 3\ \text{pF}$$
$$|y_{12}| = 80\times 10^{-6}, \quad \Phi_{12} = 260^\circ$$
$$|y_{21}| = 3\times 10^{-2}, \quad \Phi_{21} = 330^\circ.$$

1. Determine the exact nature of the neutrodyne admittance y_N and calculate the y parameters of the neutrodyned transistor.

2. The external load is $R_L = 50\ \Omega$, the Q of the complete primary winding is 100 and the total primary inductance is $L_1 = 1\ \mu$H. Calculate n, the turns ratio between the secondary and one-half of the primary such that the voltage gain $v_{\bar{0}}/v_i$ has a 3 dB pass-band of 200 kHz. (Take the coupling coefficient of the transformer to be 1.)

Solution. 1. The circuit can be represented by two four-terminal networks in parallel as shown in Fig. 2. Then with the transformer formed by the two halves of the primary taken to be perfect we have

$$\begin{cases} i_1' = y_N v_1 + y_N v_2 \\ i_2' = y_N v_1 + y_N v_2 . \end{cases}$$

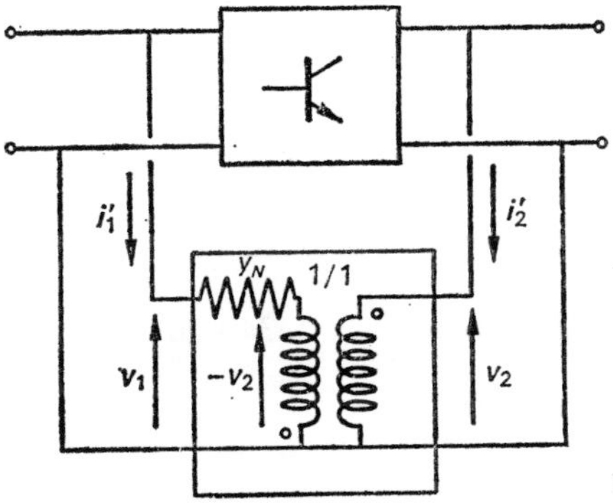

FIG. 2.

The parameters of the neutrodyned transistor are then

$$y_{ije}' = y_{ije}+y_N .$$

The neutrodyne condition is that

$$y_{12e}' = y_{12e}+y_N = 0$$

or

$$y_N = -y_{12e} = -(80\times 10^{-6} \underline{|260°})\text{.}$$

Thus

$$y_N = 80\times 10^{-6} \cos 80° + \text{j}\, 80\times 10^{-60} \sin 80°$$
$$= (71\times 9 \text{ k}\Omega)^{-1} + \text{j}\omega\, 1\times 16 \text{ pF.}$$

That is, the neutrodyne admittance is formed by a resistance of 71·9 kΩ in parallel with a capacity of 1·16 pF. This admittance is too small to affect y_{11} and y_{21} appreciably and it only modifies y_{22} slightly.

$$y_{22} = 38{\cdot}9\times 10^{-6} + \text{j}\omega\, 4{\cdot}16 \text{ pF} = g_{22} + \text{j}\omega C_{22}\text{.}$$

2. The circuit losses can be represented by a conductance in parallel with the inductance of the primary, i.e.

$$G_0 = \frac{1}{\omega LQ} = 1{\cdot}49\times 10^{-4} \text{ S}^{\dagger} \text{ (or mho)}$$

or, taken in parallel with one-half of the primary alone,

$$G_0' = 5{\cdot}95\times 10^{-4} \text{ S}$$

The equivalent circuit is shown in Fig. 3.

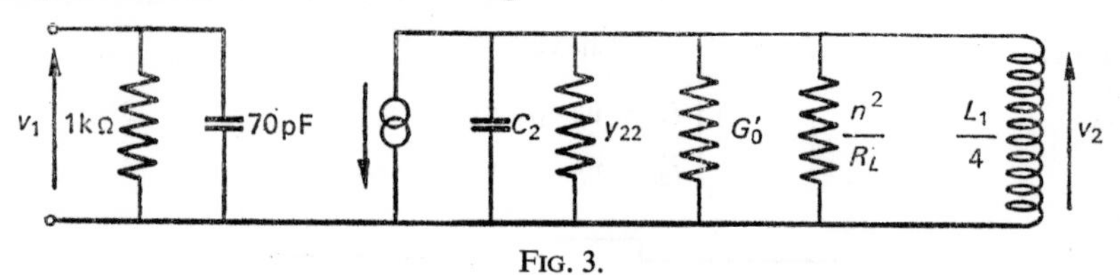

FIG. 3.

The pass-band is

$$B = \frac{f_0}{Q} = f_0 \cdot G \cdot \omega L$$

where G is, in this case, the total parallel conductance. Here

$$G = g_{22} + G_0' + \frac{n^2}{R_L} = (634 + 20\,000n^2)\times 10^{-6}\text{.}$$

The condition to be satisfied is

$$G = \frac{B}{f_0 \cdot \omega L} = \frac{0{\cdot}2}{10{\cdot}7} \cdot \frac{1}{\dfrac{10^{-6}}{4} \cdot 2\pi\times 10{\cdot}7\times 10^6} = 1{\cdot}11\times 10^{-3}\text{.}$$

Hence we obtain

$$n = \frac{1}{6{\cdot}47}\text{.}$$

† We use here S, Siemen, as the reciprocal of the ohm for the unit of conductance.

We thus have a behaviour as if the load were purely resistive being $Z_L = 1/G$, the corresponding voltage gain being

$$\frac{v_0}{v_1} = \frac{|y_{21}|}{G} \cdot n = 4{\cdot}2.$$

EXERCISE No. 19

An electrode for electrophysiological application can be represented by a generator of e.m.f. u_i and the internal impedance of a parallel RC circuit. In order to construct a system to measure u_i the circuit of Fig. 1 is used. The amplifier has a gain G which is frequency independent, a zero output impedance and an input impedance equivalent to R_i and C_i in parallel. What must be the value of C_1 if v_O is to be the image of u_i?

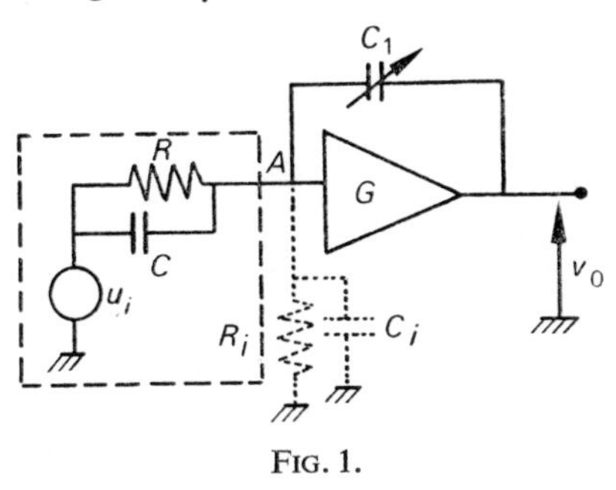

FIG. 1.

Solution. Writing the sum of the currents at A as zero we have

$$(u_i - v_a)\frac{(1+RCs)}{R} + C_1 s(G-1)\, v_a - v_a\frac{(1+R_iC_is)}{R_i} = 0$$

where v_a is the potential at A and $s = j\omega$. Then

$$v_a = u_i \frac{\dfrac{1}{R}}{\dfrac{1}{R} + \dfrac{1}{R_i} + (C+C_i)s - C_1(G-1)\,s}.$$

The output $v_O = Gv_a$ will be the image of u_i if the gain in this expression is independent of s, or

$$(C+C_i) - C_1(G-1) = 0$$

which requires

$$C_1 = \frac{C+C_i}{G-1}.$$

EXERCISE No. 20

The response of an amplifier, in both amplitude and phase, satisfies the relation

$$G(\omega) = \frac{K}{\omega_1 + j(\omega - \omega_0)}.$$

Sketch the form of $|G(\omega)|$ and $\varphi(\omega)$ in the case where $\omega_1 \ll \omega_0$.

A wave train of pulsatance ω_0 and of total length $T = 1/\omega_1$ is fed to the input of the amplifier. Using the idea of the equivalent low-pass filter, find the envelope of the high-frequency output signal.

Solution

$$|G(\omega)| = \frac{K}{\sqrt{\omega_1^2 + (\omega - \omega_0)^2}}$$

and

$$\tan \varphi = -\frac{\omega - \omega_0}{\omega_1}.$$

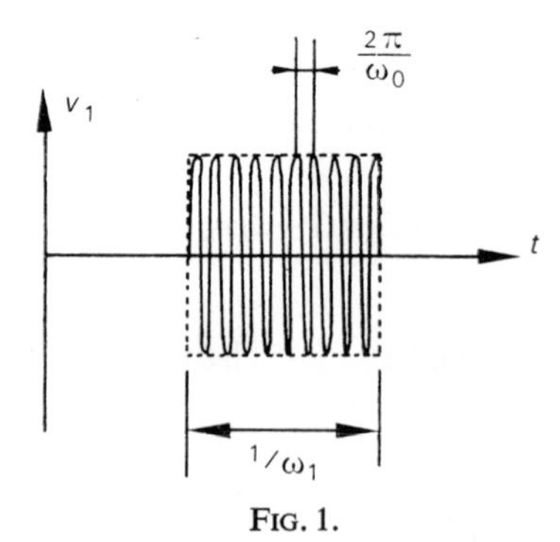

FIG. 1.

The curve of the gain against frequency is thus symmetric about the frequency ω_0 while the phase curve is antisymmetric.

For $|\omega - \omega_0| \ll \omega_1$, $|G(\omega)| = \dfrac{K}{\omega_1}$

while, for $\quad |\omega-\omega_0| \gg \omega_1, \quad |G(\omega)| = \dfrac{K}{|\omega-\omega_0|}$

and the full curve has the form of Fig. 2.

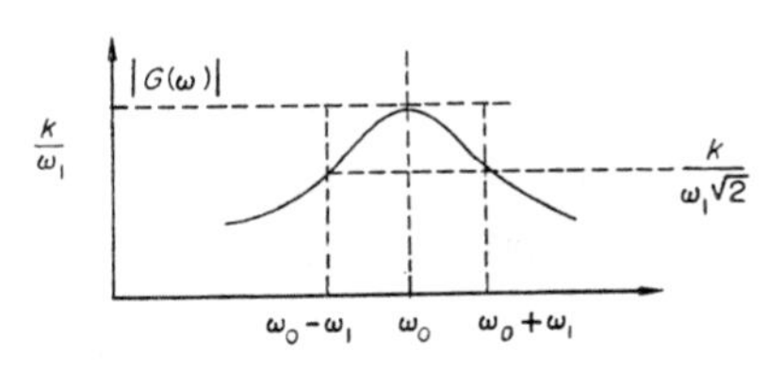

FIG. 2.

This is a case where it is possible to introduce the idea of an equivalent low-pass filter having the same form for the gain curve given, but with the central frequency transferred to $\omega_0 = 0$. This is equivalent to taking ω in place of $\omega-\omega_0$ so that

$$G_{LF}(\omega) = \frac{K}{\omega_1+j\omega} = \frac{K}{\omega_1}\cdot\frac{\omega_1}{\omega_1+j\omega}$$

which is the term K/ω_1 multiplied by the gain of an integrating RC circuit with $\omega_1 = 1/RC$.

The envelope of the response to the signal of Fig. 1 will be the same as the response of the low-pass filter when it is excited by a low frequency signal with the form of the envelope of the high frequency input signal. Now a square pulse fed to an RC circuit gives an output which follows two exponential arcs as shown in Fig. 3(a) and so the full result is that shown in Fig. 3(b).

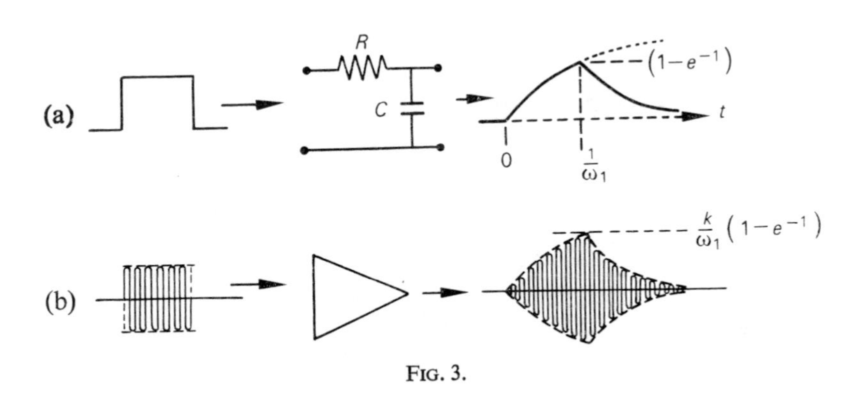

FIG. 3.

EXERCISE No. 21

1. The operational amplifier of Fig. 1 has an infinite input impedance, a zero output impedance, an infinite gain and an infinite pass-band. Calculate the gain $|G(\omega)| = v_2(\omega)/v_1(\omega)$ of the complete system and state the function performed by the circuit. Sketch the curve of

$$\log |G(\omega)| = f(\log \omega)$$

for $$R_2 = 100\ \text{k}\Omega, \quad C_1 = 1\ \mu\text{F}.$$

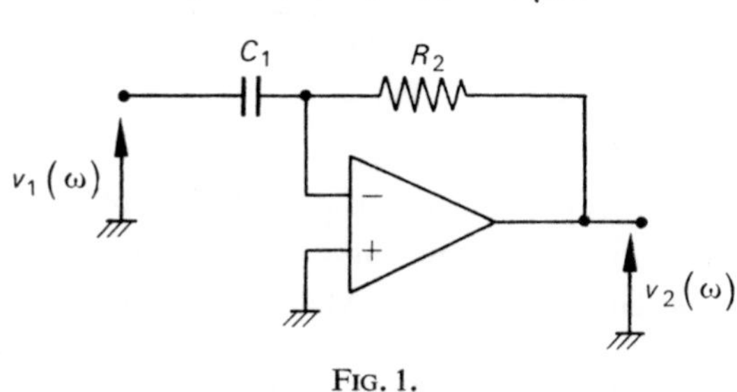

FIG. 1.

The input signal v_1 consists of a sinusoidal signal of 1 volt peak-to-peak (i.e. 1 V_{pp}) at 20 Hz on which is superimposed a noise voltage of 1 mV_{pp} which may, for simplicity, be considered to have a frequency of 500 kHz. What is the output signal $v_2(t)$ which is obtained in this case?

2. In order to avoid the inconvenient effects which appear from the above problem the circuit is modified as shown in Fig. 2. We put

$$\omega_1 = \frac{1}{R_1C_1}, \quad \omega_2 = \frac{1}{R_2C_2}$$

and choose $\omega_1 \ll \omega_2$. Calculate v_2/v_1 as a function of ω.

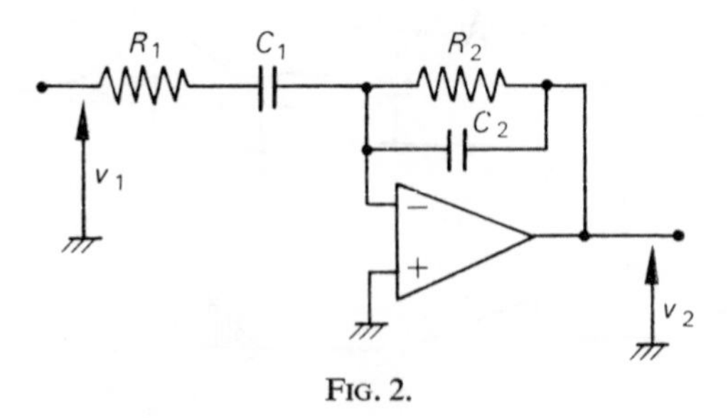

FIG. 2.

How can the representation of the circuit be simplified in the three cases where

$$\omega \ll \omega_1, \quad \omega_1 \ll \omega \ll \omega_2, \quad \omega_2 \ll \omega?$$

Deduce the asymptotic form of $|G(\omega)|$ for $R_1 = 1$ kΩ, $R_2 = 100$ kΩ, $C_1 = 1$ μF, $C_2 = 1$ nF.

What is the form of $v_2(t)$ in this case, if the signal and noise of part 2 are applied at the input?

3. Consider now that the gain of the amplifier used has a finite value G_0. Show that the gain in Fig. 1 can be written in the form

$$G = \frac{-G_0}{1+\beta(G_0+1)}$$

and calculate β. If $G_0 \gg 1$, over what frequency range will the system be a perfect differentiating circuit? Illustrate this by taking $G_0 = 10^5$.

Solution. 1. $|G(\omega)| = \omega C_1 R_2$ which corresponds to a differentiating circuit. The required curve is a straight line of slope 6 dB/octave.

At 20 Hz the gain will be $2\pi.20.10^{-1} = 4\pi$ corresponding to a sinusoidal output of 12·5 V_{pp}.

At 500 kHz the gain is $2\pi.5.10^5.10^{-1} = 10^5\pi$ which would correspond to a noise output of 314·1 V_{pp} although in fact the amplifier would be saturated. The actual signal will thus be impossible to detect and we see that a very small noise signal will block the circuit.

2. Taking the potential to be zero at the input (i.e. taking a virtual earth) the gain is

$$G(\omega) = \frac{v_2}{v_1} = \frac{-j\omega C_1 R_2}{\left(1+\frac{j\omega}{\omega_1}\right)\left(1+j\frac{\omega}{\omega_2}\right)}.$$

If $\omega \ll \omega_1$ (and hence $\omega \ll \omega_2$) the denominator can be taken as unity and

$$G(\omega) \simeq -j\omega C_1 R_2 \quad \text{or} \quad |G(\omega)| = \omega C_1 R_2 .$$

The system is again a differentiator.

If $\omega \gg \omega_1$ but $\omega \ll \omega_2$, the denominator becomes $j\omega/\omega_2$ and

$$G(\omega) \simeq -R_2/R_1$$

so that we have, effectively, an ordinary linear amplifier.

If $\omega \gg \omega_2$ (and hence $\omega \gg \omega_1$), the denominator is approximately $-\omega^2/\omega_1\omega_2$ and

$$G(\omega) \simeq \frac{1}{j\omega C_2 R_1}.$$

The system acts as an integrating circuit.

With the figures given, $\omega_1 = 10^3$ and $\omega_2 = 10^4$. Then

$$\text{for} \quad \omega = 10 \quad |G(\omega)| = 1 \quad \text{(first approximation valid)}$$

$$\text{for} \quad \omega = 10^6 \quad |G(\omega)| = 1 \quad \text{(third approximation valid)}$$

and hence the asymptotic curve is as shown in Fig. 3.

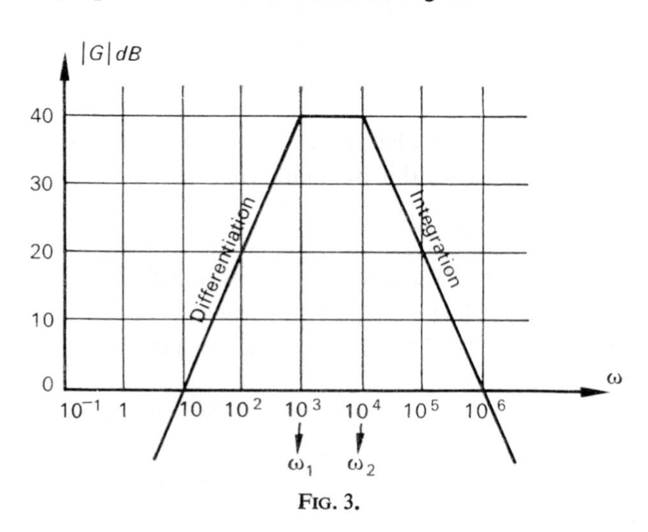

FIG. 3.

For the 20 Hz signal, $\omega = 125{\cdot}6$
and

$$|G(20)| \simeq 12{\cdot}56$$

so that v_2 will contain 20 Hz component of 12·56 V_{pp}. For the noise at 500 Hz, $\omega = 3{\cdot}14 \times 10^6$ and

$$|G(500\ \text{k})| \simeq 0{\cdot}32.$$

The noise output is thus only 0·32 mV_{pp}.

(It may be stated generally that a differentiating circuit will only operate efficiently if the band of operation is limited to that which is strictly necessary.)

3. If v_1' is the voltage at the input, we can write

$$v_2 = -G_0 v_1'$$

and

$$\frac{v_2}{v_1} = \frac{-G_0}{1+\dfrac{Z_1}{Z_2}(1+G_0)} \quad \text{with} \quad Z_1 = \frac{1}{j\omega C_1}, \quad Z_2 = R_2.$$

Thus

$$\beta = \frac{1}{j\omega C_1 R_2}.$$

If $G_0 \gg 1$ we can write

$$G = \frac{-G_0}{1+\dfrac{G_0}{j\omega C_1 R_2}}.$$

The gain will be proportional to ω if $G_0/\omega C_1 R_2 \gg 1$ which requires that

$$\omega \ll \frac{G_0}{C_1 R_2}.$$

For $G_0 = 10^5$ this requires $\omega \ll 10^6$.

EXERCISE No. 22

The amplifier used in the system of Fig. 1 has a gain slightly less than unity, i.e.

$$A = 1-\varepsilon$$

while its input impedance is infinite, its output impedance is zero and the impedance of the circuit R_0C_0 is very large and can be neglected. Show that the network between M and N can be represented by a self-inductance and give the expression for the quality factor Q. For what frequency is Q a maximum? As a numerical example take $\varepsilon = 1\%$, $R = 5$ kΩ, $R_0 = 10$ MΩ, $C_0 = 0{\cdot}1$ μF and $\omega = 10$ rad/sec.

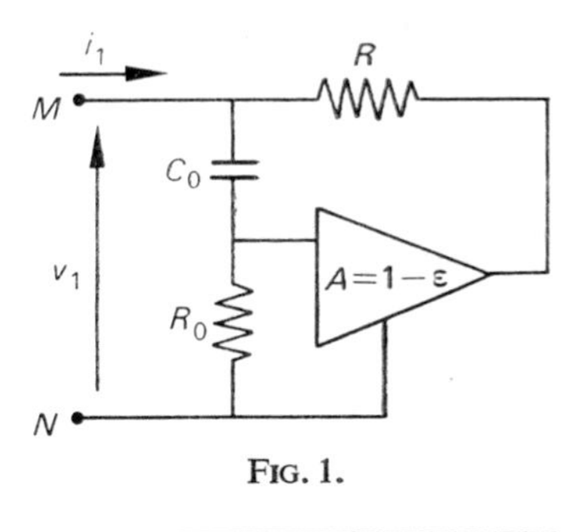

FIG. 1.

Solution. The potential divider formed by R_0 and C_0 gives an input voltage to the amplifier of

$$v_0 = v_1 \frac{j\omega C_0 R_0}{1+j\omega C_0 R_0}.$$

The output voltage is

$$v_2 = v_0(1-\varepsilon) = Av_0.$$

If the current through the R_0C_0 combination is neglected, the input current will be

$$i_1 = \frac{v_1 - v_2}{R} = \frac{v_1}{R}\left(1 - \frac{j\omega C_0 R_0 A}{1 + j\omega C_0 R_0}\right)$$

$$= \frac{v_1}{R}\left(\frac{1 + jx\varepsilon}{1 + jx}\right) \quad \text{where} \quad x = R_0 C_0 \omega = \frac{\omega}{\omega_0}.$$

The input impedance is thus

$$Z_i = \frac{v_1}{i_1} = R\left[\frac{1 + \varepsilon x^2}{1 + \varepsilon^2 x^2} + \frac{jx(1 - \varepsilon)}{1 + \varepsilon^2 x^2}\right].$$

This corresponds to a resistance

$$R' = \frac{1 + \varepsilon x^2}{1 + \varepsilon^2 x^2} R$$

in series with a self inductance

$$L = \frac{R(1 - \varepsilon) R_0 C_0}{1 + \varepsilon^2 x^2} \simeq R \cdot \frac{R_0 C_0}{1 + \varepsilon^2 x^2}.$$

The quality factor is

$$Q = \frac{\omega L}{R} = \frac{(1 - \varepsilon)x}{1 + \varepsilon x^2} \simeq \frac{x}{1 + \varepsilon x^2}.$$

This is a maximum for $x = 1/\sqrt{\varepsilon}$ when

$$Q_{\max} = \frac{1}{2\sqrt{\varepsilon}}.$$

Numerical example:

$$\omega_0 = \frac{1}{R_0 C_0} = 1 \quad x = \frac{\omega}{\omega_0} = 10.$$

Then

$$L = 5000 \text{ henrys}, \quad R = 10 \text{ k}\Omega$$

and

$$Q = 5.$$

(For the given value of ε this is, in fact, the value $Q_{\max}$.)

The inductance simulated is thus large but with a very small Q.

EXERCISE No. 23

Synthesis of Transfer Functions with Integrator Circuits.

If the operational amplifiers used are perfect, with infinite gain and input impedance, output impedance zero and pass-band infinitely large, calculate the output voltage for the circuits shown in Figs. 1 and 2 below.

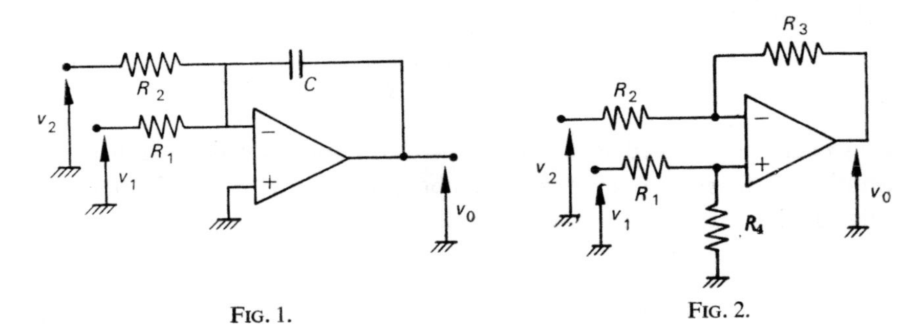

FIG. 1. FIG. 2.

Show how the transfer function

$$H(S) = \frac{v_0(s)}{v_i(s)} = -\frac{2s+1}{6s+1}$$

can be synthesized with the aid of two operational amplifiers.

Solution. Since the gain is infinite, the differential input voltage must be infinitely small. Then, in the circuit of Fig. 1, v_1 must be effectively zero (i.e. we have a virtual earth). Also, since the input current of the amplifier will be zero, we can write at the junction of R_1, R_2 and C

$$\frac{v_1}{R_1}+\frac{v_2}{R_2}+v_0Cs = 0 \qquad (s = j\omega).$$

From this

$$v_0 = -\frac{1}{Cs}\left(\frac{v_1}{R_1}+\frac{v_2}{R_2}\right).$$

For Fig. 2 the input voltage at the positive terminal will be $v_1R_4/(R_1+R_4)$. Then, for the input current to be zero, at the junction R_2R_3 we have

$$\frac{v_2-v_1\dfrac{R_4}{R_1+R_4}}{R_2}+\frac{v_0-v_1\dfrac{R_4}{R_1+R_4}}{R_3} = 0$$

and

$$v_0 = \left(\frac{R_2+R_3}{R_2}\right)\left(\frac{R_4}{R_1+R_4}\right)v_1 - \frac{R_3}{R_2}v_2$$

The condition

$$H(s) = \frac{v_2}{v_1} = -\frac{2s+1}{6s+1}$$

can also be written as

$$v_2 = -\frac{v_1}{3} - \frac{v_1+v_2}{6s}. \tag{1}$$

Thus v_2 can be obtained as the sum of two terms, one of which results from an integration. Reference to the results just obtained indicates a combination of the circuits of Figs. 1 and 2 as shown in Fig. 3.

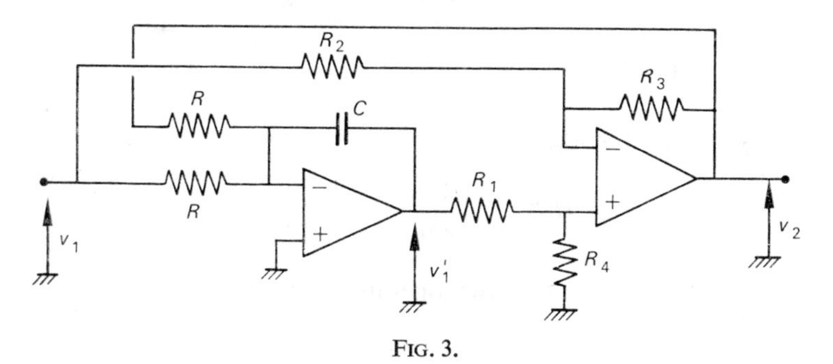

FIG. 3.

The conditions that equation (1) is satisfied are that

$$\frac{R_3}{R_2} = \frac{1}{3}, \quad RC = 1$$

$$\left(1+\frac{R_3}{R_2}\right)\left(\frac{R_4}{R_1+R_4}\right) = \frac{1}{6} \quad \text{from which} \quad \frac{R_1}{R_4} = 7.$$

We see then that

$$v_1' = -\frac{v_1+v_2}{s}$$

and

$$v_2 = \frac{1}{6}v_1' - \frac{1}{3}v_1 = -\frac{v_1}{3} - \frac{v_1+v_2}{6s}.$$

(This method can be generalized to higher order transfer functions.)

EXERCISE No. 24

Consider the transfer function

$$\frac{v_2}{v_1} = \frac{1}{8} \cdot \frac{s}{s^2 + \frac{1}{8}s + 36}.$$

What is the nature and the characteristics of the corresponding filter? How can such a filter be formed with the aid of perfect operational amplifiers used as integrators and adders?

Solution. This is a second order transfer function corresponding to a pass-band as in Exercise 14, Chapter 1. We can write

resonance pulsatance	$\omega_0 = 6$
quality factor	$Q = 48$
maximum gain (on resonance)	$= 1.$

Dividing through the expression by s^2 gives

$$\frac{v_2}{v_1} = \frac{\frac{1}{8s}}{1 + \frac{1}{8s} + \frac{36}{s^2}}$$

or

$$v_2 = -\frac{1}{s}\left[-\frac{1}{8}v_1 + \frac{1}{8}v_2 - \left(-\frac{36}{s}v_2\right)\right] \tag{1}$$

which can be obtained with only three operational amplifiers as shown in Fig. 1.

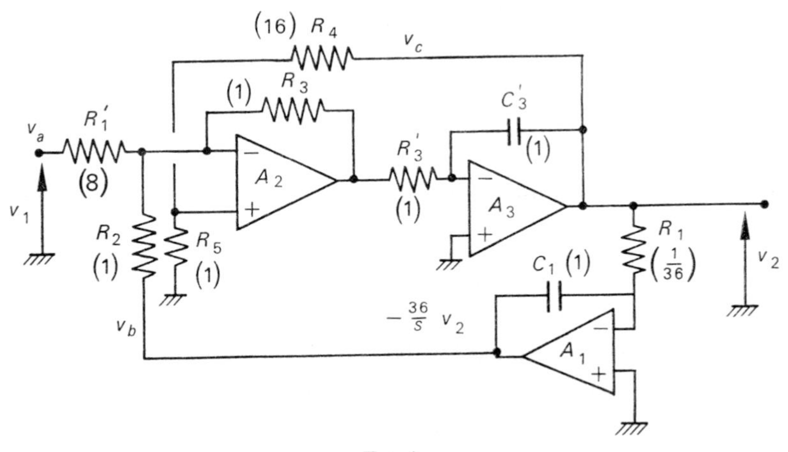

FIG. 1.

Here A_1 gives the function $-(36/s)v_2$ requiring $R_1C_1 = 1/36$. A_2 is used to produce the addition of the three terms in the bracket of equation (1). It is easy to show that

$$v_2' = -v_a\frac{R_3}{R_1'} - v_b\frac{R_3}{R_2} + \frac{R_3R_5}{R_4+R_5}\left(\frac{1}{R_1'}+\frac{1}{R_2}+\frac{1}{R_3}\right)v_c$$

so that the necessary condition is

$$\frac{R_3}{R_1'} = \frac{1}{8}, \quad \frac{R_3}{R_2} = 1 \quad \text{and} \quad \frac{R_3R_5}{R_4+R_5}\left(\frac{1}{R_1'}+\frac{1}{R_2}+\frac{1}{R_3}\right) = \frac{1}{8}$$

which requires, in relative magnitudes,

$$R_3 = 1, \quad R_1' = 8, \quad R_2 = 1 \quad \text{and} \quad \frac{R_4}{R_5} = 16.$$

The final integration is produced by A_3 with $R_3'C_3' = 1$.

EXERCISE No. 25

The gyrator

In Fig. 1 below the two gyrators are symmetrical with the same gyration resistance R. Draw the circuit for the equivalent four terminal network made up of classical R, C and L elements and having the same transfer function.

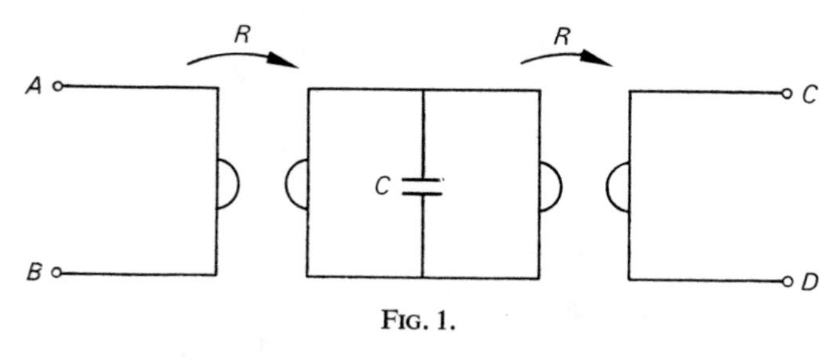

FIG. 1.

Solution. A gyrator is a four-terminal network having an impedance matrix of the form

$$(Z) = \begin{pmatrix} 0 & -R \\ R & 0 \end{pmatrix}.$$

That is, the impedance is inverted for the two directions and the input impedance is related to the load impedance as

$$Z_i = \frac{R}{Z_L}.$$

The second gyrator of the system of Fig. 1 has an infinite load impedance and, therefore, a zero input impedance which, effectively, short circuits C. If the output CD is floating, therefore, the input impedance at AB will be infinite.

If CD is short-circuited so that the input impedance of the second gyrator is infinite, the input impedance at AB is given by

$$Z_{AB} = R^2 j\omega C = j\omega(R^2C)$$

which is equivalent to a self-inductance $L = R^2C$. The equivalent circuit is thus that shown in Fig. 2.

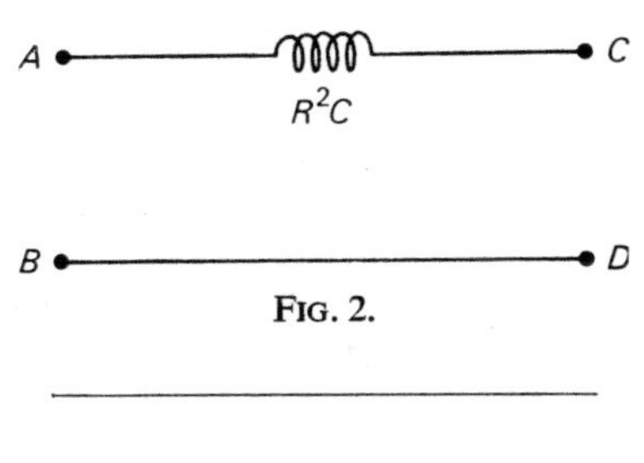

FIG. 2.

EXERCISE No. 26

The Yanagisawa Synthesis

1. The operational amplifier used in the circuit of Fig. 1 is perfect, i.e. $G = \infty$, $Z_i = \infty$, $Z_o = 0$. Calculate the input impedance Z_{AB} between A and B. Deduce the circuit for a negative impedance converter for the current (an I.NIC).

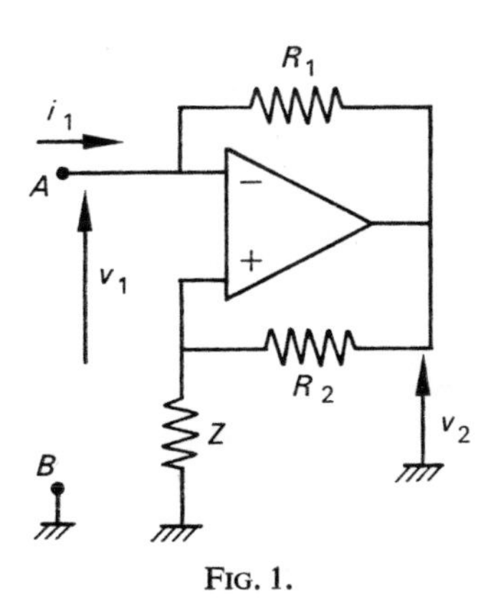

FIG. 1.

If the I.NIC has to have a unit transformation coefficient, what relationship must exist between R_1 and R_2?

2. Consider the circuit of Fig. 2 proposed by Yanagisawa using an I.NIC with coefficient unity. Calculate the transfer function v_2/v_1 in terms of the circuit admittances.

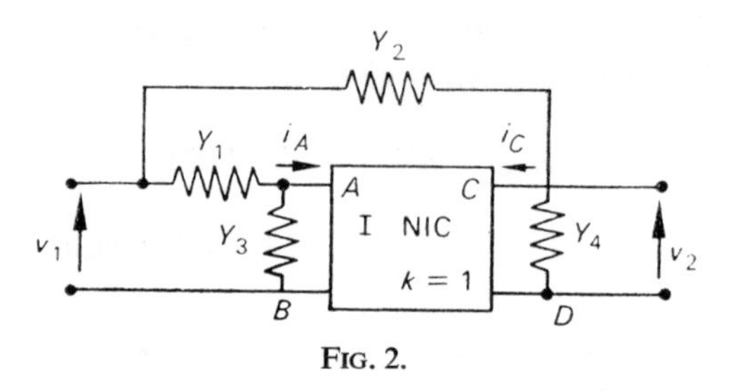

FIG. 2.

3. Knowing that any transfer function $A(s)/B(s)$ can be written as

$$H(s) = \frac{A(s)}{B(s)} = \frac{1}{1+\dfrac{B(s)-A(s)}{A(s)}} = \frac{1}{1+\dfrac{\dfrac{B(s)-A(s)}{Q(s)}}{\dfrac{A(s)}{Q(s)}}}$$

where $Q(s)$ is a polynomial in s which does not differ in degree by more than one from $A(s)$ or $B(s)$, the quantities $(A-B)/Q$ and A/Q may be considered as input functions of either impedance or admittance.

Taking note of this fact and writing $Q(s) = s+1$, express (Y_3-Y_4) and (Y_1-Y_2) in terms of s such that the system of Fig. 2 has a transfer function

$$H(s) = \frac{s-1}{s^2+2s+2}.$$

4. Write the expression (Y_3-Y_4) in the form

$$As+B+\frac{Cs}{s+\gamma}$$

and find A, B, C and γ.

Write (Y_1-Y_2) in the form

$$A'+\frac{B's}{s+\beta}$$

and find A', B' and β.

Hence deduce the circuit for the Yanagisawa filter which will give the required transfer function and give the normalized values of the components.

Solution. 1. The input impedance of the amplifier is infinite and so all the input current goes through R_1. Then

$$i_1 = \frac{v_1 - v_2}{R_1}.$$

Since the amplifier gain is infinite, the two inputs must be at the same potential v_1 so that

$$v_1 = v_2 \frac{\mathbf{Z}}{R_2 + \mathbf{Z}}.$$

Thus

$$i_1 = \frac{-R_2}{ZR_1} \cdot v_1$$

from which

$$Z_{AB} = -\frac{R_1}{R_2} . Z.$$

The circuit is thus an I.NIC with a coefficient

$$k = \frac{R_1}{R_2}.$$

(There will be the same potential on the load Z as at the inputs.)

This coefficient k will be unity if $R_1 = R_2$.

2. In Fig. 2 $\quad v_{AB} = v_{CD} \quad$ (I.NIC)

and $\quad i_A = i_C \quad$ (since $k = 1$).

Taking the condition for a current node we see that

$$\frac{v_2}{v_1} = \frac{Y_1 - Y_2}{Y_1 - Y_2 + Y_3 - Y_4}.$$

3.

$$H(s) = \frac{1}{1 + \dfrac{Y_3 - Y_4}{Y_1 - Y_2}}$$

for the circuit of Fig. 2. We require

$$\frac{A(s)}{B(s)} = \frac{s-1}{s^2 + 2s + 2}$$

from which we can take

$$\frac{B(s) - A(s)}{Q(s)} = \frac{s^2 + s + 3}{s+1} = Y_3 - Y_4$$

and

$$\frac{A(s)}{Q(s)} = \frac{s-1}{s+1} = Y_1 - Y_2.$$

4. If we write

$$\frac{s^2+s+3}{s+1} = s+3-\frac{3s}{s+1}$$

we can identify $Y_3 = s+3, \quad Y_4 = \dfrac{3s}{s+1} = \dfrac{1}{\dfrac{1}{3}+\dfrac{1}{3}s}.$

Similarly

$$\frac{s-1}{s+1} = \frac{2s}{s+1} - 1$$

so that we can identify

$$Y_1 = \frac{2s}{s+1} = \frac{1}{\dfrac{1}{2}+\dfrac{1}{2s}} \quad \text{and} \quad Y_2 = 1.$$

Thus the required circuit is that shown in Fig. 3.

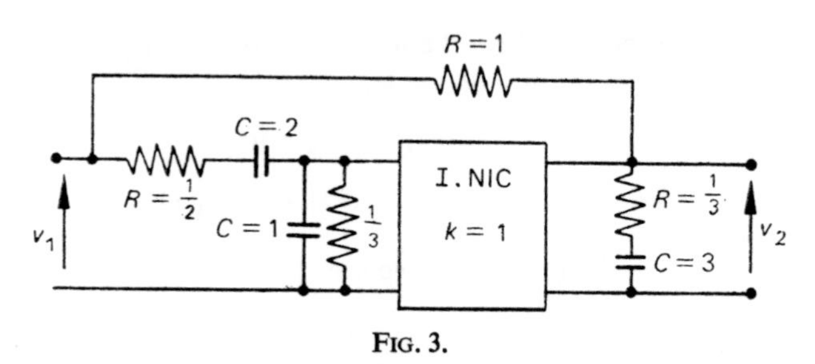

FIG. 3.

Chapter 3

Feedback and Stability

EXERCISE No. 1

An amplifier with nominal gain G_0 is fed by a source of e.m.f. 0·028 V_{rms} and negligible internal resistance. The output voltage is 36 V_{rms} and the second harmonic distortion is measured to be 7%.

1. By means of a feedback network a fraction 1·2% of the output is returned to the input such that βG_0 in the classical feedback expression is real and positive. What will be the new output voltage if the input is unchanged?

2. If we require the amplifier to give an output of 36 V_{rms} with 1% of second harmonic what must be the new level of feedback and what must be the new input voltage?

Solution. 1. The nominal voltage gain is

$$G_0 = \frac{V_2}{V_1} = \frac{36}{0{\cdot}028} = 1285.$$

With negative feedback

$$G = \frac{G_0}{1+\beta G_0} = 78{\cdot}3.$$

Then, with $V_1 = 0{\cdot}028\ \mathrm{V_{rms}}$, $V_2 = V_1 \times G = 2{\cdot}2\ \mathrm{V_{rms}}$.

2. With negative feedback the distortion is reduced by a factor $1+\beta G_0$ so that

$$1\% = \frac{7\%}{1+\beta G_0}$$

and $\beta G_0 = 6$ or $\beta \simeq 0{\cdot}5\%$.

In this case the gain is

$$G = \frac{1285}{7} = 183{\cdot}5.$$

To give 36 V_{rms} output will then require an input

$$V_1' = \frac{36}{183{\cdot}5} = 0{\cdot}19\ \mathrm{V_{rms}}.$$

EXERCISE No. 2

Show that the three-stage amplifier of Fig. 1, for which the field effect transistors F_1, F_2 and F_3 are identical, can oscillate if the internal resistance of the H.T. supply, r, is not negligible.

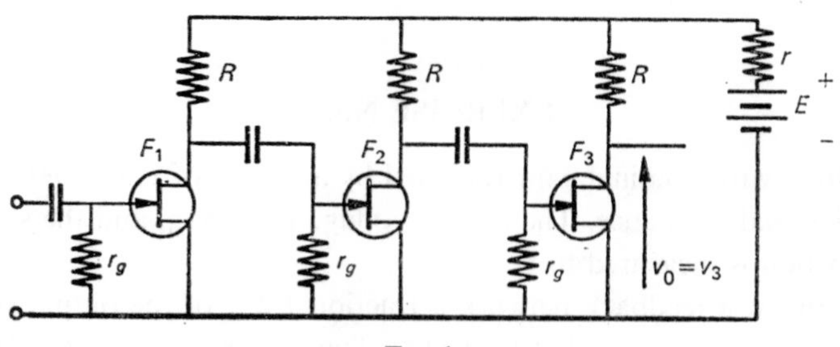

FIG. 1.

Take g to be the slope of the F.E.T.'s and assume r_g, the gate resistance, is very much larger than the load R.

As a numerical example calculate the maximum allowable value for r if g is 4 mA/V and R is 10 kΩ.

Solution. If u_0 is the input voltage and v_1, v_2, v_3 are the alternating drain potentials of F_1, F_2 and F_3, the currents of the three F.E.T.'s will be

$$i_{a1} = gu_0, \qquad i_{a2} = gv_1 \quad \text{and} \quad i_{a3} = gv_2.$$

The potential at the common point of the loads will be

$$E - r(I_{a1} + I_{a2} + I_{a3}).$$

Taking the alternating components only, the F.E.T. potentials are then

$$\begin{aligned} v_3 &= -rg(u_0 + v_1 + v_2) - Rgv_2 \\ v_2 &= -rg(u_0 + v_1 + v_2) - Rgv_1 \\ v_1 &= -rg(u_0 + v_1 + v_2) - Rgu_0 \end{aligned}$$

from which

$$v_3 = \frac{A}{\Delta} u_0$$

with Δ as the determinant

$$\Delta = \begin{vmatrix} 1 & g(r+R) & rg \\ 0 & 1+rg & g(r+R) \\ 0 & rg & 1+rg \end{vmatrix} = 1 + 2rg - rRg^2.$$

If Δ is zero the system can oscillate and this requires

$$g = \frac{1}{R}\left[1+\sqrt{1+\frac{R}{r}}\right].$$

or

$$r = \frac{1}{g(Rg-2)}.$$

In the numerical example this requires

$$r \leqslant 6{\cdot}58\,\Omega.$$

EXERCISE No. 3

The Nyquist Criterion

An amplifier of nominal gain G_0 has feedback through a loop network of gain β. The open loop gain is such that

$$\beta G_0 = \frac{A}{(1+RCs)^2}.$$

Study the stability of the system as a function of A using the Nyquist criterion.

Solution. An amplifier which has feedback will be stable if, when ω varies from $-\infty$ to $+\infty$, the point representing the open loop gain $\beta(\omega)\,G_0(\omega)$ describes a locus in the complex plane which does not enclose the point -1.

If we put $1/RC = \omega_0$ so that $\theta = \text{arc tan}\ \omega/\omega_0$, we have

$$\beta G_0 = \frac{A}{\left(1+j\dfrac{\omega}{\omega_0}\right)^3} = \frac{Ae^{-j3\theta}}{\left[1+\left(\dfrac{\omega}{\omega_0}\right)^2\right]^{3/2}}.$$

For $\quad \omega = 0, \quad \theta = 0, \quad \beta G_0 = A.$

For $\quad 3\theta = -\pi, \quad \theta = -\dfrac{\pi}{3}, \quad \dfrac{\omega}{\omega_0} = -\sqrt{3}, \quad \beta G_0 = \dfrac{A}{8}\,e^{-j\pi}.$

For $\quad \omega \to \infty, \quad \beta G_0 \to 0, \quad 3\theta \to -\dfrac{3\pi}{2}.$

The form of the Nyquist diagram is thus as shown in Fig. 1.

We see that the system is:

$$\text{stable if} \quad A < 8$$

$$\text{unstable if } A > 8$$

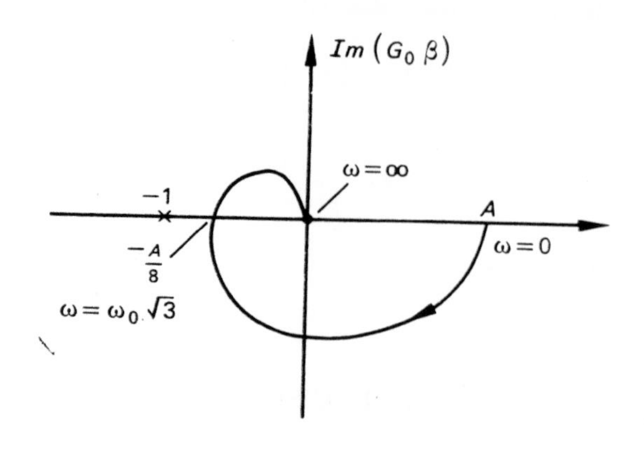

FIG. 1.

EXERCISE No. 4

Stability and the Operational Amplifier

An operational amplifier has an infinite input impedance and its gain varies with frequency as shown in Fig. 1.

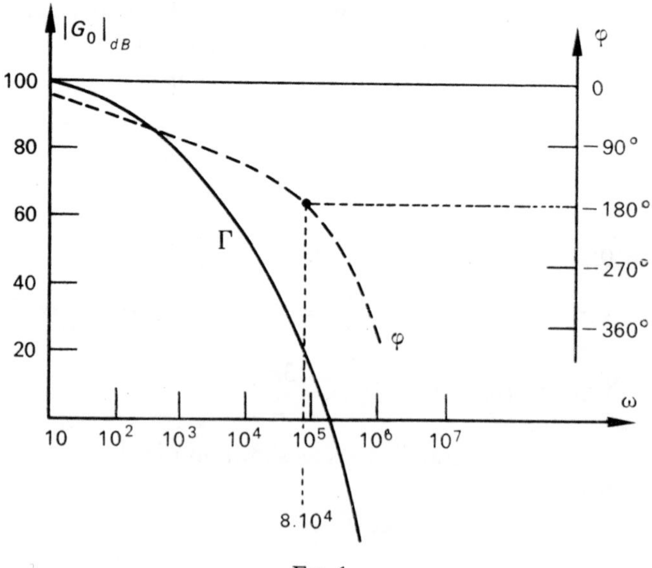

FIG. 1.

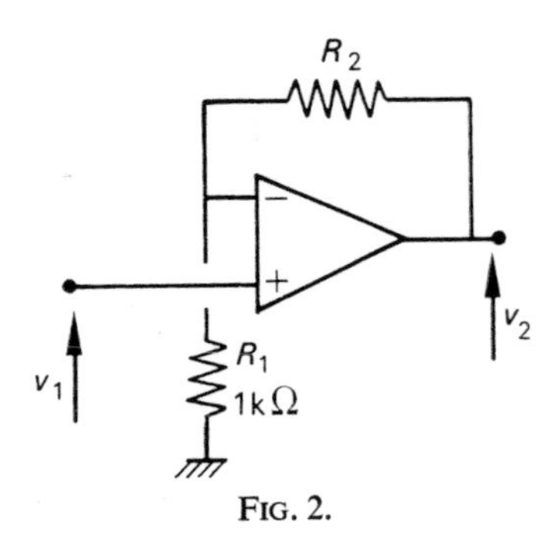

FIG. 2.

1. The amplifier is used in the circuit of Fig. 2 to produce a positive gain amplifier. What is the maximum value of R_2 for the system to be stable?

2. The resistance R_2 is now replaced by a capacitance C. Give the expressions for v_2/v_1 as a function of frequency in the case where the amplifier gain is (a) infinite, (b) given by Fig. 1.

3. Use a graphical method to find if the circuit is stable for (a) $C = 0{\cdot}3\ \mu F$ anp (b) $C = 0{\cdot}3$ nF.

Solution. 1. The system gain can be written

$$G = \frac{G_0}{1+\beta G_0}$$

with $\beta = R_1/(R_1+R_2)$ being real.

The modulus and the phase of the amplifier gain vary in a regular manner and so the Nyquist diagram will have a simple form. It follows that the system will be stable if, at the frequency where the phase of βG_0 reaches 180°, the modulus of the open loop gain is less than 1. If we draw on the same diagram as the gain curve the gain $|1/\beta|$ will cut the curve Γ at a point M with pulsatance ω_M where $|G_0| = |1/\beta|$ or

$$|\beta(\omega_M)\,G_0(\omega_M)| = 1.$$

For $\qquad \omega < \omega_M, \quad |G_0| > \left|\frac{1}{\beta}\right|, \quad |\beta G_0| > 1.$

For $\qquad \omega > \omega_M \quad |\beta G_0| < 1.$

On the other hand, if N is the point on Γ for which the phase of βG_0, which is here equivalent to the phase of G_0, is 180°, there will be stability for the frequency ω_N at N if $|\beta(\omega_N)\,G_0(\omega_N)| < 1$. That is, if N is to the right of M.

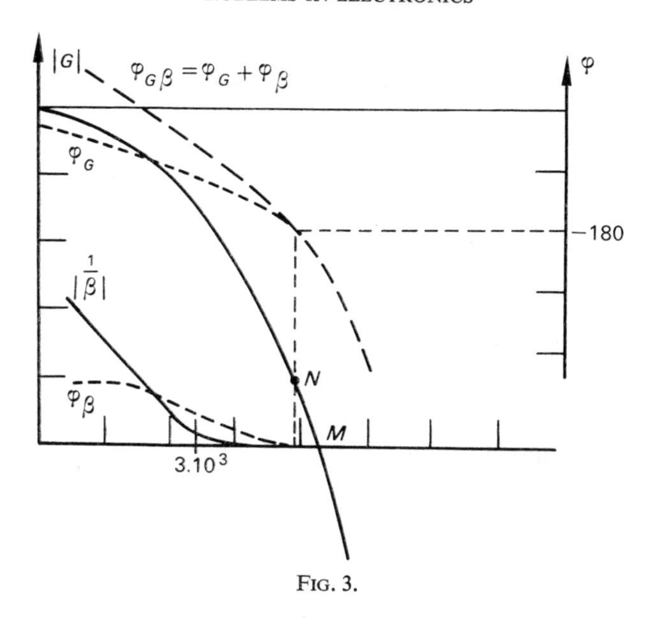

FIG. 3.

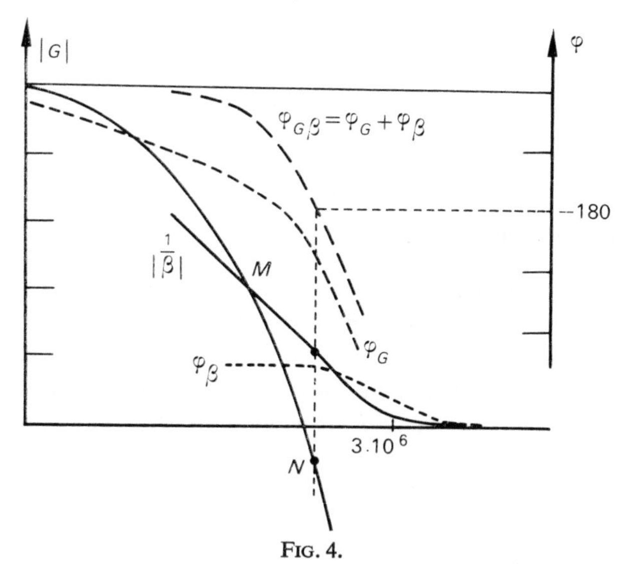

FIG. 4.

We see that here φ is 180° for $\omega \simeq 8\times10^4$ for which the gain is 20 dB. Thus we require

$$\beta = \frac{R_1}{R_1+R_2} = \frac{1}{10}$$

the least value of R_2 is $9R_1$ or 9 kΩ.

2. (a) If G_0 is infinite

$$\frac{v_2}{v_1} = 1+\frac{1}{j\omega CR_1}.$$

(b) For the finite gain we have the same expression as in part 1 with

$$\beta = \frac{R_1}{R_1+\dfrac{1}{j\omega C}} = \frac{j\omega CR_1}{1+j\omega CR_1} = \frac{j\dfrac{\omega}{\omega_1}}{1+\dfrac{j\omega}{\omega_1}}$$

and $$\frac{1}{\beta} = 1-j\frac{\omega_1}{\omega} \quad \text{where} \quad \omega_1 = \frac{1}{R_1C}.$$

3. In order to study the stability in this case we look for the value of frequency for which βG_0 has a phase of 180° and the value of $|\beta G_0|$ for this frequency.

$$\beta = \frac{1}{\sqrt{1+\dfrac{\omega_1^2}{\omega^2}}}\, e^{j\phi_B} \quad \text{with} \quad \varphi_\beta = \text{arc tan}\,\frac{\omega_1}{\omega}.$$

On the same graph as $|G_0(\omega)|$ we draw $1/|\beta|$ and $\varphi_\beta(\omega)$ with the two values of ω_1 taken as (a) $3{\cdot}3\times10^3$ and (b) $3{\cdot}3\times10^6$ rad/sec. The results obtained are shown respectively on Figs. 3 and 4.

(a) For 0·3 μF there is instability for the pulsatance ω_M for which the total phase angle is π which is found in a region where $|\beta G_0| > 1$, i.e. N to the left of M.

(b) For 0·3 nF there is always stability. (This demonstrates in a general fashion that an integrating circuit with a large time constant cannot be obtained if the operational amplifier is not unconditionally stable.)

EXERCISE No. 5

It is required to use the circuit of Fig. 1 to produce a low frequency oscillator. Give the equivalent a.c. circuit formed by the series-parallel association of two four-terminal networks (i.e. in series at the input and in parallel at the output).

Calculate the h_{ij} parameters of the feedback circuit assuming that the transformer is perfect ($k = 1$).

For what turns ratio n will the circuit oscillate and what will be the value of the oscillation frequency?

What is the lowest value of load R_L, for a given m, for oscillation to be impossible?

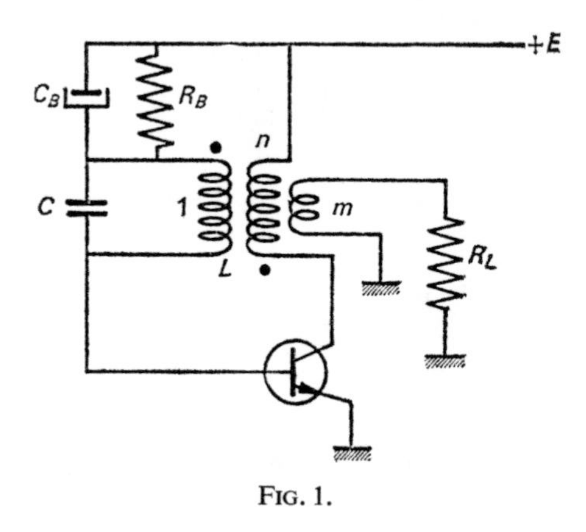

FIG. 1.

Solution. Since C_B is an electrolytic condenser, it will have a large value and the top of the *LC* circuit will be earthed for a.c. Then, taking note of the direction of the feedback winding, we can draw the equivalent circuit of Fig. 2.

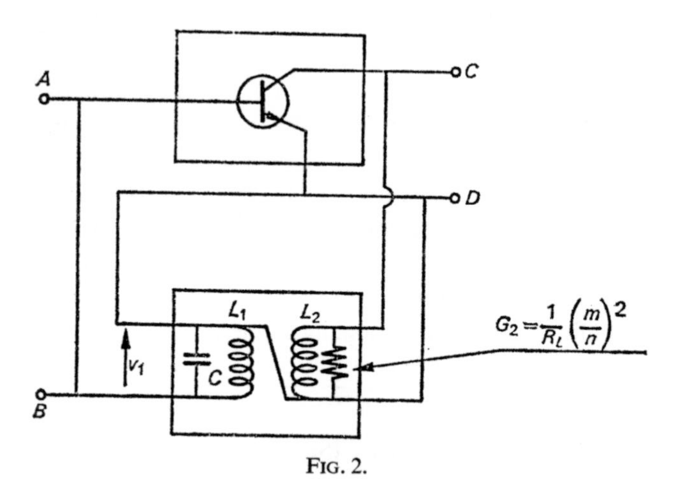

FIG. 2.

Let us calculate first h'_{11} and h'_{21} with the secondary short-circuited. With the transformer perfect the resistance at the primary is zero and $v_1 = 0$ so that $h'_{11} = 0$. In this case all the primary current goes through L_1 since the impedance of C is not zero. Then $i_2 = -i_1/n$ and $h'_{21} = -1/n$. The other two parameters are obtained on taking $i_1 = 0$. If we then consider the secondary side only we have the circuit of Fig. 3, from which

$$h'_{22} = \mathrm{j}\frac{\omega C}{n^2} + \frac{1}{n^2 \mathrm{j}\omega L_1} + G_2 .$$

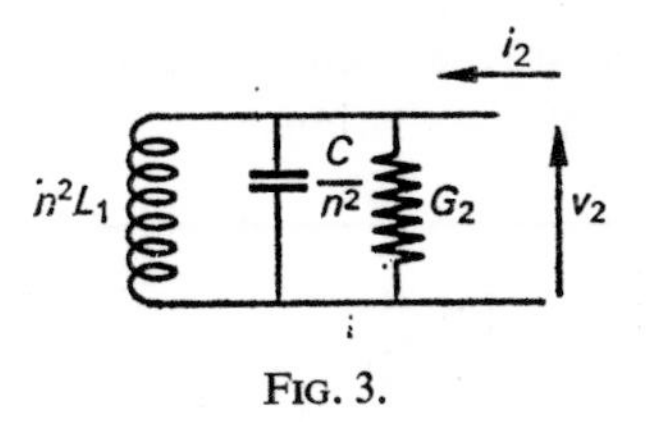

FIG. 3.

Further, because the transformer is perfect

$$v_2 = +nv_1 \quad \text{so that} \quad h'_{21} = +\frac{1}{n}.$$

The network *ABCD* has parameters $H_{ij} = h_{ij}+h'_{ij}$ and the oscillation condition is

$$\Delta H = 0$$

or

$$h_{11e}\left(h_{22e}+G_2+\frac{j\omega C}{n^2}+\frac{1}{n^2 j\omega L_1}\right)-\left(\frac{1}{n}+h_{12e}\right)\left(h_{21e}-\frac{1}{n}\right) = 0.$$

Putting $h_{22}+G_2 = G'_2$ and separating the real and imaginary parts gives

$$\omega = \frac{1}{\sqrt{L_1 C}}$$

and, neglecting h_{12} ($\sim 10^{-4}$) with respect to $1/n$

$$h_{11e}G'_2-\frac{1}{n}\left(\beta-\frac{1}{n}\right) = 0$$

and

$$n = \frac{\beta\pm\sqrt{\beta^2-4h_{11e}G'_2}}{2h_{11e}G'_2}.$$

Oscillation can occur for $\beta^2-4h_{11e}G'_2 > 0$, which requires

$$R_L > \left(\frac{m}{n}\right)^2 \frac{1}{\dfrac{\beta^2}{4h_{11e}}-h_{22e}}.$$

EXERCISE No. 6

An amplifier with infinite input impedance and zero output impedance, supplied by a signal $u_0 = U_0 \sin \omega t$ gives an output

$$v_1 = G\left(1-\frac{u_0}{A}-\frac{u_0}{B}\right)^2 u_0.$$

1. Calculate, in terms of U_0, the gain at frequency ω.

2. If the amplifier is used in the oscillator circuit of Fig. 1 what will be the frequency and amplitude of the oscillations?

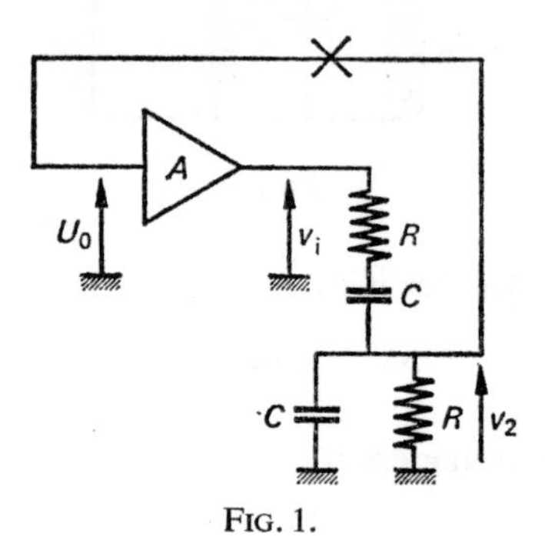

FIG. 1.

Solution. 1.

$$v_1 = G\left(1-\frac{U_0 \sin \omega t}{A}-\frac{U_0^2 \sin^2 \omega t}{B}\right)U_0 \sin \omega t.$$

The terms in $\sin^2 \omega t$ only contribute to the second harmonic, while

$$\frac{G}{B}\cdot U_0 \sin^3\omega t = \frac{3}{4}\frac{GU_0^3}{B}\sin \omega t-\frac{GU_0^3}{4B}\sin^3 \omega t.$$

The terms in ω in v_1 are then

$$v_{1\omega} = GU_0\left(1-\frac{3}{4}\frac{U_0^2}{B}\right)\sin \omega t.$$

The gain at the frequency ω is therefore

$$G_\omega = G\left(1-\frac{3}{4}\frac{U_0^2}{B}\right)$$

which is a real quantity.

2. Since the gain of the amplifier is real, oscillation will occur for a frequency where the feedback filter re-injects at the input a voltage in phase with (or in opposition to) the output v_1. Now

$$\frac{v_2}{v_1} = \frac{\omega CR}{3\omega CR+j(\omega^2C^2R^2-1)},$$

which is real and positive for $\omega^2C^2R^2 = 1$ or $\omega = 1/RC$. At this frequency

$$\frac{v_2}{v_1} = \frac{1}{3}.$$

The oscillation condition is

$$v_2 = u_0$$

and so

$$\frac{v_1}{3} = u_0$$

or

$$\frac{1}{3}G\left(1-\frac{3U_0^2}{4B}\right)U_0 = U_0$$

from which the amplitude of oscillation is

$$U_0 = 2\sqrt{B\left(\frac{1}{3}-\frac{1}{G}\right)}.$$

EXERCISE No. 7

In the frequency changer circuit shown in Fig. 1 below, we assume that the frequencies considered are sufficiently low for the current gain to be taken as real and the input characteristic to be taken as that of the base-emitter diode given by

$$I_B = I_{BO}(e^{V_B/\psi}-1)$$

$$\psi = 25 \text{ mV}, \quad I_{BO} = 10^{-7} \text{ A}.$$

The input voltage, $a \cos \omega t$, is applied in series with the local oscillator voltage $b \cos \Omega t$ in the base circuit. The working point

$$V_{BE} = V_M, \quad I_B = I_M$$

can be adjusted by means of the bias system R_1R_2. The output voltage at the frequency $(\omega+\Omega)$ is taken from the collector terminal.

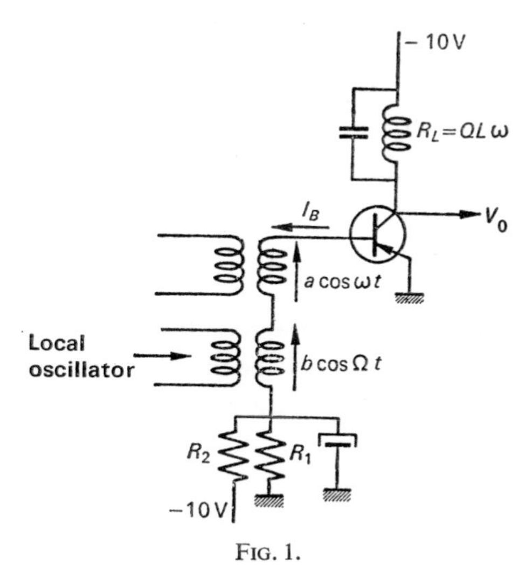

FIG. 1.

By considering a Taylor expansion to third order calculate the conversion factor

$$g_{mc} = \frac{i_2(\Omega+\omega)}{V_1(\omega)}.$$

Solution

$$I_B = I_{BO}(e^{V_B/\psi}-1),$$

so, with a base voltage V_M+v, we have

$$I_B(V_M+v) = I_{BO}(e^{(V_M+v)/\psi}-1).$$

This gives the change in current as

$$i_B = I_B(V_M+v)-I_B(V_M) = vI'_B(V_M)+\frac{v^2}{2!}I''_B(V_M)+\frac{v^3}{3!}I'''_B(V_M)$$

$$= v\,\frac{I_B(V_M)+I_{BO}}{\psi}+\frac{v^2}{2}\,\frac{I_{BM}+I_{BO}}{\psi^2}+\frac{v^3}{6}\,\frac{I_B(V_M)+I_{BO}}{\psi^3}$$

or

$$\frac{i_B}{I_B(V_M)+I_{BO}} = \frac{1}{\psi}\left(v+\frac{v^2}{2\psi}+\frac{v^3}{6\psi^2}\right). \tag{1}$$

The applied a.c. base potential is

$$v = a \sin \omega t+b \sin \Omega t.$$

With this voltage substituted in (1) the only term which gives a frequency $(\omega+\Omega)$ is that in v^2 from which we have

$$v^2 \rightarrow ab \cos (\omega+\Omega)t,$$

so that

$$i_B(\omega+\Omega) = \frac{I_B(V_M)+I_{BO}}{\psi^2}\,ab \cos (\omega+\Omega)t$$

and

$$g_{mc} = \frac{I_B(V_M)+I_{BO}}{\psi^2}\cdot b\beta.$$

We find the conversion factor to depend on the working point and to be linear in the local signal voltage. This will only be the exact case for b small, as we have considered, but is not generally the case in practice.

Chapter 4

Non-Linear Circuits and Switching

EXERCISE No. 1

Consider the monostable circuit of Fig. 1.

1. What value of R_{B_2} is required for the transistor T_2 to be saturated with no input? (Take $R_{B_2} \gg R_2$).

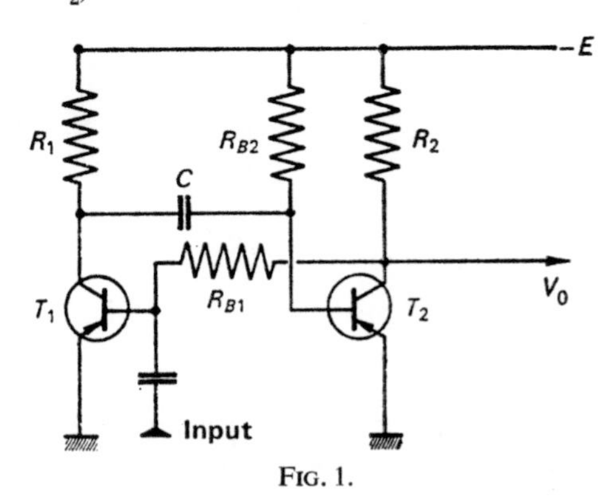

FIG. 1.

2. What value of R_{B_1} is required for T_1 to be saturated when T_2 is cut-off?

3. A short negative pulse is applied at the input which briefly switches on T_1. What is the sequence of events which follows? (How, in the Abraham and Bloch multivibrator would the charge and discharge of the capacitor C be studied?) What is the duration of the square pulse delivered to the output?

4. After what period from the end of an input pulse will the system be sensitive to a further pulse?
(Take the base-emitter voltage of a conducting transistor to be zero.)

Solution. 1. Neglecting V_{BE}, the base current of T_2 is

$$I_{B_2} = \frac{-E}{R_{B_2}}$$

from which the normal collector current is

$$\frac{-\beta E}{R_{B_2}}.$$

The maximum collector current is

$$\frac{-E}{R_2}$$

and there will be saturation if

$$\left|-\frac{E}{R_2}\right| < \left|-\frac{E}{R_{B_2}}\right|,$$

or

$$R_{B_2} < \beta R_2.$$

2. Using the same reasoning we have

$$R_{B_1}+R_2 < \beta R_1$$

or, effectively,

$$R_{B_1} < \beta R_1.$$

3. In the quiescent state the potentials are those of Fig. 2 and the capacity C is charged to the potential E.

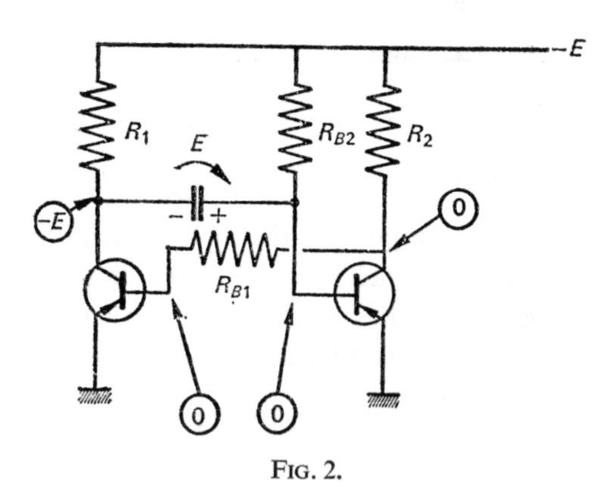

FIG. 2.

If the pulse makes T_1 conduct, its collector potential will drop to zero. The charge on C cannot change instantaneously, the potential across C remains, so that the base of T_2 goes to $+E$ which cuts off this transistor. Immediately after the end of the pulse therefore, T_2 will be cut-off and, following part 2 above, T_1 will conduct. The charging circuit for C is that of Fig. 3.

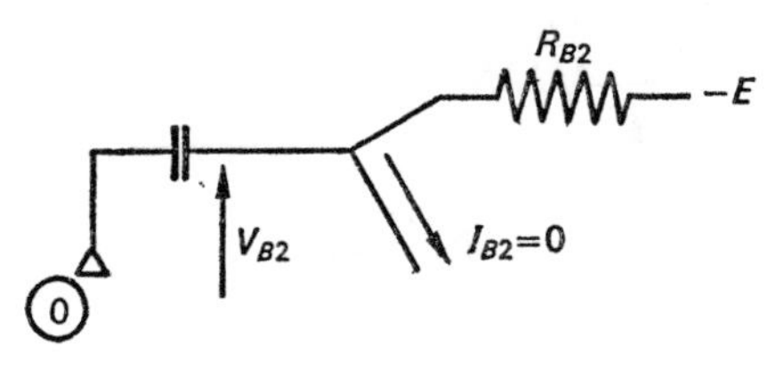

FIG. 3.

The potential V_{B_2} then changes as

$$V_{B_2}+E = 2E\, e^{-t/R_{B_2}C}.$$

Then T_2 switches to the conducting state when the initial state is reached, i.e.

$$V_{B_2} \to 0,$$

which occurs when

$$t = R_{B_2}C \ln 2.$$

The voltage V_2 is thus a negative square pulse of length t.

4. Immediately after the "swing-back", the capacitor C will be completely discharged and we begin from Fig. 4. To regain the initial state it is necessary for the potential to build up again to the value E. The total time for this process is three times the time constant and so the recovery time is

$$T_r \simeq 3R_1C.$$

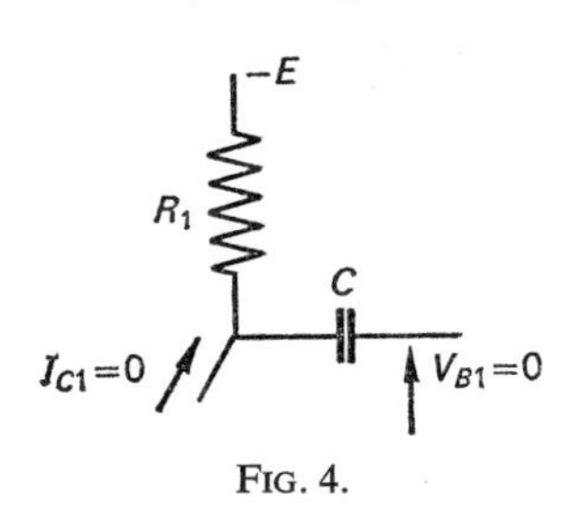

FIG. 4.

EXERCISE No. 2

Generation of a Symmetric Saw Tooth Wave

The system of Fig. 1 has a perfect operational amplifier for which the output voltage is limited by saturation to $\pm V_0$. The input voltage varies linearly with time from $-V_1$ to $+V_1$ where $V_1 \geqslant V_0$. How does v_2 vary with time?

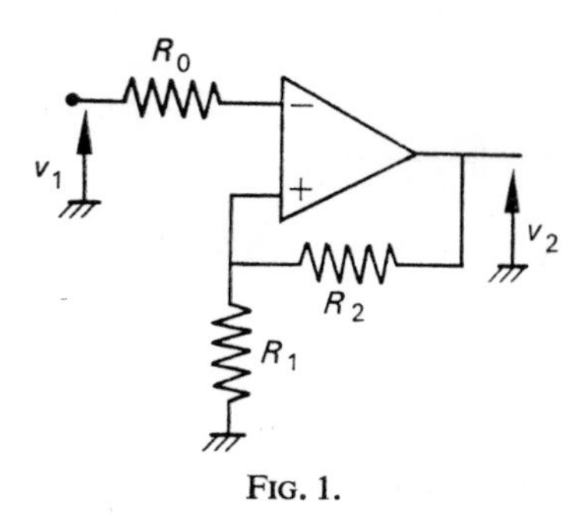

FIG. 1.

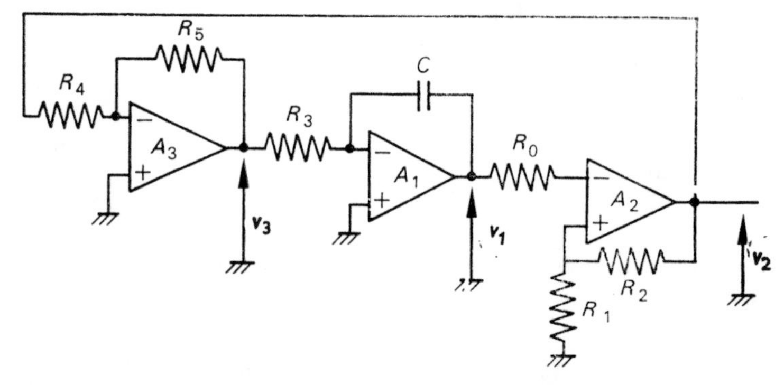

FIG. 2.

From this result find the waveform generated by the oscillator circuit shown in Fig. 2 for $V_0 = 15$ V, $R_1 = R_2 = R_3 = 10$ kΩ, $C = 1$ μF and $R_4 = R_5$.

Solution. If v_1 is very negative the operational amplifier is saturated and $v_2 =$ $= +V_0$ so that the input terminal + has a potential

$$V_L = +V_0 \cdot \frac{R_1}{R_1+R_2}$$

due to the divider R_1R_2. This will remain so as long as $v_1 < V_L$. As soon as v_1 passes the value V_L, the operational amplifier acts as a comparator of infinite gain to change v_2 instantly to $-V_0$. This state is stable, since the input terminal – is now at $-V_L$, as long as $v > -V_L$. The system thus acts as a Schmitt trigger with a large hysteresis as is seen in Fig. 3.

The circuit of Fig. 2 is, in effect, an oscillator. If the potential at the point A_2 is $+V_0$ (i.e. the comparator is switched to the "high" position) the potential at A_3 will be $-V_0$ since, with $R_4 = R_5$, there is a gain of -1 for the first amplifier. This voltage is integrated by R_3C so that v_1 increases linearly as

$$v_1 = v_{10} - \frac{1}{R_3C}\int_0^t v_3(t)\,dt = v_{10} + \frac{v_0}{R_3C}.t,$$

where v_{10} is the initial value of v_1. When v_1 reaches $+V_L$, v_2 changes abruptly to $-V_0$ and v_3 to $+V_0$ so that v_1 begins to decrease linearly until the system switches again when $v_1 = -V_L$ and so on.

The waveform of v_3 is thus a symmetric saw-tooth with a period twice that necessary to change from $-V_L$ to $+V_L$ or

$$T = 2\times 2V_L.\frac{R_3C}{v_0} = 4\frac{R_1}{R_1+R_2}.R_3C,$$

or, numerically, $V_L = 7{\cdot}5$ volts and $T = 2\times 10^{-2}$ sec.

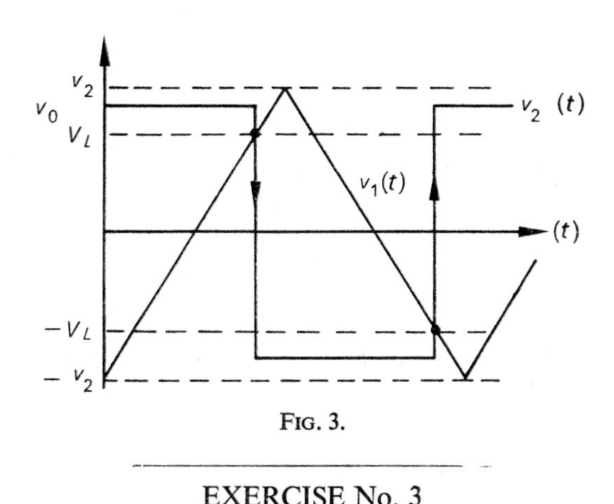

FIG. 3.

EXERCISE No. 3

Diode Circuits

We consider that the diodes in the following question have a characteristic such that

$I = 0$ for $V < V_0$ (where $V_0 \simeq 0{\cdot}6$ volt for silicon)

and resistance $= 0$ for $V > V_0$

1. Circuits with diodes and resistances.

If a sinusoidal voltage of amplitude 10 volts is applied to the input terminals *AE* of the networks of Figs. 1 to 4 what will be the form of the output across *BE*? What would be the response of Fig. 3 if, for $V > V_0$, there were a finite diode resistance?

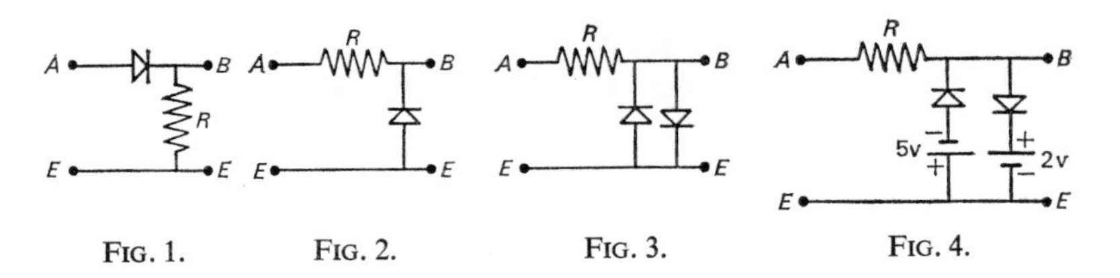

Fig. 1. Fig. 2. Fig. 3. Fig. 4.

The input signal across $A'E$ in Fig. 5 is a symmetric saw-tooth of 10 volts amplitude as shown. What will be the output signal?

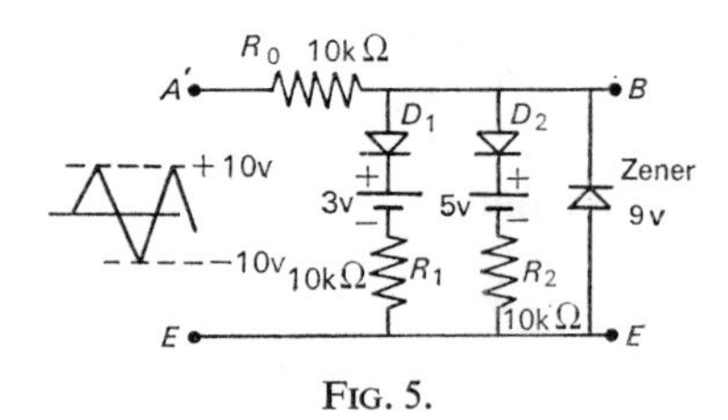

Fig. 5.

2. Diodes and capacitors.

A 10 volt sinusoidal signal is applied at $t = 0$ across AE of the circuits of Figs. 6 to 8. What will be the output waveforms?

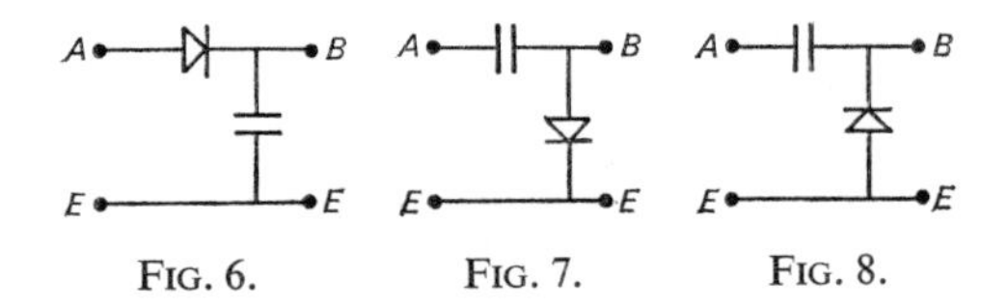

Fig. 6. Fig. 7. Fig. 8.

3. Diode–capacitor–resistor circuits.

A square wave of amplitude 10 volts is applied to the input AE of the circuits of Figs. 9 and 10. Show the form of the output signal as a function of time.

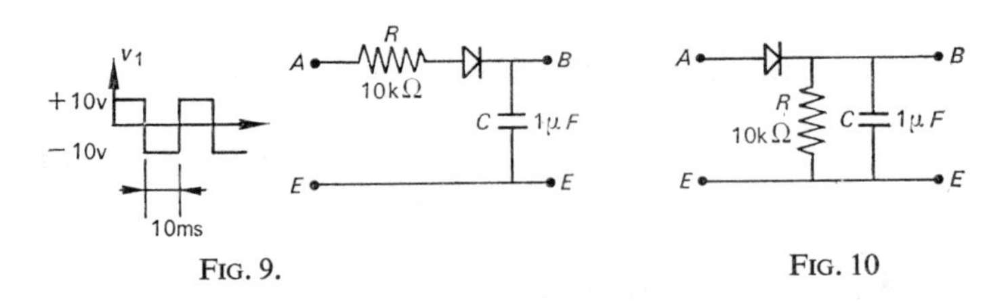

Fig. 9. Fig. 10

Solution. 1. For the circuits of Figs. 1 and 2 the output has approximately a half-wave rectified form as shown in Figs. 11 and 12 respectively, the cut-off being at $+V_0$ in the first case and $-V_0$ in the second.

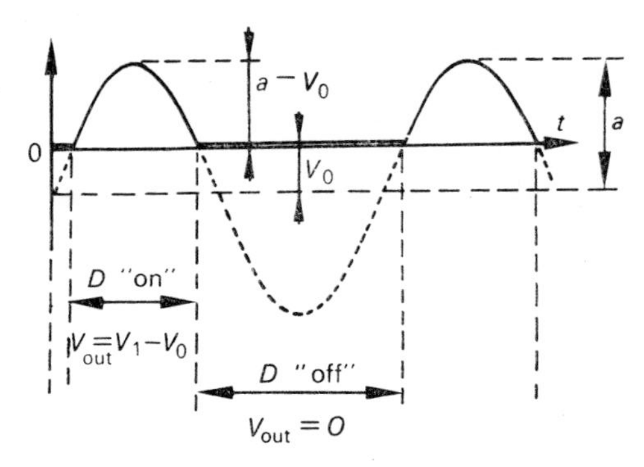

FIG. 11.

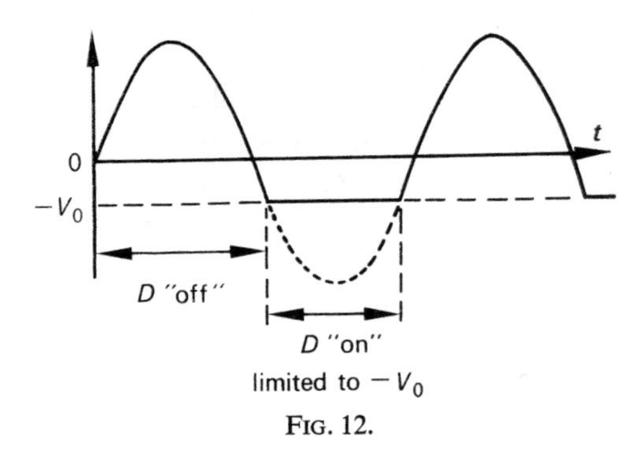

FIG. 12.

In the third circuit (Fig. 3) the voltage is limited at both $+V_0$ and $-V_0$ as in Fig. 13(a)

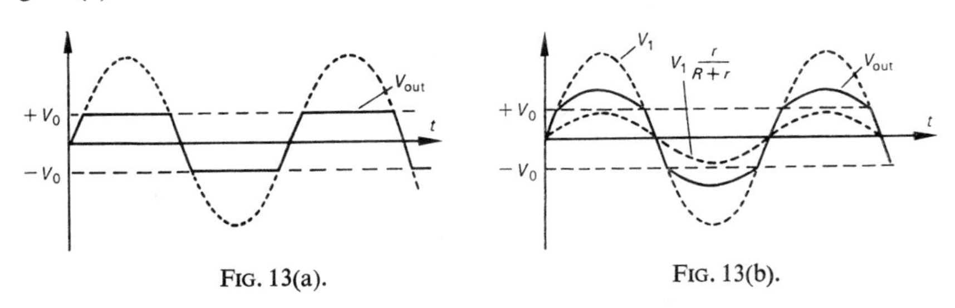

FIG. 13(a). FIG. 13(b).

If we take the resistance at $V > V_0$ to be r for each diode, the output voltage for $V > V_0$ is no longer constant. The circuit will behave as a potential divided between r and R with the resultant output of Fig. 13(b).

In the case of Fig. 4 the limit voltages are simply displaced from the values in Fig. 13(a), since the diodes do not conduct except when $V_1 > V_0+2$ in the one case and $V_1 < -V_0-5$ in the other. The result is shown in Fig. 14.

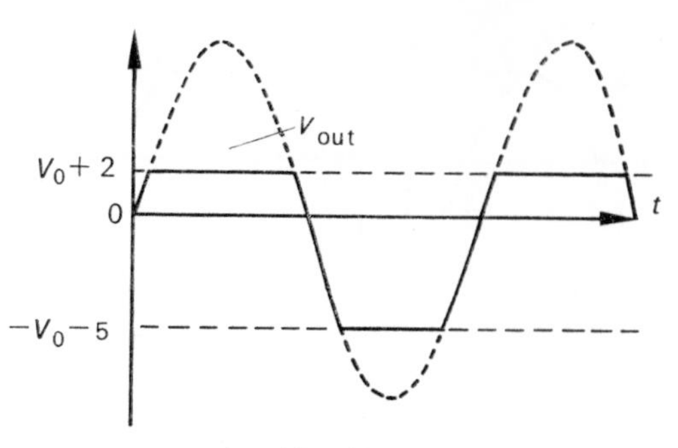

FIG. 14.

The circuit of Fig. 5 behaves as a waveform shaper. For $0 < V_1 < V_0+3$ all the diodes are cut off; $V_{out} = V_1$. For $V_1 > V_0+3$ the diode D_1 conducts, the two resistances R_0 and R_1 constitute a potential divider and V_{out} varies half as fast as V_1, as long as $V_{out} < V_0+5$ which occurs for $V_1 = V_1'$.

If $V_1 > V_1'$, $V_{out} > V_0+5$, D_2 then conducts so that the resistance R_2 is introduced into the circuit and the slope becomes 1/3. The Zener diode has no effect until V_{out} is 9 volts, which never occurs in this case. On the other hand, acting as a normal diode, the Zener will limit the voltage at $-V_0$ negative. The resultant waveform is shown in Fig. 15.

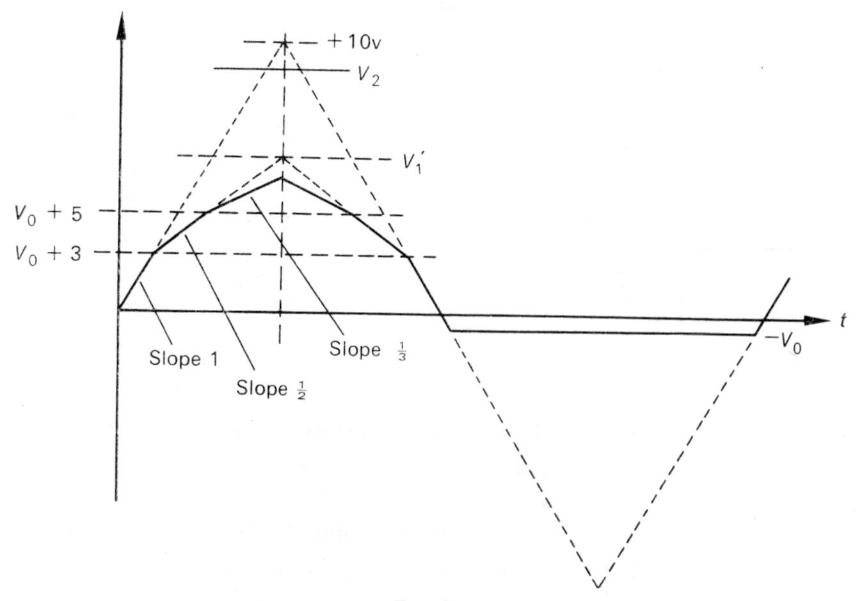

FIG. 15.

2. The input signal commences at $t = 0$ so that the capacitor is uncharged for $T < 0$. The circuit of Fig. 6 detects the peak voltages since, as V_1 increases, the diode conducts charging C but the capacitor cannot discharge when V_1 decreases. Fig. 16 shows the variation of the potential across C which would be observed if the amplitude of V_1 were modulated in the manner indicated.

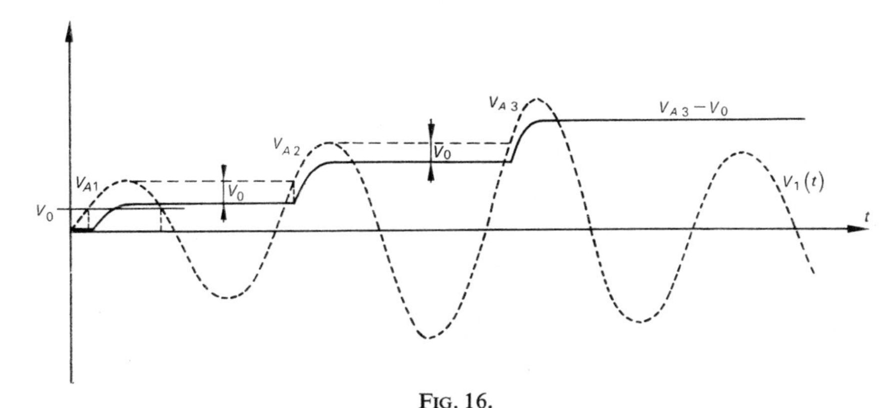

FIG. 16.

At the first positive peak $V_{out} = V_{A1} - V_0$. The capacitor holds this potential and then increases again for $V_1 > (V_{A1} - V_0) + V_0$. The output voltage reaches $V_{A2} - V_0$ and then remains constant as long as $V_1 < (V_{A2} - V_0) + V_0$ and so on.

In the case of Fig. 7, the increase in the input voltage is registered at B up to the point where the diode conducts. From there V_{out} remains constant at $V_{out} = V_0$ and C charges up.

At the peak of the wave the voltage of C is $V_A - V_0$; V_1 then decreases and the diode is cut-off, C maintains the constant potential $(V_A - V_0)$, since it cannot discharge and $V_{out} = V_1 - (V_A - V_0)$. The diode cannot conduct, since V_1 will never be greater than V_A and the output voltage will be a sine wave displaced relative to V_1 as shown in Fig. 17.

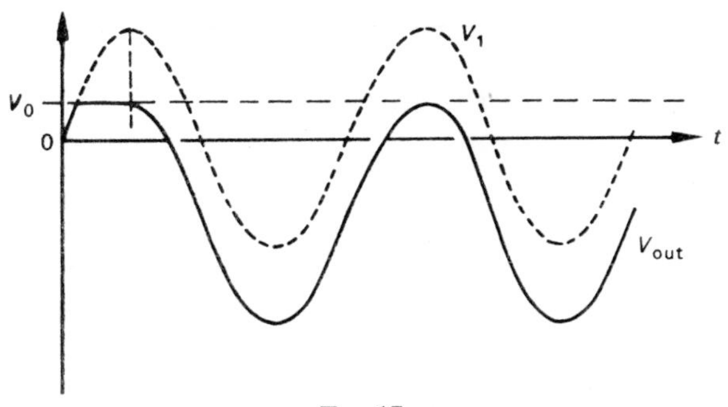

FIG. 17.

Fig. 8 gives the same result with the sign reversed.

3. Diode-resistance-capacitance circuits.

Since, in Fig. 9, the input voltage is greater than V_0, when this voltage is positive the diode will conduct and C will charge up through R such that

$$v_{out} = (10 - V_0)(1 - e^{-t/RC}).$$

With $RC = 10^{-2}$ s, $V_0 = 0{\cdot}6$ V the output voltage at the end of 10 ms will be

$$v_{out} = 9{\cdot}4(1 - e^{-1}) = 5{\cdot}9 \text{ V}.$$

V_1 now drops abruptly to -10 V so that the diode is cut-off and C keeps its charge. As soon as V_1 again goes positive, the charge on C increases and, at the end of the next positive half cycle,

$$v_{out} = 9{\cdot}4(1 - e^{-2}) = 8{\cdot}3 \text{ V}.$$

The output voltage thus follows Fig. 18.

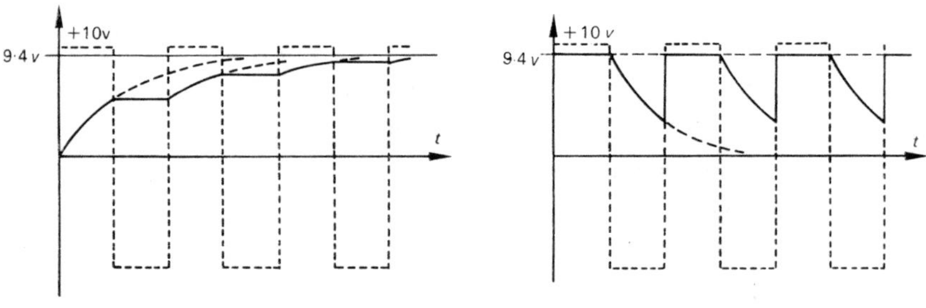

FIG. 18. FIG. 19.

Finally, in the case of Fig. 10, the capacitor charges through the diode while the effective voltage is $(10 - 0{\cdot}6)$ V and then discharges slowly towards zero through R, while V_1 is negative, as $e^{-t/RC}$ as shown in Fig. 19.

EXERCISE No. 4

The Functional Amplifier

It is proposed to use the functional amplifier technique (the basic circuit of which is shown in Fig. 1) to obtain a circuit for which the output voltage is proportional to the square root of the input

$$V_2 = -k\sqrt{V_1} \qquad V_1 > 0.$$

1. $Z(I)$ is a non-linear impedance which can be defined in terms of an $I-V$ characteristic. What must be the form of this characteristic if the required variation of V_2 with V_1 is to be obtained for $R = 10^4\,\Omega$ and $|V_2| = 5$ V for $V_1 = 5$ V?

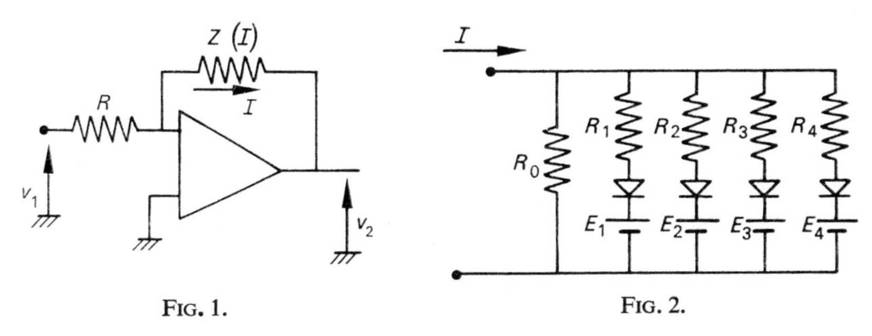

FIG. 1. FIG. 2.

2. It is possible to obtain $Z(I)$ with a waveform shaping system such as that shown in Fig. 2. If we take the diodes to be ideal (i.e. $R = 0$ for $V > 0$, $R = \infty$ for $V < 0$) the $I-V$ characteristic between $V = 0$ and $V = 5$ V will approach an ideal form if the curve is drawn tangential to the lines at $V = 1$ V, 2 V, 3 V etc. In this case, when account is taken of the curvature of the real diode characteristics, the actual and theoretical curves will be very close. Calculate R_0, R_1, R_2, R_3 and R_4 and find E_1, E_2, E_3, E_4 by a graphical method.

Solution. 1. The input voltage at the negative terminal of the operational amplifier is zero (fictional earth) and so the voltage across Z will be $-V_2$. Also $V_1 = RI$. Thus, for the $I-V$ characteristic

$$V = -V_2 = k\sqrt{V_1} = k\sqrt{RI},$$

or

$$I = \frac{V^2}{k^2R}.$$

The value of k required is given, since $k^2 = 5$ and thus

$$I = \frac{V^2}{5\times 10^4}.$$

2. The tangent at the origin, $V = 0$, must be horizontal and hence $R_0 = \infty$.

For $V > E_1$, the diode D_1 will conduct and the slope of the curve will be equal to R_1. For this to be tangential to the curve at $V = 1$ volt requires that

$$R_1 = \left(\frac{dV}{dI}\right)_{1\text{ volt}} = \frac{k^2R}{2\times 1\text{ volt}} = 25\text{ k}\Omega.$$

The voltage E_1 is determined by the intersection of the tangent with the V-axis, i.e. $E = 0{\cdot}5$ volt.

When $V = E_2$, D_2 begins to conduct and the slope is due to R_1 and R_2 in parallel. This must be the slope at $V = 2$ volts, or

$$\frac{R_1 R_2}{R_1 + R_2} = \left(\frac{dV}{dI}\right)_2 = \frac{25\ \text{k}\Omega}{2}$$

or $R_2 = 25\ \text{k}\Omega$.

This will give the tangent to the curve at 2 volts, provided that E_2 is the voltage at the intersection of the two tangents at 1 volt and 2 volts, as shown in Fig. 3. Thus $E_2 = 1{\cdot}5$ V.

Similarly $R_3 = R_2 = 25\ \text{k}\Omega$ and $E_3 = 2{\cdot}5$ V being the voltage at the intersection of the points 2 volts and 3 volts.

Finally $R_4 = R_3 = 25\ \text{k}\Omega$, $E_4 = 3{\cdot}5$ V.

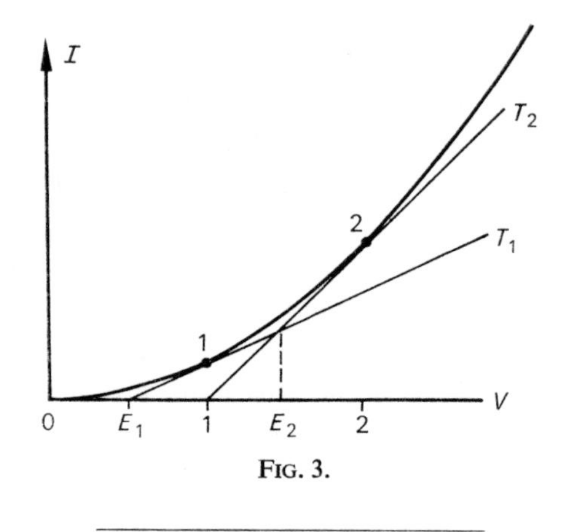

FIG. 3.

EXERCISE No. 5

The UJT Relaxation Oscillator

The unijunction transistor (UJT) which is shown in the relaxation circuit of Fig. 1 has the following characteristics when the supply voltage is 20 volts:

Peak current	$I_p = 1\ \mu\text{A}$
Peak voltage	$V_p = \eta E = 14$ volts ($\eta = 0{\cdot}7$)
Valley current	$I_v = 10$ mA
Valley voltage	$V_v = 2$ volts.

The three parts of the characteristic can be represented by straight lines as shown in Fig. 2, the slope of the third part (for $I > I_v$) corresponding to a resistance of 100 Ω.

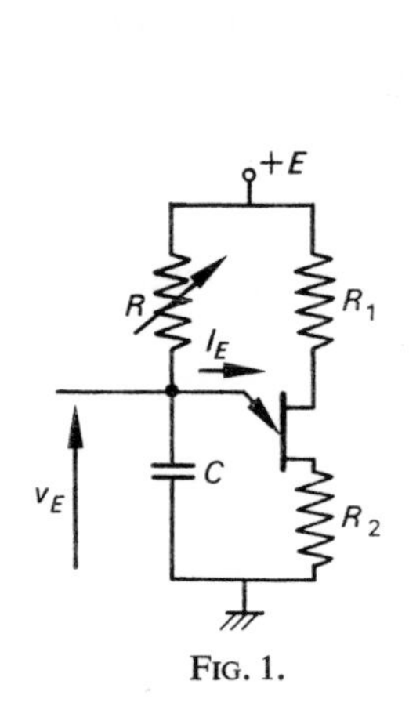

FIG. 1.

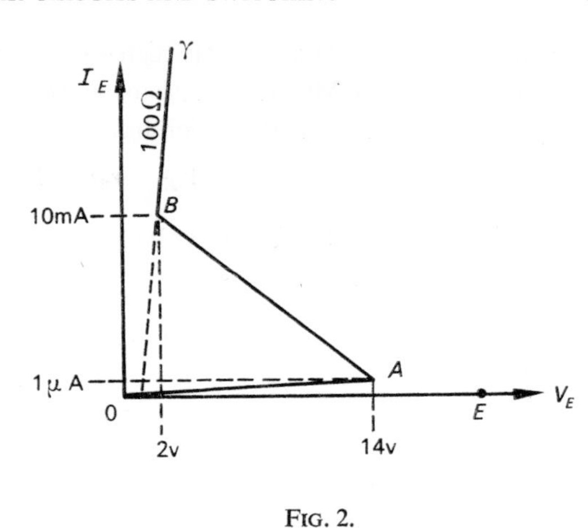

FIG. 2.

If the drop in potential across R_1 and R_2 can be ignored, what are the limits on R for the system to oscillate? Calculate the recurrence frequency in terms of R and C.

Solution. Initially, with the capacitor uncharged, the working point of the UJT will be at 0 on the branch OA. The UJT will then appear as an equivalent resistance

$$r_1 = \frac{V_p}{I_p} = 14\ \text{M}\Omega.$$

The equivalent circuit then is that of Fig. 3. The capacitor C now charges to the voltage defined by the potential divider Rr_1 with a time constant $(R \,||\, r_1)C$ or

$$V_E = E\frac{r_1}{R+r_1}\left[1-\exp\left(\frac{-t}{\dfrac{r_1RC}{R+r_1}}\right)\right].$$

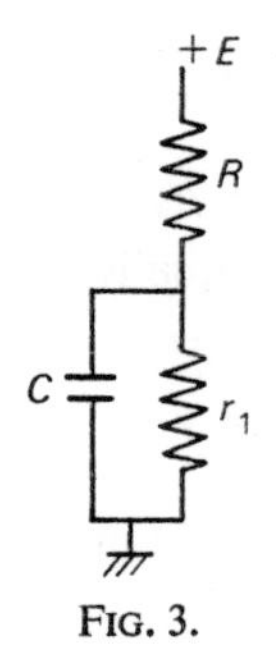

FIG. 3.

When the working point reaches A (which is only possible if $E.r_1/R+r_1 > \eta E = V_p$ or $R < r_1(1-\eta)/\eta = 6\,\text{M}\Omega$) the transistor will conduct heavily so that the working point jumps to the branch BC which has the equation

$$V_E = r_2 I_E + V_0.$$

V_0 is the intercept of BC with the voltage axis. The new equivalent circuit is given in Fig. 4.

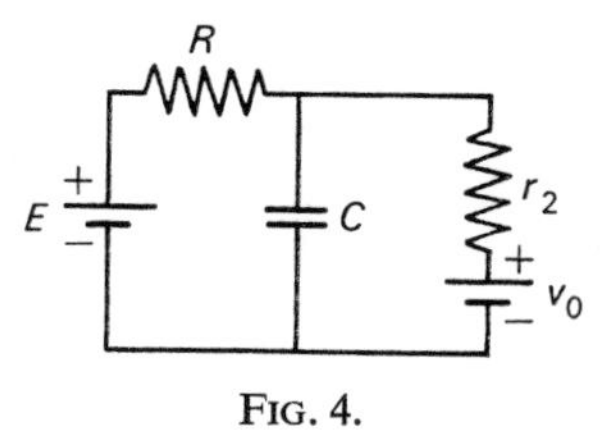

FIG. 4.

V_E now decreases exponentially towards a limit

$$V_L = \frac{r_2 E + R V_0}{r_2 + R}$$

with the time constant

$$(R \,||\, r_2)C,$$

so that

$$V_v - V_L = (V_p - V_L)\exp\left(\frac{-t}{\dfrac{r_2 RC}{R+r_2}}\right).$$

The transistor then cuts off when the working point arrives at B, which will be after a time t_1 such that

$$t_1 = \frac{r_2 RC}{R+r_2}\ln\frac{V_p - V_L}{V_v - V_L}.$$

This cut-off can only occur if $V_L < V_v$ or

$$R > r_2 . \frac{E - V_v}{V_v - V_0}$$

which imposes a second condition on R.

(Note that the first condition can equally be written as

$$R < \frac{E - \eta E}{\eta E/r} = \frac{E - V_p}{I_p},$$

which represents the slope of the line EA in Fig. 2. Similarly the second condition is equivalent to $R > (E - V_v)/I_v$.)

The UJT can only oscillate, therefore, if the load-line of the emitter

$$V_E = E - RI_E$$

cuts the characteristic curve in the region AB where the resistance is negative. Numerically we see that

$$1800\,\Omega < R < 6\text{ M}\Omega.$$

The oscillator period is found by adding together the times for the two exponential processes as V_E varies between V_p and V_v. That is

$$T = \frac{r_1RC}{R+r_1}\ln\frac{r_1(E-V_v)-V_vR}{r_1(E-V_p)-V_pR}+\frac{r_2RC}{R+r_2}\ln\frac{V_p-V_L}{V_v-V_L}.$$

EXERCISE No. 6

De Morgan's Theorems

Use de Morgan's theorems to put the boolean expression

$$Z = A\bar{B}+B\bar{C}A$$

into a form which can be obtained by means of a circuit using only

(i) nand gates
(ii) nor gates.

(The variables ABC are available but not their complements.)

Solution. (i) We use the theorem in the form

$$\overline{\alpha\beta} = \bar{\alpha}+\bar{\beta},$$

$$\text{or}\quad \alpha+\beta = \overline{\bar{\alpha}\bar{\beta}}.$$

Then, writing

$$Z = \overline{\overline{A\bar{B}}\;\overline{B\bar{C}A}}$$

this expression can be represented by the circuit of Fig. 1.

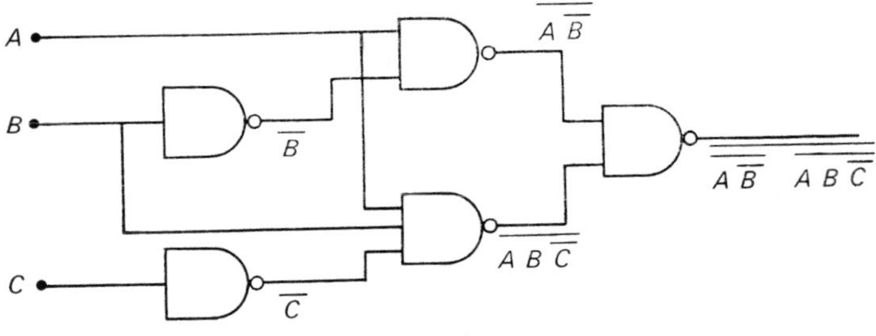

FIG. 1.

(ii) If we use the theorem $\alpha\beta = \overline{\bar{\alpha}+\bar{\beta}}$

we can put

$$Z = \overline{\bar{A}+B}+\overline{\bar{A}+\bar{B}+C}$$

for which one possible circuit is

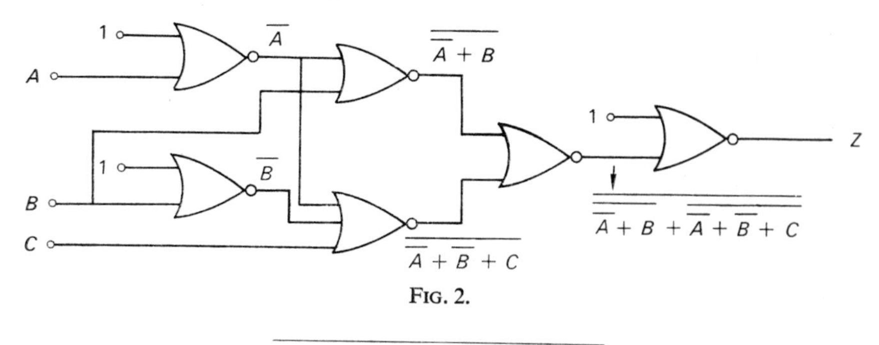

FIG. 2.

EXERCISE No. 7

Karnaugh Diagram

Use a Karnaugh diagram to simplify the expression

$$X = A\bar{C}D+\bar{B}\bar{C}D+\bar{A}BD+ABCD.$$

Solution. Since each variable can have two values, there are sixteen possible states for the system which correspond to the sixteen cases on a Karnaugh diagram.

The most elementary, if not the most rapid, method is to construct a "truth" table in which a figure 1 is used for the cases where A, B, C, D or $X = 1$. This gives the following table

A	B	C	D	X	A	B	C	D	X
0	0	0	0	0	1	0	0	0	0
			1	1				1	1
		1	0	0			1	0	0
			1	0				1	0
	1	0	0	0		1	0	0	0
			1	1				1	1
		1	0	0			1	0	0
			1	1				1	1

from which

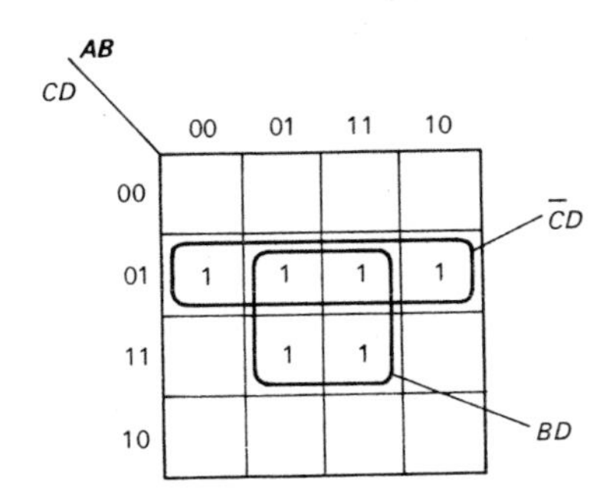

We see two 4th order loops leading to the result

$$X = BD+\bar{C}D = D(B+\bar{C}).$$

EXERCISE No. 8

A decade counter is formed by placing four bistable units B_1, B_2, B_3 and B_4 in series with their outputs Q_1, Q_2, Q_3 and Q_4 feeding to an associated circuit which returns the system to zero at the 10th pulse. The first unit B_1 switches its state with each pulse and the state 0 corresponds to $Q_1 = Q_2 = Q_3 = Q_4 = 0$. Show on a Karnaugh diagram the 10 states corresponding to the decade.

The display of the ten figures is given by a special tube consisting of seven segments which may each light up independently on the application of an external voltage (the +1 logic). To display the figure 5, for example, the segments *a*, *c*, *d*, *f* and *g* are illuminated as shown in Fig. 1. If the ten figures have the form shown in this diagram, what is the boolean function $A(Q_1, Q_2, Q_3, Q_4)$ associated with the voltage which must be applied to the segment "*a*" if it is to light up as required for the figures 0, 2, 3, 5, 7, 8, 9? Represent this function on a Karnaugh diagram. It will be noted that certain cases in the diagram correspond to states which never arise. From this obtain the simplest expression for *A*.

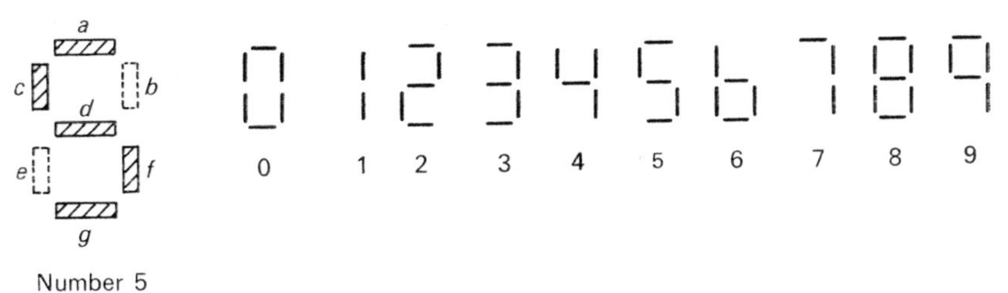

FIG. 1.

Solution. Table I shows the values of Q_1 etc. for each state and, from this, the Karnaugh diagram of Table II may be obtained.

TABLE I

State	Q_1	Q_2	Q_3	Q_4
0	0	0	0	0
1	1	0	0	0
2	0	1	0	0
3	1	1	0	0
4	0	0	1	0
5	1	0	1	0
6	0	1	1	0
7	1	1	1	0
8	0	0	0	1
9	1	0	0	1

TABLE II

Q_3Q_4 \ Q_1Q_2	00	01	11	10
00	0	2	3	1
01	8	×	×	9
11	×	×	×	×
10	4	6	7	5

For the segment "*a*" to be illuminated in the states 0, 2, 3, 5, 7, 8, 9, it follows that

$$A = \bar{Q}_1\bar{Q}_2\bar{Q}_3\bar{Q}_4+\bar{Q}_1Q_2\bar{Q}_3\bar{Q}_4+Q_1Q_2\bar{Q}_3\bar{Q}_4+Q_1\bar{Q}_2Q_3\bar{Q}_4$$
$$+Q_1Q_2Q_3\bar{Q}_4+\bar{Q}_1\bar{Q}_2\bar{Q}_3Q_4+Q_1\bar{Q}_2\bar{Q}_3Q_4.$$

The cases marked with a cross in Table II are never used and one can put either 1 or 0 in these positions to give the maximum order loop. Thus, for example, with four 4th order loops we can write Table III and

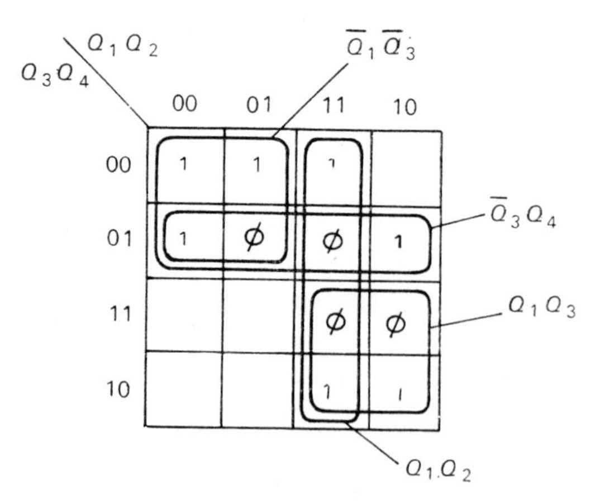

$$A = Q_1Q_2+Q_1Q_3+\bar{Q}_1\bar{Q}_3+\bar{Q}_3Q_4$$

with the sign Ø here signifying 0 or 1.

Chapter 5

Noise

EXERCISE No. 1

What is the noise factor of a network formed by a parallel resistance R_p which is supplied from an internal resistance ϱ as shown in Fig. 1?

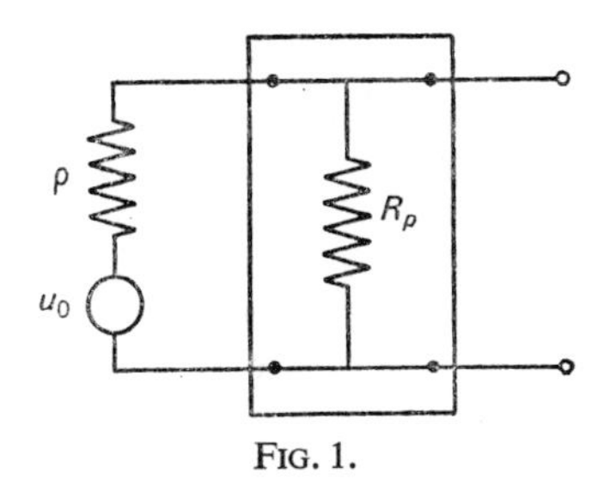

FIG. 1.

Solution. The noise factor is

$$F = \frac{\dfrac{\text{Available input signal power}}{\text{Available input noise power}}}{\dfrac{\text{Available output signal power}}{\text{Available output noise power}}}.$$

In the present case the input noise is that of the resistance ϱ, i.e. kTB.

The available signal power is $u_0^2/4\varrho$. The combination of the network and the source has an output impedance

$$\frac{R_p\varrho}{R_p+\varrho}$$

and an e.m.f.

$$u_0\frac{R_p}{R_p+\varrho}.$$

The available output power of the signal is thus

$$\frac{u_0^2R_p}{(R_p+\varrho)^2}\times\frac{1}{4}\frac{R_p+\varrho}{R_p\varrho}=\frac{u_0^2R_p}{4\varrho(R_p+\varrho)}.$$

The output noise is that of a resistance

$$\frac{R_p \varrho}{R_p + \varrho}$$

which is also kTB as can be seen from Fig. 2.

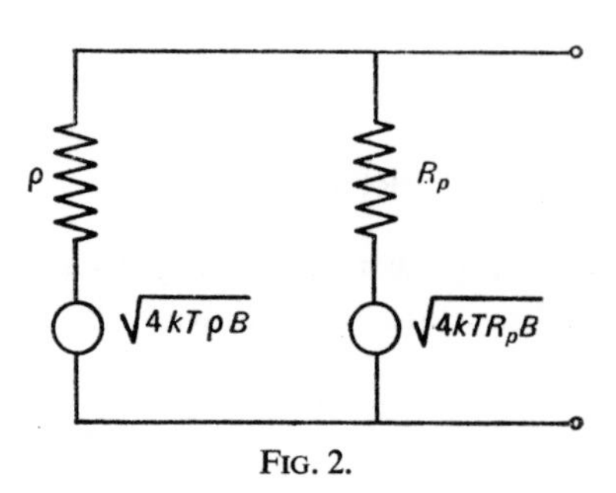

FIG. 2.

Thus, following our definition above

$$F = 1 + \frac{\varrho}{R_p}.$$

EXERCISE No. 2

A triode amplifier with internal resistance R_a and noise resistance R_N is fed from a source with internal resistance ϱ_g.

1. Calculate the noise factor of the system.

2. To decrease the gain, the signal is attenuated by a potential divider formed by the resistances R_1 and R_2 of Fig. 1. What is the new noise factor? What is this noise factor for $R_1 = R_2 = R$?

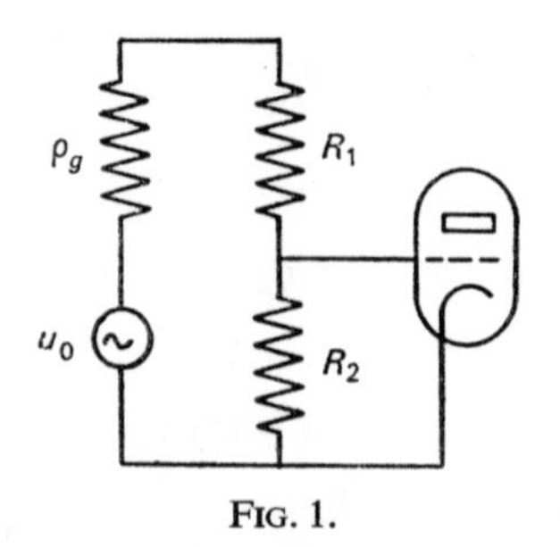

FIG. 1.

Solution. 1. In the absence of R_N we would have a perfect amplifier. Hence we can replace the system by the circuit of Fig. 2 which, following the previous exercise, gives a noise factor

$$F = 1+\frac{R_N}{\varrho_g}.$$

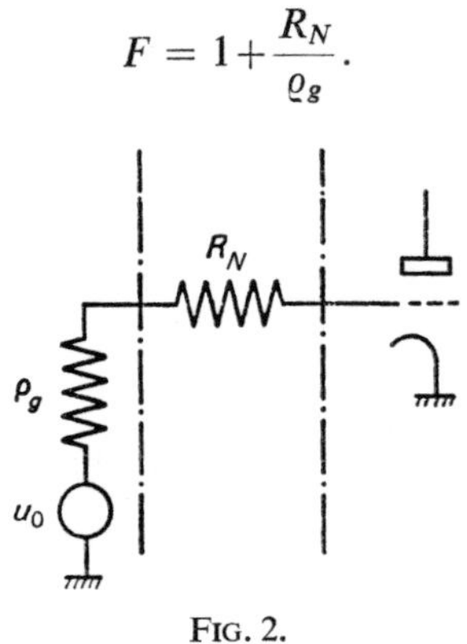

FIG. 2.

2. Following the above reasoning it is simply necessary to calculate the noise of the circuit of Fig. 3. The noise power at both the input and output will be kTB and the noise factor becomes

$$F = \frac{1}{G_p}$$

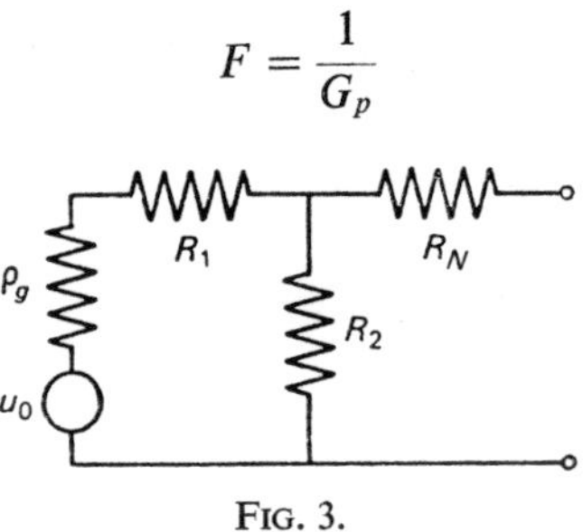

FIG. 3.

where G_p is the power gain. The whole system can be considered as a source of e.m.f.

$$u_0\frac{R_2}{R_1+\varrho_g+R_2}$$

with internal resistance

$$\frac{R_2(R_1+\varrho_g)}{R_2+R_1+\varrho_g}+R_N.$$

The available output power is thus

$$P_{uo} = \frac{u_0^2\dfrac{R_2^2}{(R_1+\varrho_g+R_2)^2}}{4\left[R_N+\dfrac{R_2(R_1+\varrho_g)}{R_2+R_1+\varrho_g}\right]},$$

while the available input power is

$$P_{ui} = \frac{u_0^2}{4\varrho_g}.$$

Then

$$G_p = \frac{R_2^2\varrho_g}{[R_N(R_1+\varrho_g+R_2)+R_2(R_1+\varrho_g)](R_1+\varrho_g+R_2)}$$

from which

$$F = 1+\frac{R_1}{\varrho_g}+\frac{R_1+\varrho_g}{R_2}+\frac{R_1}{\varrho_g}\cdot\frac{R_1+\varrho_g}{R_2}+\frac{R_N}{\varrho_g}\left(1+\frac{R_1+\varrho_g}{R_2}\right)^2.$$

For $R_1 = R_2 = R$

$$F = 3+\frac{2R}{\varrho_g}+\frac{\varrho_g}{R}+\frac{R_N}{\varrho_g}\left(2+\frac{\varrho_g}{R}\right)^2.$$

EXERCISE No. 3

From the point of view of noise, a triode can be represented as a perfect valve together with a noise resistance R_N in series with the grid and a resistance R_C, representing the real part of the H.F. input impedance in series with it and at a temperature some five times room temperature. The schematic system is represented in Fig. 1.

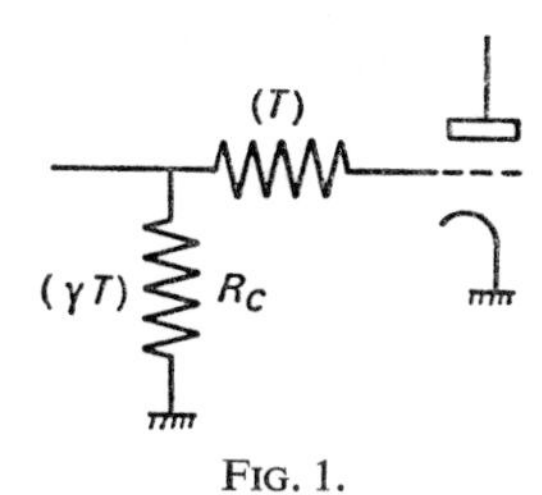

FIG. 1.

Calculate the noise factor F for a common-grid amplifier stage where the source impedance ϱ_g is optimized to give a minimum value of F.

Find F for the case where the amplification factor μ is 40, the resistance R_a of the valve is 10 kΩ, $R_C = 100$ kΩ and $R_N = 500$ Ω.

What is the value of F in the general case when the impedances at the input are matched?

Solution. With the noise resistances included the equivalent circuit becomes as in Fig. 2. The corresponding characteristic equation of the triode is

$$iR_a = \mu u+v,$$

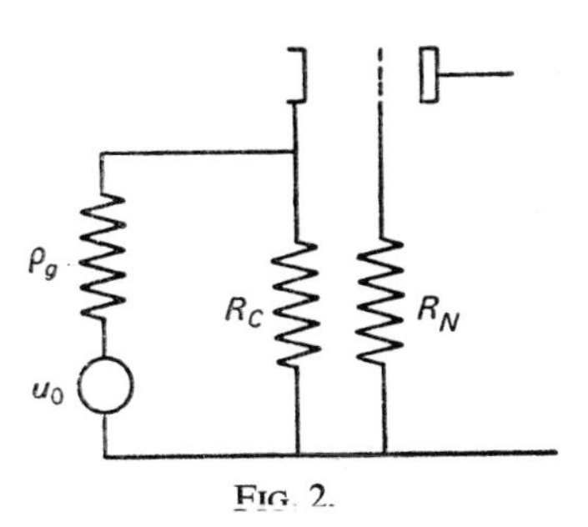

FIG. 2.

with

$$u = \Delta(V_g - V_K) = \Delta(V_g - V_E) + \Delta(V_E - V_K) = -\Delta(V_K - V_E)$$

and

$$v = \Delta(V_A - V_K) = \Delta(V_A - V_E) + \Delta(V_E - V_K) = v_{AE} + u$$

where V_E is the earth potential. We now find the equivalent generator of the system to find the gain and the available power output from the amplifier stage.

The impedance is

$$R_a + \frac{\varrho_g R_c}{R_c + \varrho_g}(\mu+1)$$

(where we note that Thévenin's theorem cannot be applied) and the e.m.f. is

$$u_0 \frac{R_c}{R_c + \varrho_g}(\mu+1).$$

The avialable output power is

$$P_{uo} = \frac{u_0^2(\mu+1)^2 \left(\dfrac{R_c}{R_c + \varrho_g}\right)^2}{4\left(R_a + \dfrac{\varrho_g R_c(\mu+1)}{\varrho_g + R_c}\right)}$$

while that at the input is

$$P_{ui} = \frac{u_0^2}{4\varrho_g}.$$

The available power gain is

$$G_p = \frac{\varrho_g(\mu+1)^2 \left(\dfrac{R_c}{R_c + \varrho_g}\right)^2}{R_a + \dfrac{\varrho_g R_c(\mu+1)}{\varrho_g + R_c}}.$$

To calculate the output noise power we note that the system contains three noise sources and make a direct calculation; i_1 and i_2 are taken as the currents through ϱ_g and R_c^1 respectively. Then the triode characteristic is $R_a i = \mu u + v$ with

$$u = \Delta(V_g - V_K) = e_N - e_g - \varrho_g i_1$$
$$v = -v_{AE} - e_g - \varrho_g i_1.$$

Also

$$i = i_1 + i_2$$

and

$$e_g + \varrho_g i_1 - e_c - R_c i_2 = 0$$

From these five equations the e.m.f. due to the output noise is found as

$$\mu e_N - (\mu+1) e_g + \frac{\varrho_g}{\varrho_g + R_C}(\mu+1)(e_g - e_c).$$

With the output impedance already calculated above and all terms $\overline{e_i e_{j \neq i}} = 0$ the output power of the noise can be taken as

$$P_{No} = \mu^2 \overline{e_N^2} + \left(\frac{\varrho_g}{\varrho_g + R_c}\right)^2 (\mu+1)^2 \overline{e_c^2} + (\mu+1)^2 \left(\frac{R_c}{R_c + \varrho_g}\right)^2 \overline{e_g^2} \Big/ 4\left[R_a + \frac{\varrho_g R_c}{\varrho_g + R_c}(\mu+1)\right].$$

The noise factor is given by $\quad F = \dfrac{1}{G_p} \cdot \dfrac{P_{No}}{kTB} \quad$ or,

$$F = \frac{\mu^2 R_N + \left(\dfrac{\varrho_g}{\varrho_g + R_c}\right)^2 (\mu+1)^2 . 5R_c + (\mu+1)^2 \left(\dfrac{R_c}{R_c + \varrho_g}\right)^2 \varrho_g}{\varrho_g (\mu+1)^2 \left(\dfrac{R_c}{R_c + \varrho_g}\right)^2}$$

$$= 1 + \frac{5\varrho_g}{R_c} + \left(\frac{\mu}{\mu+1}\right)^2 \left(1 + \frac{\varrho_g}{R_c}\right)^2 \frac{R_N}{\varrho_g}.$$

This may be separated into terms in ϱ_g and $1/\varrho_g$ and terms independent of ϱ_g to give

$$F = 1 + 2\left(\frac{\mu}{\mu+1}\right)^2 \frac{R_N}{R_c} + \left(\frac{\mu}{\mu+1}\right)^2 \frac{R_N}{\varrho_g} + \frac{\dfrac{\varrho_g}{R_c}}{\left[5 + \left(\dfrac{\mu}{\mu+1}\right)^2 \dfrac{R_N}{R_c}\right]}$$

with $\mu \gg 1$, so that $(\mu/(\mu+1))^2 \simeq 1$ this becomes

$$F = 1 + \frac{2R_N}{R_c} + \frac{R_N}{\varrho_g} + \frac{\dfrac{\varrho_g}{R_c^2}}{(5R_c + R_N)}.$$

The optimum value of F will occur when $\partial F/\partial \varrho_g = 0$. Taking

$$F = F_0 + \frac{A}{\varrho_g} + \frac{\varrho_g}{B}$$

$$\frac{\partial F}{\partial \varrho_g} = -\frac{A}{\varrho_g^2} + \frac{1}{B}$$

which is zero for $\varrho_{g\,\text{opt}} = \sqrt{AB}$. In the present case this gives

$$\varrho_{g\,\text{opt}} = R_c \sqrt{\frac{R_N}{5R_c + R_N}}.$$

Putting the numerical values given we see

$$\varrho_{g\,\text{opt}} = 10^5 \sqrt{\frac{500}{5.10^5 + 500}} = 3159\ \Omega.$$

The input impedance of the amplifier stage is, with the large value of R_c given, effectively the inverse of the mutual conductance. There will thus be matching for

$$\varrho_g = \frac{R_a}{\mu}.$$

Then

$$F_{\text{matched}} = 1 + 2\frac{R_N}{R_c} + \mu\frac{R_N}{R_a} + \frac{(5R_c + R_N)R_a}{\mu R_c^2}.$$

The numerical values give

$$F_{\text{matched}} = 3{\cdot}126 \quad \text{(or about 5 dB)},$$

while, for the optimum value of ϱ_g,

$$F_{\min} = 1 + 2\frac{R_N}{R_c} + 2\frac{R_N(5R_c + R_N)}{R_c^2} = 1{\cdot}31 \text{ (or about } 1{\cdot}2 \text{ dB)}$$

EXERCISE No. 4

1. A real four-terminal network can be replaced by a perfect network with the same characteristics with an associated current source and an associated voltage source which represent the noise output. (These two noise sources are supposed to be uncorrelated.)

Calculate the noise factor of such a system, as indicated in Fig. 1, when it has an input from a source of internal impedance ϱ_g. (To simplify the calculation take $h_{12} = 0$.)

Put

$$\overline{e_n^2} = 4kTB\ R_s$$

$$\overline{i_n^2} = \frac{4kTB}{R_p}.$$

What value of the internal impedance ϱ_g will give the lowest noise?

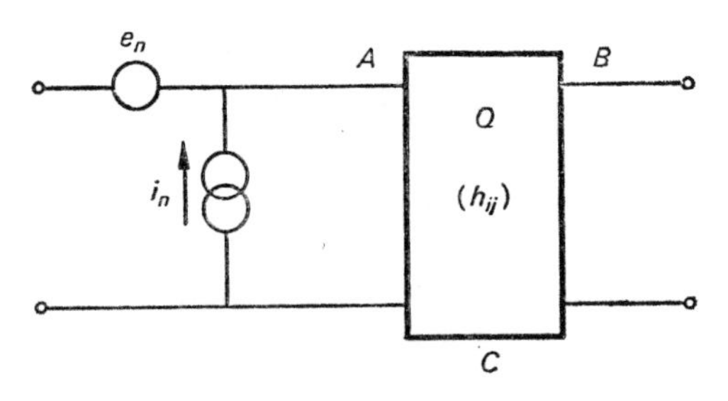

FIG. 1.

2. If the network has actually three terminals AB and C, calculate the new noise factor when the system is used with A earthed as in Fig. 2.

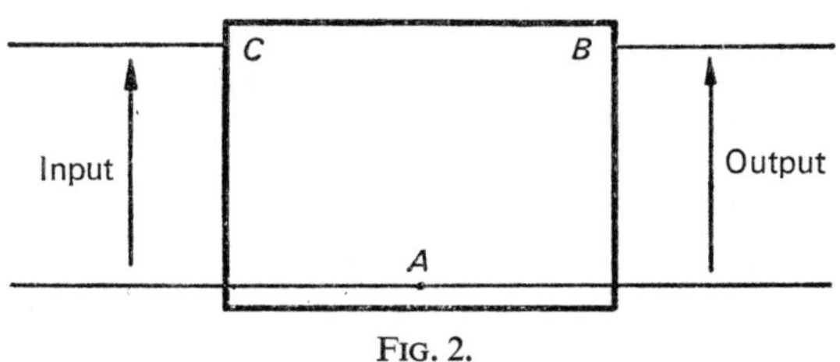

FIG. 2.

Express F as a function of the conductance $g = h_{21}/h_{11}$ and h_{22} in the simple case where $h_{22} \ll 1/\varrho_g$ and $h_{21} \gg 1$. If, further, $h_{22} \ll g$, what is the optimum value of ϱ_g?

Solution. 1. If h_{12} is zero, the output impedance of the network is simply $1/h_{22}$ and the open circuit voltage gain is

$$G_v = \frac{\beta \times 1/h_{22}}{h_{11}} = \frac{\beta}{h_{11}h_{22}}. \tag{1}$$

The available power output is then

$$P_{uo} = \frac{\beta^2 v_1^2}{h_{11}^2 h_{22}^2} \times \frac{1}{4 \times \dfrac{1}{h_{22}}} = \frac{\beta^2 v_1^2}{4h_{11}^2 h_{22}}.$$

The input impedance is h_{11} so that

$$v_1 = u_0 \frac{h_{11}}{h_{11}+\varrho_g}$$

where u_0 is the e.m.f. of the generator of internal impedance ϱ_g. Thus

$$P_{uo} = \frac{u_0^2}{4} \frac{\beta^2}{h_{22}(h_{11}+\varrho_g)^2}$$

from which the available power gain is

$$G_{pu} = \frac{\beta^2 \varrho_g}{h_{22}(h_{11}+\varrho_g)^2}.$$

With the equivalent circuit transformed as shown in Fig. 3 the voltage for the noise output is

$$v_1 = \frac{h_{11}\varrho_g}{h_{11}+\varrho_g} \cdot \left(i_n + \frac{e_n+e_g}{\varrho_g}\right).$$

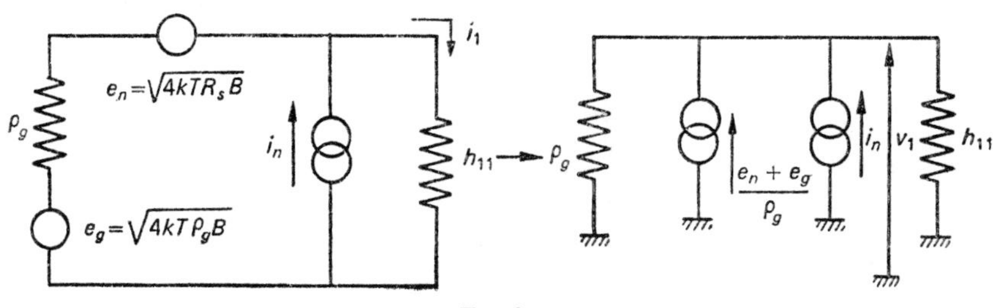

FIG. 3.

Thus, from (1) above, the available noise output power is

$$P_{nuo} = \frac{\beta^2 \varrho_g^2}{4h_{22}(h_{11}+\varrho_g)^2} \left(i_n + \frac{e_n+e_g}{\varrho_g}\right)^2$$

or, developing the bracket and allowing for the absence of correlation

$$P_{nuo} = \frac{\beta^2 \varrho_g^2}{h_{22}(h_{11}+\varrho_g)^2} \left(\frac{1}{R_p} + \frac{R_s}{\varrho_g^2} + \frac{1}{\varrho_g}\right) kTB.$$

The noise factor is obtained immediately as

$$F = \frac{1}{G_{pu}} \frac{P_{nuo}}{kTB} = 1 + \frac{\varrho_g}{R_p} + \frac{R_s}{\varrho_g}.$$

F is a minimum for $\varrho_g = \sqrt{R_s R_p}$ (see the preceding exercise) and

$$F_{\min} = 1 + 2\sqrt{R_s/R_p}.$$

2. With respect to the signal, and with $h_{12} = 0$, the equivalent circuit is that of Fig. 4 which is analogous with that of Exercise 3 with

$$\frac{1}{h_{22}} \equiv R_a, \quad R_c \equiv h_{11}, \quad g = \frac{h_{21}}{h_{11}} \equiv \frac{\mu}{R_a}.$$

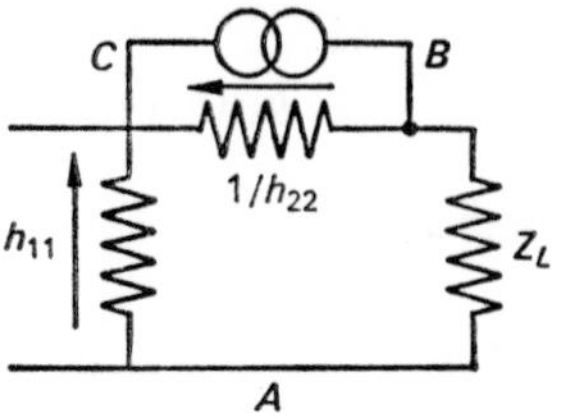

FIG. 4.

The output impedance is thus

$$Z_{\text{out}} = \frac{1}{h_{22}} + \frac{\varrho_g h_{11}}{\varrho_g + h_{11}} \left(1 + \frac{h_{21}}{h_{11} h_{22}}\right)$$

and the e.m.f. of the equivalent generator is

$$u_0 \frac{h_{11}}{h_{11} + \varrho_g} \left(1 + \frac{h_{21}}{h_{11} h_{22}}\right).$$

From this the gain in available power is given by

$$\frac{1}{G_{pu}} = \frac{\varrho_g}{h_{11}\varrho_g(1+\mu)} + \frac{h_{11} + \varrho_g}{h^2_{11} h_{22} \varrho_g (1+\mu)^2} \quad \text{with} \quad \mu = \frac{h_{21}}{h_{11} h_{22}}.$$

The circuit for the noise is that of Fig. 5.

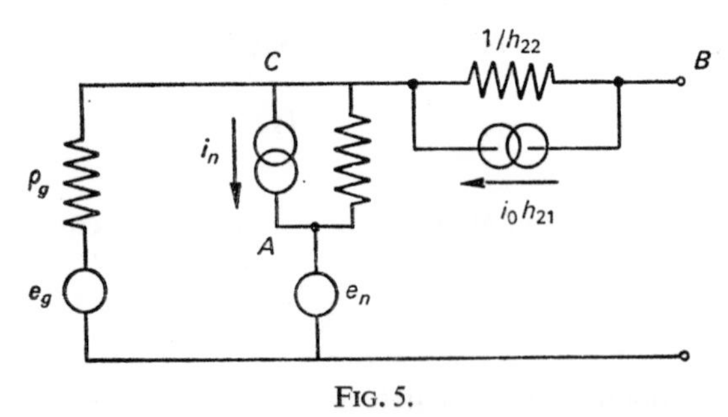

FIG. 5.

If we put $v_A - v_C = v_1$, we can write for the point C

$$\frac{e_g - e_n + v_1}{\varrho_g} - i_n + \frac{v_1}{h_{11}} + \frac{h_{21}}{h_{11}} v_1 + (v_2 - e_n + v_1) h_{22} = 0$$

with v_2 as the output voltage.

At B, with open-circuit output,

$$(e_n - v_1 - v_2)h_{22} - \frac{h_{21}}{h_{11}} \cdot v_1 = 0.$$

From the two equations we get

$$v_2 = \frac{\varrho_g h_{22} - h_{21}}{h_{22}(h_{11} + \varrho_g)} \cdot e_n + \frac{h_{22}h_{11} + h_{21}}{h_{22}(h_{11} + \varrho_g)} e_g - \frac{(h_{22}h_{11} + h_{21})\varrho_g}{h_{22}(h_{11} + \varrho_g)} \cdot i_n.$$

Since the output impedance is the same for both the signal and the noise (the two noise sources being fictional in this context) the available noise output power is

$$P_{nou} = \frac{(\varrho_g h_{22} - h_{21})^2 \overline{e_n^2} + (h_{22}h_{11} + h_{21})^2 \overline{e_g^2} + \varrho_g^2 (h_{22}h_{11} + h_{21})^2 \overline{i_n^2}}{4h_{22}(h_{11} + \varrho_g)\,[\varrho_g + h_{11} + \varrho_g h_{11} h_{22}(1 + \mu)]}.$$

The required noise factor is thus

$$F = \frac{1}{G_{pu}} \cdot \frac{P_{nuo}}{kTB} = \frac{(\varrho_g h_{22} - h_{21})^2 R_s + (h_{22}h_{11} + h_{21})^2 \varrho_g + \dfrac{\varrho_g^2}{R_p}(h_{22}h_{11} + h_{21})^2}{h_{11}^2 (1 + \mu)^2 \varrho_g h_{22}^2}.$$

If $\varrho_g \ll 1/h_{22}$ we can neglect $\varrho_g h_{22}$ compared with h_{21} and, taking h_{21}^2 and h_{11}^2 as factors in the numerator and denominator respectively,

$$F = \left(\frac{g}{h_{22} + g}\right)^2 \left[\left(1 + \frac{h_{22}}{g}\right)^2 + \frac{R_s}{\varrho_g} + \frac{\varrho_g}{R_p}\left(1 + \frac{h_{22}}{g}\right)^2\right].$$

If further, $h_{22} \ll g$ then

$$F \simeq 1 + \frac{R_s}{\varrho_g} + \frac{\varrho_g}{R_p},$$

which is the same as in the circuit with C earthed and which has, therefore, the same optimum value for ϱ_g.

EXERCISE No. 5

A field effect transistor (with common source connection) has a noise factor of 5 dB at 20 Hz when the source impedance is 500 kΩ. The noise factor is a minimum of 0·5 dB for $\varrho_g = 10$ MΩ.

Calculate the noise factor with common gate connection if the transistor is supplied from a generator of internal impedance 500 Ω and the conductance of the F.E.T. is $g = 2$ mA/V.

Solution. The noise of an F.E.T. can be represented by a voltage source in series and a current source in parallel as shown in Fig. 1.

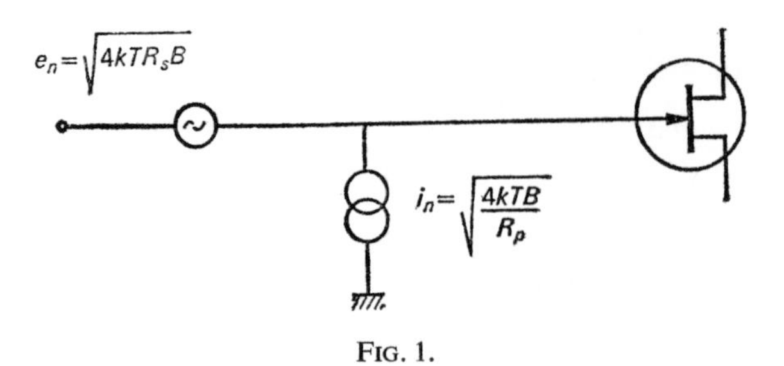

FIG. 1.

The minimum noise factor is

$$F_{\min} = 1+2\sqrt{\frac{R_s}{R_p}} \quad \text{for} \quad \varrho_g = \sqrt{R_s R_p}\,.$$

With the data given above we see that

$$\frac{R_s}{R_p} = \frac{1}{4}(F_{\min}-1)^2 = 3{\cdot}65\times 10^{-3}$$

$$R_s R_p = \varrho_{g\,\text{opt}}^2 = 10^{14}\,\Omega^2$$

from which $R_s \simeq 604\,\text{k}\Omega$, $R_p \simeq 165\,\text{M}\Omega$.

The field effect transistor is a typical case of a network for which $h_{12} = 0$ at low frequencies (here 20 Hz) and $h_{22} \ll g$. Under these conditions the expression for F is the same as in the normal circuit arrangement so that, for $\varrho_g = 500\,\Omega$,

$$F = 1+\frac{\varrho_g}{R_p}+\frac{R_s}{\varrho_g} \simeq 1+\frac{R_s}{\varrho_g} = 1208 \text{ or } 30{\cdot}8 \text{ dB}.$$

Notation

δ : skin depth
ε_0 : permittivity of free space
ε_r : relative permittivity
ε^* : complex permittivity $\varepsilon^* = \varepsilon' - j\varepsilon''$
λ_0 : free-space wavelength
λ : wavelength in a general medium
μ_0 : permeability of free space
μ_r : relative permeability
μ^* : complex permeability $\mu^* = \mu' - j\mu''$
σ : conductivity
φ : loss angle
c_0 : speed of light, $c_0 = (\varepsilon_0\mu_0)^{-1/2}$

Transmission Lines

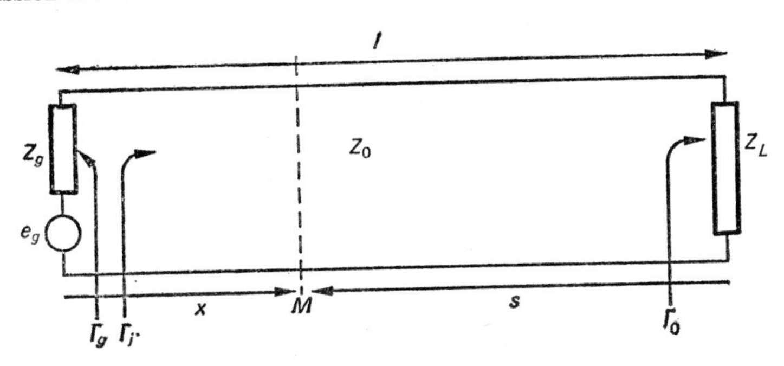

v^+, i^+ : complex voltage and current associated with the incident wave
v^-, i^- : complex voltage and current associated with the reflected wave
P_i or P_o : incident power
γ : propagation constant for a wave travelling along the line, $\gamma = \alpha + j\beta$
α : attenuation coefficient
β : phase constant, $\beta = \omega/v = 2\pi/\lambda$
v : phase velocity
$\Gamma_0 = |\Gamma_0|\, e^{j\theta_0}$: reflection coefficient at the end of the line

Γ	: reflection coefficient at a point along the line
Γ_i	: reflection coefficient at the line input
$S = \left\lvert \dfrac{v_{max}}{v_{min}} \right\rvert$	: standing wave ratio ($S > 1$)
Z_L	: load impedance
Z_0	: characteristic impedance
z_L	: reduced load impedance $z_L = Z_t/Z_0$

Wave Guides

H_x	: component of the electromagnetic field in complex notation
$\mathcal{H}_x$	: component of the electromagnetic field in real notation
λ	: wavelength in an infinite medium
β	: phase constant in an infinite medium, $\beta = 2\pi/\lambda$
λ_g	: guide wavelength
β_g	: phase constant in a waveguide, $\beta_g = 2\pi/\lambda_g$
λ_c	: cut-off wavelength
β_c	: cut-off phase constant $\beta_c = 2\pi/\lambda_c$
η_0	: impedance of free space, $\eta_0 = \sqrt{\mu_0/\varepsilon_0}$
Z_{TM}	: impedance for a TM mode
Z_{TE}	: impedance for a TE mode

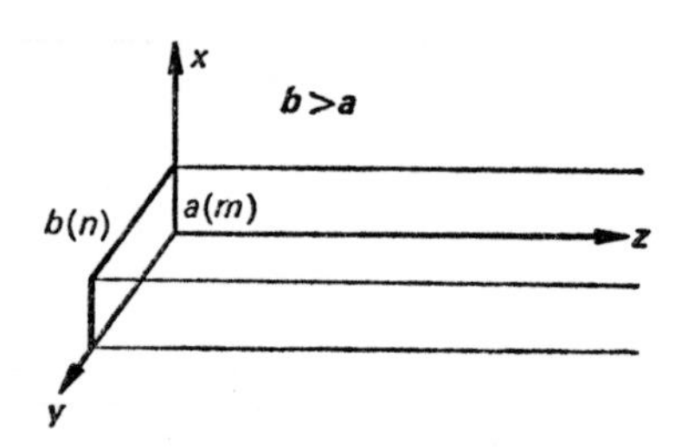

Enumeration of the TE_{mn} and TM_{mn} modes in a rectangular guide. Thus the fundamental mode is that designated as TE_{01}.

Chapter 6

Transmission Lines

Transmission Lines

EXERCISE No. 1

Consider a coaxial cable in which the radii of the inner and outer conductors are respectively $a = 0{\cdot}65$ mm and $b = 2{\cdot}35$ mm. The space between the conductors is filled with polythene of relative permittivity $\varepsilon_r = 2{\cdot}26$. Calculate the capacity C and inductance L per unit length of the cable and hence deduce its characteristic impedance and the phase velocity of electromagnetic waves in the line.

Solution. The capacity per unit length of a coaxial line is

$$C = \frac{2\pi\varepsilon_0\varepsilon_r}{\ln \dfrac{b}{a}},$$

with

$$\varepsilon_0 = \frac{1}{36\pi}\, 10^{-9}\ \text{F/m}$$

and

$$\ln \frac{b}{a} = 2{\cdot}3 \log_{10} \frac{b}{a},$$

The capacity in pF/m is then

$$C_{(\text{pF/m})} = \frac{24\varepsilon_r}{\log_{10} \dfrac{b}{a}} = 97{\cdot}5\ \text{pF/m}.$$

The inductance per unit length is

$$L = \frac{\mu_0}{2\pi} \ln \frac{b}{a}.$$

With

$$\mu_0 = 4\pi \times 10^{-7}\ \text{H/m}$$

and

$$L_{(\mu\text{H/m})} = 0{\cdot}46 \log_{10} \frac{b}{a} = 0{\cdot}256\ \mu\text{H/m},$$

this gives

$$Z_0 = \sqrt{\frac{L}{C}} = 51{\cdot}3\ \Omega$$

and

$$v = \frac{1}{\sqrt{LC}} = \frac{1}{\sqrt{\mu_0 \varepsilon_0 \varepsilon_r}} = \frac{c_0}{\sqrt{\varepsilon_r}} = 0{\cdot}66\ c_0 = 200\,000 \text{ km/s}$$

EXERCISE No. 2

Consider that the coaxial line in the preceding exercise is not perfect.

1. Calculate the resistance R per unit length due to the skin effect for frequencies greater than 1 MHz if the conductors are of copper with conductivity $\sigma = 5{\cdot}8 \times 10^7$ S/m and permeability $\mu = \mu_0$.

2. Calculate the conductance G per unit length due to losses in the dielectric which may be represented by a loss angle φ.

3. Calculate R and G for $f = 100$ MHz and $f = 3000$ MHz given that, for polythene, $\tan \varphi = 2 \times 10^{-4}$ at 1 MHz and 5×10^{-4} at 3000 MHz.

Solution. 1. Because of the skin effect, at high frequencies the current is localized in a region close to the surface of a conductor. This localization may be represented by the skin depth given by

$$\delta = \sqrt{\frac{2}{\omega \mu \sigma}}$$

which, for copper at a frequency of 1 MHz, gives

$$\delta \simeq 50\ \mu.$$

The resistance per unit length of the outer conductor is

$$R_o = \frac{1}{\sigma} \frac{l}{S_o} \quad \text{with} \quad l = 1 \text{ m} \quad \text{and} \quad S_o = 2\pi b \delta.$$

The resistance per unit length of the inner conductor is

$$R_i = \frac{1}{\sigma} \frac{l}{S_i} \quad \text{with} \quad l = 1 \text{ m}, \quad S_i = 2\pi a \delta$$

so that

$$R = R_i + R_o = \frac{1}{2\pi\sigma\delta}\left(\frac{1}{b} + \frac{1}{a}\right)$$

with

$$\frac{1}{2\pi\sigma\delta} = \frac{1}{2\pi\sigma}\sqrt{\pi\mu_0 f\sigma} = \frac{1}{2\pi}\sqrt{\frac{\pi\mu_0}{\sigma}}\sqrt{f} = 4{\cdot}15\times10^{-8}\sqrt{f}\ \Omega,$$

$$R = 4{\cdot}15\times10^{-8}\left[\frac{1}{b}+\frac{1}{a}\right]\sqrt{f}\ \Omega/\text{m}.$$

2. The conductance per unit length is due to losses in the dielectric, that is to say, to the complex permittivity

$$\varepsilon^* = \varepsilon_0\varepsilon_r(1-\text{j}\tan\varphi).$$

In the presence of such losses the capacity C of a condenser filled with the dielectric of permittivity $\varepsilon = \varepsilon_0\varepsilon_r$ becomes

$$C^* = C(1-\text{j}\tan\varphi).$$

Its admittance is then

$$Y = \text{j}\omega C^* = \text{j}\omega C(1-\text{j}\tan\varphi)$$

so that

$$Y = \omega C\tan\varphi+\text{j}\omega C = G+\text{j}\omega C.$$

The presence of dielectric losses thus gives rise to the conductance

$$G = \omega C\tan\varphi$$

which may be calculated as

$$G = \frac{2\pi\varepsilon_0\varepsilon_r}{\ln\dfrac{b}{a}}\times2\pi f\tan\varphi = 151\frac{\varepsilon_r f\tan\varphi}{\log_{10}\dfrac{b}{a}}\ \text{pS/m}.$$

3. For

$$f = 100\ \text{MHz},\quad R = 0{\cdot}815\ \Omega/\text{m};$$
$$f = 3000\ \text{MHz},\quad R = 4{\cdot}47\ \Omega/\text{m}.$$

Assuming that $\tan\varphi$ varies linearly with f gives

$$\tan\varphi = 2{\cdot}1\times10^{-4}\ \text{at 100 MHz},\quad G = 1{\cdot}25\times10^{-5}\ \text{S/m},$$
$$\tan\varphi = 5\times10^{-4}\ \text{at 3000 MHz},\quad G = 9{\cdot}15\times10^{-4}\ \text{S/m}.$$

EXERCISE No. 3

The parameters of a coaxial line are R, G, L, C. In the case where $\omega L \gg R$ and $\omega C \gg G$, show that the propagation constant can be written in the form $\gamma = \alpha_1+\alpha_2+\text{j}\beta$ and find the form of the variation of α_1 and α_2 with frequency.

Assuming that $\tan\varphi$ is independent of frequency, at what frequency will the dielectric losses become equal to the conduction losses? (Take $\tan\varphi = 5\times10^{-4}$, the conductivity of the copper conductors as $5{\cdot}8\times10^{7}$ S/m and the respective radii as $a = 1{\cdot}29$ mm, $b = 4{\cdot}7$ mm.)

Solution. If the propagation constant is taken in its normal form

$$\gamma = \sqrt{(R+j\omega L)(G+j\omega C)} = j\omega\sqrt{LC}\sqrt{\left(1+\frac{R}{j\omega L}\right)\left(1+\frac{G}{j\omega C}\right)}$$

then, on expanding the terms to second order,

$$\gamma = j\omega\sqrt{LC}\left(1+\frac{R}{2j\omega L}+\frac{R^2}{8\omega^2L^2}\right)\left(1+\frac{G}{2j\omega C}+\frac{G^2}{8\omega^2C^2}\right)$$

$$= \frac{R}{2}\sqrt{\frac{C}{L}}+\frac{G}{2}\sqrt{\frac{L}{C}}+j\omega\sqrt{LC}\left(1+\frac{1}{2}\left(\frac{R}{2\omega L}-\frac{G}{2\omega C}\right)^2\right).$$

Hence,

$$\beta = \sqrt{LC}\ (G \text{ and } R \text{ giving second order corrections})$$

$$\alpha_1 = \frac{R}{2}\sqrt{\frac{C}{L}}\ \text{(losses due to the metal conductors)}$$

$$\alpha_2 = \frac{G}{2}\sqrt{\frac{L}{C}}\ \text{(losses due to the dielectric).}$$

From Exercises 1 and 2 we have

$$\sqrt{\frac{L}{C}} = \frac{1}{2\pi}\sqrt{\frac{\mu_0}{\varepsilon_0\varepsilon_r}}\ln\frac{b}{a},$$

$$R = \frac{1}{2\pi}\sqrt{\frac{\pi\mu}{\sigma}}\left(\frac{1}{a}+\frac{1}{b}\right)\sqrt{f}, \quad \text{with} \quad \mu \simeq \mu_0,$$

and

$$G = \omega C\tan\varphi,$$

so that

$$\alpha_1 = \frac{1}{2\ln\dfrac{b}{a}}\sqrt{\frac{\pi\varepsilon_0\varepsilon_r}{\sigma}}\left[\frac{1}{a}+\frac{1}{b}\right]\sqrt{f}$$

$$\alpha_2 = \pi\sqrt{\mu_0\varepsilon_r\varepsilon_0}\tan\varphi f.$$

If it is assumed that $\tan\varphi$ is independent of frequency, α_1 and α_2 will be equal if

$$\sqrt{f} = \frac{1}{\sqrt{\pi\mu_0\sigma}}\cdot\frac{1}{\tan\varphi}\cdot\frac{1}{2\ln\dfrac{b}{a}}\left[\frac{1}{a}+\frac{1}{b}\right].$$

If $\tan\varphi = 5\times10^{-4}$, $b = 4{\cdot}7$ mm and $a = 1{\cdot}29$ mm, this gives

$$\sqrt{f} = 5{\cdot}05\times10^4 \quad \text{or} \quad f = 2550 \text{ MHz.}$$

EXERCISE No. 4

A line consisting of two parallel wires in air has the characteristic parameters, for each kilometre of length,

$$R = 2\,\Omega, \qquad L = 3 \text{ mH},$$
$$G = 1{\cdot}5\times10^{-7} \text{ S}, \qquad C = 12 \text{ nF}.$$

1. Calculate the values at 500 Hz of:

   the characteristic impedance Z_0;

   the propagation constant $\gamma = \alpha + j\beta$;

   the phase velocity v.

2. Assuming that there is no variation of loss angle with frequency, calculate the above quantities for $f = 2000$ Hz.

Solution 1.

$$\omega = 2\pi f = 3{\cdot}14\times10^3 \text{ rad/s}$$
$$\omega L = 9{\cdot}42\,\Omega/\text{km}, \qquad \omega C = 37{\cdot}7\times10^{-6} \text{ S/km}$$
$$Z_0 = \sqrt{\frac{R+j\omega L}{G+j\omega C}}$$

with

$$R+j\omega L = 2+j9{\cdot}42 = 9{\cdot}65\,\underline{/78^\circ}\ \Omega/\text{km}$$
$$G+j\omega C = (0{\cdot}15+j37{\cdot}7)10^{-6} \simeq 37{\cdot}7\times10^{-6}\,\underline{/90^\circ} \text{ S/km},$$

so that

$$Z_0^2 = 2{\cdot}56\times10^5\,\underline{/-12^\circ}$$
$$Z_0 = 5{\cdot}06\times10^2\,\underline{/-6^\circ} \simeq (503-j53)\Omega$$
$$\gamma^2 = (R+j\omega L)(G+j\omega C) \simeq 3{\cdot}64\times10^{-4}\,\underline{/168^\circ}$$
$$\gamma = 1{\cdot}91\times10^{-2}\,\underline{/84^\circ} = (0{\cdot}198+j1{\cdot}894)\times10^{-2}.$$

Then

$$\alpha = 1{\cdot}98\times10^{-3} \text{ neper/km}$$
$$\beta = 1{\cdot}894\times10^{-2} \text{ rad/m.}$$

The relation $\beta = \frac{\omega}{v}$ gives $v = \frac{3{\cdot}14\times10^3}{1{\cdot}894\times10^{-2}} = 165\,800$ km/s.

2. For $f = 2000$ MHz

$$\omega L = 37{\cdot}68 \gg R, \qquad \omega C \gg G,$$

so that

$$Z_0 = \sqrt{\frac{L}{C}} = 500\ \Omega$$

$$\gamma = j\beta \quad \text{where} \quad \beta = \omega\sqrt{LC} = 7{\cdot}5\times10^{-2}\ \text{rad/km}$$

and

$$v = \frac{1}{\sqrt{LC}} = 166\,600\ \text{km/s}.$$

EXERCISE No. 5

A load $Z_L = (50+j162{\cdot}5)\,\Omega$ is placed at the end of a lossless line of characteristic impedance $Z_0 = 125\,\Omega$. Calculate the modulus and phase angle of the reflection coefficient at this load and deduce the standing wave ratio.

Solution. The reflection coefficient at the end of a line is

$$\Gamma_0 = \frac{Z_L - Z_0}{Z_L + Z_0}$$

or, with reduced impedances

$$\Gamma_0 = \frac{z_L - 1}{z_L + 1}.$$

Here

$$z_L = \frac{50+j\,162{\cdot}5}{125} = 0{\cdot}4+j\,1{\cdot}3$$

so that

$$\Gamma_0 = |\Gamma_0|\,e^{j\theta_0} = \frac{-0{\cdot}6+j\,1{\cdot}3}{1{\cdot}4+j\,1{\cdot}3},$$

$$|\Gamma_0| = 0{\cdot}75, \qquad \theta_0 = 72°.$$

The standing wave ratio is given by

$$S = \frac{1+|\Gamma_0|}{1-|\Gamma_0|} = \frac{1{\cdot}75}{0{\cdot}25} = 7.$$

EXERCISE No. 6

What relation exists between the standing wave ratio S and the reduced reactance $x = X/Z_0$ when a load $Z_L = Z_0+jX$ is placed at the end of a lossless line?

Solution. The terminal impedance is $Z_L = Z_0+jX$ giving a reduced impedance

$$z_L = \frac{Z_L}{Z_0} = 1+jx.$$

The reflection coefficient at this impedance placed at the end of the line is

$$\Gamma_0 = \frac{z_L-1}{z_L+1} = \frac{jx}{2+jx}$$

and its modulus is

$$|\Gamma_0| = \frac{|x|}{\sqrt{4+x^2}}. \tag{1}$$

The standing wave ratio (V.S.W.R.) is given by

$$S = \frac{1+|\Gamma|}{1-|\Gamma|}, \quad \text{from which} \quad |\Gamma| = \frac{S-1}{S+1} = |\Gamma_0|,$$

Thus, on substituting from equation (1),

$$\frac{|x|}{\sqrt{4+x^2}} = \frac{S-1}{S+1}$$

or

$$x^2 = \frac{(S-1)^2}{(S+1)^2}(4+x^2),$$

which gives

$$x^2\left[1-\frac{(S-1)^2}{(S+1)^2}\right] = 4\frac{(S-1)^2}{(S+1)^2},$$

$$Sx^2 = (S-1)^2,$$

$$|x| = \frac{S-1}{\sqrt{S}}.$$

EXERCISE No. 7

A lossless line of characteristic impedance $Z_0 = 50\,\Omega$ is terminated by an unknown impedance Z_L. The distance between two adjacent voltage minima is $d = 8$ cm, the V.S.W.R. is $S = 2$ and the first voltage minimum is situated at $s_m = 1{\cdot}5$ cm from the load. What is the value of Z_L?

Solution $d = \lambda/2 = 8$ cm. Thus $\lambda = 16$ cm,

At a voltage minimum, the impedance is given by $Z_m = Z_0/S$. By the formula for the transformation of the impedance we can relate Z_m and Z_L such that

$$Z_m = \frac{Z_0}{S} = Z_0 \frac{Z_L + jZ_0 \tan \beta s_m}{Z_0 + jZ_L \tan \beta s_m}$$

with $\beta = 2\pi/\lambda$.

The reduced terminal impedance is then calculated as

$$z_L = \frac{Z_L}{Z_0} = \frac{\frac{1}{S} - j \tan \beta s_m}{1 - j\frac{1}{S} \tan \beta s_m} = \frac{1 - jS \tan \beta s_m}{S - j \tan \beta s_m}.$$

Since $$\beta s_m = \frac{2\pi}{16} \times 0{\cdot}15 \simeq 0{\cdot}59, \quad \tan \beta S_m = 0{\cdot}67$$

and

$$z_L = \frac{1 - j1{\cdot}34}{2 - j0{\cdot}67} \simeq 0{\cdot}65 - j0{\cdot}45,$$

so that

$$Z_L = Z_0 z_L = (32{\cdot}5 - j\,22{\cdot}5)\,\Omega.$$

EXERCISE No. 8

A lossless line with characteristic impedance $Z_0 = 50\ \Omega$ and propagation constant $\gamma = j\beta$ is terminated by an impedance Z_L as shown in Fig. 1.

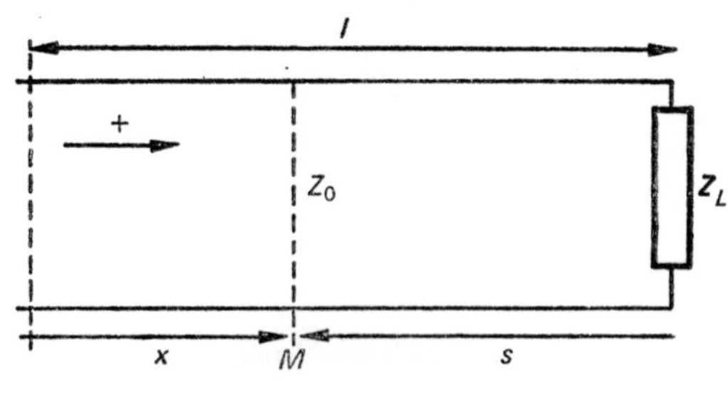

FIG. 1.

Let v^+ and i^+, v^- and i^- be the complex voltage and current of the incident and reflected waves respectively and $\Gamma_0 = |\Gamma_0| e^{j\theta_0}$ be the reflection coefficient from the end of the line.

1. Express the ratios

$$\left|\frac{v}{v^+}\right|^2, \quad \left|\frac{v}{v^+}\right|, \quad \left|\frac{i}{i^+}\right|^2 \quad \text{and} \quad \left|\frac{i}{i^+}\right|$$

in terms of $|\Gamma_0|$, θ_0, and $\theta = \beta s$, if v and i are respectively the complex total voltage and complex total current at a point M distance s from Z_L.

2. Determine the positions of the voltage maxima and minima.

Solution. 1. If the point M is at a distance x from the input to the line, then

$$v = v^+ + v^- = v_1 \, e^{-j\beta x} + v_2 \, e^{+j\beta x},$$

$$i = i^+ + i^- = \frac{1}{Z_0}(v_1 \, e^{-j\beta x} - v_2 \, e^{+j\beta x}).$$

On putting $s = l - x$ this gives

$$v(s) = v_1 \, e^{-j\beta l} \, e^{+j\beta s} + v_2 \, e^{+j\beta l} \, e^{-j\beta s}.$$

At the end of the line $s = 0$ and

$$v(0) = v_1 \, e^{-j\beta l} + v_2 \, e^{+j\beta l} = v^+(l)\,(1+\Gamma_0),$$

with

$$\Gamma_0 = \frac{v_1}{v_2} \, e^{2j\beta l}.$$

Thus

$$v(s) = v_1 \, e^{-j\beta l}(e^{+j\beta s} + \Gamma_0 \, e^{-j\beta s}),$$

$$v(s)\, v^*(s) = |v|^2 = v_1 v_1^* [e^{+j\beta s} + \Gamma_0 \, e^{-j\beta s}]\,[e^{-j\beta s} + \Gamma_0^* \, e^{+j\beta s}]$$

$$v_1 v_1^* = v^+(v^+)^* = |v^+|^2,$$

and

$$\left|\frac{v}{v^+}\right|^2 = 1 + |\Gamma_0|^2 + |\Gamma_0|\,[e^{-j(\theta_0 - 2\beta s)} + e^{+j(\theta_0 - 2\beta s)}],$$

$$= 1 + |\Gamma_0|^2 + 2|\Gamma_0| \cos(\theta_0 - 2\theta), \quad \text{with} \quad \theta = \beta s;$$

$$\left|\frac{v}{v^+}\right| = \sqrt{1 + |\Gamma_0|^2 + 2|\Gamma_0| \cos(\theta_0 - 2\theta)}.$$

Similarly

$$\left|\frac{i}{i^+}\right|^2 = 1 + |\Gamma_0|^2 - 2|\Gamma_0| \cos(\theta_0 - 2\theta),$$

$$\left|\frac{i}{i^+}\right| = \sqrt{1 + |\Gamma_0|^2 - 2|\Gamma_0| \cos(\theta_0 - 2\theta)}.$$

2. The quantity $|v|$ is a maximum when $\theta_0 - 2\theta = \pm n.2\pi$. Thus

$$|v| = |v^+|\,(1+|\Gamma_0|),$$

so that

$$s_M = \left(\frac{\theta_0}{4\pi}\right)\lambda \pm n\frac{\lambda}{2};$$

At this position $|i| = |i^+|\,(1-|\Gamma_0|)$ and the current is a minimum. The quantity $|v|$ is a minimum for $\theta_0 - 2\theta = \pm(2n+1)\pi$ so that

$$|v| = |v^+|\,(1-|\Gamma_0|)$$

and

$$s_m = \left(\frac{\theta_0}{4\pi}\right)\lambda \pm \left(n+\frac{1}{2}\right)\frac{\lambda}{2}.$$

EXERCISE No. 9

A slightly lossy transmission line has a characteristic impedance $Z_0 = 100\ \Omega$ and a length $2{\cdot}4\ \lambda$ (where λ is the wavelength of the radiation considered). The V.S.W.R. at the position of the load is $S_L = 4$ and at the input is $S_i = 3$, while the distance from the load to the first voltage minimum is $0{\cdot}1\ \lambda$. Find:

1. The total attenuation (in nepers) produced by the line;
2. The load impedance;
3. The input impedance.

Solution. 1. The reflection coefficient at the load is

$$\Gamma_0 = |\Gamma_0|\, e^{j\theta_0}.$$

with

$$|\Gamma_0| = \frac{S_L - 1}{S_L + 1} = \frac{3}{5} = 0{\cdot}6.$$

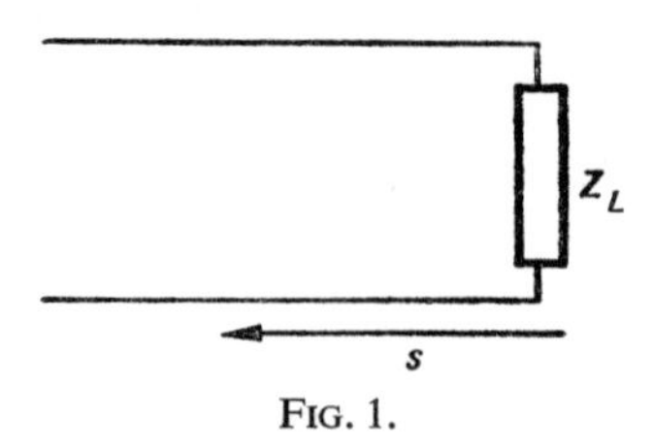

FIG. 1.

The reflection coefficient at the input is

$$\Gamma_i = \Gamma_0\, e^{-2\gamma s},$$

from which

$$|\Gamma_i| = |\Gamma_0|\, e^{-2\alpha s},$$

with

$$|\Gamma_i| = \frac{S_i-1}{S_i+1} = \frac{2}{4} = \frac{1}{2}$$

$$2\alpha s = \ln\left|\frac{\Gamma_0}{\Gamma_i}\right| = \ln 1{\cdot}2,$$

so that $\alpha s = \frac{1}{2}\ln 1{\cdot}2 = 1{\cdot}15 \log 1{\cdot}2 = 0{\cdot}09$ neper.
(From this one can calculate $\alpha = 0{\cdot}09/2{\cdot}4\lambda = 0{\cdot}0375$ neper/wavelength.)

2. The distance of the first minimum from the load is

$$s_m = 0{\cdot}1\,\lambda$$

and the attenuation $\alpha s_m = 0{\cdot}00375$ neper may be neglected.

Then

$$\theta_0 - \frac{4\pi s_m}{\lambda} = -\pi,$$

from which

$$\theta_0 = -0{\cdot}6\pi = -1{\cdot}885 \text{ radian}$$
$$\Gamma_0 = 0{\cdot}6\, e^{-j0{\cdot}6\pi} = -0{\cdot}6(0{\cdot}309+j\,0{\cdot}95) = -(0{\cdot}185+j\,0{\cdot}57).$$

Also

$$z_L = \frac{1+\Gamma_L}{1-\Gamma_L} = 0{\cdot}37-j\,0{\cdot}66$$

and

$$Z_L = (37-j\,66)\,\Omega.$$

3. $\Gamma_i = |\Gamma_0|\, e^{j\theta_0}\, e^{-2\alpha s}\, e^{-2j\beta s} = |\Gamma_i|\, e^{j(\theta_0-2\beta s)}$

with

$$2\beta s = \frac{4\pi}{\lambda}\cdot 2{\cdot}4\,\lambda = 9{\cdot}6\pi \quad \text{and} \quad \theta_0 - 2\beta s = -10{\cdot}2\pi.$$

Then

$$\Gamma_i = |\Gamma_i|\, e^{-j0.2\pi} = 0{\cdot}404-j\,0{\cdot}293 \simeq 0{\cdot}4-j\,0{\cdot}3$$

and

$$z_i = \frac{1{\cdot}4-j\,0{\cdot}3}{0{\cdot}6+j\,0{\cdot}3} = 1{\cdot}67-j\,1{\cdot}33,$$

or

$$Z_i = (167-j\,133)\,\Omega.$$

EXERCISE No. 10

A lossless line of impedance $Z_0 = 50\,\Omega$ is terminated by a load $Z_L = (7{\cdot}5 - \mathrm{j}\,25)\,\Omega$. The frequency of the radiation to be used is 3000 MHz, the phase velocity being $0{\cdot}66\, c_0$, where c_0 is the velocity of light.

1. What is the distance of the first voltage minimum from the load?
2. At what distance from Z_L is the first voltage maximum situated and what is the reflection coefficient at this point?

Solution. 1. The reduced load impedance is

$$z_L = \frac{Z_L}{Z_0} = 0{\cdot}15 - \mathrm{j}\,0{\cdot}5.$$

The reflection coefficient at the load is

$$\Gamma_0 = \frac{z_L - 1}{z_L + 1} = -\frac{(0{\cdot}85 + \mathrm{j}\,0{\cdot}5)}{(1{\cdot}15 - \mathrm{j}\,0{\cdot}5)} = |\Gamma_0|\, \mathrm{e}^{\mathrm{j}\theta_0}.$$

Thus

$$|\Gamma_0|^2 = \frac{(0{\cdot}85)^2 + (0{\cdot}5)^2}{(1{\cdot}15)^2 + (0{\cdot}5)^2} \quad \text{or} \quad |\Gamma_0| = 0{\cdot}78$$

and

$$\theta_0 = \pi + \varphi_1 - \varphi_2 \quad \text{with} \quad \tan\varphi_1 = \frac{0{\cdot}5}{0{\cdot}85},$$

$$\tan\varphi_2 = -\frac{0{\cdot}5}{1{\cdot}15}.$$

Then

$$\theta_0 = 1{\cdot}3\,\pi \text{ radians}$$

and

$$s_m = \frac{\theta_0}{4\pi} \cdot \lambda \pm \frac{\lambda}{4} = (0{\cdot}325 \pm 0{\cdot}25)\lambda.$$

The first minimum is situated at $s_m = 0{\cdot}075\,\lambda$. Then, with

$$v = 0{\cdot}66 c_0 = 200\,000 \text{ km/s},$$

the wavelength is

$$\lambda = \frac{v}{f} = \frac{2 \times 10^8}{3 \times 10^9} = 0{\cdot}066 \text{ m}$$

and

$$s_m = 0{\cdot}495 \text{ cm}.$$

2. The first maximum is at

$$s_M = s_m + \frac{\lambda}{4} = 2{\cdot}155 \text{ cm};$$

and at this point

$$\Gamma = \Gamma_0 \, e^{-2j\beta s_M} = |\Gamma_0| \, e^{j(\theta_0 - 2\beta s_M)}.$$

Since $2\beta s_M = \theta_0$

$$\Gamma = |\Gamma_0|.$$

EXERCISE No. 11

Measurements of the V.S.W.R. along a transmission line are usually made with the aid of "square law" crystal which gives a current $i = a|v|^2$ where a is a constant and v is the voltage at the point of measurement.

1. If the maximum and minimum values of i are $i_M = 50$ μA and $i_m = 20$ μA what is the value of S, the standing wave ratio?

The square law $i = a|v|^2$ is only valid for low powers. When $|v_{\max}|$ is large, the following method is used.

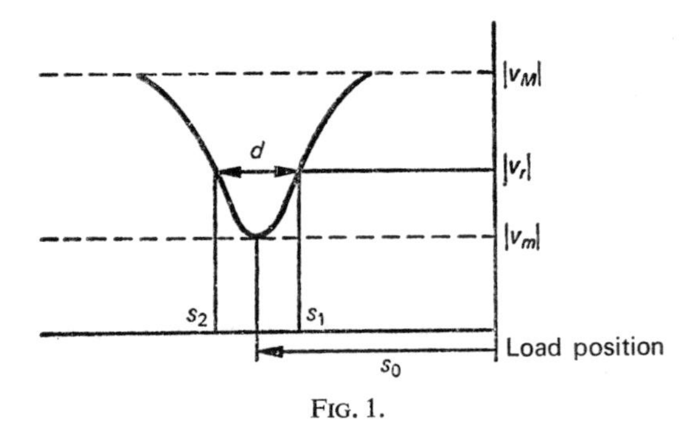

FIG. 1.

The distance d is found between two points on the axis s_1 and s_2 which are symmetric about the minimum position s_0. That is, two points are found with the same voltage $|v_r|$, between $|v_{\max}|$ and $|v_{\min}|$, for which the law $i \propto |v|^2$ is valid. If the wavelength λ, the distance d and the ratio $R = |v_r/v_m|$ are known, the V.S.W.R. can be found.

2. The reflection coefficient from the end of the line is $\Gamma_0 = |\Gamma_0| \, e^{j\theta_0}$. Calculate the current i measured at $s = s_0$ and $s = s_2$ and the ratio k^2 in terms of $|\Gamma_0|$, $\beta = 2\pi/\lambda$ and d.

3. Express S in terms of d, k and λ and find this relation for (a) $k^2 = 2$, (b) $\pi d/\lambda \ll 1$.

Solution. 1. The currents at the maxima and minima are

$$i_M = a|v_M|^2,$$
$$i_m = a|v_m|^2,$$

so that

$$S = \left|\frac{v_M}{v_m}\right| = \sqrt{\frac{i_M}{i_m}} = 1{\cdot}58.$$

2. At $s = s_0$, we have

$$i(s_0) = a|v(s_0)|^2 = a|v^+|^2[1+|\Gamma_0|^2+2|\Gamma_0| \cos(2\beta s_0-\theta_0)]$$

and at $s = s_2$

$$i(s_2) = a|v^+|^2[1+|\Gamma_0|^2+2|\Gamma_0| \cos(2\beta s_2-\theta_0)].$$

For a voltage minimum

$$2\beta s_0-\theta_0 = \pm(2n+1)\pi, \quad \text{so that} \quad -\theta_0 = \pm(2n+1)\pi-2\beta s_0$$

and at $s = s_2$

$$2\beta s_2-\theta_0 = 2\beta s_2-2\beta s_0\pm(2n+1)\pi.$$

Now

$$s_2-s_0 = d/2,$$

and therefore

$$i(s_2) = a|v^+|^2[1+|\Gamma_0|^2-2|\Gamma_0| \cos \beta d],$$
$$i(s_0) = a|v^+|^2[1-|\Gamma_0|]^2,$$

and

$$k^2 = \left|\frac{v(s_2)}{v(s_0)}\right|^2 = \frac{i(s_2)}{i(s_0)} = \frac{1+|\Gamma_0|^2-2|\Gamma_0| \cos \beta d}{[1-|\Gamma_0|]^2}.$$

3. Since

$$S = \frac{1+|\Gamma_0|}{1-|\Gamma_0|}, \quad \text{one has} \quad |\Gamma_0| = \frac{S-1}{S+1},$$

so that

$$[1-|\Gamma_0|]^2 = \frac{4}{(S+1)^2},$$
$$1+|\Gamma_0|^2 = \frac{2(S^2+1)}{(S+1)^2},$$

from which

$$k^2 = \frac{(S^2+1)-(S^2-1)\cos\beta d}{2} = \frac{S^2(1-\cos\beta d)+1+\cos\beta d}{2}$$

But

$$\cos\beta d = \cos 2\pi\frac{d}{\lambda} \quad \text{gives} \quad 1+\cos\beta d = 2\cos^2\frac{\pi d}{\lambda},$$

so that

$$k^2 = S^2\sin^2\frac{\pi d}{\lambda}+\cos^2\frac{\pi d}{\lambda}$$

$$S = \sqrt{\frac{k^2-\cos^2\frac{\pi d}{\lambda}}{\sin^2\frac{\pi d}{\lambda}}}.$$

If (a) $k^2 = 2$ we have

$$S = \sqrt{1+\frac{1}{\sin^2\frac{\pi d}{\lambda}}}.$$

while if (b) $(\pi d/\lambda) \ll 1$

$$S \simeq \frac{\lambda}{\pi d}\sqrt{k^2-1}.$$

EXERCISE No. 12

The expression for the potential and current at a point on a transmission line.

A line of length l and characteristic impedance Z_0 is supplied by a generator of e.m.f. e_g and internal impedance Z_g and is terminated by a load impedance Z_L. The reflection coefficients at the planes $x = 0$ and $x = l$ are Γ_g and Γ_0 respectively and the propagation constant is $\gamma = \alpha+j\beta$.

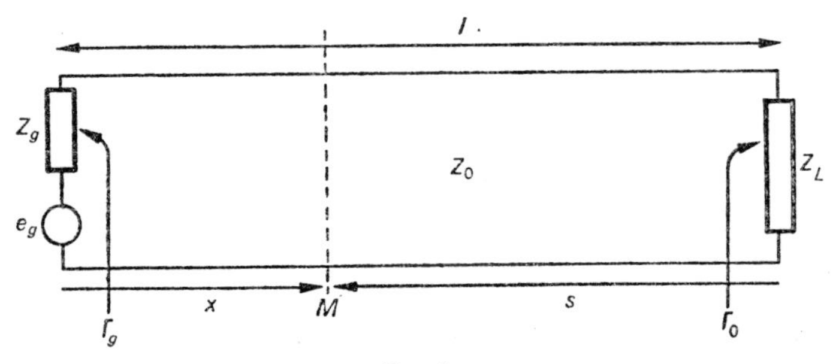

FIG. 1.

At a point M on the line, a distance x from the generator, the voltage and current are given by

$$v(x) = v_1\,e^{-\gamma x} + v_2\,e^{+\gamma x},$$

$$i(x) = \frac{1}{Z_0}[v_1\,e^{-\gamma x} - v_2\,e^{+\gamma x}].$$

It is proposed to calculate v_1 and v_2 in terms of the boundary conditions imposed by the load and source impedances.

1. At the instant $t = 0$ the generator is connected to the line at $x = 0$. What is the impedance seen by the generator? What is the voltage $v_{i0}(0)$ applied to the input at this instant?

2. Let $v_{i0}(x)$ be the voltage of the incident wave at M and $v_{r0}(x)$ the voltage of the wave at M obtained by reflection of $v_{i0}(x)$ at the load. Give expressions for $v_{i0}(x)$ and $v_{r0}(x)$ in terms of $v_{i0}(0)$, Γ_0 and γl.

3. When the wave v_{r0} is reflected at $x = 0$ it gives rise to an effective incident wave v_{i1}. Express $v_{r1}(x)$ in terms of $v_{i0}(x)$, Γ_0, Γ_g and γl.

4. Deduce:

(a) $v_{in}(x)$ and $v_{rn}(x)$, the voltages after n reflections;

(b) the potential $v(x)$ in terms of e_g, Z_0, Γ_g, Γ_0, γl;

(c) the value of the quantities v_1 and v_2.

Solution. 1. At $t = 0$ the existence of Z_L is unknown to the generator which see in effect, an infinite line of impedance Z_0. The equivalent circuit is then as in Fig. 2

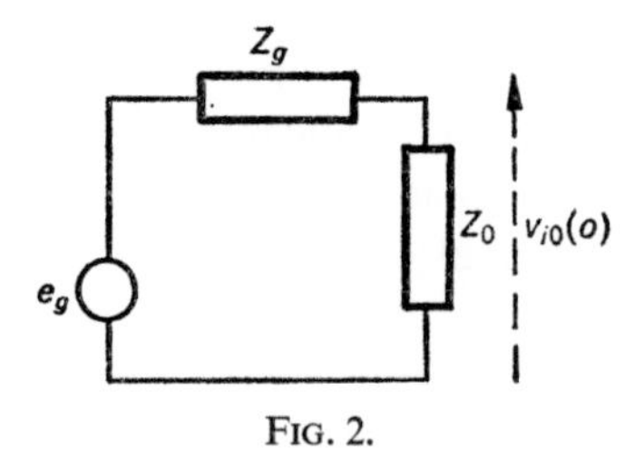

FIG. 2.

from which

$$v_{i0}(0) = e_g \frac{Z_0}{Z_0 + Z_g}. \tag{1}$$

2. At M we have

$$v_{i0}(x) = v_{i0}(0)\,e^{-\gamma x}$$

which gives at $x = l$,

$$v_{i0}(l) = v_{i0}(0)\,e^{-\gamma l}. \tag{2}$$

If Γ_0 is the reflection coefficient at the load Z_L we have

$$v_{r0}(l) = \Gamma_0 v_{i0}(l) = \Gamma_0 v_{i0}(0)\,e^{-\gamma l}$$

and

$$v_{r0}(x) = v_{r0}(l)\,e^{-\gamma s} \quad \text{where} \quad s = l-x,$$

so that

$$v_{r0}(x) = \Gamma_0 v_{i0}(0)\,e^{-2\gamma l}\,e^{+\gamma x}. \tag{3}$$

3. At $x = 0$

$$v_{r0}(0) = \Gamma_0 v_{i0}(0)\,e^{-2\gamma l}.$$

After reflection at Z_g the reflected wave is

$$v_{i1}(0) = \Gamma_g v_{r0}(0) = \Gamma_0 \Gamma_g v_{i0}(0)\,e^{-2\gamma l},$$

where

$$\Gamma_g = \frac{Z_g - Z_0}{Z_g + Z_0},$$

from which

$$v_{i1}(x) = \Gamma_0 \Gamma_g v_{i0}(0)\,e^{-2\gamma l}e^{-\gamma x}$$

$$v_{i1}(l) = \Gamma_0 \Gamma_g v_{i0}(0)\,e^{-3\gamma l} \tag{4}$$

$$v_{r1}(l) = \Gamma_0^2 \Gamma_g v_{i0}(0)\,e^{-3\gamma l}.$$

Thus

$$v_{r1}(x) = \Gamma_0^2 \Gamma_g v_{i0}(0)\,e^{-4\gamma l}e^{+\gamma x}. \tag{5}$$

4. From 2 and 4 the incident wave after n reflections is

$$v_{in}(x) = v_{in-1}(x)\Gamma_0 \Gamma_g e^{-2\gamma l}$$

while the corresponding reflected wave is

$$v_{rn}(x) = v_{rn-1}(x)\Gamma_0 \Gamma_g e^{-2\gamma l}$$

Repeating this result for $(n-1)$, $(n-2)$, etc. gives

$$v_{in}(x) = v_{i0}(x)[1+\Gamma_0\Gamma_g e^{-2\gamma l}+(\Gamma_0\Gamma_g e^{-2\gamma l})^2+\ldots]$$

$$v_{rn}(x) = v_{r0}(x)[1+\Gamma_0\Gamma_g e^{-2\gamma l}+(\Gamma_0\Gamma_g e^{-2\gamma l})^2+\ldots];$$

ro m which

$$v_n(x) = v_{in}(x)+v_{rn}(x),$$

$$= [v_{i0}(x)+v_{r0}(x)]\,[1+\Gamma_0\Gamma_g e^{-2\gamma l}+\ldots]$$

$$v(x) = v_n(x) \quad \text{when} \quad n \to \infty,$$

and

$$v(x) = \frac{v_{i0}(x)+v_{r0}(0)}{1-\Gamma_0\Gamma_g e^{-2\gamma l}}.$$

The voltage distribution in the line is uniquely determined by the first incident wave and the first reflected wave, the subsequent reflections being included through the term

$$[1-\Gamma_0\Gamma_g e^{-2\gamma l}]^{-1},$$

Thus we have

$$v(x) = v_{i0}(0)\,\frac{e^{-\gamma x}+\Gamma_0 e^{-2\gamma l}e^{+\gamma x}}{1-\Gamma_0\Gamma_g e^{-2\gamma l}},$$

from which

$$v_1 = e_g\,\frac{Z_0}{Z_0+Z_g}\,\frac{1}{1-\Gamma_0\Gamma_g e^{-2\gamma l}},$$

$$v_2 = e_g\,\frac{Z_0}{Z_0+Z_g}\,\frac{\Gamma_0 e^{-2\gamma l}}{1-\Gamma_0\Gamma_g e^{-2\gamma l}}.$$

EXERCISE No. 13

A transmission line of characteristic impedance Z_0 and length l is terminated by a load Z_L. If v_L and v_i are the voltages at the load and input respectively, find the ratio v_L/v_i in terms of Z_0, Z_L and γl where $\gamma = \alpha+j\beta$ is the propagation constant.

Solution. The voltages at a point x on the line is

$$v(x) = Ae^{-\gamma x}+Be^{+\gamma x}.$$

At the ends of the line

$$v(0) = A+B = v_i,$$
$$v(l) = Ae^{-\gamma l}+Be^{+\gamma l};$$

and

$$\Gamma_0 = \frac{Z_L-Z_0}{Z_L+Z_0} = \frac{Be^{\gamma l}}{Ae^{-\gamma l}}.$$

Thus

$$v_L = Ae^{-\gamma l}\left(1+\frac{Z_L-Z_0}{Z_L+Z_0}\right) = \frac{2Z_L}{Z_L+Z_0}\,Ae^{-\gamma l}$$

$$v_i = A\left(1+\frac{B}{A}\right) = A\left(1+\frac{Z_L-Z_0}{Z_L+Z_0}\,e^{-2\gamma l}\right)$$

and

$$\frac{v_L}{v_i} = \frac{2Z_L e^{-\gamma l}}{Z_L+Z_0+(Z_L-Z_0)\,e^{-2\gamma l}}$$

$$= \frac{Z_L}{Z_L\cosh\gamma l+Z_0\sinh\gamma l}.$$

EXERCISE No. 14

Considering the transmission line shown in Fig. 1, calculate the parameters (u, Z_i) of the equivalent circuit shown in Fig. 2, the characteristic impedance of the line being Z_0.

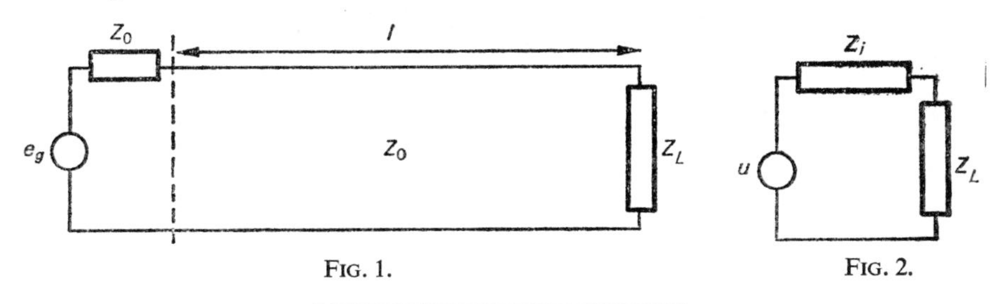

FIG. 1. FIG. 2.

Solution. This case requires the application of Thévenin's theorem.

(a) The generator can be replaced effectively by its internal impedance Z_0. Thus the impedance seen from the load is Z_0 so that

$$Z_0 = Z_i.$$

(b) The load Z_L is replace d by an infinite resistance. The voltage at the output is then u as seen in Fig. 3.

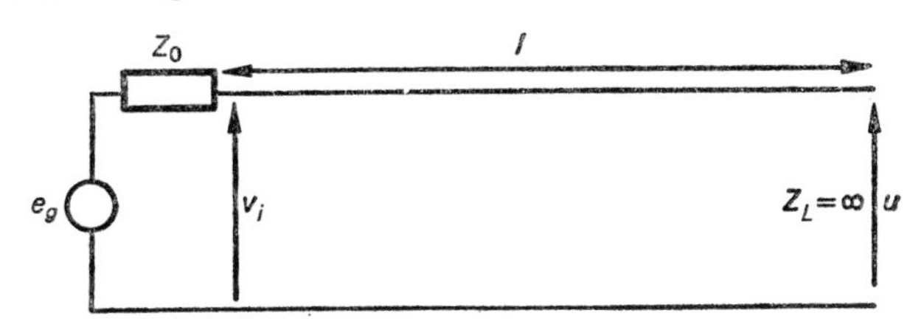

FIG. 3.

The impedance seen in the input plane is

$$Z_i = \frac{Z_0}{\tanh \gamma l},$$

from which

$$v_i = \frac{Z_i}{Z_0+Z_i} \cdot e_g = \frac{Z_0}{Z_0+Z_0 \tanh \gamma l} e_g = \frac{e_g}{1+\tanh \gamma l}.$$

Now it was shown in the preceding exercise that

$$\frac{v_L}{v_i} = \frac{Z_L}{Z_L \cosh \gamma l + Z_0 \sinh \gamma l}.$$

In the present case $Z_L = \infty$ and $v_L = u$.

Thus

$$u = \frac{v_i}{\cosh \gamma l} = \frac{e_g}{1+\tanh \gamma l} \frac{1}{\cosh \gamma l} = \frac{e_g}{\cosh \gamma l + \sinh \gamma l} = e_g \mathrm{e}^{-\gamma l}.$$

Note: The voltage across Z_L is

$$v_L = u \frac{Z_L}{Z_L + Z_0} = e_g \mathrm{e}^{-\gamma l} \frac{Z_L}{Z_L + Z_0}.$$

Chapter 7

Power Matching

EXERCISE No. 1

Efficiency of a transmission line

Consider a length l of a transmission line with a constant of propagation $\gamma = \alpha + j\beta$ and a characteristic impedance Z_0 which is real. If the efficiency of the line is defined as the ratio k between output power P_L delivered at the load and the input power P_i, calculate this efficiency as a function of Γ_0, the reflection coefficient at the load. Under what condition does k have its maximum value k_m?

In the particular case where $Z_L = R_L$, express k in terms of k_m and the standing wave ratio S.

Calculate k for the case where $\alpha = 1{\cdot}4$ dB/m, $l = 2$ m and $R_L = 3Z_0$.

Solution. The reflection coefficient at the line input is

$$\Gamma_i = \Gamma_0 e^{-2\gamma l}.$$

The input power transmitted along the line is given by

$$P_i = P_0(1-|\Gamma_i|^2) = P_0(1-|\Gamma_0|^2 e^{-\gamma 4l})$$

where P_0 is the power supplied at the input. The power arriving at the load is

$$P_0' = P_0 e^{-2\alpha l}$$

and the power delivered to the load is

$$P_L = P_0'(1-|\Gamma_0|^2),$$

so that

$$k = \frac{P_L}{P_i} = \frac{P_0'(1-|\Gamma_0|^2)}{P_0(1-|\Gamma_0|^2 e^{-4\alpha l})} = e^{-2\alpha l}\frac{1-|\Gamma_0|^2}{1-|\Gamma_0|^2 e^{-4\alpha l}},$$

The efficiency of the line is a maximum when $Z_L = Z_0$ so that $\Gamma_0 = 0$ and

$$k = k_m = e^{-2\alpha l}.$$

Then

$$k = k_m \frac{1-|\Gamma_0|^2}{1-|\Gamma_0|^2 k_m}.$$

If $Z_L = R_L$,

$$|\Gamma_0|^2 = \left(\frac{R_L - Z_0}{R_L + Z_0}\right)^2.$$

If $R_L > Z_0$, $S = R_L/Z_0$ and if $R_L < Z_0$, $S = Z_0/R_L$. In both cases

$$|\Gamma_0|^2 = \left(\frac{S-1}{S+1}\right)^2 = \left(\frac{1/S-1}{1/S+1}\right)^2.$$

Thus

$$k = k_m \frac{4S}{(S+1)^2-(S-1)^2k_m^2}.$$

If $l = 2$ m and $\alpha = 1{\cdot}4$ dB/m:

$$2{\cdot}8 = 10 \log \frac{1}{k_m}, \quad \text{or} \quad \frac{1}{k_m} = 1{\cdot}9,\ k_m = 0{\cdot}527.$$

For $S = 1$, $k = k_m = 0{\cdot}527$, an efficiency of 52·7%

For $S = 3$, $R_L/Z_0 = 3$, $k = 0{\cdot}423$, an efficiency of 42·3%.

It will be seen that it is important for a transmission line to be matched. Where there are no losses, $k = k_m = 1$ and the efficiency is 100% when the line does not dissipate energy; but in this case $P_i = P_L = P_0(1-|\Gamma_0|^2)$ and it is still advantageous to match the line.

EXERCISE No. 2

Consider the coaxial cable shown in Fig. 1 where the inner conductor is held by rings of dielectric of thickness t separated by a distance d, the relative permittivity being ε_r. A load is placed at the position T which would be matched to the empty cable. What relation must exist between d, t and ε_r if there is to be no reflection from the plane E, the line being assumed lossless?

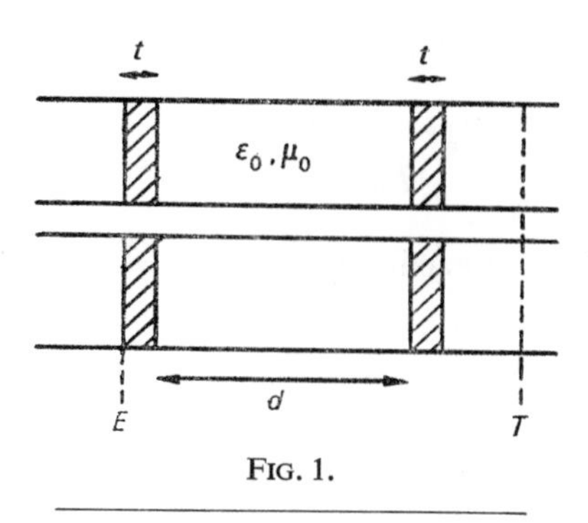

FIG. 1.

Solution. Let Z_0 and Z_1 be the characteristic impedances of the cable empty and filled with the dielectric respectively, and let β_0 and β_1 be the respective propagation constants. Then

$$Z_1 = \frac{Z_0}{\sqrt{\varepsilon_r}}, \quad \beta_1 = \beta_0\sqrt{\varepsilon_r}.$$

What is required is that the line should be matched in the plane E when there is a matched load at T. Since the system is symmetric it is possible to calculate the impedance seen from T and from E at the plane π shown in Fig. 2.

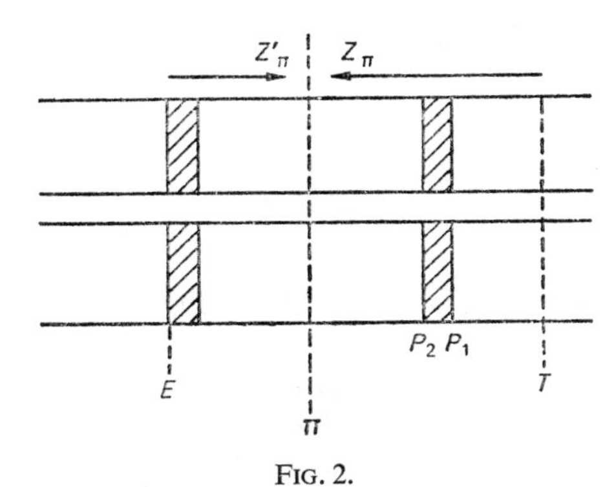

FIG. 2.

The impedance seen at P_1 is Z_0 which becomes, at P_2,

$$Z(t) = Z_1 \frac{Z_0 + jZ_1 \tan \beta_1 t}{Z_1 + jZ_0 \tan \beta_1 t}.$$

The transfer of this impedance to the position π gives

$$Z_\pi = Z_0 \frac{Z(t) + jZ_0 \tan \beta_0 \dfrac{d}{2}}{Z_0 + jZ(t) \tan \beta_0 \dfrac{d}{2}}.$$

The impedance seen at π from E will have the same form but with t replaced by $-t$, d by $-d$. That is

$$Z'_\pi = Z_0 \frac{Z(-t) - jZ_0 \tan \beta_0 \dfrac{d}{2}}{Z_0 - jZ(-t) \tan \beta_0 \dfrac{d}{2}},$$

where, for matching, we must have

$$Z'_\pi = Z_\pi. \tag{1}$$

Let us put

$$\theta_0 = \tan\frac{\beta_0 d}{2} \quad \text{and} \quad \theta_1 = \tan \beta_1 t.$$

On reducing Z'_π and Z_π to a common denominator, equation (1) becomes

$$Z_0[Z(t)-Z(-t)](1-\theta_0^2)-2j\theta_0[Z(t)\,Z(-t)-Z_0^2] = 0, \tag{2}$$

where

$$Z(t)-Z(-t) = Z_1\frac{Z_0+jZ_1\theta_1}{Z_1+jZ_0\theta_1}-Z_1\frac{Z_0-jZ_1\theta_1}{Z_1-jZ_0\theta_1}$$

$$= 2jZ_1\theta_1\frac{Z_1^2-Z_0^2}{Z_1^2+Z_0^2\theta_1^2}$$

and

$$Z(t)\,Z(-t) = Z_1^2\frac{Z_0+Z_1^2\theta_1^2}{Z_1+Z_0\theta_1^2}.$$

Substituting in (2) gives

$$Z_0Z_1\theta_1(Z_1^2-Z_0^2)(1-\theta_0^2) = \theta_0\theta_1^2(Z_1^4-Z_0^4)$$

or

$$\frac{Z_0Z_1}{Z_1^2+Z_0^2} = \frac{\theta_1\theta_0}{1-\theta_0^2}.$$

Substituting for $Z_1 = Z_0/\sqrt{\varepsilon}$ and $2\theta_0/(1-\theta_0^2) = \tan\beta_0 d$ gives

$$\frac{2\sqrt{\varepsilon_r}}{1+\varepsilon_r} = \tan\beta_1 t . \tan\beta_0 d,$$

which gives, finally,

$$\frac{2\sqrt{\varepsilon_r}}{1+\varepsilon_r} = \tan\beta_0\sqrt{\varepsilon_r}.t\,\tan\beta_0 d.$$

EXERCISE No. 3

If it is required to match a load $Z_L = R_L+jX_L$ to a line with characteristic impedance $Z_0 = R_0$, this is usually accomplished by means of a line of length l (ideally lossless) of characteristic impedance $Z_1 = R_1$ as shown in Fig. 1.

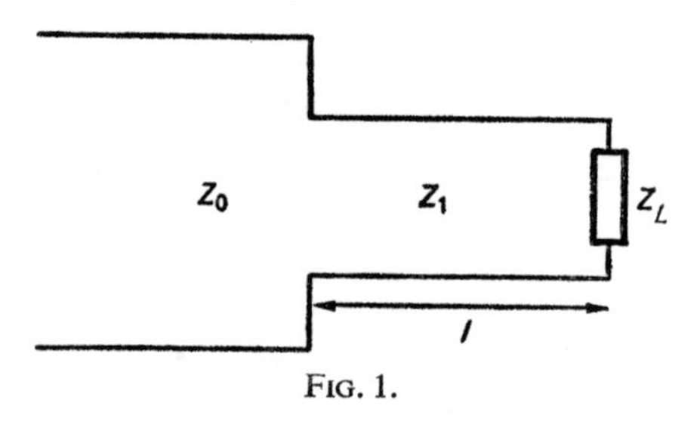

FIG. 1.

Calculate the values of Z_1 and l to produce matching. Is it always possible to achieve such a match?

For numerical examples consider that:

$$Z_0 = 75\ \Omega, \quad Z_L = (18+j25)\ \Omega$$
$$Z_L = (80+j45)\ \Omega$$
$$Z_L = (18+j45)\ \Omega.$$

Solution. The input impedance of the line of length l is

$$Z_i = Z_1 \frac{Z_L+jZ_1 \tan \dfrac{2\pi l}{\lambda}}{Z_1+jZ_L \tan \dfrac{2\pi l}{\lambda}}.$$

It is required that $Z_i = Z_0$. Putting $Z_L = R_L+jX_L$ and $\tan 2\pi l/\lambda = \theta$ this gives

$$Z_0Z_1+j(R_L+jX_L)\,\theta . Z_0 = Z_1R_L+jX_LZ_1+j\theta Z_1^2.$$

Equating the real and imaginary parts separately gives two equations

$$Z_0Z_1-X_LZ_0\theta = Z_1R_L$$
$$\theta R_LZ_0 = X_LZ_1+\theta Z_1^2,$$

so that it is required that

$$Z_1 = \sqrt{Z_0R_L+\frac{X_L^2Z_0}{R_L-Z_0}},$$
$$\theta = Z_1\frac{(Z_0-R_L)}{X_LZ_0}.$$

Since Z_1 is taken to be real, there is only a solution if

$$Z_0R_L+\frac{X_L^2Z_0}{R_L-Z_0} > 0.$$

(a) For $R_L > Z_0$ this condition is always satisfied

(b) For $R_L < Z_0$, we need $Z_0R_L > \dfrac{X_L^2Z_0}{Z_0-R_L}$,

from which

$$Z_0R_L > |Z_L|^2.$$

Numerical example.

1.
$$Z_L = (18+j25)\Omega \quad \text{and} \quad R_L < Z_0$$
$$|Z_L|^2 = (18)^2+(25)^2 = 949$$
$$Z_0R_L = 1350 > |Z_L|^2$$

i.e. there is a possible solution.

From the expressions for Z and θ we find

$$Z_1 \simeq 23\ \Omega, \quad \theta = 0{\cdot}7 \quad \text{or} \quad \frac{2\pi l}{\lambda} = 0{\cdot}611+n\pi$$

and
$$l = 0{\cdot}097\lambda+\frac{n\lambda}{2}.$$

2.
$$Z_L = (80+j45)\ \Omega \quad \text{so that} \quad R_L > Z_0$$
$$Z_1 = 190\ \Omega$$
$$\theta = -0{\cdot}283 \quad \text{and} \quad l = 0{\cdot}455\lambda+\frac{n\lambda}{2}.$$

3.
$$Z_L = (18+j45)\ \Omega, \quad R_L < Z_0$$
$$|Z_L|^2 = 2349$$
$$R_LZ_0 = 1350,$$

so that
$$R_LZ_0 < |Z_L|^2 \quad \text{and matching is impossible.}$$

EXERCISE No. 4

A lossless line of characteristic impedance $Z_0 = 50\ \Omega$ and terminated in a load of impedance Z_L is supplied by a generator giving an incident power of $P_i = 500$ mW.

1. If the V.S.W.R. is 1·4 what is the power reaching the load?
2. What are the values of the maximum and minimum impedances along the line?
3. What are the values of the maximum and minimum voltage and current?

Solution. 1. The power transmitted by the line is related to the incident power by

$$P_t = P_i-P_r = P_i(1-|\Gamma|^2),$$

P_r being the reflected power and Γ the reflection coefficient. Now

$$S = \frac{1+|\Gamma|}{1-|\Gamma|} \quad \text{so that} \quad |\Gamma| = \frac{S-1}{S+1},$$

and

$$P_t = P_i \frac{4S}{(S+1)^2} = 486 \text{ mW}.$$

2. $$Z_M = SZ_0 = 70\,\Omega; \quad Z_m = \frac{Z_0}{S} = 35{\cdot}7\,\Omega.$$

3. $$P_t = \frac{V_M^2}{SZ_0},$$

where V_m is maximum r.m.s. voltage. Then

$$V_M = \sqrt{P_t S Z_0} = 5{\cdot}83 \text{ V}_{\text{rms}}$$

and

$$V_m = \frac{V_M}{S} = 4{\cdot}16 \text{ V}_{\text{rms}}.$$

Then

$$I_M = \frac{V_M}{Z_0} = \frac{5{\cdot}8}{50} = 117 \text{ mA}_{\text{rms}}$$

$$I_m = \frac{I_M}{S} = 83 \text{ mA}_{\text{rms}}.$$

EXERCISE No. 5

A lossless 75 Ω line terminates in a load $Z_L = (51{\cdot}5+\text{j}45)\,\Omega$ and the power supplied by a generator is 150 mW. What is the current through the load?

Solution. The power transmitted to the load is $P_t = P_i(1-|\Gamma|^2)$ with

$$\Gamma = \frac{Z_L - Z_0}{Z_L + Z_0} = \frac{-23{\cdot}5+\text{j}45}{126{\cdot}5+\text{j}45} \quad \text{and} \quad |\Gamma|^2 = 0{\cdot}14.$$

Then

$$P_t = 150 \times 0{\cdot}86 = 129 \text{ mW}.$$

With this power dissipated in the load

$$P_t = I_L^2 R_L,$$

where R_L is the real part of Z_L. Then

$$I_L = \sqrt{P_t/R_L} = 50 \text{ mA}_{\text{rms}}.$$

EXERCISE No. 6

A lossless 50 Ω line is terminated in a matched load and a conductance G is placed in parallel with the line.

1. If one quarter of the incident power is reflected, what is the value of G?
2. How much power is dissipated in the matched load when the incident power is 40 mW?
3. What is the attenuation of the system in decibels?

Solution. 1. As will be seen from Fig. 1 the total admittance in the plane of G is $Y = G+Y_0$. The reflection coefficient is thus

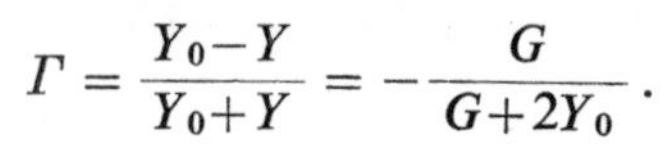

$$\Gamma = \frac{Y_0-Y}{Y_0+Y} = -\frac{G}{G+2Y_0}.$$

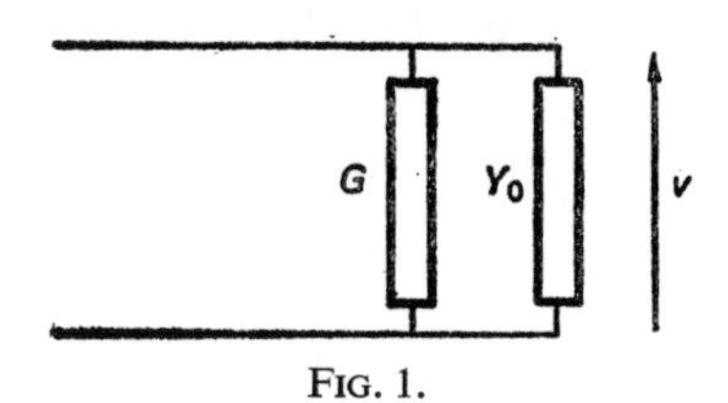

FIG. 1.

The ratio of the reflected and incident powers is then

$$\frac{P_r}{P_i} = |\Gamma|^2 = \left(\frac{G}{G+2Y_0}\right)^2 = \frac{1}{4}$$

from which

$$G = 2Y_0 = 4\times 10^{-2}\ \mathrm{S}.$$

2. The transmitted power is 30 mW and this is dissipated in G and Y_0. Thus

$$P_t = \tfrac{1}{2}(G+Y_0)\,v^2,$$
$$P_{Y_0} = \tfrac{1}{2}Y_0 v^2,$$

and

$$P_{Y_0} = \frac{Y_0}{G+Y_0}P_t = \frac{P_t}{3} = 10\ \mathrm{mW}.$$

3. The attenuation of the system is

$$\alpha = 10\log\frac{P_i}{P_{Y_0}} = 10\log 4 = 6\ \mathrm{dB}.$$

EXERCISE No. 7

The attenuating system shown in Fig. 1 consists of two resistances R_1 and R_2 in parallel across the line at a separation of a quarter of a wavelength. Calculate R_1 and R_2 knowing that:

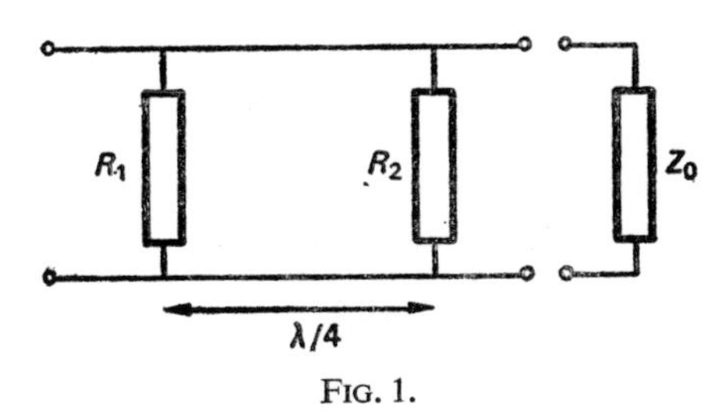

FIG. 1.

(a) The line is lossless, has a characteristic impedance of 50 Ω and is terminated in a matched load; (b) the system is matched in the plane of R_1 and; (c) the total attenuation of the system is 10 dB.

Solution. The admittance in the plane of R_2 is $Y = G_2 + Y_0$ or, in terms of reduced quantities,

$$y = g_2 + 1.$$

The total reduced admittance in the plane of R_1 is

$$y' = g_1 + \frac{1}{g_2+1}$$

since a quarter-wave line transforms a reduced admittance into its inverse. There will be matching if $y' = 1$ or

$$g_1 = \frac{g_2}{g_2+1}.$$

The incident power P_i is then completely transmitted and is absorbed in G_1, G_2 and Y_0 since the line is lossless. The power dissipated in G_2 and Y_0 is given by

$$P = P_i \frac{\dfrac{1}{g_2+1}}{g_1 + \dfrac{1}{g_2+1}} = P_i \frac{1}{g_2+1}.$$

The power dissipated in Y_0 is

$$P_{Y_0} = P\frac{1}{g_2+1} = P_i\frac{1}{(g_2+1)^2}$$

and the total attenuation is

$$\alpha = 10\log\frac{P_i}{P_{Y_0}} = 10\log(g_2+1)^2,$$

so that, for $\alpha = 10$ dB

$$g_2+1 = \sqrt{10},$$

$$g_2 = \frac{G_2}{Y_0} = \frac{Z_0}{R_2} = 2{\cdot}16, \quad \text{or} \quad R_2 = 23\ \Omega.$$

Because

$$g_1 = \frac{g_2}{g_2+1}, \qquad R_1 = 73\ \Omega.$$

EXERCISE No. 8

A lossless 50 Ω line is terminated by a matched load. Between the source and the load, three resistances, R_1, R_2 and R_1 respectively, are placed in the positions π_1, π_2 and π_3 which are separated by quarter wavelengths as shown in Fig. 1.

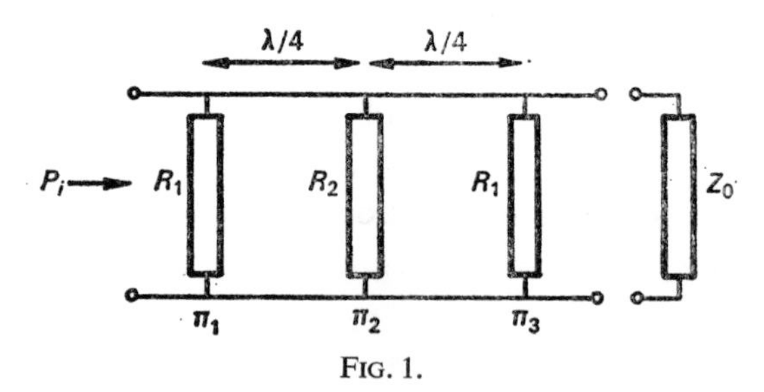

FIG. 1.

The resistances are chosen so that there is matching in the plane π_1 and one third of the incident power of 0·9 W is dissipated in the resistance R_1 situated at π_1.

1. Calculate R_1 and R_2.
2. What is the power dissipated in R_2 and in the resistance R_1 at π_3?
3. What is the power reaching Z_0?

Solution. If we use the property of a quarter-wave line as an impedance transformer we see that the reduced admittances at π_3, π_2 and π_1 are, respectively,

$$y_{\pi_3} = g_1+1, \qquad y_{\pi_2} = g_2+\frac{1}{g_1+1};$$

$$y_{\pi_1} = g_1+\frac{1}{g_2+\dfrac{1}{g_1+1}}.$$

For matching in π_1 we require $y_{\pi_1} = 1$ or

$$1 = g_1+\frac{1}{g_2+\dfrac{1}{g_1+1}}.$$

from which we find

$$\frac{1}{1-g_1} = g_2+\frac{1}{g_1+1}$$

and

$$g_2 = \frac{2g_1}{1-g_1^2}.$$

The power dissipated in the resistance R_1 at π_1 is related to the incident power P_i by

$$P_{R_1} = P_i\frac{g_1}{y_{\pi_1}} = P_i g_1,$$

so that

$$\frac{P_{R_1}}{P_i} = g_1 = \frac{1}{3}.$$

Because

$$g_1 = \tfrac{1}{3}, \qquad g_2 = \tfrac{3}{4};$$

and

$$R_1 = 3\,Z_0 = 150\,\Omega,$$

$$R_2 = \tfrac{4}{3}\,Z_0 = 66{\cdot}6\,\Omega.$$

2. The power transmitted beyond π_1 is 600 mW and the power dissipation in R_2 is

$$P_{R_2} = 600\frac{g_2}{y_{\pi_2}} = 600\,g_2(1-g_1) = 300\text{ mW}.$$

Of the 300 mW which passes beyond π_2 the part dissipated in R_1 at π_3 is

$$P_{R_1} = 300\frac{g_1}{y_{\pi_3}} = 300\frac{g_1}{g_1+1} = 75 \text{ mW}.$$

3. The remaining 225 mW will be dissipated in the matched load.

EXERCISE No. 9

A load impedance $Z_L = (20+j\,52)\Omega$ is used to terminate a lossless $80\,\Omega$ line. At a distance of $1{\cdot}592\,\lambda$ another impedance $Z = (40+j\,32)\Omega$ is placed in parallel across the line. If the incident power is $P_0 = 100$ mW what is the power received at the load?

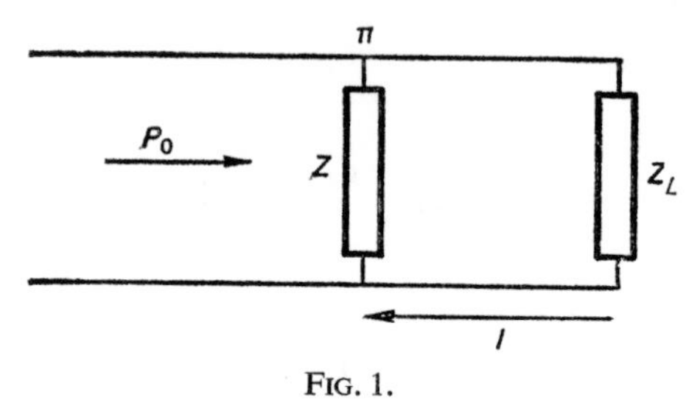

FIG. 1.

Solution. First method (see Fig. 2). Considering the total current (incident + reflected) the power reaching Z_L will be

$$P_{Z_L} = \tfrac{1}{2}\,(\text{real part of } v'i'^*),$$

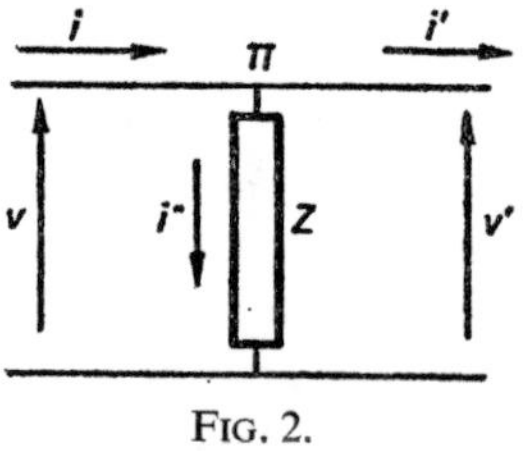

FIG. 2.

since the line is lossless. Now $v' = v$ and $i' = i-i'' = i-Yv$, so that

$$P_{Z_L} = \tfrac{1}{2}\,\text{Re}\,(vi^*)-\tfrac{1}{2}\,\text{Re}\,(Y^*vv^*).$$

Also $v = v_i(1+\Gamma)$ where v_i is the potential of the incident wave approaching the plane π and Γ the reflection coefficient at π and

$$i = \frac{v_i}{Z_0}(1-\Gamma).$$

Thus

$$P_{Z_L} = \frac{|v_i|^2}{2Z_0}(1-|\Gamma|^2) - \frac{1}{2}\,\mathrm{Re}\,[(G-\mathrm{j}B)\,|v_i|^2\,|1+\Gamma|^2],$$

$$= P_0(1-|\Gamma|^2) - \frac{1}{2}\,\mathrm{Re}\left[\frac{G-\mathrm{j}B}{Y_0}\,\frac{|v_i|^2}{Z_0}\,|1+\Gamma|^2\right].$$

Finally

$$P_{Z_L} = P_0(1-|\Gamma|^2 - g\,|1+\Gamma^2).$$

where $P_0 = 100$ mW.

For the calculation of Γ it is necessary to know the impedance at the plane π. The impedance transformed from the load Z_L is, in reduced terms,

$$z_L' = \frac{z_L + \mathrm{j}\tan\dfrac{2\pi l}{\lambda}}{1+\mathrm{j}z_L\tan\dfrac{2\pi l}{\lambda}}.$$

Since $z_L = 0{\cdot}25+\mathrm{j}\,0{\cdot}65$, this gives $z_L' = 1+2\mathrm{j}$ and

$$y_L' = 0{\cdot}2-\mathrm{j}\,0{\cdot}4,$$

so that the total admittance at π is

$$y_\pi = y + y_L'.$$

Now

$$y = \frac{1}{z} = (0{\cdot}5+\mathrm{j}\,0{\cdot}4)^{-1} \simeq 1{\cdot}2-\mathrm{j}$$

and

$$y_\pi = 1{\cdot}4-\mathrm{j}\,1{\cdot}4$$

Then

$$\Gamma = \frac{1-y_\pi}{1+y_\pi}.$$

gives

$$|\Gamma|^2 = 0{\cdot}274.$$

Taking

$$|1+\Gamma|^2 = (1+\Gamma)(1+\Gamma)^* = 1+\Gamma+\Gamma^*+|\Gamma|^2$$

with $\Gamma+\Gamma^*$ = twice real part of $\Gamma = 2\times -0{\cdot}378$
the expression for the power at the load gives

$$P_{Z_L} = 0{\cdot}105\,P_0 = 10{\cdot}5\text{ mW}.$$

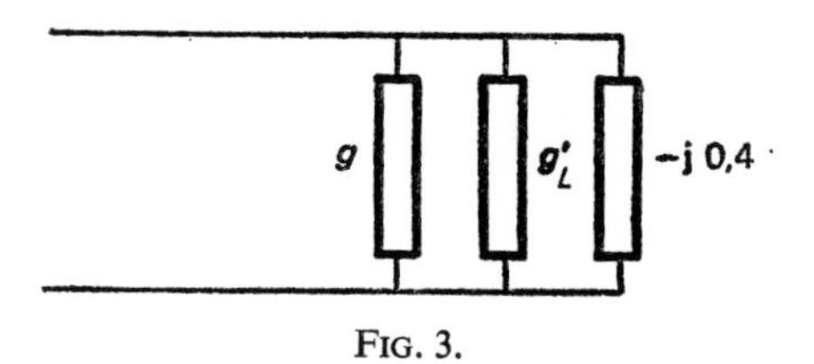

FIG. 3.

Second method. The equivalent circuit shown in Fig. 3 allows the calculation of y_L' as seen at π. The power at the combined impedance Z and Z_L is

$$P_t = P_0(1-|\Gamma|^2),$$

which is divided between the two conductances g and g_L' so that

$$P_{g_L'} = P_0(1-|\Gamma|^2)\frac{g_L'}{g+g_L'} = 10{\cdot}4 \text{ mW}$$

and, as the line is lossless, $P_{g_L'} = P_{Z_L}$.

The second method, although quicker, does not differ in principle from the first since one may show that

$$P_0(1-|\Gamma|^2)\frac{g_L'}{g+g_L'} = P_0(1-|\Gamma|^2)-gP_0|1+\Gamma|^2.$$

Chapter 8

Use of the Smith Chart

EXERCISE No. 1

A $75\,\Omega$ line is terminated by an impedance Z_L. With the aid of a Smith chart find the reflection coefficient and standing wave ratio due to the load when it takes the following values.

$$Z_1 = 225\,\Omega, \quad Z_2 = 75\,\Omega, \quad Z_3 = 15\,\Omega,$$
$$Z_4 = j45\,\Omega, \quad Z_5 = -j225\,\Omega, \quad Z_6 = (45+j120)\,\Omega.$$

Solution. The Smith chart makes use of normalized co-ordinates; in the present case normalized relative to $Z_0 = 75\,\Omega$.

(a) The reduced impedance corresponding to $Z_L = Z_1 = 225\,\Omega$ is

$$z_1 = Z_1/Z_0 = 3.$$

On the Smith chart z_1 is represented by the point A situated at the intersection of the real axis and the circle of radius $r = 3$. The modulus of the reflection coefficient is given

either directly by the scale graduated in $|\Gamma_0|$,

or by the ratio OA/R_c where R_c is the radius of the chart.

The phase of the reflection coefficient is given by the angle between OA and OJ.

The V.S.W.R. S is given

either by the scale graduated in S,

or by the scale of the reduced resistance along the axis OJ.

Taking these readings for z_1 gives

$$|\Gamma_0| = 0{\cdot}5, \quad \theta_0 = 0, \quad S = r = 3.$$

(b) $Z_2 = 75\,\Omega$, $z_2 = 1$ is represented by O, the centre of the chart, so that

$$|\Gamma_0| = 0, \quad \theta_0 = 0, \quad S = 1.$$

(The line is matched.)

(c) $Z_3 = 15\,\Omega$, $z_3 = 0{\cdot}2$, i.e. the point B.

Then

$$|\Gamma_0| = 0{\cdot}66, \quad \theta_0 = (\boldsymbol{OJ}, \boldsymbol{OB}) = \pi,$$

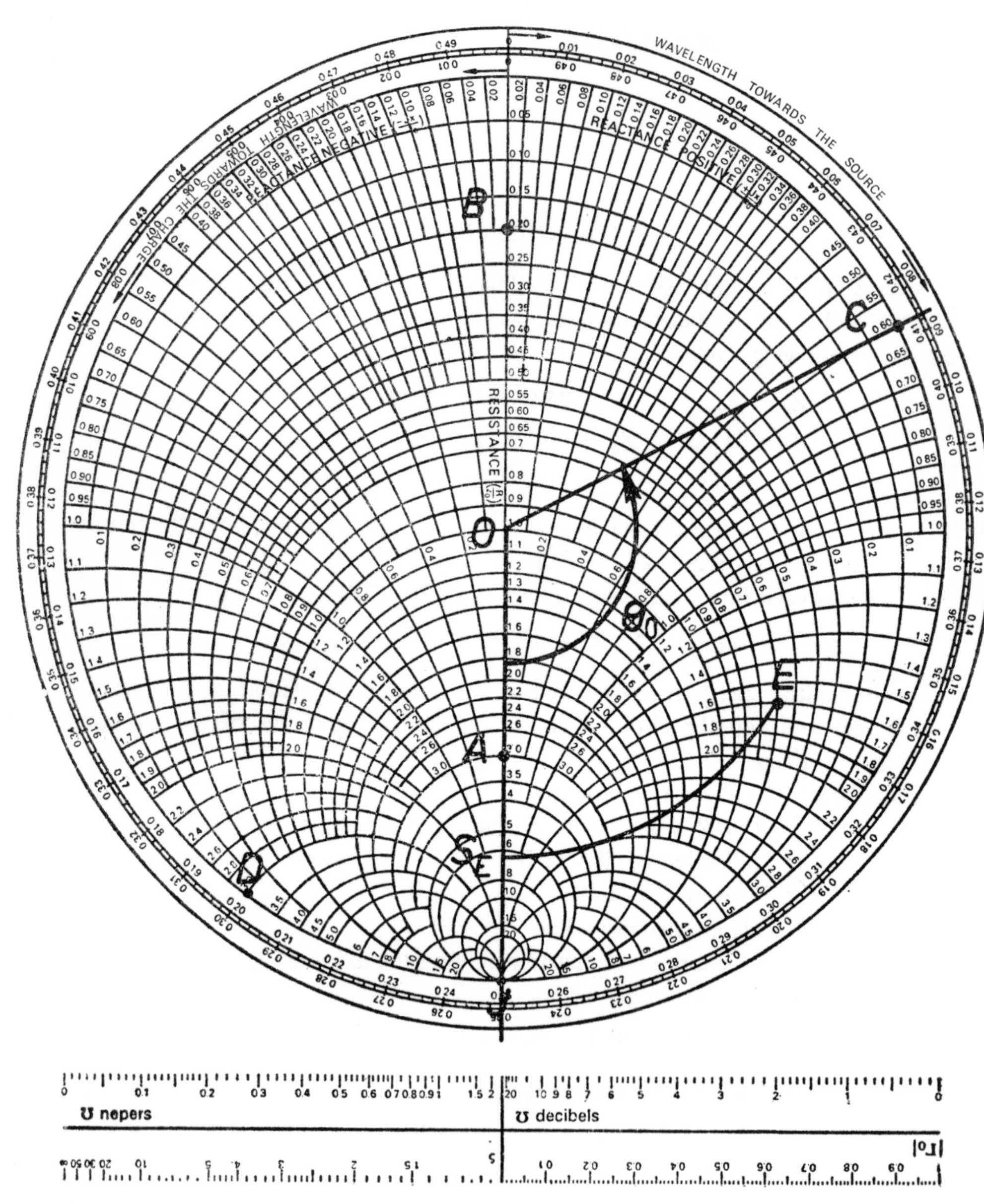
WAVELENGTH TOWARDS THE SOURCE
WAVELENGTH TOWARDS THE CHARGE
REACTANCE NEGATIVE
REACTANCE POSITIVE
RESISTANCE
B
C
O
A
D
E
ʊ nepers
ʊ decibels
|Γ₀|

FIG. 1.

To find S the distance OB is measured off along OJ and the position along the scale of reduced resistance gives $S = 5$.

(d) $Z_4 = \mathrm{j}\,45\,\Omega$, $z_4 = \mathrm{j}\,0{\cdot}6$ which is the point C on the circle $r = 0$ for a positive reactance. In this case

$$|\Gamma_0| = 1 \quad \text{and} \quad S = \infty$$

and θ_0 is the angle $(\boldsymbol{OJ}, \boldsymbol{OC})$. This angle is determined most rapidly from the external scale of the circle on which $0{\cdot}5$ represents 2π radians. Then

$$\theta_0 = 4\pi(0{\cdot}25 - 0{\cdot}086) = 2{\cdot}06 \text{ radians} \simeq 119°.$$

(e) $Z_5 = -\mathrm{j}225\ \Omega$, $z_5 = -\mathrm{j}3$. This is the point D corresponding to a negative reactance and

$$|\Gamma_0| = 1, \quad S = \infty, \quad \theta_0 = (\boldsymbol{OJ}, \boldsymbol{OD}) = -0{\cdot}05 \times 4\pi = -36°.$$

(f) $Z_6 = (45 + \mathrm{j}120)\ \Omega$, $z_6 = 0{\cdot}6 + \mathrm{j}1{\cdot}6$. This is the point E corresponding to the intersection of the circle passing through $0{\cdot}6$ on the resistance axis and the reactive circle $1{\cdot}6$.

Then

$$|\Gamma_0| = |OE|/R_c = 0{\cdot}73 \quad \text{and} \quad \theta_0 = (\boldsymbol{OJ}, \boldsymbol{OE}) = 59°.$$

When $|OE|$ is measured off along OJ one finds $S = 6{\cdot}4$.

EXERCISE No. 2

A lossless transmission line with air as the dielectric and a characteristic impedance of $50\,\Omega$ has a length $l = 34$ cm terminated by a resistance $R = 12{\cdot}5\,\Omega$. Calculate the parameters of the equivalent series circuit at the input of this line for $f = 150$ MHz and also for the parallel equivalent circuit.

Solution. For $f = 150$ MHz the wavelength in air is

$$\lambda = \frac{3 \times 10^8}{1{\cdot}5 \times 10^8} = 2\,\text{m},$$

so that

$$\frac{l}{\lambda} = \frac{0{\cdot}34}{2} = 0{\cdot}17.$$

Since

$$r = \frac{R}{Z_0} = \frac{12{\cdot}5}{50} = 0{\cdot}25$$

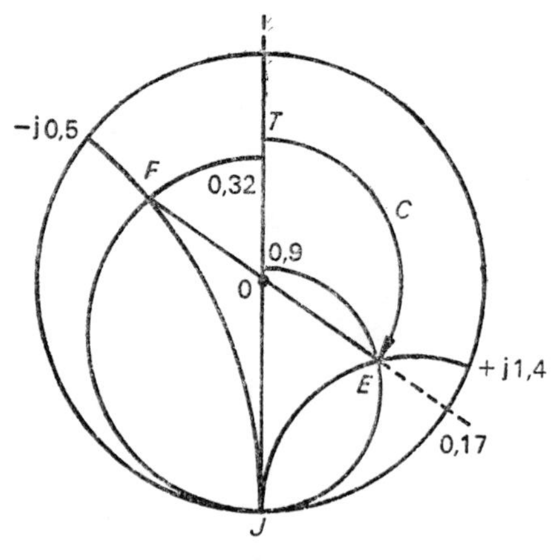

FIG. 1.

we start at the point T on the chart corresponding to $r = 0{\cdot}25$ and draw about O the circle C of constant V.S.W.R. as shown in Fig. 1.

Following the circle C through the angle $(\boldsymbol{OT}, \boldsymbol{OE}) = 0{\cdot}17$ on the outer circle we can find the reduced input impedance represented by the point E as

$$z = 0{\cdot}9 + \mathrm{j}1{\cdot}4$$

$$Z = (45 + \mathrm{j}70)\,\Omega.$$

The equivalent circuit would consist of a resistance $R = 45\ \Omega$ and an inductance

$$L = \frac{70}{\omega} = \frac{70}{2\pi \times 1{\cdot}5 \times 10^8} = 7{\cdot}45 \times 10^{-8}\,\mathrm{H}.$$

The reduced input admittance is given by the point F diametrically opposite to E with $OF = OE$. Then

$$y = 0{\cdot}32 - \mathrm{j}0{\cdot}5,$$

or

$$Y = \frac{1}{50}(0{\cdot}32 - \mathrm{j}0{\cdot}5)\,S = \frac{1}{R'} - \mathrm{j}\frac{1}{\omega L'}.$$

The equivalent parallel circuit at the input would thus consist of

$$R' = 156\ \Omega \text{ in parallel with } L' = 0{\cdot}106\ \mu\mathrm{H}.$$

EXERCISE No. 3

A load $Z_L = (75+j75)\,\Omega$ is placed as the termination of a 75 Ω lossless line. Calculate the impedance $Z(s)$ seen at the distance

$$s = \frac{\lambda}{8},\quad \frac{\lambda}{4},\quad 0{\cdot}45\lambda,\quad \frac{\lambda}{2},\quad \lambda,$$

from the load, λ being the wavelength of the signal. At what distance from the load is this impedance real?

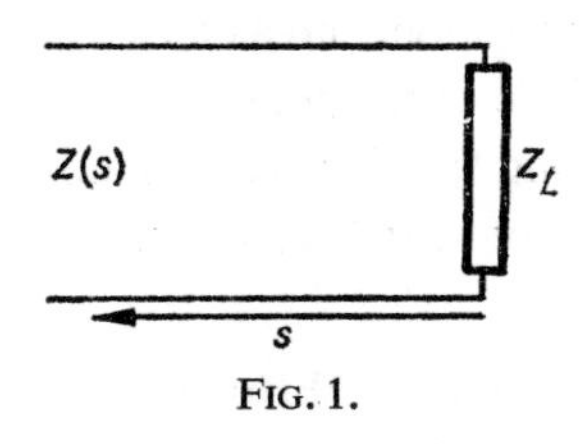

FIG. 1.

Solution. The reduced load impedance is

$$z_L = \frac{Z_L}{Z_0} = 1+j,$$

which is represented on the chart of Fig. 2 by the point T.

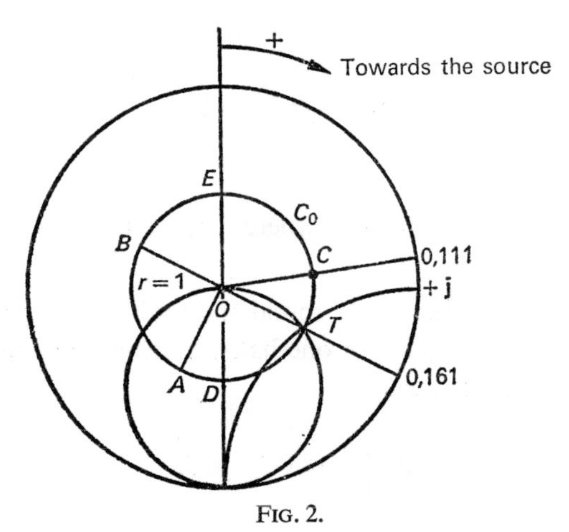

FIG. 2.

If the line is lossless, the modulus of the reflection coefficient will be constant and, at all points on the line

$$|\Gamma| = |\Gamma_0|.$$

On the chart

$$|\Gamma_0| = \frac{OT}{R_c}.$$

A displacement along the transmission line corresponds to a displacement on the chart around the circle C_0 with centre O and radius OT. At s

$$\Gamma(s) = \Gamma_0\, e^{-2j\beta s} = |\Gamma_0|\, e^{j(\theta_0 - 2\beta s)}$$

where $\beta = 2\pi/\lambda$.

Thus, when s increases, $\theta_0 - 2\beta s$ decreases and a displacement of position from the load towards the source corresponds to a clockwise movement along C_0 from T. The mark on the external circle corresponding to the point T is 0·161.

(a) The point A corresponding to $z(\lambda/8)$ is situated on C_0 and, since $(\boldsymbol{OT}, \boldsymbol{OA}) =$ 0·125, the effective distances from the load are related as

$$\frac{s_A}{\lambda} = \frac{s_T}{\lambda} + 0{\cdot}125 = 0{\cdot}161 + 0{\cdot}125 = 0{\cdot}286.$$

Reading from the chart gives

$$z\left(\frac{\lambda}{8}\right) = 2 - j$$

and

$$Z\left(\frac{\lambda}{8}\right) = (150 - 75j)\ \Omega.$$

(b) For

$$2\beta s_0 = 4\pi \frac{s_0}{\lambda} = 2\pi$$

we have

$$\Gamma(s_0) = \Gamma_0, \quad \text{where} \quad \frac{s_0}{\lambda} = 0{\cdot}5.$$

It is seen that one complete circle round the Smith's chart corresponds to a displacement along the waveguide of one half-wavelength and 0·25λ corresponds to a half-circle of the chart. Then

$$z\left(\frac{\lambda}{4}\right) = 0{\cdot}5 - j0{\cdot}5,$$

being at the point B symmetrically opposite to T through $\boldsymbol{O}$.

$$Z\left(\frac{\lambda}{4}\right) = (37{\cdot}5 - j37{\cdot}5)\ \Omega.$$

(c) The point C which represents $Z(0{\cdot}45\lambda)$ is found from $(OT, OC) = 0{\cdot}45$ so that

$$\frac{s_C}{\lambda} = \frac{s_T}{\lambda} + 0{\cdot}45 = 0{\cdot}5 + 0{\cdot}1111.$$

Reading off the chart gives

$$z(0{\cdot}45\lambda) = 0{\cdot}6 + \mathrm{j}0{\cdot}65$$

$$Z(0{\cdot}45\lambda) = (45 + \mathrm{j}48{\cdot}75)\,\Omega.$$

(d) For $s = \lambda/2, \lambda, n\lambda/2$ for all integers n, $Z(s) = Z_L$.

(e) The value of $Z(s)$ is real at the two points where the circle C_0 cuts the real axis.

At D, $\quad z(s) = 2{\cdot}6 \quad$ so that $\quad Z = R = 195\,\Omega$

and $$\frac{s}{\lambda} = \frac{s_0}{\lambda} - 0{\cdot}161 = 0{\cdot}089 + \frac{n}{2}.$$

At E, $\quad z(s') \simeq 0{\cdot}38 \left(\text{actually } \dfrac{1}{2{\cdot}6}\right) \quad$ and $\quad Z(s') = R' = 28{\cdot}8\,\Omega$

$$\frac{s'}{\lambda} = \left(0{\cdot}089 + \frac{n}{2}\right) + 0{\cdot}25 = 0{\cdot}339 + \frac{n}{2}.$$

Note: at D, $z(s) = S$ and the V.S.W.R. on the line is 2·6.

EXERCISE No.4

What length of a 50Ω lossless line will have an input admittance of $Y_i = \mathrm{j} \times 10^{-2}$ S when it is terminated in a short circuit? Take the working wavelength as 1 m.

Solution. The reduced input admittance of a short circuited line is

$$y_i = -\mathrm{j} \cot \frac{2\pi l}{\lambda}.$$

Using the Smith's chart for the admittance, $y_i = \mathrm{j}0{\cdot}5$ is represented by the point E in Fig. 1. The terminating admittance is infinite (point J on the chart). In going from the point J to the point E in the clockwise direction (i.e. towards the source) one moves along a length of line given by

$$\frac{l}{\lambda} = \frac{s_E}{\lambda} - \frac{s_J}{\lambda} = 0{\cdot}074 + 0{\cdot}25 = 0{\cdot}324,$$

so that $l = 32{\cdot}4$ cm$+ n \times 50$ cm.

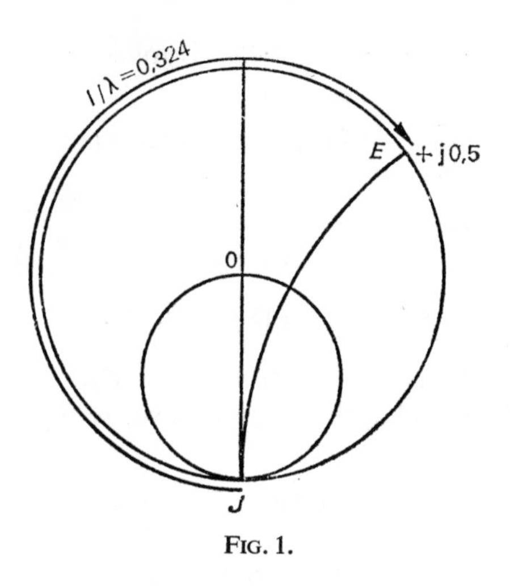

FIG. 1.

EXERCISE No. 5

A lossy 50 Ω line is terminated in a load $Z_L = (35+j75)\,\Omega$. What is the impedance seen at a distance $s_0 = 0{\cdot}375\lambda$ from the load if the attenuation per wavelength is $\alpha = 0{\cdot}27$ neper?

Solution. In Fig. 1 the point T represents $z_L = 0{\cdot}7+j1{\cdot}5$. The change in phase over the length s_0 is given by the angle

$$(\boldsymbol{OT}, \boldsymbol{OA}) = 0{\cdot}375.$$

Then

$$\frac{s_A}{\lambda} = \frac{s_T}{\lambda}+0{\cdot}375 = 0{\cdot}166+0{\cdot}375 = 0{\cdot}541.$$

The point A would represent $z(s_0)$ in the absence of attenuation. Here, however,

$$\Gamma(s_0) = |\Gamma_0|\, e^{j(\theta_0-2\beta s_0)}\, e^{-2\alpha s_0}.$$

The attenuation thus reduces the modulus of the reflection coefficient. This variation of $|\Gamma|$ is found from the scale graduated in nepers as follows:

The distance corresponding to the radius OA (or OT) is marked off along the attenuation scale. In this case the point A reads $u_0 = 0{\cdot}2$ nepers.

The total attenuation $\alpha s_0 = 0{\cdot}27\times0{\cdot}375 = 0{\cdot}1$ neper is added to u_0 to give the value $u_1 = 0{\cdot}3$ which corresponds to the point B.

The radius OB represents the new modulus of the reflection coefficient

$$|\Gamma(s_0)| = \frac{OB}{R_c}.$$

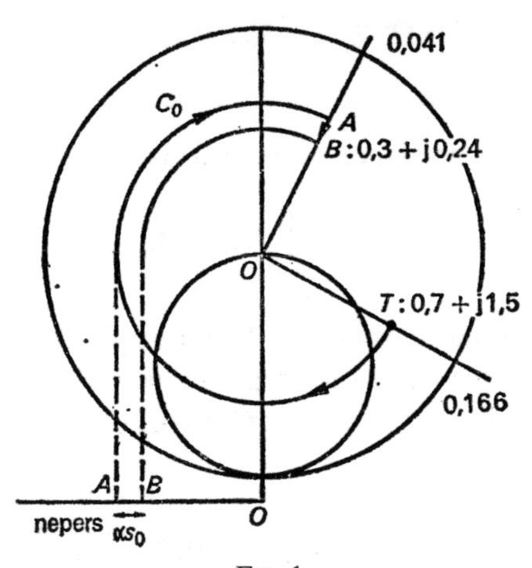

FIG. 1.

The point representing the impedance $z(s_0)$ will be that shown as B, the point of intersection of the circle of radius OB and the radius OA.

This point gives

$$z(s_0) = 0{\cdot}3+j0{\cdot}24,$$

$$Z(s_0) = (15+j12)\,\Omega.$$

The complete displacement from T to B would follow a logarithmic spiral. However, we have taken this displacement in two steps: a displacement with zero attenuation from T to A around the circle C_0, and a displacement at constant phase angle from A to B.

EXERCISE No. 6

Calculate the input impedence of the system shown in Fig. 1 in the case where the lines L_1 and L_2 are lossless.

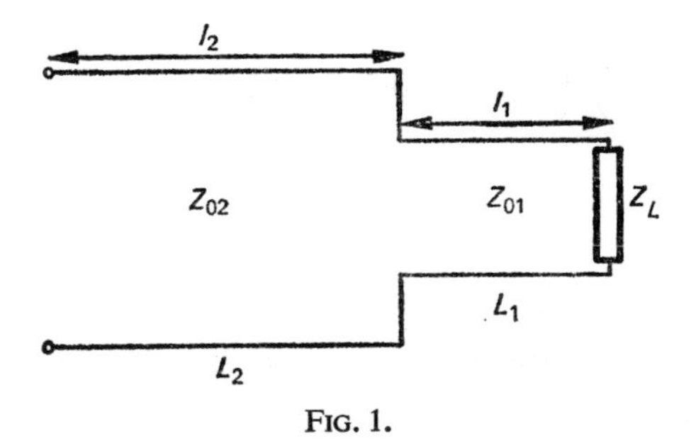

FIG. 1.

Take $Z_{01} = 50\ \Omega$, $Z_{02} = 125\ \Omega$, $Z_L = (50+j150)\ \Omega$, $l_1 = 0{\cdot}077\lambda$, $l_2 = 0{\cdot}291\lambda$.

Solution. The reduced impedance of the load is

$$z_L = \frac{Z_L}{Z_{01}} = 1+3j,$$

which is represented on the chart of Fig. 2 by the point T on the circle C_1.

A displacement along the line L_1 towards the source by an amount $l_1/\lambda = 0{\cdot}077$ corresponds to a clockwise rotation along C_1 to the point A such that $(\boldsymbol{OT}, \boldsymbol{OA}) = 0{\cdot}077$ and the angle at A is

$$\frac{s_T}{\lambda}+0{\cdot}077 = 0{\cdot}205+0{\cdot}077 = 0{\cdot}282.$$

Then $z(l_1) \simeq 2-j4$.

The line L_2 is thus terminated by an impedance $Z(l_1) = Z_{01}(2-j4)$.

In terms of the impedance Z_{02} of L_2 this is equivalent to a load impedance

$$z'(l_1) = \frac{Z_{01}}{Z_{02}}(2-j4) = \frac{2}{5}(2-j4) = 0{\cdot}8-j1{\cdot}6,$$

which is represented by the point B.

(Note that this point B can be found from A by displacement along the circle of constant ratio $R/X = -2$ corresponding to the phase angle $\varphi = 63°30'$, since the change in the reference impedance from Z_{01} to Z_{02} does not change the phase of $z(l_1)$ and $z'(l_1)$. Thus

$$\tan\varphi\,\big(z(l_1)\big) = -\frac{4}{2} = -\frac{1{\cdot}6}{0{\cdot}8} = \tan\varphi\,\big(z'(l_1)\big).$$

In going around the circle C by a distance corresponding to $0{\cdot}291\lambda$ from the point B we arrive at the point C such that the angle at C is

$$\frac{s_B}{\lambda}+0{\cdot}291 = 0{\cdot}327+0{\cdot}291 = 0{\cdot}618.$$

Thus

$$z_c = 0{\cdot}35+j0{\cdot}85$$

and the input impedance to the system is

$$Z = 125(0{\cdot}35+j0{\cdot}85) = (43{\cdot}75+j106{\cdot}25)\ \Omega.$$

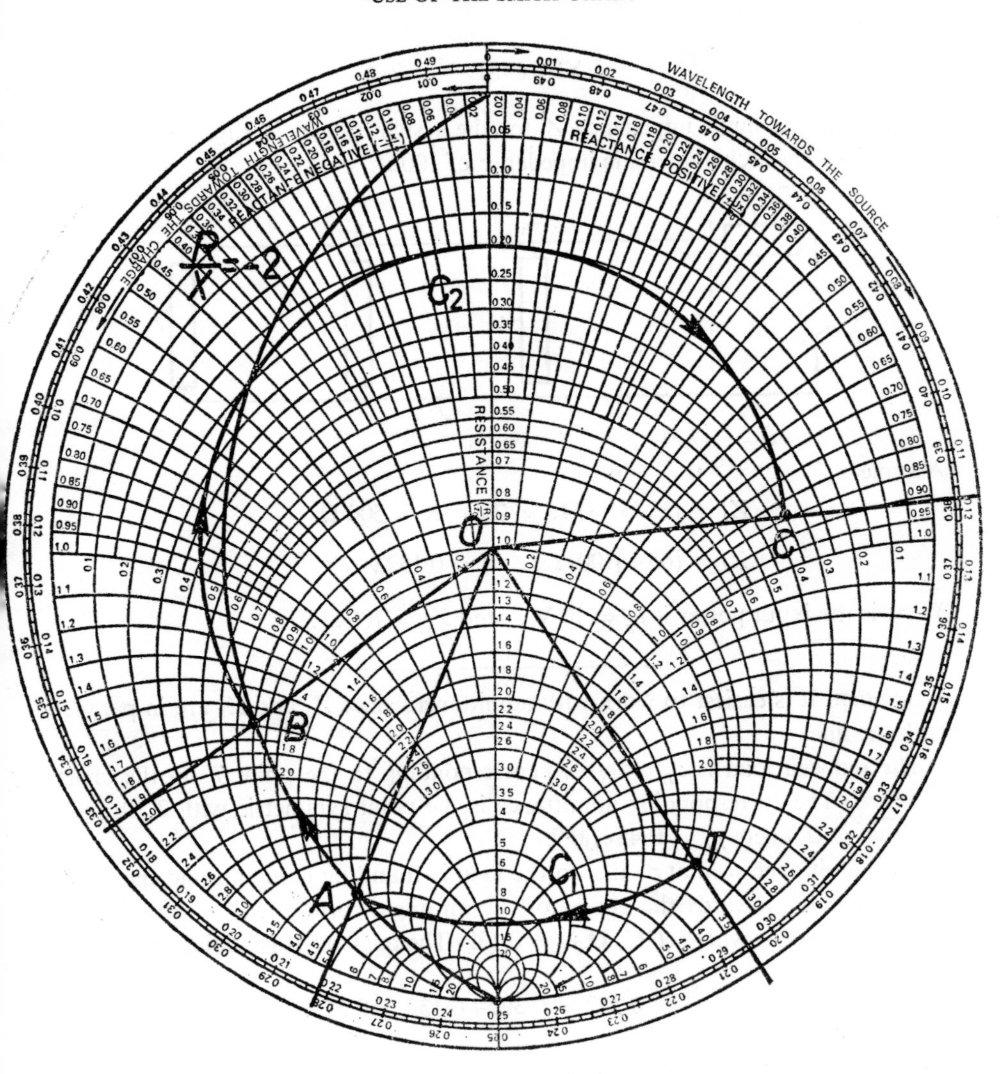
WAVELENGTH TOWARDS THE SOURCE
WAVELENGTH TOWARDS THE CHARGE
REACTANCE POSITIVE
REACTANCE NEGATIVE
RESISTANCE
$\frac{R}{X} = -2$
C_2
C_1
O
G
B
A
ʊ nepers
ʊ decibels

FIG. 2.

EXERCISE No. 7

Calculate the input impedance Z_i for the system shown in Fig. 1 where each of the lines has a characteristic impedance of 100 Ω.

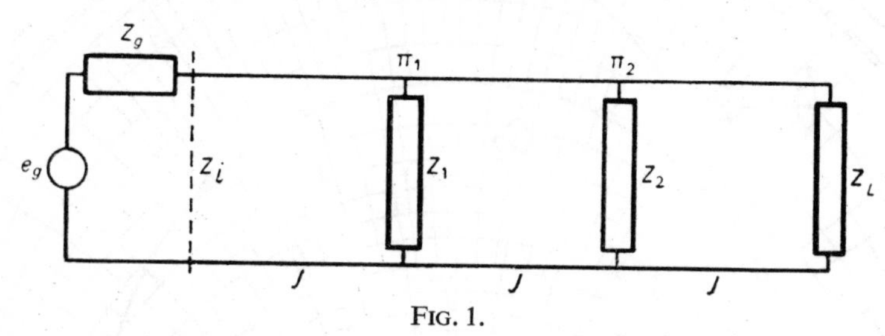

FIG. 1.

Assume

$$Z_1 = jX_1 = j87\ \Omega, \quad Z_2 = R = 500\ \Omega,$$

$$Z_L = (65+j100)\ \Omega, \quad l = 0{\cdot}132\lambda$$

What is the input power if $e_g = 10\ V_{rms}$ and $Z_g = 100\ \Omega$?

Solution. It is easiest to work with admittances since Z_1 and Z_2 are in parallel with the line.

The reduced load is $z_L = 0{\cdot}65+j$ corresponding to the point T on the chart in Fig. 2. The corresponding admittance is the point symmetric with T through O—that is U where $y_L = 0{\cdot}45-j0{\cdot}7$.

The admittance seen at π_2 due to the load is found by going around the circle C_0 through U, towards the generator, by an angle corresponding to $0{\cdot}132\lambda$ to reach the point A where $(\boldsymbol{OU}, \boldsymbol{OA}) = 0{\cdot}132$. The angle at A is equivalent to

$$\frac{s_u}{\lambda}+0{\cdot}132 = 0{\cdot}392+0{\cdot}132 = 0{\cdot}524.$$

The admittance is $y = 0{\cdot}3+j0{\cdot}14$ and the total admittance at π_2 is

$$y_{\pi_2} = y+\frac{1}{z_2} \quad \text{with} \quad z_2 = 5$$

so that

$$y_{\pi_2} = 0{\cdot}3+j0{\cdot}14+0{\cdot}2 = 0{\cdot}5+j0{\cdot}14 \quad \text{(point } B\text{)}.$$

A displacement equivalent to $0{\cdot}132\ \lambda$ around the circle C_1 from B gives the admittance seen at π_1. That is, at the point C,

$$y' = 1{\cdot}1+j0{\cdot}75$$

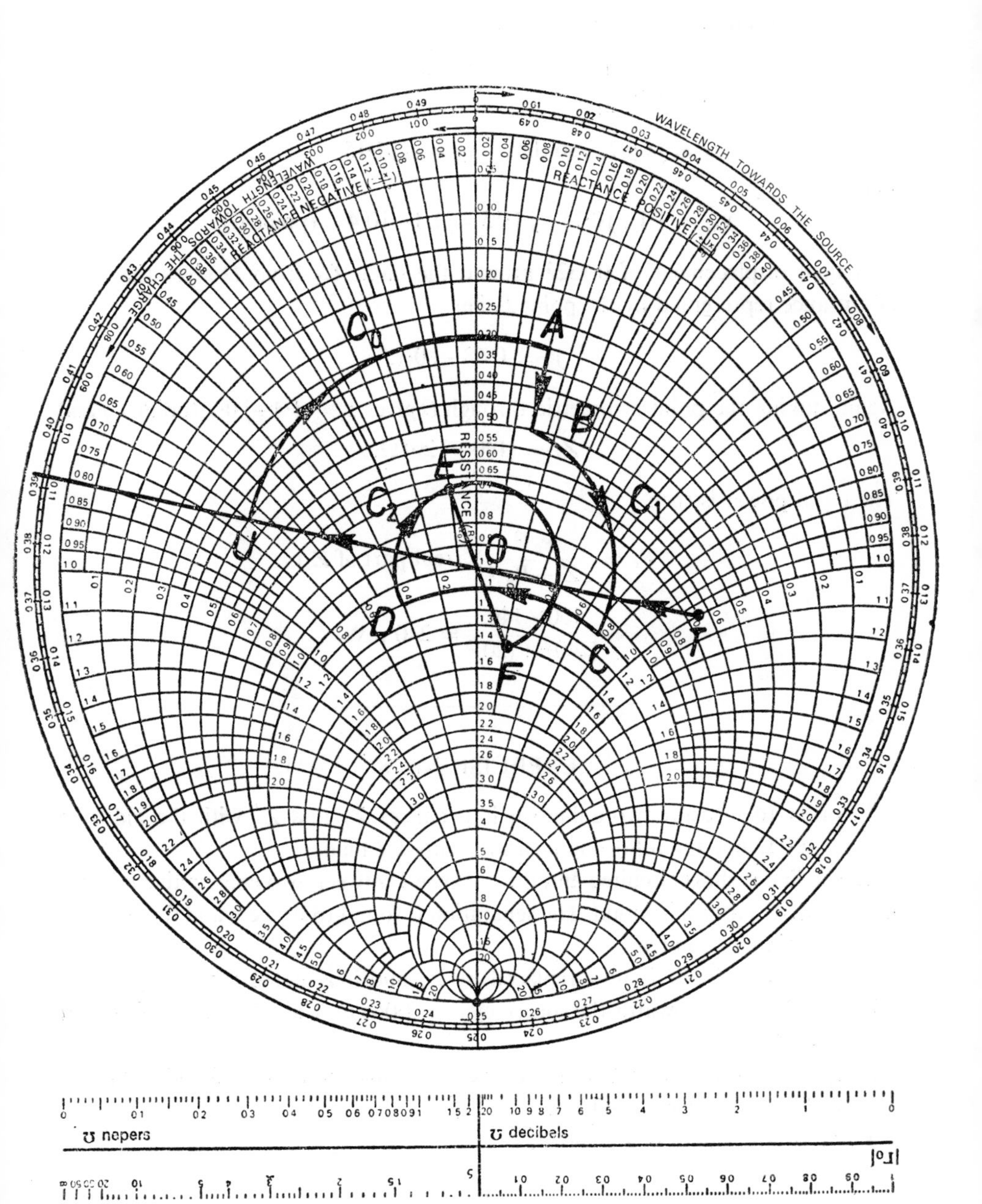
WAVELENGTH TOWARDS THE SOURCE
WAVELENGTH TOWARDS THE CHARGE
REACTANCE NEGATIVE
REACTANCE POSITIVE
RESISTANCE
A
B
C
C0
C1
C2
D
E
F
J
O
υ nepers
υ decibels
|ρL|

FIG. 2.

and the total admittance at π_1 is

$$y_{\pi_1} = y' + \frac{1}{z_1}$$

$$z_1 = \mathrm{j}\,0{\cdot}87 \quad \text{so that} \quad y_1 = -\mathrm{j}\,1{\cdot}15$$

and

$$y_{\pi_1} = 1{\cdot}1 + \mathrm{j}0{\cdot}75 - \mathrm{j}\,1{\cdot}15 = 1{\cdot}1 - \mathrm{j}0{\cdot}4 \quad \text{(point } D\text{)}.$$

A final displacement of $0{\cdot}132\lambda$ around the circle C_2 from D gives the input admittance represented by the point E, i.e.

$$y_i = 0{\cdot}68 - \mathrm{j}\,0{\cdot}1.$$

Again, taking the point symmetric through O gives $z_i = 1{\cdot}4 + \mathrm{j}\,0{\cdot}2$ (point F) and

$$Z_i = (140 + \mathrm{j}\,20)\,\Omega.$$

The input power is then

$$P_i = \frac{|e_g|^2}{|Z_i + R_g|^2} \times (\text{real part of } Z_i)$$

$$= \frac{100 \times 140}{(240)^2 + (20)^2} = 240 \text{ mW}.$$

EXERCISE No. 8

In order to measure two unknown impedances Z_L and Z_L' with the aid of a lossless 100 Ω line the following measurements are made.

1. The line is short circuited and the distance l between two consecutive minima in the V.S.W. pattern is found. The separation is found to be 5 cm and the position of one minimum is noted.

2. The short circuit is now replaced by Z_L and it is noted that the previous minimum is shifted towards the source by $d = 1{\cdot}04$ cm, while the S.W.R. is found to be $S = 2$.

3. Z_L is replaced by Z_L' when the value of S is found to be $S' = 3$ and d changes to $d' = 1{\cdot}65$ cm towards the load.

What are the values of Z_L and Z_L'?

Solution. 1. With the line short circuited the voltage distribution is of the form shown in Fig. 1a. The minima are, in fact, zero and are separated by $\lambda/2$ so that $\lambda = 10$ cm.

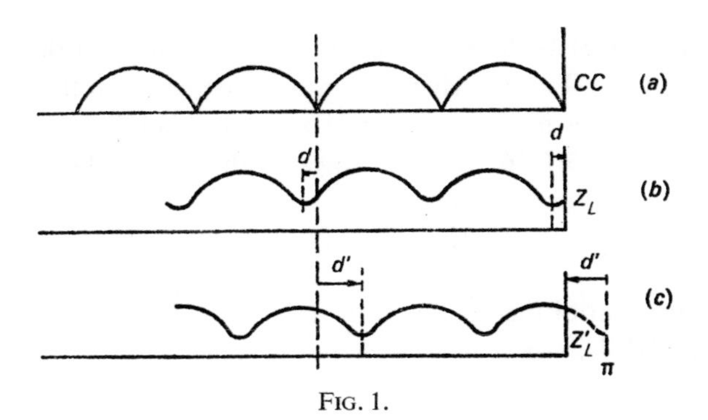

FIG. 1.

2. With the line terminated in Z_L and $S = 2$, the locus of the points on the Smith chart which represent the impedances along the line must be the circle about O of radius $r = 2$. The terminating impedance is necessarily on this circle which is shown in Fig. 2. At a voltage minimum the reduced impedance is real and equal to $z_m = 1/S$ so that here $z_m = 0{\cdot}5$, which is the point on the real axis.

Since the minimum is displaced by $d = 1{\cdot}04$ cm *towards the source*, the voltage distribution will be as shown in Fig. 1b. The first minimum is at d from the load and is represented by the point m.

To find z_L it is sufficient to move round the circle radius Om by an amount $d/\lambda = 0{\cdot}104$ *towards the load.* This gives the point T such that

$$z_L = 0{\cdot}7 - \mathrm{j}\,0{\cdot}5 \quad \text{and} \quad Z_L = (70 - \mathrm{j}\,50)\,\Omega.$$

Briefly we see that a displacement d of the minimum *towards the source* on the line corresponds to a displacement d/λ *towards the load* on the chart.

3. With the load terminated in Z'_L the locus of the points for $S = 3$ cuts the

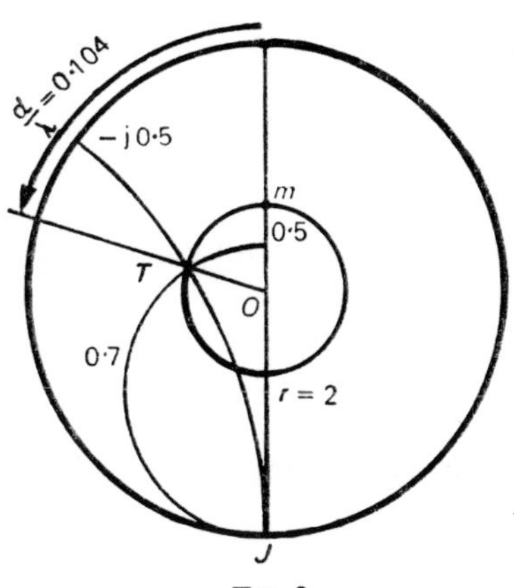

FIG. 2.

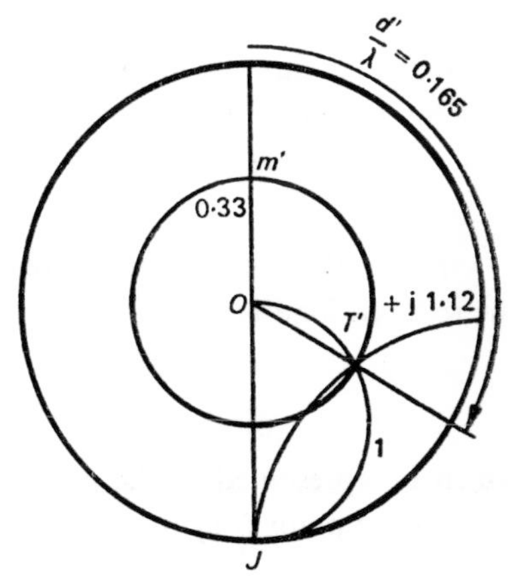

FIG. 3.

real axis at m' ($r = 1/3$) as shown in Fig. 3. The distribution of the voltage signal is of the form shown in Fig. 1c, where the minimum nearest to the load is, in fact, virtually situated, as shown, in the plane π. To find the load z_L' the circle is followed through an angle $d'/\lambda = 0{\cdot}165$ from m' towards the source. This gives the point T' such that

$$z_L' = 1 + \mathrm{j}\,1{\cdot}12, \quad \text{or} \quad Z_L' = (100 + \mathrm{j}\,112)$$

Thus a displacement of the minimum by a distance d' *towards the load* requires a displacement of d'/λ *towards the source* on the chart.

EXERCISE No. 9

A load of $Z_L = (25 - \mathrm{j}\,75)\ \Omega$ is placed as the termination of a lossless $100\ \Omega$ line, the wavelength of the signal being $\lambda = 30$ cm. To match this load to the line, use is made of a stub composed of a variable length l of the lossless line short-circuated at its termination. This stub is placed in parallel with the line at a distance s from the load as is shown in Fig. 1.

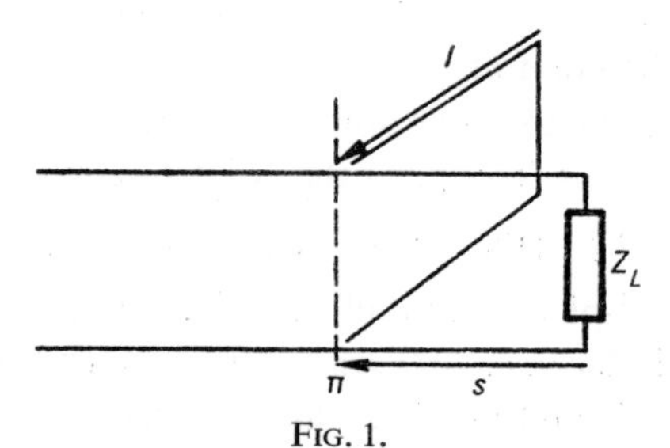

FIG. 1.

Calculate, with the aid of the Smith's chart, the values of l and s for matching. Is there more than one solution?

Solution. Because the stub is placed in parallel we will work with admittances, the reduced load is

$$z_L = \frac{Z_L}{Z_0} = 0{\cdot}35 - \mathrm{j}\,0{\cdot}75,$$

which is represented as the point T of Fig. 3. The corresponding reduced admittance is the point U symmetric through O where

$$y_L = 0{\cdot}4 + \mathrm{j}\,1{\cdot}2.$$

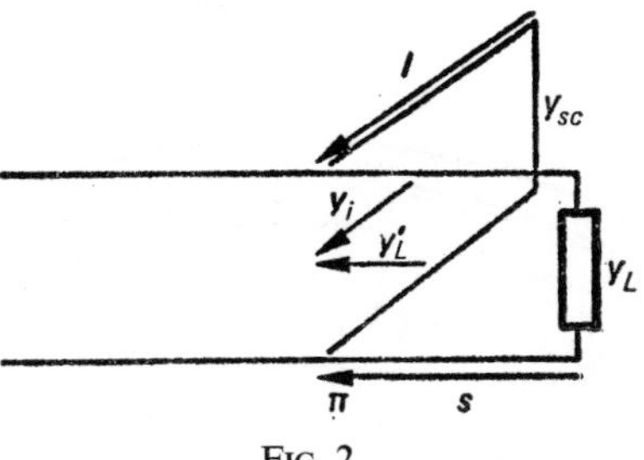

FIG. 2.

The displacement from $Z_L(y_L)$ towards the plane π corresponds to a displacement around the circle C (centred on O and of radius OU). The admittance due to y_L but in the plane π will be (see Fig. 2)

$$y_L' = g' + jb'.$$

The input admittance of a short-circuited line of length l is

$$y_i = -j \cot \beta l = jb_i.$$

The condition for matching in the plane π is that $y_\pi = y_0 = 1$ (or $Z_\pi = Z_0$) or

$$g' = 1, \qquad b' + b_i = 0.$$

The stub must thus be placed in a position where the reduced admittance due to Z_L is of the form $y_L' = 1 + jb'$. That is, the point on the Smith chart must correspond to the intersection of the circle $g = 1$ and the circle C. There are seen to be two possible solutions.

First case (chart Fig. 3):

At P_1 we have

$$y_L' = 1 + j\,2{\cdot}1.$$

(a) The distance s is determined by the angle ($\boldsymbol{OU}$, $\boldsymbol{OP_1}$) with the help of the scale on the outer circle. This angle between U and P_1 corresponds to a movement along the line from the load towards the source of

$$\frac{s_1}{\lambda} = 0{\cdot}19 - 0{\cdot}144 = 0{\cdot}046.$$

Because a complete circuit of the chart brings us back to P_1, there are a family of distances s_1 with $y_L' = 1 + j\,2{\cdot}1$, i.e.

$$s_1 = 0{\cdot}046\,\lambda + n\frac{\lambda}{2}$$

$$s_1 = 1{\cdot}38 \text{ cm} + n \times 15 \text{ cm}$$

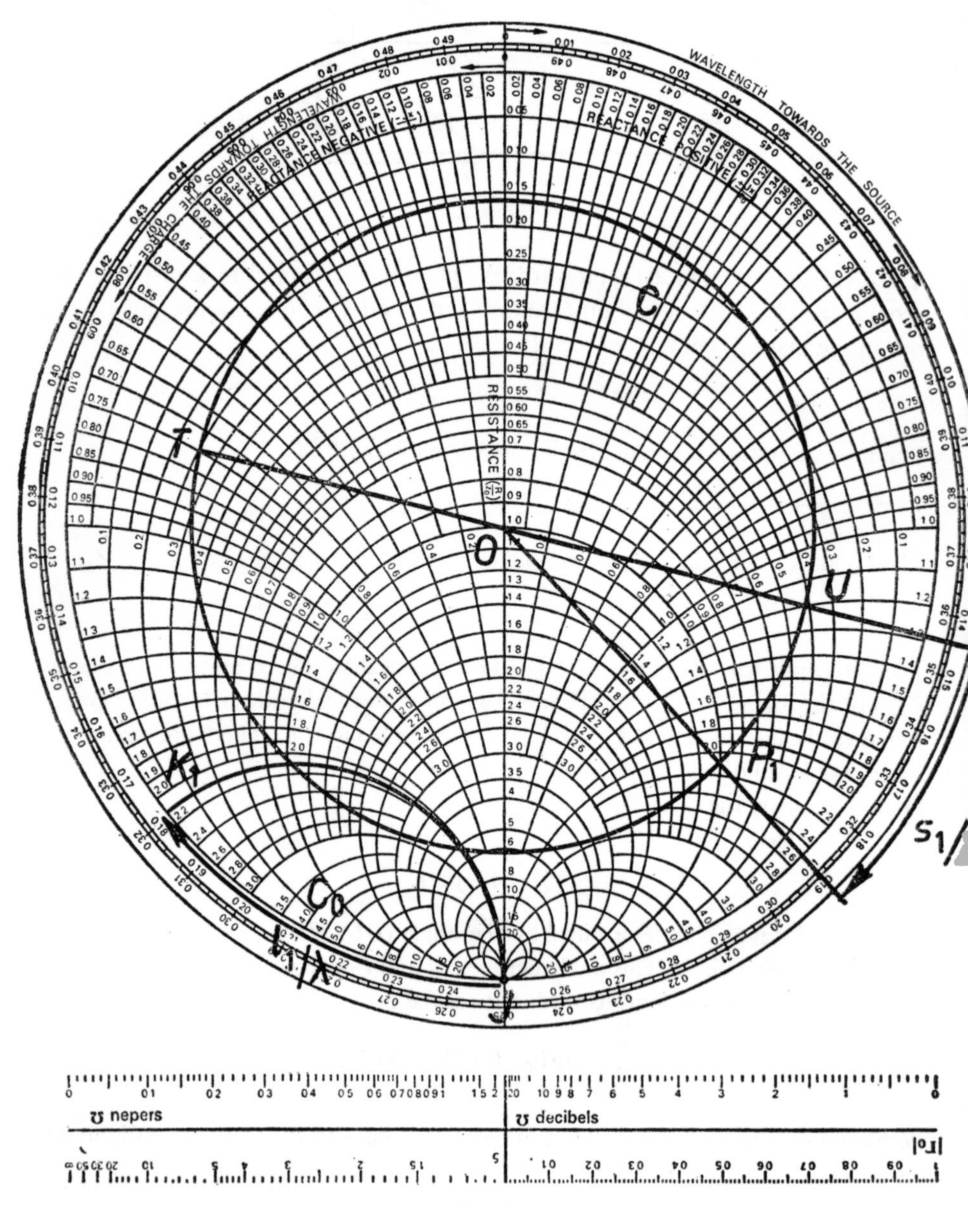

FIG. 3.

(b) The total admittance at the plane π is

$$y_\pi = 1+j\,2{\cdot}1+j\,b_i.$$

The input admittance of the stub must thus be $y_i = -j\,2{\cdot}1$, which is represented on the chart by the point K_1.

The problem has thus become that shown in Fig. 4: gives the input admittance of the stub, what is its length?

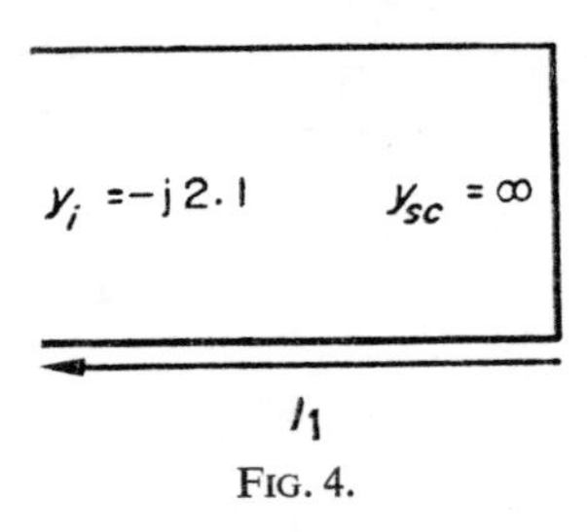

FIG. 4.

When l varies, the locus of the input admittance is the outer circle $C_0(g = 0)$. Since the short-circuit is represented by the point J, the angle ($\boldsymbol{OJ}$, $\boldsymbol{OK_1}$) allows l_1/λ to be determined. Thus

$$\frac{l_1}{\lambda} = 0{\cdot}321-0{\cdot}25 = 0{\cdot}071,$$

$$l_1 = 0{\cdot}071\,\lambda+n\frac{\lambda}{2}$$

$$l_1 = 2{\cdot}13\text{ cm}+n\times 15\text{ cm}.$$

(Although there are an infinite number of solutions, convenience will normally require $l_1 = 2{\cdot}13$ cm).

Second case: (chart Fig. 5). At P_2, $y_L' = 1-j\,2{\cdot}1$.

(a) s_2/λ is given by the angle ($\boldsymbol{OU}$, $\boldsymbol{OP_2}$), so that

$$\frac{s_2}{\lambda} = 0{\cdot}31-0{\cdot}144 = 0{\cdot}166,$$

$$s_2 = 4{\cdot}98\text{ cm}+n\times 15\text{ cm}.$$

(b) The input admittance must be $y_i = j\,2{\cdot}1$ (the point K_2). l_2/λ is given by the angle ($\boldsymbol{OJ}$, $\boldsymbol{OK_2}$) so that

$$\frac{l_2}{\lambda} = 0{\cdot}25+0{\cdot}179 = 0{\cdot}1429$$

$$l_2 = 12{\cdot}87\text{ cm}+n\times 15\text{ cm}.$$

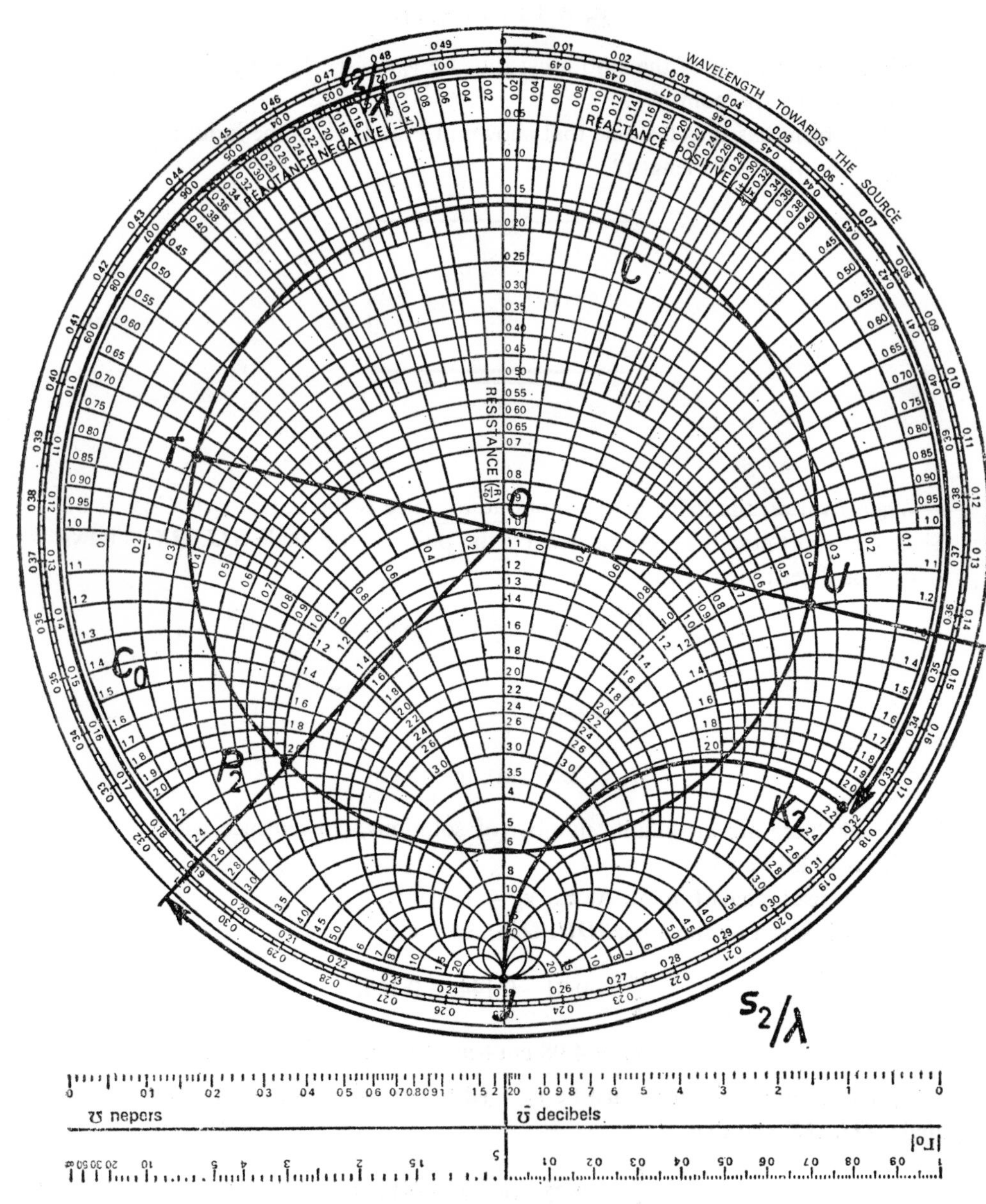
WAVELENGTH TOWARDS THE SOURCE
REACTANCE NEGATIVE
REACTANCE POSITIVE
RESISTANCE
C
C_0
T
O
U
P_2
K_2
l_2/λ
s_2/λ
nepers
decibels
$|\Gamma_0|$

Fig. 5.

Chapter 9

Free and Guided Waves

EXERCISE No. 1

Consider a sine wave propagated along the z-axis in the positive direction in a dielectric medium ($\sigma = 0$) with a propagation constant β_g. Show that for a transverse electric mode (TE) the transverse components of the electromagnetic field can be expressed in terms of the derivatives of H_z with respect to x and y.

Solution. For propagation along the z-axis we can write

$$\partial/\partial z = -j\beta_g.$$

In the case of a TE mode we have, of course, $E_z = 0$. Under these conditions Maxwell's equations (in complex notation) give **curl** $\boldsymbol{H} = j\omega\varepsilon \boldsymbol{E}$. Thus

$$\begin{vmatrix} \boldsymbol{i} & \boldsymbol{j} & \boldsymbol{k} \\ \dfrac{\partial}{\partial x} & \dfrac{\partial}{\partial y} & -j\beta_g \\ H_x & H_y & H_z \end{vmatrix} = j\omega\varepsilon \begin{vmatrix} E_x \\ E_y \\ 0 \end{vmatrix}$$

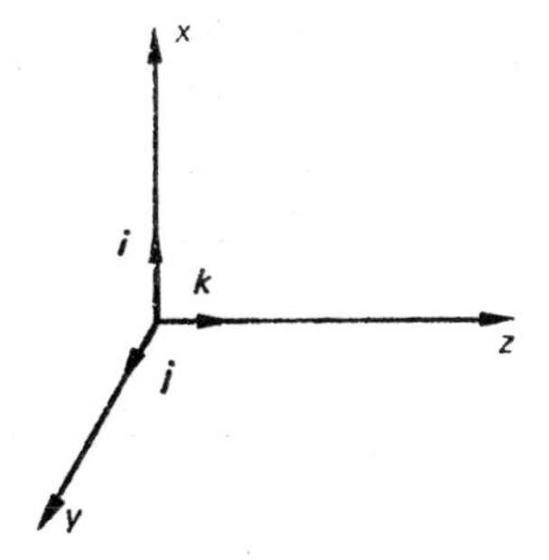

FIG. 1.

so that

$$\frac{\partial H_z}{\partial y} + j\beta_g H_y = j\omega\varepsilon E_x \tag{1}$$

$$-\frac{\partial H_z}{\partial x} - j\beta_g H_x = j\omega\varepsilon E_y \tag{2}$$

$$\frac{\partial H_y}{\partial x} - \frac{\partial H_x}{\partial y} = 0. \tag{3}$$

Similarly $\mathbf{curl}\, \boldsymbol{E} = -j\omega\mu \boldsymbol{H}$ gives

$$+j\beta_g E_y = -j\omega\mu H_x \tag{4}$$

$$-j\beta_g E_x = -j\omega\mu H_y \tag{5}$$

$$\frac{\partial E_y}{\partial x} - \frac{\partial E_x}{\partial y} = -j\omega\mu H_z. \tag{6}$$

From (4) and (5)

$$\frac{E_x}{H_y} = -\frac{E_y}{H_x} = \frac{\omega\mu}{\beta_g} = Z_{\mathrm{TE}} \tag{7}$$

where Z_{TE} is the impedance for the TE mode.

Substituting from (5) in (1) gives

$$\frac{\partial H_z}{\partial y} = j\left(\frac{\omega^2\varepsilon\mu}{\beta_g} - \beta_g\right) H_y,$$

so that

$$H_y = \frac{-j\beta_g}{\beta^2 - \beta_g^2}\frac{\partial H_z}{\partial y}, \quad \text{with} \quad \beta^2 = \omega^2\varepsilon\mu.$$

Substituting from (4) in (2) gives

$$H_x = \frac{-j\beta_g}{\beta^2 - \beta_g^2}\frac{\partial H_z}{\partial x}.$$

From (7) we have

$$E_x = \frac{-j\omega\mu}{\beta^2 - \beta_g^2}\frac{\partial H_z}{\partial y},$$

$$E_y = \frac{j\omega\mu}{\beta^2 - \beta_g^2}\frac{\partial H_z}{\partial x},$$

so that

$$\boldsymbol{H}_T = \frac{-j\beta_g}{\beta^2 - \beta_g^2}\,\mathbf{grad}_T H_z,$$

$$\boldsymbol{E}_T = \frac{j\omega\mu}{\beta^2 - \beta_g^2}\,\boldsymbol{k} \wedge \mathbf{grad}_T H_z.$$

where the suffix T indicates the transverse vector (e.g. $\boldsymbol{H}_T = \boldsymbol{i}H_x + \boldsymbol{j}H_y$). Then

$$\boldsymbol{E}_T . \boldsymbol{H}_T = 0.$$

EXERCISE No. 2

Consider that the wave of the preceding exercise is propagated between two perfectly conducting parallel sheets in the yOz plane at $x = 0$ and $x = d$ as shown in Fig. 1. Obtain expressions for all of the field components in real notation.

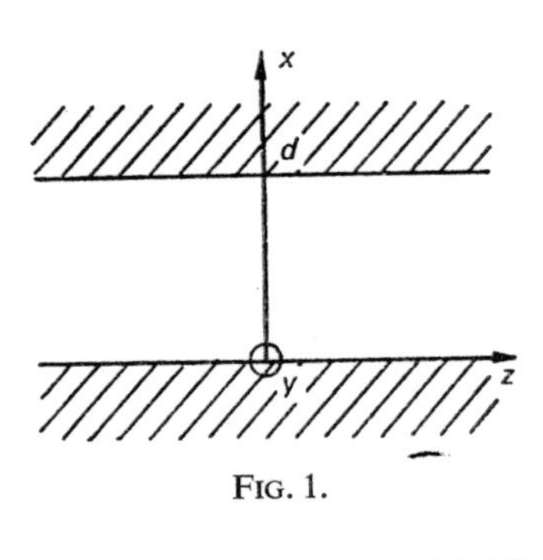

FIG. 1.

Solution. Since the conducting planes extend indefinitely in the directions $+y$ and $-y$, the field components cannot depend on y. Thus:

$$E_x = 0, \quad H_y = 0.$$

There remain

$$H_x = \frac{-j\beta_g}{\beta^2-\beta_g^2}\frac{\partial H_z}{\partial x}, \quad E_y = \frac{j\omega\mu}{\beta^2-\beta_g^2}\frac{\partial H_z}{\partial x}.$$

We calculate H_z from the relation

$$\mathbf{curl}\,(\mathbf{curl}\,\boldsymbol{H}) = \mathbf{grad}\,(\mathrm{div}\,\boldsymbol{H}) - \Delta\boldsymbol{H} = j\omega\varepsilon\,\mathbf{curl}\,\boldsymbol{E} = \omega^2\varepsilon\mu\boldsymbol{H}$$

where

$$\Delta = \frac{\partial^2}{\partial x^2}+\frac{\partial^2}{\partial y^2}+\frac{\partial^2}{\partial z^2}.$$

As div $\boldsymbol{H} = 0$ this gives

$$\Delta\boldsymbol{H}+\omega^2\varepsilon\mu\boldsymbol{H} = 0;$$

so that, for H_z,

$$\frac{\partial^2 H_z}{\partial x^2}+\frac{\partial^2 H_z}{\partial y^2}+\frac{\partial^2 H_z}{\partial z^2}+\omega^2\varepsilon\mu H_z = 0,$$

with

$$\frac{\partial^2 H_z}{\partial y^2} = 0, \quad \frac{\partial^2 H_z}{\partial z^2} = -\beta_g^2$$

Thus

$$\frac{\partial^2 H_z}{\partial x^2} = -(\beta^2-\beta_g^2)\, H_z,$$

$$H_z = A\cos\sqrt{(\beta^2-\beta_g^2)}\,x + B\sin\sqrt{(\beta^2-\beta_g^2)}\,x.$$

Now, by the normal boundary conditions, $\boldsymbol{H}_t$ must be parallel to the planes of the conductors. This condition is satisfied for H_z at both $x = 0$ and $x = d$, while

$$H_x = \frac{-\mathrm{j}\beta_g}{\beta^2-\beta_g^2}\frac{\partial H_z}{\partial x}$$

$$= \frac{-\mathrm{j}\beta_g}{\beta^2-\beta_g^2} \times [-A\sqrt{(\beta^2-\beta_g^2)}\sin\sqrt{(\beta^2-\beta_g^2)}\,x + B\sqrt{(\beta^2-\beta_g^2)}\cos\sqrt{(\beta^2-\beta_g^2)}\,x].$$

and we must have

$$H_x(0) = H_x(d) = 0,$$

so that

$$B = 0 \quad \text{and} \quad \sin\sqrt{(\beta^2-\beta_g^2)}\,d = 0,$$

which requires

$$\sqrt{(\beta^2-\beta_g^2)}\,d = p\pi.$$

where p is any integer. The complex expressions for the components are then

$$H_z = A\cos p\pi\frac{x}{d}\,\mathrm{e}^{-\mathrm{j}\beta_g z},$$

$$H_x = \mathrm{j}\beta_g\frac{d}{p\pi}A\sin p\pi\frac{x}{d}\,\mathrm{e}^{-\mathrm{j}\beta_g z}$$

$$E_y = -\mathrm{j}\omega\mu\frac{d}{p\pi}A\sin p\pi\frac{x}{d}\,\mathrm{e}^{-\mathrm{j}\beta_g z}.$$

On multiplying by $\mathrm{e}^{\mathrm{j}\omega t}$ and taking the real parts, this gives:

$$\mathcal{H}_z = A\cos p\pi\frac{x}{d}\cos(\omega t-\beta_g z),$$

$$\mathcal{H}_x = -A\beta_g\frac{d}{p\pi}\sin p\pi\frac{x}{d}\sin(\omega t-\beta_g z),$$

$$\mathcal{E}_y = A\omega\mu\frac{d}{p\pi}\sin p\pi\frac{x}{d}\sin(\omega t-\beta_g z).$$

With

$$\beta_g = \sqrt{\beta^2 - p\frac{\pi}{d}}\;.$$

EXERCISE No. 3

A waveguide is filled with a dielectric of permittivity ε and permeability μ for which the propagation constant, as an infinite medium, is $\beta = 2\pi/\lambda$. The guide is assumed to be perfect (i.e. the walls are made from a metal of infinite conductivity).

An electromagnetic wave is propagated along the guide, with propagation constant $\beta_g = 2\pi/\lambda_g$, in the z-direction.

1. What relation exists between β and β_g and what factors affect this relationship?

2. What relation exists between the phase velocity v_p and the group velocity v_g in the guide?

3. If the impedance of the guide is defined as the ratio $\left|\frac{E_T}{H_T}\right|$ between the transverse electric and magnetic components, and Z_{TE} and Z_{TM} are the values of this impedance for a TE_{mn} and a TM_{mn} mode respectively, calculate Z_{TE} and Z_{TM} in terms of λ_g and λ. (Show that $Z_{TE}.Z_{TM}$ = constant.)

4. Assuming that it would be possible to vary the frequency of the electromagnetic wave from 0 to infinity, what would be the corresponding range of β_g, v_p, v_g, Z_{TE} and Z_{TM}?

Solution. 1. For the dielectric (ε, μ) Maxwell's equations may be written

$$\textbf{curl}\ \boldsymbol{E} = -j\omega\mu\boldsymbol{H}, \quad \textbf{curl}\ \boldsymbol{H} = j\omega\varepsilon\boldsymbol{E},$$

which gives

$$\textbf{curl}\ (\textbf{curl}\ \boldsymbol{E}) = -\Delta E = \omega^2\varepsilon\mu\boldsymbol{E}.$$

In order to study the propagation of the wave in the guide, it is necessary to study the propagation equation

$$\Delta\boldsymbol{E}+\omega^2\varepsilon\mu\boldsymbol{E} = 0.$$

which, for propagation along the z-direction gives

$$\Delta_T\boldsymbol{E}+[\omega^2\varepsilon\mu-\beta_g^2]\,\boldsymbol{E} = 0.$$

Δ_T being, as before, the Laplacian for the transverse co-ordinates (x, y). For propagation in free space

$$\Delta_T\boldsymbol{E} = 0,$$

and

$$\beta_g^2 = \beta^2 = \omega^2 \varepsilon \mu.$$

In a waveguide this is replaced by

$$\beta^2 - \beta_g^2 = \beta_c^2,$$

where β_c depends on the boundary conditions and

$$\beta_c = \frac{2\pi}{\lambda_c},$$

λ_c being the cut-off wavelength.

2. $v_p = \dfrac{\omega}{\beta_g}, \quad v_g = \dfrac{d\omega}{d\beta_g},$

with

$$\beta_g^2 = \beta^2 - \beta_c^2 = \omega^2 \varepsilon \mu - \beta_c^2,$$

so that

$$\frac{d\beta_g}{d\omega} = \frac{\beta \sqrt{\varepsilon\mu}}{\beta_g} = \frac{\lambda_g}{\lambda} \cdot \frac{1}{v}.$$

Thus

$$v_g = v \frac{\lambda}{\lambda_g},$$

where v is the velocity in an infinite medium of dielectric (ε, μ). We see that

$$v_p v_g = \frac{\beta_g}{\beta \sqrt{\varepsilon\mu}} \cdot \frac{\omega}{\beta_g} = v^2.$$

3. In cartesian co-ordinates, we have for the TE_{mn} mode (i.e. for $E_z = 0$)

$$\textbf{curl}\, \boldsymbol{E} = \begin{vmatrix} \boldsymbol{i} & \boldsymbol{j} & \boldsymbol{k} \\ \dfrac{\partial}{\partial x} & \dfrac{\partial}{\partial y} & -j\beta_g \\ E_x & E_y & 0 \end{vmatrix} = -j\omega\mu \begin{vmatrix} H_x \\ H_y \\ H_z \end{vmatrix};$$

from which

$$j\beta_g E_y = -j\omega\mu H_x,$$

$$-j\beta_g E_x = -j\omega\mu H_y,$$

or

$$\frac{E_x}{H_y} = \frac{\omega\mu}{\beta_g} = -\frac{E_y}{H_x}.$$

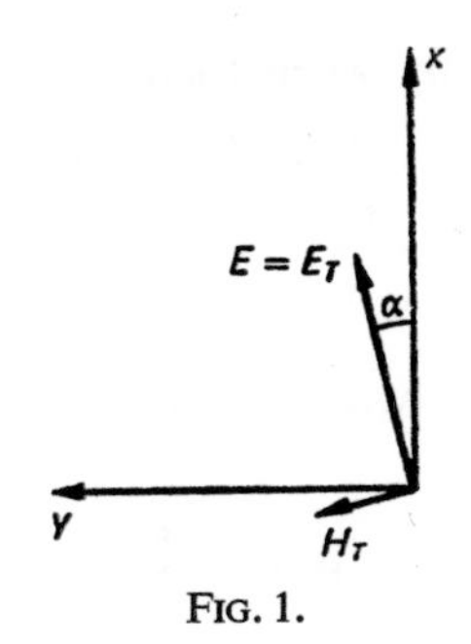

FIG. 1.

Fig. 1 shows the orientation of the transverse electric and magnetic fields in the $x-y$ plane. Thus

$$\left|\frac{E_T}{H_T}\right| = \left|\frac{E_T}{H_T}\right| \frac{\cos\alpha}{\cos\alpha} = \frac{E_x}{H_y} = \frac{\omega\mu}{\beta_g},$$

$$Z_{\text{TE}} = \frac{\omega\mu}{\beta_g} = \sqrt{\frac{\mu}{\varepsilon}}\frac{\lambda_g}{\lambda}.$$

Similarly we can show

$$Z_{\text{TM}} = \sqrt{\frac{\mu}{\varepsilon}}\frac{\lambda}{\lambda_g},$$

so that

$$Z_{\text{TE}}Z_{\text{TM}} = \frac{\mu}{\varepsilon}.$$

4. $\beta_g^2 = \beta^2 - \beta_c^2$, so that $\beta_g = \sqrt{1-(\beta_c/\beta)^2}$

(a) For $\beta > \beta_c$ (i.e. $f > f_c$, the cut-off frequency) β_g will vary from 0 to β when f varies from f_c to infinity.

(b) For $\beta < \beta_c$ we have

$$\beta_g = \mathrm{j}\beta\sqrt{\left(\frac{\beta_c}{\beta}\right)^2 - 1} = \mathrm{j}\beta_g'.$$

The exponential in the propagation equation $e^{\pm \mathrm{j}\beta_g z}$ becomes $e^{\pm\beta_g' z}$, of which only the term $e^{-\beta_g' z}$ has meaning. This latter term shows that propagation is replaced by attenuation—we have an evanascent mode.

$$v_p = \frac{\omega}{\beta_g} = \frac{1}{\sqrt{\varepsilon\mu}} \frac{1}{\sqrt{1-\left(\frac{\beta_c}{\beta}\right)^2}}$$

v_p is seen to have meaning only when there is propagation, i.e. when $\beta > \beta_c$. Then v_p varies from infinity to $v = 1/\sqrt{\varepsilon\mu}$ when f varies from f_c to infinity.

$$v_g = \frac{1}{\sqrt{\varepsilon\mu}} \frac{\beta_g}{\beta} = \frac{1}{\sqrt{\varepsilon\mu}} \sqrt{1-\left(\frac{\beta_c}{\beta}\right)^2}$$

varies from 0 to v when f varies from f_c to infinity.

$Z_{TE} = \sqrt{\frac{\mu}{\varepsilon}} \frac{\beta}{\beta_g}$ varies from infinity to $\eta = \sqrt{\frac{\mu}{\varepsilon}}$ (the impedance of the infinite medium).

$Z_{TM} = \sqrt{\frac{\mu}{\varepsilon}} \frac{\beta_g}{\beta}$ varies from zero to $\eta = \sqrt{\frac{\mu}{\varepsilon}}$ as f varies from f_c to infinity.

EXERCISE No. 4

The cross-section of the lossless rectangular waveguide of Fig. 1 has the dimensions

$$b = 7{\cdot}214 \text{ cm}, \quad a = 3{\cdot}404 \text{ cm}.$$

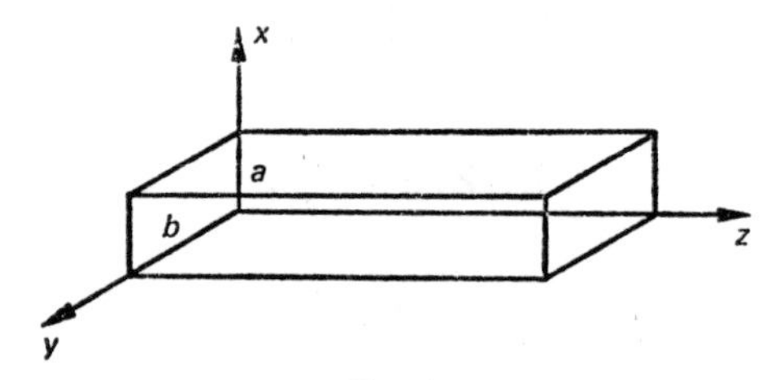

FIG. 1.

Calculate the propagation coefficient in the guide, and the corresponding phase velocity, for each of the modes TE_{01}, TE_{10}, TE_{11}, TE_{20} at a frequency of 5 GHz.

Solution. The propagation coefficient in the guide, β_g, is given by

$$\beta_g^2 = \beta_0^2 - \beta_c^2,$$

with

$$\beta_0^2 = \left(\frac{2\pi}{\lambda_0}\right)^2 = 4\pi^2 f^2 \varepsilon_0 \mu_0,$$

$$\beta_c^2 = \left(\frac{2\pi}{\lambda_c}\right)^2 = \left(\frac{m\pi}{a}\right)^2 + \left(\frac{n\pi}{b}\right)^2 \text{ for a } T_{mn} \text{ mode.}$$

Then

$$\beta_g = \frac{2\pi}{\lambda_g} = \frac{2\pi}{\lambda_c}\sqrt{\left(\frac{\lambda_c}{\lambda_0}\right)^2 - 1}.$$

In the present case

$$f = 5 \text{ GHz} \quad \text{and} \quad \lambda_0 = \frac{3\times 10^{10}}{5\times 10^9} = 6 \text{ cm}.$$

(a) For the TE_{01} mode

$$(\lambda_c)_{01} = 2b = 14{\cdot}428 \text{ cm},$$

so that

$$\beta_g = \frac{2\pi}{(\lambda_c)_{01}}\sqrt{(2{\cdot}404)^2 - 1} = 0{\cdot}954 \text{ rad/cm}$$

corresponding to a wavelength

$$\lambda_g = 6{\cdot}6 \text{ cm}.$$

The phase velocity is then

$$v_p = \frac{2\pi f}{\beta_g} = c_0\frac{\lambda_g}{\lambda} = 1{\cdot}1 c_0 = 330\,000 \text{ km/s}.$$

(b) For the TE_{10} mode, $(\lambda_c)_{10} = 2a = 6{\cdot}808$ cm and

$$\beta_g = \frac{2\pi}{(\lambda)_{10}}\sqrt{(1{\cdot}134)^2 - 1} = 0{\cdot}495 \text{ rad/cm},$$

so that

$$\lambda_g = 12{\cdot}7 \text{ cm},$$

$$v_p = 2{\cdot}12 c_0 = 636\,000 \text{ km/s}.$$

(c) For the TE_{11} mode

$$(\lambda_c)_{11} = \frac{2ab}{\sqrt{a^2+b^2}} = 6{\cdot}25 \text{ cm}$$

$$\beta_g = \frac{2\pi}{(\lambda_c)_{11}}\sqrt{(1{\cdot}025)^2 - 1} = 0{\cdot}284 \text{ rad/cm},$$

$$\lambda_g = 22 \text{ cm}, \quad v_p = 3{\cdot}68\, c_0 = 1\,100\,000 \text{ km/s}.$$

It is seen that, as λ_c increases towards λ, the guide wavelength and the phase velocity are both increased.

(d) In the TE_{20} mode with $(\lambda_c)_{20} = a$,

$$\beta_g = \frac{2\pi}{a}\sqrt{\left(\frac{a}{\lambda_0}\right)^2 - 1},$$

where, in this case, $a < \lambda_0$. Then

$$\beta_g = +j\frac{2\pi}{a}\sqrt{1-\left(\frac{a}{\lambda_0}\right)^2} = +j\beta'_g$$

The term $e^{\pm j\beta_g z}$ becomes $e^{\pm\beta'_g z}$ with only the term $e^{-\beta'_g z}$ having a meaning for a wave propagating in the positive z-direction. There is thus an evanescent mode with

$$\alpha = \beta'_g = \frac{2\pi}{a}\sqrt{1-(0{\cdot}568)^2} = 1{\cdot}51 \text{ neper/cm.}$$

EXERCISE No. 5

Remembering that, in a rectangular waveguide propagating the TE_{mn} mode, the z-component of the magnetic field is, in complex notation,

$$H_z = H_0 \cos m\frac{\pi x}{a} \cos n\frac{\pi y}{b} e^{-j\beta_g z}.$$

1. Give the expressions for the field components for the TE_{01} mode;
2. Calculate the power carried by the wave.

Solution. 1. For the TE_{01} mode; $m = 0$, $n = 1$. The H_z component is then

$$H_z = H_0 \cos\frac{\pi y}{b} e^{-j\beta_g z}.$$

From Exercise 1 we know that all the transverse components of the electromagnetic field can be expressed as functions of H_z to give, in this case,

$$H_x = \frac{-j\beta_g}{\beta^2-\beta_g^2}\frac{\partial H_z}{\partial x} = 0,$$

$$H_y = \frac{+j\beta_g}{\beta^2-\beta_g^2} H_0 \frac{\pi}{b} \sin\frac{\pi y}{b} e^{-j\beta_g z},$$

$$E_x = -\frac{j\omega\mu}{\beta^2-\beta_g^2}\frac{\partial H_z}{\partial y} = \frac{j\omega\mu}{\beta^2-\beta_g^2} H_0 \frac{\pi}{b} \sin\frac{\pi y}{b} e^{-j\beta_g z},$$

$$E_y \propto \frac{\partial H_z}{\partial x} = 0.$$

We also know that

$$\beta_c = \frac{2\pi}{\lambda_c} = \sqrt{\left(\frac{m\pi}{a}\right)^2+\left(\frac{n\pi}{b}\right)^2}$$

$$\beta_g^2+\beta_c^2 = \beta^2$$

where λ_c is the cut-off wavelength. Then

$$\beta_c = \frac{2\pi}{\lambda_c} = \frac{\pi}{b} \quad \text{or} \quad \lambda_c = 2b.$$

The components become

$$H_z = H_0 \cos \frac{\pi y}{b} e^{-j\beta_g z},$$

$$H_y = j\beta_g \frac{b}{\pi} H_0 \sin \frac{\pi y}{b} e^{-j\beta_g z},$$

$$E_x = j\omega\mu \frac{b}{\pi} H_0 \sin \frac{\pi y}{b} e^{-j\beta_g z}.$$

2. The power carried by the wave is

$$P = \frac{1}{2} \int_S (E \wedge H^*) dS,$$

where S is the guide cross-section.

$$P = \frac{1}{2} \int_S E_x H_y^* \, dS = \frac{1}{2} \omega\mu\beta_g \left(\frac{b}{\pi}\right)^2 H_0^2 \int_0^a dx \int_0^a \sin^2 \frac{\pi y}{b} \, dy$$

$$P = \frac{1}{2} \omega\mu\beta_g \left(\frac{b}{\pi}\right)^2 H_0^2 a \frac{b}{2}$$

or

$$P = \frac{ab}{4} \sqrt{\frac{\mu}{\varepsilon}} \frac{\beta\beta_g}{\beta_c^2} H_0^2.$$

EXERCISE No. 6

A rectangular guide with dimensions $a = 1{\cdot}012$ cm, $b = 2{\cdot}286$ cm propagate a wave in the TE_{01} mode with a frequency of 9400 MHz and an incident power $P_0 = 15$ mW.

Calculate the real components of the electromagnetic field if the guide dielectric is air.

Solution

$$(\lambda_c)_{01} = 2b = 4{\cdot}572 \text{ cm},$$

$$\lambda_0 = \frac{c_0}{f} = \frac{3 \times 10^{10}}{9{\cdot}4 \times 10^9} = 3{\cdot}19 \text{ cm},$$

from which

$$\lambda_g = \frac{\lambda_c}{\sqrt{\left(\frac{\lambda_c}{\lambda_0}\right)^2 - 1}} = \frac{4{\cdot}572}{\sqrt{(1{\cdot}43)^2 - 1}} = 4{\cdot}45 \text{ cm.}$$

The power in the wave (cf. Exercise 5) is

$$P = \frac{ab}{4}\sqrt{\frac{\mu_0}{\varepsilon_0}}\,\frac{\lambda_c^2}{\lambda\lambda_g}H_0^2,$$

and

$$H_0 = 0{\cdot}684 \text{ A/m};$$

Also

$$H_z = H_0 \cos\frac{\pi y}{b}\,e^{-j\beta_g z},$$

$$H_y = j\beta_g\frac{b}{\pi}H_0 \sin\frac{\pi y}{b}\,e^{-j\beta_g z},$$

$$E_x = j\omega\mu\frac{b}{\pi}H_0 \sin\frac{\pi y}{b}\,e^{-j\beta_g z}.$$

Multiplying by $e^{j\omega t}$ and taking the real parts gives

$$\begin{aligned}
\mathscr{H}_z &= 0{\cdot}684 \cos 1{\cdot}37y \cos(\omega t - 1{\cdot}412z) \text{ A/m},\\
\mathscr{H}_y &= -0{\cdot}702 \sin 1{\cdot}37y \sin(\omega t - 1{\cdot}142z) \text{ A/m},\\
\mathscr{E}_x &= -369{\cdot}5 \sin 1{\cdot}37y \sin(\omega t - 1{\cdot}412z) \text{ V/m},
\end{aligned}$$

where y and z are in centimetres.

EXERCISE No. 7

Consider a section of waveguide with internal dimensions $a = 0{\cdot}284$ cm and $b = 0{\cdot}568$ cm.

1. Classify the different cut-off frequencies for the modes TE_{mn} and TM_{mn}, which can propagate in the guide, up to a frequency $f_c = 4(f_c)_{01}$.
2. What is the range of frequencies for which it is possible to propagate only the TE_{01} mode?
3. What is the range for the TE_{10} mode alone?

Solution. 1. Here $b/a = 2$. The cut-off wavelength is given by

$$\lambda_c = \frac{2}{\sqrt{\left(\frac{m}{a}\right)^2 + \left(\frac{n}{b}\right)^2}} \quad \text{or} \quad f_c = \frac{c}{2}\sqrt{\left(\frac{m}{a}\right)^2 + \left(\frac{n}{b}\right)^2},$$

these formulae being valid for both TE_{mn} and TM_{mn} modes.

The guide fundamental is that mode with the lowest cut-off frequency, i.e. the TE_{01} mode. For this mode

$$(f_c)_{01} = \frac{c_0}{2b} = 26\,400 \text{ MHz} = 26{\cdot}4 \text{ GHz}.$$

The cut-off frequency for the general TE_{on} mode can be written;

$$(f_c)_{on} = n\frac{c_0}{2b} = n(f_c)_{01}.$$

while for the TE_{mo} modes

$$(f_c) = m\frac{c_0}{2a} = m\frac{c_0}{b}, \quad \text{since} \quad b = 2a,$$

or

$$(f_c)_{mo} = m(f_c)_{02}.$$

If we take into account the fact that $b = 2a$ we obtain a general formula

$$(f_c)_{mn} = \frac{c}{2b}\sqrt{(2m)^2+n^2} = (f_c)_{01}\sqrt{(2m)^2+n^2}.$$

Those modes with frequencies less than $4(f_c)_{01}$ are then given when

$$(2m)^2+n^2 \leqslant 16;$$

$$m = 1 \quad n = 1 : (f_c)_{11} = (f_c)_{01}\sqrt{5};$$

$$m = 1 \quad n = 2 : (f_c)_{12} = (f_c)_{01}\sqrt{8};$$

$$m = 1 \quad n = 3 : (f_c)_{13} = (f_c)_{01}\sqrt{13}.$$

For the TM_{mn} mode

$$H_z = 0, \quad E_z \propto \sin m\frac{\pi x}{a} \sin n\frac{\pi y}{b};$$

which does not exist for $m = 0$ or $n = 0$. The first transverse magnetic mode is TM_{11}. We thus get the scheme of frequencies given in Fig. 1.

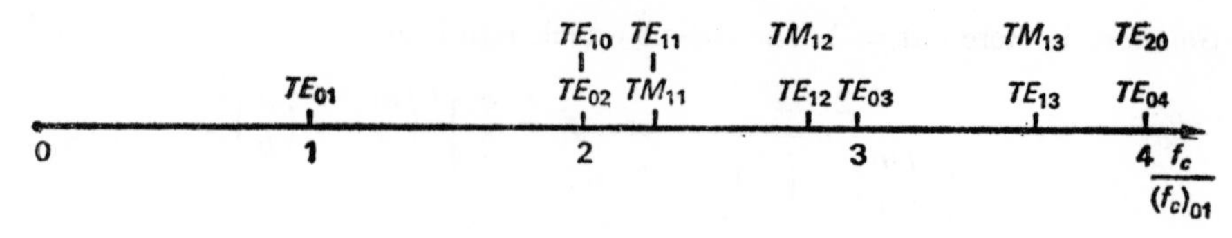

FIG. 1.

2. If the TE_{01} mode is to be propagated alone we must have

$$(f_c)_{01} \leqslant f \leqslant (f_c)_{02},$$

or

$$(f_c)_{01} \leqslant f \leqslant 2(f_c)_{01},$$

$$26{\cdot}4 \text{ GHz} \leqslant f \leqslant 52{\cdot}8 \text{ GHz}.$$

3. It is impossible to propagate the mode TE_{10} by itself since when

$$(f_c)_{10} < f < (f_c)_{11},$$

the modes TE_{01} and TE_{02} can propagate as well.

EXERCISE No. 8

What range of frequency can b: used if it is required to propagate the single mode TE_{01} along a rectangular guide of dimensions $a = 1{\cdot}016$ and $b = 2{\cdot}286$ cm? Calculate the variation of the guide wavelength over this frequency range.

Solution. 1. The cut-off frequency for the TE_{01} mode is

$$(f_c)_{01} = \frac{c_0}{2b} = 6556 \text{ MHz}.$$

We can write the general equation for the cut-off frequencies in the form

$$(f_c)_{mn} = \frac{c_0}{2b}\sqrt{(2{\cdot}25m)^2+n^2}\,,$$

since

$$\frac{b}{a} = 2{\cdot}25.$$

The next cut-off frequency to $(f_c)_{01}$ is

$$(f_c)_{02} = 2(f_c)_{01} < (f_c)_{10} = 2{\cdot}25(f_c)_{01},$$

which gives the range of frequency for TE_{01} alone as

$$6556 \text{ MHz} \leq f \leq 13\,112 \text{ MHz}.$$

2. With

$$\beta_c^2 + \beta_g^2 = \beta_0^2$$

or

$$\frac{1}{\lambda_c^2} + \frac{1}{\lambda_g^2} = \frac{1}{\lambda_0^2},$$

we have

$$\lambda_g = \frac{\lambda_c}{\sqrt{\left(\frac{\lambda_c}{\lambda_0}\right)^2 - 1}} = \frac{2b}{\sqrt{\left(\frac{f}{f_c}\right)^2 - 1}}.$$

Then, for

$$\begin{aligned} f &\to f_c, & \lambda_g &\to \infty; \\ f &= 1{\cdot}1 f_c, & \lambda_g &= 9{\cdot}98 \text{ cm}; \\ f &= 2 f_c, & \lambda_g &= \frac{2b}{\sqrt{3}} = 2{\cdot}64 \text{ cm}. \end{aligned}$$

We see that the guide wavelength varies drastically in the neighbourhood of the cut-off frequency.

EXERCISE No. 9

In order that a rectangular guide of dimensions $a = 1{\cdot}016$ cm and $b = 2{\cdot}286$ cm can propagate a TE_{01} mode at 5000 MHz, the guide is filled with a dielectric of relative permittivity ε_r.

1. What are the limits on ε_r if it is required that only the TE_{01} mode shall propagate?

2. Taking $\varepsilon_r = 2{\cdot}25$, calculate the guide wavelength, the phase velocity and the attenuation for a dielectric loss tangent $\tan \varphi = 10^{-3}$.

Solution. 1. In the absence of the dielectric the propagation constant β_g is given by

$$\beta_g^2 = \beta_0^2 - \beta_c^2 \tag{1}$$

with

$$\beta_0^2 = \omega^2 \varepsilon_0 \mu_0 = \left(\frac{2\pi}{\lambda_0}\right)^2,$$

which gives

$$\lambda_g = \frac{\lambda_c}{\sqrt{\left(\frac{\lambda_c}{\lambda_0}\right)^2 - 1}}.$$

There will be propagation if

$$\lambda_c > \lambda_0.$$

In the present case, where the mode is TE_{01}, $\lambda_c = 2b$ and there will be propagation if

$$\lambda_0 < 2b = 4 \cdot 572 \text{ cm}.$$

However

$$\lambda_0 = \frac{2 \cdot 10^{10}}{5 \cdot 10^9} = 6 \text{ cm}.$$

and propagation is not allowed. With the dielectric in the guide equation (1) becomes

$$\beta_g^2 = \beta^2 - \beta_c^g,$$

with

$$\beta^2 = \omega^2 \varepsilon_0 \varepsilon_r \mu_0 = \left(\frac{2\pi}{\lambda}\right)^2$$

and

$$\lambda = \frac{\lambda_0}{\sqrt{\varepsilon_r}} \quad \text{and} \quad \lambda_g = \frac{\lambda_c}{\sqrt{\left(\frac{\lambda_c}{\lambda}\right)^2 - 1}}.$$

There will not be propagation for $\lambda < \lambda_c$ or if $\lambda_0/\sqrt{\varepsilon_r} < 2b$. In addition, if we require that there shall be no higher modes propagated,

$$(\lambda_c)_{02} < \lambda < (\lambda_c)_{01},$$

the mode immediately above the fundamental (TE_{01}) being the TE_{02} mode with cut-off wavelength $(\lambda_c)_{02} = b$. Thus we require

$$b < \frac{\lambda_0}{\sqrt{\varepsilon_r}} < 2b,$$

so that

$$\frac{\lambda_0}{2b} < \sqrt{\varepsilon_r} < \frac{\lambda_0}{b},$$

$$\left(\frac{\lambda_0}{2b}\right)^2 < \varepsilon_r < \left(\frac{\lambda_0}{b}\right)^2,$$

which, in this case, gives

$$1 \cdot 72 < \varepsilon_r < 6 \cdot 88.$$

2. For $\varepsilon_r = 2 \cdot 25$ the wavelength in the infinite medium is $\lambda = \dfrac{6}{\sqrt{2 \cdot 25}} = 4$ cm.

Thus

$$\lambda_g = \frac{\lambda_c}{\sqrt{(1\cdot 14)^2 - 1}} = 8\cdot 35 \text{ cm};$$

Since φ is small, tan $\varphi \simeq \varphi$ and the complex dielectric permittivity is

$$\varepsilon^* = \varepsilon_0 \varepsilon_r (1 - j\varphi).$$

which gives the complex propagation constant

$$\beta_g^{*2} = \omega^2 \varepsilon_0 \mu_0 \varepsilon_r (1 - j\varphi) - \beta_c^2 = \beta_g^2 - j\varphi\beta^2.$$

Then

$$\beta_g^* = \beta_g \sqrt{1 - j\varphi\left(\frac{\beta}{\beta_g}\right)^2},$$

with

$$\frac{\beta}{\beta_g} = \frac{\lambda_g}{\lambda} = \frac{8\cdot 35}{4}, \quad \varphi = 10^{-3};$$

and finally

$$\beta_g^* \approx \beta_g \left[1 - j\frac{\varphi}{\lambda}\left(\frac{\lambda_g}{\lambda}\right)^2\right]$$

The propagation term $e^{-j\beta_g^* z}$ becomes

$$e^{-j\beta_g z} e^{-\alpha z},$$

with

$$\alpha = \beta_g \frac{\varphi}{2}\left(\frac{\lambda_g}{\lambda}\right)^2 = \pi\varphi\frac{\lambda_g}{\lambda^2} = 0\cdot 166 \times 10^{-2} \text{ neper/m}$$

and

$$v_p = \frac{\omega}{\beta_g} = 417\,000 \text{ km/s}.$$

EXERCISE No. 10

An electromagnetic wave is propagated along a waveguide of conductivity σ.

1. Give the general formula which allows the attenuation due to the joule losses to be calculated.

2. Calculate this attenuation for the TE_{01} mode propagating along a rectangular guide.

3. As a numerical example, take $a = 1\cdot 016$ cm, $b = 2\cdot 286$ cm, $\sigma = 5\cdot 8 \times 10^7$ S/m $f = 9600$ MHz.

Solution. 1. The attenuation coefficient is given by

$$2\alpha = \frac{\text{Power dissipated per unit length}}{\text{Power transmitted}} = \frac{P_d}{P}$$

with

$$P = \tfrac{1}{2} \int_S (\boldsymbol{E} \wedge \boldsymbol{H}^*) \, \mathrm{d}\boldsymbol{S}.$$

The power dissipated by the Joule heating per unit length is

$$P_d = \tfrac{1}{2} R_s \int_C ii^* \, \mathrm{d}l$$

where R_s is the resistance of a parallelapiped of unit base area and width δ, δ being the skin depth, the integral is taken round the guide circumference C as shown in Fig. 1.

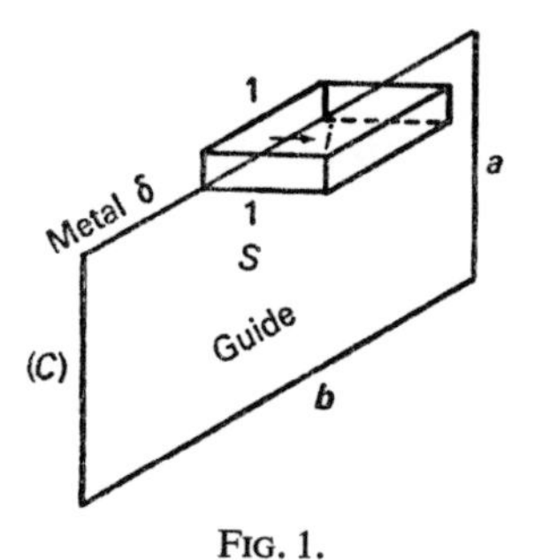

FIG. 1.

Now $|i| = |H_t|$ where H_t is the magnetic field close to, and tangential to, the guide surface. Thus

$$P = \tfrac{1}{2} R_S \int_C |H_t|^2 \, \mathrm{d}l,$$

so that

$$\alpha = \frac{1}{2} \frac{R_S \int_C |H_t|^2 \, \mathrm{d}l}{\int_S (\boldsymbol{E} \wedge \boldsymbol{H}^*) \, \mathrm{d}\boldsymbol{S}},$$

where the denominator integral is over the guide cross-section S.

Note: The expressions for the field components are taken to be those for the case where $\sigma = \infty$, any change due to the finite conductivity being generally negligible.

2. The expressions for the field components have been given in Exercise 5

$$H_x = 0, \quad H_y = j\beta_g \frac{b}{\pi} H_0 \sin \frac{\pi y}{b} e^{-j\beta_g z}$$

$$H_z = H_0 \cos \frac{\pi y}{b} e^{-j\beta_g z},$$

$$\int_C |H_t|^2 \, dl = 2 \int_0^a |H_z|^2 \, dx + 2 \int_a^b [|H_z|^2 + |H_y|^2] \, dy$$

$$= 2H_0^2 a + 2 \int_0^b H_0^2 \cos^2 \frac{\pi y}{b} \, dy + 2\beta_g^2 \frac{b^2}{\pi^2} H_0^2 \int_c^b \sin^2 \frac{\pi y}{b} \, dy$$

$$= H_0^2 \left[2a + b + \left(\frac{\beta_g}{\beta_c} \right)^2 b \right]$$

$$= H_0^2 \left[2a + b \left(1 + \frac{\beta_g^2}{\beta_c^2} \right) \right] = H_0^2 \left[2a + b \frac{\beta^2}{\beta_c^2} \right].$$

Also from Exercise 5

$$\frac{1}{2} \int_S (E \wedge H^*) \, dS = \frac{ab}{4} \sqrt{\frac{\mu_0}{\varepsilon_0}} \frac{\lambda_c^2}{\lambda \lambda_g} H_0^2,$$

from which

$$\alpha = \frac{1}{2} R_S \frac{2a + b \dfrac{\lambda_c^2}{\lambda^2}}{\dfrac{ab}{2} \sqrt{\dfrac{\mu_0}{\varepsilon_0}} \dfrac{\lambda_c^2}{\lambda \lambda_g}} = \frac{R_S}{\eta_0 a} \frac{1 + \dfrac{2a}{b} \left(\dfrac{\lambda}{\lambda_c} \right)^2}{\sqrt{1 - \left(\dfrac{\lambda}{\lambda_c} \right)^2}},$$

$$\lambda = \frac{c_0}{f} = \frac{3 \times 10^{10}}{9 \cdot 6 \times 10^9} = 3 \cdot 12 \text{ cm}$$

$$\frac{\lambda}{\lambda_c} = \frac{\lambda}{2b} = \frac{3 \cdot 12}{4 \cdot 572} = 0 \cdot 684$$

and

$$\alpha \approx 1 \cdot 3 \times 10^{-2} \text{ neper/m}$$

EXERCISE No. 11

A cylindrical waveguide has a radius $a = 5$ cm. What is the cut-off wavelength for the propagation of the TE_{01} mode? If the frequency is 5000 MHz, what is the guide wavelength?

Solution. 1. The cut-off wavelength for the TE_{01} mode is here given by

$$\frac{2\pi}{\lambda_c} = \frac{x'_{01}}{a}$$

where x'_{01} is the first zero of the derivative of the Bessel function J_0, i.e.

$$J'_0(x'_{01}) = 0.$$

Thus

$$x'_{01} = 3{\cdot}83,$$

and

$$\lambda_c = \frac{2\pi a}{x'_{01}} = 0{\cdot}82\times 2a,$$

$$\lambda_c = 8{\cdot}2 \text{ cm}.$$

2. $\lambda = \dfrac{c_0}{f} = \dfrac{3\times 10^{10}}{5\times 10^{9}} = 6$ cm,

so that

$$\lambda_g = \frac{\lambda}{\sqrt{1-\left(\frac{\lambda}{\lambda_c}\right)^2}} = 8{\cdot}8 \text{ cm}.$$

EXERCISE No. 12

A cylindrical guide has radius 3·6 cm. What are the cut-off wavelengths of the first six modes which can be propagated in the guide?

Solution. For the TE_{nm} mode the cut-off wavelength is given by

$$\frac{2\pi a}{\lambda_c} = x'_{nm},$$

x'_{nm} being the mth zero of the derivative of the nth order Bessel function J_n.

For the TM_{nm} mode, on the other hand, we have the cut-off wavelengths

$$\frac{2\pi a}{\lambda_c} = x_{nm}.$$

where x_{nm} is the mth zero of J_n.

For the lowest order TE modes this gives

$$x'_{01} = 3{\cdot}832, \quad \text{giving} \quad (f'_c)_{01} = \frac{c_0}{2\pi a}.3{\cdot}832 = 5081 \text{ MHz},$$

$$x'_{11} = 1{\cdot}84, \quad \text{giving} \quad (f'_c)_{11} = \frac{c_0}{2\pi a}.1{\cdot}84 = 2440 \text{ MHz},$$

$$x'_{21} = 3{\cdot}054, \quad \text{giving} \quad (f'_c)_{21} = \frac{c_0}{2\pi a}.3{\cdot}054 = 4049 \text{ MHz},$$

$$x'_{31} = 4{\cdot}2, \quad \text{giving} \quad (f'_c)_{31} = \frac{c_0}{2\pi a}.4{\cdot}2 = 5569 \text{ MHz},$$

and for the lowest TM modes

$$x_{01} = 2{\cdot}405, \quad \text{giving} \quad (f_c)_{01} = 3189 \text{ MHz}$$
$$x_{11} = 3{\cdot}832, \quad \text{giving} \quad (f_c)_{11} = 5081 \text{ MHz}$$
$$x_{21} = 5{\cdot}136, \quad \text{giving} \quad (f_c)_{21} = 6810 \text{ MHz}.$$

The distribution of frequencies is thus that shown in Fig. 1.

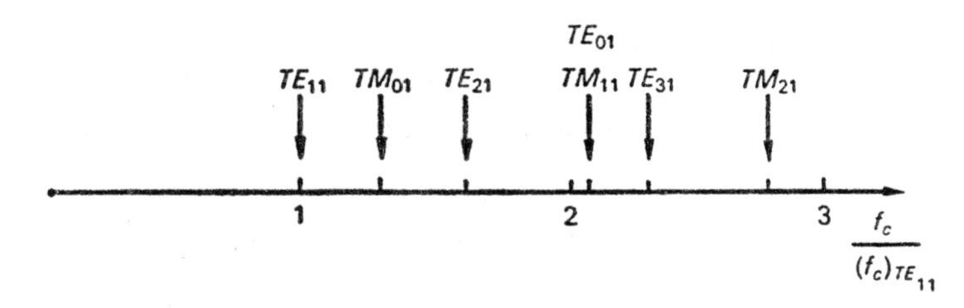

FIG. 1.

Chapter 10

The Resonant Cavity

EXERCISE No. 1

Consider a cylindrical lossless guide of radius a in which a sinusoidal electromagnetic wave is propagated along the z-axis.

1. What are the field components for the TE_{01} mode?
2. If a cavity is formed by placing two perfectly conducting planes in the guide at $z = 0$ and $z = h$, what are the field components in this cavity?
3. Calculate the instantaneous stored electrical and magnetic energies in this cavity. Calculate the total stored energy and draw a conclusion.

Solution. 1. For a TE_{mn} mode

$$H_z = H_0 J_n(\beta_c r) \cos n\theta \, e^{-j\beta_g z},$$

with

$$\beta_c a = x'_{nm} \quad \text{where} \quad x'_{nm} \text{ is the } m\text{th zero of } J'_n.$$

Then

$$H_z = H_0 J_0 \left(x'_{01} \frac{r}{a} \right) e^{-j\beta_g z},$$

with

$$x'_{01} = 3{\cdot}83.$$

The other field components can be deduced from H_z as

$$\boldsymbol{E}_T = \frac{j\omega\mu}{\beta_c^2} \boldsymbol{k} \wedge \mathbf{grad}_T H_z,$$

so that

$$E_r^+ = -\frac{j\omega\mu}{\beta_c^2} \frac{1}{r} \frac{\partial H_z}{\partial \theta} = 0,$$

$$E_\theta^+ = \frac{j\omega\mu}{\beta_c^2} \frac{\partial H_z}{\partial r} = \frac{j\omega\mu}{\beta_c} J'_0 \left(x'_{01} \frac{r}{a} \right) e^{-j\beta_g z}.$$

For the magnetic field

$$H_T = \frac{-j\beta_g}{\beta_c^2}\,\mathbf{grad}_T H_z,$$

and

$$H_\theta^+ = 0,$$

$$H_r^+ = \frac{-j\beta_g}{\beta_c} H_0 J_0'\left(x_{01}' \frac{r}{a}\right) e^{-j\beta_g z}.$$

Thus, for the TE_{01} mode

$$E_\theta^+ = \frac{j\omega\mu}{\beta_c} H_0 J_0'\left(x_{01}' \frac{r}{a}\right) e^{-j\beta_g z},$$

$$H_r^+ = \frac{-j\beta_g}{\beta_c} H_0 J_0'\left(x_{01}' \frac{r}{a}\right) e^{-j\beta_g z},$$

$$H_z^+ = H_0 J_0\left(x_{01}' \frac{r}{a}\right) e^{-j\beta_g z},$$

$$E_r^+ = H_\theta^+ = 0, \quad \beta_g^2 + \beta_c^2 = \beta^2.$$

2. The total field in a cavity consists of the fields due to the incident wave propagating towards $z > 0$ and a wave reflected at the plane $z = h$ propagating towards $z < 0$. Thus

$$(H_z)_{\text{total}} = H_z^+ + H_z^- = H_0 J_0 e^{-j\beta_g z} + H_0' J_0 e^{+j\beta_g z},$$

which must satisfy the boundary conditions

$$H_z = 0 \quad \text{at} \quad z = 0 \quad \text{and} \quad z = h,$$

so that

$$H_0 + H_0' = 0 \quad \text{at} \quad z = 0;$$

$$\sin \beta_g h = 0 \quad \text{at} \quad z = h.$$

From this we get the resonance condition

$$\beta_g h = p\pi,$$

or, substituting for β_g

$$\left(\frac{p\pi}{h}\right)^2 + \left(\frac{x_{01}'}{a}\right)^2 = \omega^2 \varepsilon\mu.$$

For $0 < z < h$ (the region of the cavity)

$$H_z = -2jH_0J_0 \sin p\pi \frac{z}{h},$$

$$E_\theta = \frac{j\omega\mu}{\beta_c} H_0J_0' e^{-j\beta_g z} + \frac{j\omega\mu}{\beta_c} H_0'J_0' e^{+j\beta_g z}$$

$$= 2\frac{\omega\mu}{\beta_c} H_0J_0' \sin p\pi \frac{z}{h},$$

$$H_r = \frac{-j\beta_g}{\beta_c} H_0J_0' e^{-j\beta_g z} + \frac{j\beta_g}{\beta_c} H_0'J_0' e^{+j\beta_g z},$$

$$H_r = \frac{-2j\beta_g}{\beta_c} H_0J_0' \cos p \frac{\pi z}{h}.$$

If we multiply the field components by $e^{j\omega t}$, we can take the real parts to give

$$\mathcal{H}_r = \frac{2\beta_g}{\beta_c} H_0J_0'\left(x_{01}'\frac{r}{a}\right) \cos p\pi \frac{z}{h} \sin \omega t,$$

$$\mathcal{H}_z = 2H_0J_0\left(x_{01}'\frac{r}{a}\right) \sin p\pi \frac{z}{h} \sin \omega t,$$

$$\mathcal{E}_\theta = \frac{2\omega\mu}{\beta_c} H_0J_0'\left(x_{01}'\frac{r}{a}\right) \sin p\pi \frac{z}{h} \cos \omega t.$$

3. The instantaneous magnetic energy stored in the cavity is

$$W_m(t) = \frac{1}{2}\int_V \mu[H_r^2 + H_z^2]\,dV$$

$$= \frac{\mu}{2}\sin^2 \omega t \int_0^a \int_0^h \frac{4\beta_g^2}{\beta_c^2} H_0^2 J_0'^2 \cos^2 p\pi \frac{z}{h}\, 2\pi r\, dr\, dz$$

$$+ \frac{\mu}{2}\sin^2 \omega t \int_0^a \int_0^h 4H_0^2 J_0^2 \sin^2 p\pi \frac{z}{h}\, 2\pi r\, dr\, dz$$

$$= \frac{\mu}{2}\sin^2 \omega t \frac{4\beta_g^2}{\beta_c^2} H_0^2 \frac{h}{2} \cdot 2\pi \int_0^a J_0'^2\left(x_{01}'\frac{r}{a}\right) r\, dr$$

$$+ \frac{\mu}{2}\sin^2 \omega t\, 4H_0^2 \frac{h}{2}\, 2\pi \int_0^a J_0^2\left(x_{01}'\frac{r}{a}\right) r\, dr.$$

Now

$$J'_0 = -J_1,$$

$$\int_0^a J_1^2\left(x'_{01}\frac{r}{a}\right) r\, \mathrm{d}r = -\frac{a^2}{2} J_0(x'_{01})\, J_2('_{01}x),$$

because

$$J_1(x'_{01}) = J'_0(x'_{01}) = 0.$$

Further

$$\frac{2}{x'_{01}} J_1(x'_{01}) = 0 = J_2(x'_{01}) + J_0(x'_{01}),$$

so that

$$-J_2(x'_{01}) = J_0(x'_{01}),$$

and hence

$$\int_0^a J_1^2\left(x'_{01}\frac{r}{a}\right) r\, \mathrm{d}r = \frac{a^2}{2} J_1^2(x'_{01})$$

On the other hand

$$\int_0^a J_0^2\left(x'_{01}\frac{r}{a}\right) r\, \mathrm{d}r = \frac{a^2}{2} J_0^2(x'_{01}),$$

so that

$$W_m(t) = \mu H_0^2 h\pi a^2 J_0^2(x'_{01}) \left[\left(\frac{\beta_g^2}{\beta_c^2}\right) + 1\right] \sin^2 \omega t,$$

and finally

$$W_m(t) = \mu H_0^2(\pi a^2 h)\, J_0^2(x'_{01}) \left(\frac{\beta}{\beta_c}\right)^2 \sin^2 \omega t.$$

The stored electrical energy in the cavity is

$$W_e(t) = \frac{1}{2}\int_V \varepsilon E^2\, \mathrm{d}v$$

$$= \frac{1}{2}\frac{4\omega^2\mu^2}{\beta_c^2} H_0^2\, 2\pi\varepsilon \int_0^a J_0'^2 r\, \mathrm{d}r \int_0^h \sin^2\frac{p\pi z}{h}\, \mathrm{d}z \cdot \cos^2 \omega t,$$

and

$$W_e(t) = \frac{\omega^2\mu^2\varepsilon}{\beta_c^2} (\pi a^2 h)\, H_0^2 J_0^2(x'_{01}) \cos^2 \omega t$$

$$= \mu H_0^2(\pi a^2 h)\, J_0^2(x'_{01}) \left(\frac{\beta}{\beta_c}\right)^2 \cos^2 \omega t.$$

The total energy stored in the cavity is thus

$$W = \mu H_0^2 V J_0^2(x'_{01}) \left(\frac{\beta}{\beta_c}\right)^2.$$

which is constant, being independent of time. (Here $V = \pi a^2 h$ is the cavity volume.) The mean stored electrical energy $W_E = W/2 = W_M$, the mean stored magnetic energy $W_M = W/2 = W_E$. The total stored energy is just twice the mean stored magnetic or electrical energy.

EXERCISE No. 2

Calculate the Q-factor, Q_0, of the cavity considered in the preceding exercise.

Solution. The Q-factor (or quality factor) is given by

$$Q_0 = \omega \frac{W}{P},$$

where W is the stored energy and P is the power dissipated. In the present case the power will be dissipated by the cavity walls so that

$$P = \frac{R_S}{2} \int_S H_t H_t^* \, dS,$$

where S is taken over the whole interior cavity surface and R_S was defined in Exercise 10, Chapter 9.

The field H_t is that parallel (tangential) to the walls and is assumed to have the same form as for the perfect cavity. The power P_1 dissipated in the side walls is thus given by taking the fields at $r = a$.

$$H_r(a) = 0,$$

$$H_z(a) = -2jH_0 J_0(x'_{01}) \sin \frac{p\pi z}{h}$$

so that

$$P_1 = \frac{R_S}{2} 4H_0^2 2\pi a J_0^2(x'_{01}) \int_0^h \sin^2 \frac{p\pi z}{h} \, dz$$

$$= R_S 2\pi a h J_0^2(x'_{01}) H_0^2.$$

If P_2 is the power dissipated in each of the end faces at $z = 0$ and $z = h$ the fields must be taken as, for example:

$$z = 0, \qquad H_z = 0, \qquad H_r = -2j\frac{\beta_g}{\beta_c} H_0 J_0'\left(x_{01}' \frac{r}{a}\right),$$

from which

$$P_2 = \frac{R_S}{2} 4\left(\frac{\beta_g}{\beta_c}\right)^2 H_0^2 2\pi \int_0^a J_0'^2\left(x_{01}' \frac{r}{a}\right) r\,dr,$$

$$= 2\pi a^2 R_S \left(\frac{\beta_g}{\beta_c}\right)^2 H_0^2 J_0^2(x_{01}').$$

The total dissipated power is thus

$$P = P_1 + 2P_2 = 2\pi R_S H_0^2 J_0^2(x_{01}')\left[ah + 2a^2\left(\frac{\beta_g}{\beta_c}\right)^2\right].$$

Now let us calculate

$$Q_0 = \omega\frac{W}{P};$$

W was calculated above as

$$W = \mu H_0^2 \pi a^2 h J_0^2(x_{01}')\left(\frac{\beta}{\beta_c}\right)^2,$$

so that

$$Q_0 = \frac{\omega\mu}{2R_S}\,\frac{ah(\beta/\beta_c)^2}{h + 2a(\beta_g/\beta_c)^2},$$

where we can write

$$\frac{\omega\mu}{2R_S} = \frac{\omega\mu\sigma\delta}{2} = \frac{1}{\delta}.$$

For a TE_{01} cavity mode

$$Q_0 = \frac{1}{\delta\lambda^2}\,\frac{4\pi^2 a^3}{x_{01}'^2 + 2\pi^2\left(\frac{a}{h}\right)^3 p^2}.$$

EXERCISE No. 3

A cavity is constructed from a rectangular guide, with dimensions $a = 0{\cdot}38$ cm, $b = 0{\cdot}76$ cm, so that it oscillates in the TE_{102} mode at 50 GHz. What is the length h of the cavity if the dielectric is air?

Are there other modes with the same resonance frequency?

Solution. 1. The resonant frequency of a TE_{mnp} mode rectangular cavity is given by

$$4\pi^2 f^2 \varepsilon_0 \mu_0 = \left(m\frac{\pi}{a}\right)^2 + \left(n\frac{\pi}{b}\right)^2 + \left(p\frac{\pi}{h}\right)^2,$$

which, for the TE_{102} mode, is

$$4f^2 \varepsilon_0 \mu_0 = \frac{1}{a^2} + \frac{4}{h^2}$$

from which

$$h = \frac{1}{\sqrt{\left(\frac{f}{c_0}\right)^2 - \frac{1}{4a^2}}} = 0{\cdot}975 \text{ cm}.$$

2. Since $b = 2a$, the TE_{022} mode will have the same resonant frequency as the TE_{102} mode. We say that the two modes are degenerate.

EXERCISE No. 4

A cavity is constructed of air-filled rectangular waveguide of dimensions $a = 1{\cdot}016$ cm, $b = 2{\cdot}286$ cm. In the plane $z = 0$ the guide is closed by a perfect conductor with a small coupling hole, while the other end of the cavity is closed by a perfectly short-circuiting piston. What must be the range of movement of the piston, if the cavity is to resonate at frequencies from 9300 MHz to 10 200 MHz?

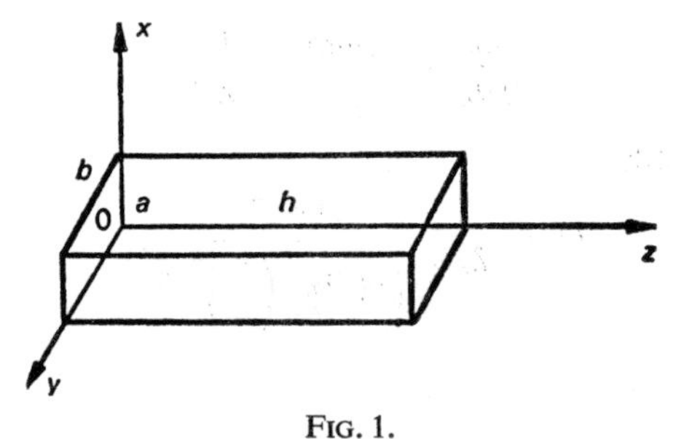

FIG. 1.

Solution. The resonant frequency of the rectangular cavity is given by

$$f_0 = c_0 \sqrt{\left(\frac{m}{2a}\right)^2 + \left(\frac{n}{2b}\right)^2 + \left(\frac{p}{2h}\right)^2}$$

where c_0 is the velocity of light and we are assuming the TE_{mnp} mode.

For the TE_{012} mode; $m = 0$, $n = 1$, $p = 2$ and

$$f_0 = c_0 \sqrt{\left(\frac{1}{2b}\right)^2 + \left(\frac{1}{h}\right)^2},$$

$$h = \frac{2b}{\sqrt{\left(\frac{2b}{\lambda_0}\right)^2 - 1}}.$$

For $f_0 = 9300$ MHz, $\lambda_0 = 3{\cdot}25$ cm, $h_1 = 4{\cdot}57$ cm,

$f_0 = 10\,200$ MHz, $\lambda_0 = 2{\cdot}94$ cm, $h_2 = 3{\cdot}84$ cm

and the range of movement required is 0·73 cm.

EXERCISE No. 5

The cavity of the preceding exercise has a Q-factor 5000 and a coupling coefficient $\alpha = 1$. The klystron oscillator works at 9600 MHz.

1. What is the cavity length when it is tuned to 9600 MHz?

2. Through what distance must the piston be moved to vary the frequency of resonance over the cavity pass band?

Solution. 1. For $f_0 = 9600$ MHz, $\lambda_0 = 3{\cdot}125$ cm and

$$h_0 = \frac{2b}{\sqrt{\left(\frac{2b}{\lambda_0}\right)^2 - 1}} = 4{\cdot}281 \text{ cm}.$$

2. When the klystron operates at f_0 and the length of the cavity is h_0 then the cavity absorption follows curve 1 of Fig. 2. The width δf at the half power points

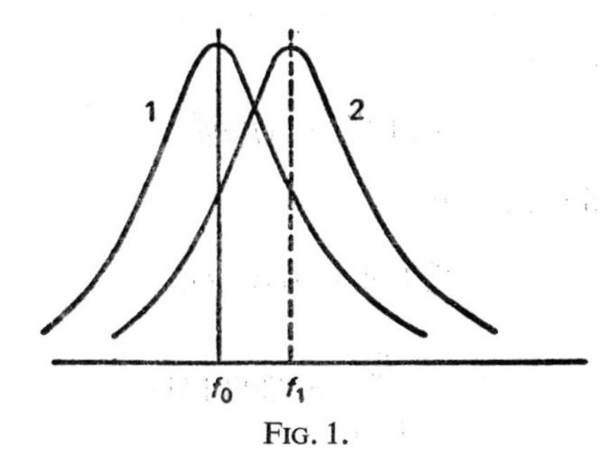

FIG. 1.

is given by

$$\frac{2\delta f}{f_0}Q_L = 1,$$

with

$$Q_L = \frac{Q_0}{1+\alpha} = 2500.$$

Then

$$(f_1 - f_0) = \frac{f_0}{2Q_L} = 1{\cdot}92 \text{ MHz}.$$

If the cavity piston is displaced so that the resonance curve is that shown as (2) in Fig. 2, then, since the source frequency remains at f_0, the power absorbed in the cavity reduces by one-half. We have

$$\mathrm{d}h = -\frac{2b}{2}\left[\left(\frac{2b}{\lambda_0}\right)^2 - 1\right]^{-3/2} \times \frac{4b^2}{c_0^2} f_0 2\, \mathrm{d}f,$$

from which

$$\mathrm{d}h = -h_0^3 \frac{f\,\mathrm{d}f}{c^2}.$$

If $\mathrm{d}f = 1{\cdot}92$ MHz, $\mathrm{d}h = -1{\cdot}6 \times 10^{-3}$ cm.

Since it can be considered that the whole resonance curve will lie within the range 4 df, a movement of 4 dh $\sim$ 1/10 mm will cover the whole curve.

EXERCISE No. 6

An air-filled cavity resonates at $f_0 = 10\,600$ MHz. What will be the resonant frequency when the cavity is filled with a dielectric of relative permittivity $\varepsilon_r = 1{\cdot}63$?

The Q-factor of the empty cavity is $Q_0 = 8200$ and the loss tangent of the dielectric is 10^{-3}. What will be the new Q-factor?

Solution. In the empty cavity

$$\beta_g^2 + \beta_c^2 = \varepsilon_0 \mu_0 \omega_0^2.$$

while in the dielectric filled cavity

$$\beta_g^2 + \beta_c^2 = \varepsilon_0 \mu_0 \varepsilon_r \omega_1^2,$$

so that

$$f_1 = \frac{f_0}{\sqrt{\varepsilon_r}} = 8300 \text{ MHz}.$$

The new Q-factor Q_0' is given by

$$\frac{1}{Q_0'} = \frac{1}{Q_0} + \frac{1}{Q_\varepsilon},$$

where $Q_\varepsilon = \cot \varphi$. Thus

$$\frac{1}{Q_0'} = \frac{1}{Q_0} + \tan \varphi = \frac{1}{8200} + 10^{-3}$$

or

$$Q_0' = 890$$

It will be noticed that $Q_0' \sim Q_\varepsilon$, the losses in the dielectric being much more important than the losses in the cavity walls.

EXERCISE No. 7

We consider a cavity coupled to two identical waveguides propagating the TE_{01} mode. The first guide (1) is matched to the microwave source while the second (2) is terminated in its characteristic impedance Z_0. The power P_0 supplied by the source is constant. (The equivalent circuit is shown in Fig. 1).

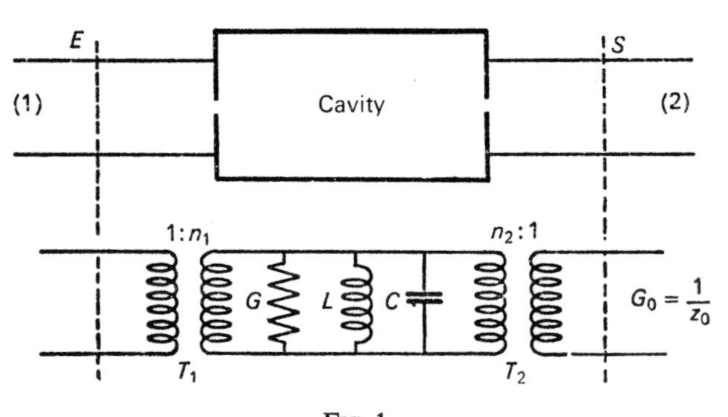

FIG. 1.

1. Find the admittance at the plane E and so deduce the equivalent circuit for this plane.

2. Taking account of the internal impedance of the source and putting

$$\alpha_1 = \frac{G_0}{n_1^2 G}, \quad \alpha_2 = \frac{G_0}{n_2^2 G},$$

calculate the loaded Q-factor, Q_L, of the cavity. Hence deduce the external Q-factors, Q_{E_1} and Q_{E_2}, and express these in terms of the equivalent circuit.

3. Calculate the power dissipated in the system as a function of P_0, α_1, α_2, Q_L and $x = \omega/\omega_0$. What is the form of this expression in the neighbourhood of resonance when $\alpha_1 = \alpha_2 = \alpha$?

4. Is it possible to have matching for both the cases $\alpha_1 \neq \alpha_2$ and $\alpha_1 = \alpha_2$?

5. Calculate the power dissipated in the admittance G_0 (i.e. the power transmitted into the guide (2).)

Solution. 1. The effective admittance across the primary of the transformer T is G_0/n_2^2. The equivalent circuit will thus be as in Fig. 2,

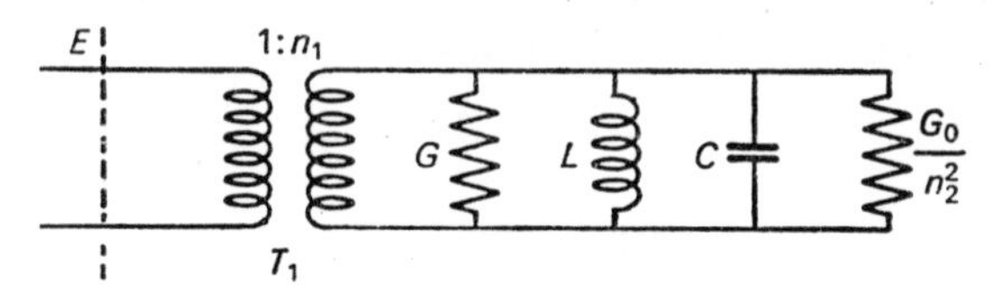

FIG. 2.

The total admittance for the secondary of T_1 is

$$Y = G + \frac{G_0}{n^2} + j\left(\omega C - \frac{1}{\omega L}\right).$$

The admittance across the primary of T is then $Y' = Yn_1^2$ and

$$Y' = n_1^2\left[G + \frac{G_0}{n_2^2} + j\left(\omega C - \frac{1}{\omega L}\right)\right]$$

The equivalent circuit for the input can thus be taken as

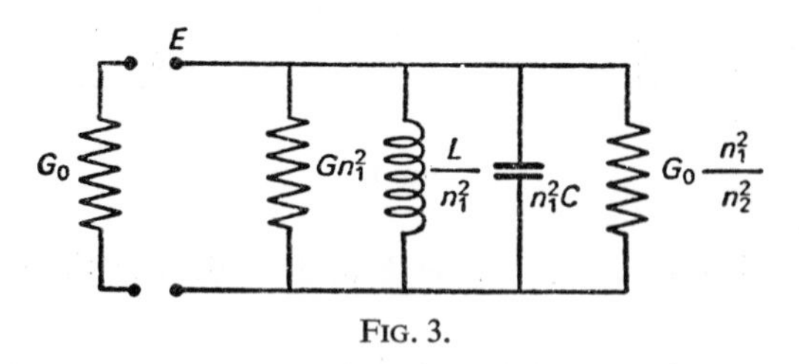

FIG. 3.

2. For the resonant circuit, including the conductance G_0 due to the generator, the Q-factor is

$$Q_L = \frac{n_1^2\omega_0 C}{G_0 + n_1^2 G + G_0\dfrac{n_1^2}{n_2^2}} = \frac{\omega_0 C}{G}\,\frac{1}{1 + \dfrac{G_0}{n_1^2 G} + \dfrac{G_0}{n_2^2 G}} = \frac{Q_0}{1+\alpha_1+\alpha_2}.$$

$Q_0 = \omega_0 C/G$ is the Q-factor of the empty cavity, so that

$$\frac{1}{Q_L} = \frac{1}{Q_0} + \frac{1}{Q_{E_1}} + \frac{1}{Q_{E_2}} = \frac{1}{Q_0} + \frac{\alpha_1}{Q_0} + \frac{\alpha_2}{Q_0}$$

and

$$Q_{E_i} = \frac{Q_0}{\alpha_i} = \frac{n_i^2 \omega_0 C}{G_0}.$$

3. The incident power is P_0 and, if the reflection coefficient in the input plane is Γ, the power dissipated in the cavity and load is

$$P = P_0(1-|\Gamma|^2);$$

$$\Gamma = \frac{1-y'}{1+y'},$$

with

$$y' = \frac{n_1^2}{G_0}\left[G + \frac{G_0}{n_2^2} + \mathrm{j}\left(\omega C - \frac{1}{\omega L}\right)\right].$$

Taking G as a factor gives

$$y' = \frac{1}{\alpha_1}\left[1+\alpha_2+\mathrm{j}Q_0\left(x-\frac{1}{x}\right)\right]$$

which gives

$$\Gamma = \frac{\alpha_1-(1+\alpha_2)-\mathrm{j}Q_0\left(x-\frac{1}{x}\right)}{\alpha_1+(1+\alpha_2)+\mathrm{j}Q_0\left(x-\frac{1}{x}\right)},$$

$$\Gamma\Gamma^* = \frac{[\alpha_1-(1+\alpha_2)]^2+Q_0^2\left(x-\frac{1}{x}\right)}{(1+\alpha_1+\alpha_2)^2+Q_0^2\left(x-\frac{1}{x}\right)^2},$$

so that

$$P_1 = \frac{4P_0\alpha_1(1+\alpha_2)}{(1+\alpha_1+\alpha_2)^2}\,\frac{1}{1+Q_L^2\left(x-\frac{1}{x}\right)^2}.$$

In the neighbourhood of the resonance

$$Q_L\left(x-\frac{1}{x}\right) \approx 2Q_L\frac{\delta\omega}{\omega_0} = u.$$

and, for $\alpha_1 = \alpha_2 = \alpha$,

$$P = \frac{4P_0\alpha(1+\alpha)}{(1+2\alpha)^2}\,\frac{1}{1+u^2}.$$

4. The system will be matched at resonance if $P = P_0, \Gamma = 0$ or $y' = 1$. That is,

$$y'(\omega_0) = \frac{1+\alpha_2}{\alpha_1} = 1.$$

Consequently, if $\alpha_1 = \alpha_2$, there can be no matching at the resonance.

5. The power which is dissipated is divided proportionately among the conductances. In the load the power dissipated is given by

$$\frac{P_2}{P} = \frac{G_0\left(\dfrac{n_1}{n_2}\right)^2}{n_1^2 G+G_0\left(\dfrac{n_1}{n_2}\right)^2} = \frac{\dfrac{G_0}{n_2^2 G}}{1+\dfrac{G_0}{n_2^2 G}} = \frac{\alpha_2}{1+\alpha_2},$$

so that

$$P_2 = \frac{4P_0\alpha_1\alpha_2}{(1+\alpha_1+\alpha_2)^2} \frac{1}{1+Q_L^2\left(x-\dfrac{1}{x}\right)^2}.$$

Where the system is matched at the resonance

$$\alpha_1 = 1+\alpha_2,$$

$$P_2 = P_0 \frac{\alpha_2}{1+\alpha_2} \frac{1}{1+Q_L^2\left(x-\dfrac{1}{x}\right)^2}.$$

EXERCISE No. 8

The end of rectangular waveguide is closed by a perfect short-circuit.

1. At what distance l from this short-circuit must another short-circuit be placed if the resulting cavity is to resonate at a frequency f

(a) in the case where the guide is lossless?

(b) in the case where there are losses in the guide but $\alpha l \ll 1$? What is the cavity conductance in each case?

2. If, in fact, the second short-circuit is an iris which acts as a negative susceptance $B = -1/\omega L$ as in Fig. 1, what is the new resonance condition in the lossless case? If, depending on the diameter of the iris, the value $|B|/Y_0$ varies between 2 and 400, what will be the range of values of l?

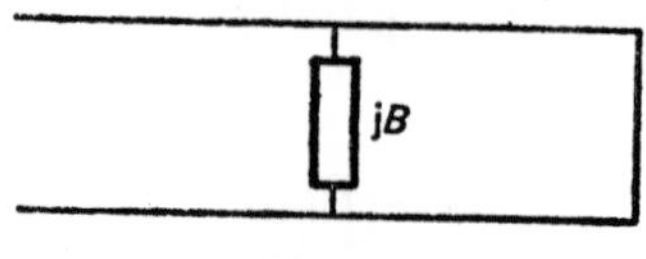

FIG. 1.

3. If we are in a region where $|B|/Y_0 \gg 1$ so that $l = p(\lambda_g/2) - \varepsilon(\lambda_g/2\pi)$ where $\alpha l \ll \varepsilon \ll 1$, what relation is there between ε and B, and what is the cavity conductance?

Solution. 1. (a) Lossless guide: the admittance due to a short-circuit at a distance l is $Y = -jY_0 \cot \beta_g l$. The second short-circuit will be placed at a position l such that $Y(l) = \infty$ or

$$\beta_g l = p\pi, \qquad l = p\frac{\lambda_g}{2}.$$

(b) When the guide is lossy

$$Y = Y_0 \coth \gamma l = Y_0 \frac{l + j \tanh \alpha l \tan \beta_g l}{\tanh \alpha l + j \tan \beta_g l}.$$

With $\alpha l \ll 1$

$$Y = Y_0 \frac{1 + j\alpha l \tan \beta_g l}{\alpha l + j \tan \beta_g l}.$$

For $\beta_g l = p\pi$ we will have $Y = Y_0/\alpha l$ which will be very large: the condition for resonance is approximately the same and, in both cases, the susceptance is zero.

2. In the case where the iris has a finite susceptance and $\alpha = 0$, the admittance in the plane of the iris will be

$$Y_t = Y + jB = j(B - Y_0 \cot \beta_g l),$$

so that, for resonance

$$\cot \beta_g l = \frac{B}{Y_0} = -\frac{|B|}{Y_0}.$$

For $\quad \dfrac{|B|}{Y_0} = 2 \quad$ we have $\quad l = 0{\cdot}425\, \lambda_g$,

while for $\quad \dfrac{|B|}{Y_0} = 400, \quad l = 0{\cdot}5\, \lambda_g$.

Thus, if $|B| \gg Y_0$, the cavity is about $\lambda_g/2$ in length.

3. If $l = p\lambda_g/2 - \varepsilon\lambda_g/2\pi$, $\beta_g l = p\pi - \varepsilon$ and

$$Y = Y_0 \frac{1 - j\alpha l\varepsilon}{\alpha l - j\varepsilon} \simeq Y_0 \frac{1}{\alpha l - j\varepsilon},$$

so that

$$Y + jB = Y_0 \frac{1}{\alpha l - j\varepsilon} + jB$$

$$= Y_0 \frac{\alpha l}{(\alpha l)^2 + \varepsilon^2} + jY_0 \frac{\varepsilon}{(\alpha l)^2 + \varepsilon^2} + jB.$$

There will be a resonance if

$$\frac{B}{Y_0} = -\frac{\varepsilon}{(\alpha l)^2+\varepsilon^2}.$$

In particular, for $\alpha l \ll \varepsilon$,

$$\frac{B}{Y_0} \simeq -\frac{1}{\varepsilon}.$$

As $B < 0$, ε is positive,

(For the case of an inductive iris the cavity will be slightly shorter than $p\lambda_g/2$.)

The cavity conductance is, finally,

$$G = Y_0 \frac{\alpha l}{(\alpha l)^2+\varepsilon^2} \simeq \frac{B^2 \alpha l}{Y_0}.$$

PART TWO

Problems

Problem No. 1

PUSH–PULL AND DISTORTION

I

It is known that the Fourier series for the anode current of a pentode used to amplify a strong signal $E_g \cos \omega t$ is of the form:

$$i_a = I_{a0}+A_0+A_1 \cos \omega t+A_2 \cos 2\,\omega t+A_3 \cos 3\,\omega t+A_4 \cos 4\,\omega t.$$

If $i_{a\,\max}$, $i_{a,\,1/2}$, $I_{a,0}$, $i_{a,\,-1/2}$, $i_{a\,\min}$ are the currents which correspond to

$$\omega t = 0, \frac{\pi}{3}, \frac{\pi}{2}, \frac{2\pi}{3}, \text{ and } \pi,$$

find A_0 to A_4 in terms of these five values.

II

Being given the static characteristics of a 6L6 pentode shown in Fig. 1, this pentode is considered to be used in a class A amplifier with:

anode potential = 250 V,
screen potential = 250 V,
bias potential = −15 V.

The load is formed by a transformer having an effective primary impedance of 2·3 kΩ.

1. Sketch the dynamic transfer characteristic of the system.

2. Calculate the distortion introduced when this tube amplifies a signal with an amplitude of 15 V. (Neglect the effect of the fourth harmonic.)

3. At what signal amplitude does the distortion fall to 6% and what power is then delivered by the transformer?

4. Two 6L6 pentodes are used in a symmetric class A_1 amplifier, each being biassed at −15 V. Sketch the composite characteristics of the amplifier.

What is the optimum load, $R_{aa}/4$, which gives the best compromise between the power output and the distortion?

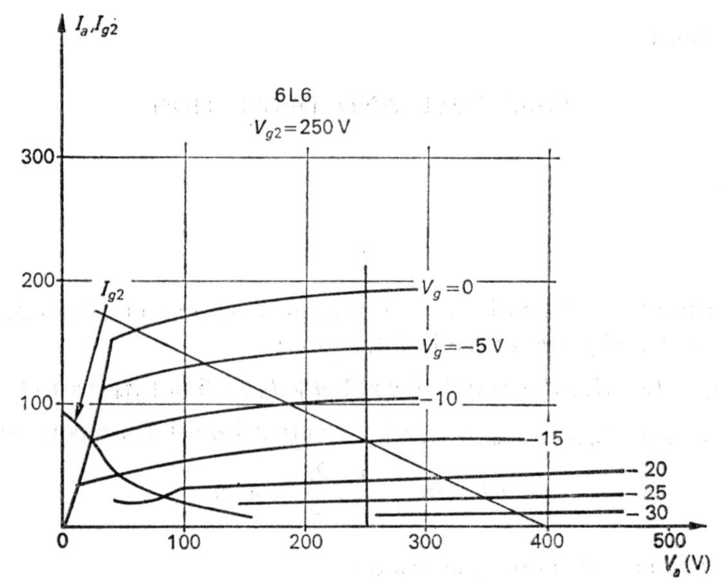

FIG. 1.

5. With an anode-to-anode load of 5600 Ω find:

(a) the distortion introduced when the signal has an amplitude of 15 V,

(b) the output power,

(c) the efficiency of the amplifier. (Ignore the power consumed by the heaters and assume that the screen current is 3·6 mA.)

6. The two valves are biassed by means of a common cathode resistance R_k. Find R_k if the bias is to be obtained in the quiescent condition and the new value of the H.T. potential for the characteristics and results to remain valid.

7. The load is actually a loud-speaker of 7 Ω resistance. What must be the transformer used if it is assumed that there is no power loss in the transformer?

8. If there is a negative feedback of 30% of the output voltage, what will be the new value of the distortion?

(Nantes, June 1962)

Solution

I

If $\omega t = 0$, then

$$i_{a\,\max} = I_{a0} + A_0 + A_1 + A_2 + A_3 + A_4. \tag{1}$$

If $\omega t = \pi/3$,

$\cos \omega t = \frac{1}{2}, \quad \cos 2\omega t = -\frac{1}{2}, \quad \cos 3\omega t = -1, \quad \cos 4\omega t = -\frac{1}{2},$ so that

$$i_{a,\,1/2} = I_{a0} + A_0 + \frac{A_1}{2} - \frac{A_2}{2} - A_3 - \frac{A_4}{2}. \tag{2}$$

If $\omega t = \pi/2$,

$$\cos \omega t = 0, \quad \cos 2\omega t = -1, \quad \cos 3\omega t = 0, \quad \cos 4\omega t = 1,$$

and

$$I_{a0} = I_{a0}+A_0-A_2+A_4. \tag{3}$$

If $\omega t = 2\pi/3$,

$$\cos \omega t = -\tfrac{1}{2}, \quad \cos 2\omega t = -\tfrac{1}{2}, \quad \cos 3\omega t = 1, \quad \cos 4\omega t = -\tfrac{1}{2},$$

and

$$i_{a,\,-1/2} = I_{a0}+A_0-\frac{A_1}{2}-\frac{A_2}{2}+A_3-\frac{A_4}{2}. \tag{4}$$

Finally, if $\omega t = \pi$,

$$\cos \omega t = -1, \quad \cos 2\omega t = 1, \quad \cos 3\omega t = -1, \quad \cos 4\omega t = 1;$$

$$i_{a\,\min} = I_{a0}+A_0-A_1+A_2-A_3+A_4. \tag{5}$$

We have thus five equations and five unknowns. The solution using determinants is long-winded and it is better to reduce the equations systematically. For example, (1)+(2)+(3)+(4)+(5) gives, on putting $B_1 = i_{a\,\max}-I_{a0}$, $B_2 = i_{a,\,1/2}-I_{a,\,0}$ etc.,

$$B_1+B_2+B_4+B_5 = 5A_0+2A_4 \tag{6}$$

where $B_3 = 0$. Also

$$(1)+(5) \rightarrow \frac{B_1+B_5}{2} = A_0+A_2+A_4 \tag{7}$$

$$\frac{(1)}{2}+(2) \rightarrow \frac{B_1}{2}+B_2 = \frac{3}{2}A_0+A_1-\frac{A_3}{2}. \tag{8}$$

The simpler set of equations is thus

$$B_1 = A_0+A_1+A_2+A_3+A_4 \tag{1}$$

$$\frac{B_1}{2}+B_2 = \frac{3}{2}A_0+A_1-\frac{A_3}{2} \tag{8}$$

$$B_1+B_2+B_4+B_5 = 5A_0+2A_4 \tag{6}$$

$$\frac{B_1+B_5}{2} = A_0+A_2+A_4 \tag{7}$$

$$0 = A_0-A_2+A_4 \tag{3}$$

(3)+(7) gives

$$\frac{B_1+B_5}{2} = 2A_0+2A_4 \tag{9}$$

and (6)−(9) gives

$$\frac{B_1}{2}+B_2+B_4+\frac{B_5}{2} = 3A_0.$$

Thus

$$A_0 = \frac{B_1+B_5}{6}+\frac{B_2+B_4}{3}$$

$$= \frac{i_{a\,\max}+i_{a\,\min}}{6}+\frac{i_{a,\,1/2}+i_{a,\,-1/2}}{3}-I_{a0}.$$

Then (9) gives

$$A_4 = \frac{i_{a\,\max}+i_{a\,\min}}{12}-\frac{i_{a,\,1/2}+i_{a,\,-1/2}}{3}+\frac{I_{a0}}{2}$$

and, substituting in the other equations, we have

$$A_1 = \tfrac{1}{3}(i_{a\,\max}-i_{a\,\min})+\tfrac{1}{3}(i_{a,\,1/2}-i_{a,\,-1/2})$$
$$A_2 = \tfrac{1}{4}(i_{a\,\max}+i_{a\,\min})-\tfrac{1}{2}I_{a0}$$
$$A_3 = \tfrac{1}{6}(i_{a\,\max}-i_{a\,\min})-\tfrac{1}{3}(i_{a,\,1/2}-i_{a,\,-1/2}).$$

II

1. The working point is defined at

$$V_a = 250\text{ V} \quad \text{and} \quad V_g = -15\text{ V}$$

which, on Fig. 1, corresponds to $i_a = 67$ mA.

The load-line through this point will have a slope

$$-\frac{1}{R_L} = -\frac{1}{2{\cdot}3\times 10^3}.$$

From this line we find the curve $I_a = f(V_g)$ shown in Fig. 2.

2. In order to find the distortion we must find from the characteristic the values of the currents $i_{a\,\min}$, $i_{a\,\max}$, $i_{a,\,1/2}$, $i_{a,\,-1/2}$.

Now $i_{a\,\max}$ corresponds to $V_g = -15+15 = 0$ V. The curve thus gives

$$i_{a\,\max} = 157{\cdot}5\text{ mA}.$$

$i_{a\,\min}$ corresponds to $V_g = -30$ V. That is, $i_{a\,\min} = 10$ mA.

$i_{a,\,1/2}$ is at $\omega t = \frac{\pi}{3}$, $\cos\omega t = \frac{1}{2}$ or $V_g = -15+\frac{15}{2} = -7{\cdot}5$ V.

Then

$$i_{a,\,1/2} = 112{\cdot}5\text{ mA}.$$

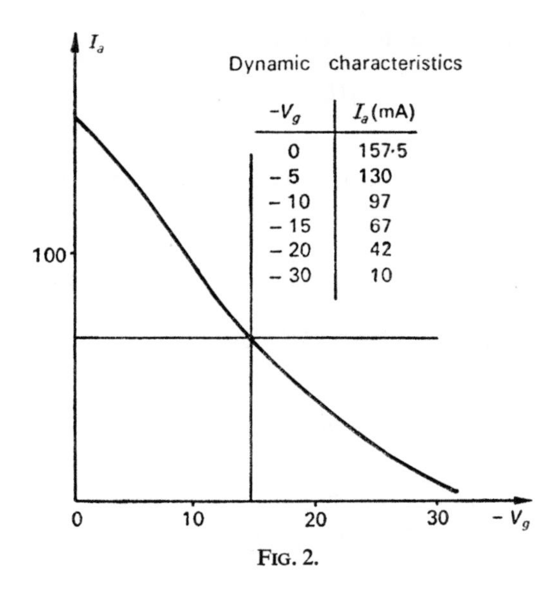

FIG. 2.

Similarly, $i_{a,\,-1/2}$ occurs at $V_g = -15-\dfrac{15}{2} = -22{\cdot}5$ V and

$$i_{a,\,-1/2} = 32{\cdot}5 \text{ mA}.$$

Using the equations of I above we have

$$A_1 = 75{\cdot}5 \text{ mA},$$
$$A_2 = 8{\cdot}5 \text{ mA},$$
$$A_3 = -2 \text{ mA},$$

from which the distortion in the second and third harmonics is

$$D_2 = \left|\frac{A_2}{A_1}\right| \to 11{\cdot}25\%,$$

$$D_3 = \left|\frac{A_3}{A_1}\right| \to 2{\cdot}65\%.$$

The total distortion is

$$D = \sqrt{D_2^2+D_3^2} = 11{\cdot}55\%.$$

3. To obtain a rigorous solution to this part it would be necessary to calculate D for different input amplitudes U_0 and then to sketch

$$D = f(U_0).$$

The graphical solution would then be found where

$$f(U_0) = 6\%.$$

In fact, for $U_0 = 5$ V, we find

$$I_{a0} = 67 \text{ mA}, \quad i_{\max} = 97{\cdot}5 \text{ mA}, \quad i_{a\,\min} = 42{\cdot}5 \text{ mA}, \quad i_{a,1/2} = 82{\cdot}5 \text{ mA},$$
$$i_{a,-1/2} = 55 \text{ mA},$$

from which,

$$A = 27{\cdot}5 \text{ mA}, \quad A_2 = 1{\cdot}5 \text{ mA} \quad \text{and} \quad A_3 = 0{\cdot}13 \text{ mA}.$$

The value of D is 5·5%, which is very close to the value 6% required. The output power will be

$$P = \frac{1}{2} \cdot \left(\frac{V_{\max} - V_{\min}}{2}\right)\left(\frac{I_{a\,\max} - I_{a\,\min}}{2}\right)$$

with $V_{\max} = 302{\cdot}5$ V and $V_{\min} = 185{\cdot}5$ V as read from the load-line. Then

$$P = 0{\cdot}825 \text{ W}.$$

4. We want the composite characteristics of the push–pull system. This is obtained by placing the characteristics of the two valves head-to-tail as shown in Fig. 3, the two being symmetric about the point $V_a = 250$ V on the voltage axis.

The points M_1 and M_2 on the two curves $V_g = 15$ V are the quiescent points of the two valves. If an input signal of $+15$ V is fed to the grid of the first valve P_1, the potential of this grid becomes $-15+15 = 0$, while the grid of P_2 becomes, because of the push–pull connection, $-15-15 = -30$ V.

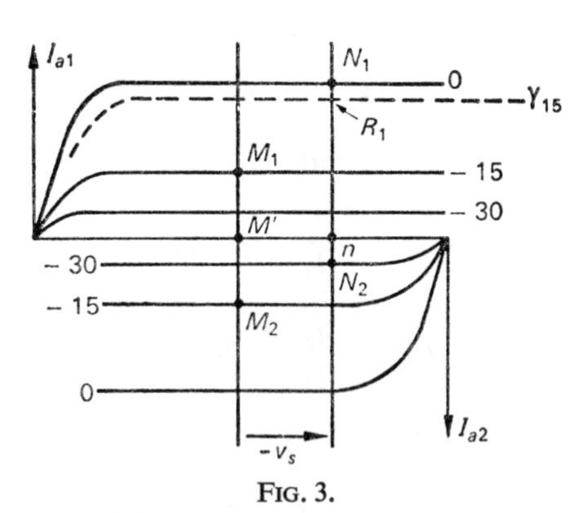

FIG. 3.

The instantaneous voltage on the anode of P_1 depends on the load, but, if it takes a value $(250\text{ V}+v_S)$, the potential on the anode of P_2 will necessarily be $(250\text{ V}-v_S)$ because of the perfect coupling between the two half-windings of the transformer primary. The working point of P_1 will be at N_1 and of P_2 at N_2.

The composite characteristic of the system is

$$i = f(v_S)$$

which is the useful current given and, for constant exciting voltage,

$$i = i_{a_1} - i_{a_2} = f(v_S) \text{ constant excitation.}$$

A point R_1 on this curve will be obtained by taking the difference of the lengths

$$nN_1 - nN_2.$$

It is thus sufficient to take the points R from Fig. 3 for the various values of v_S. The consideration of exciting voltages 15 V, 10 V, 5 V, etc., then gives the series of curves shown in Fig. 4.

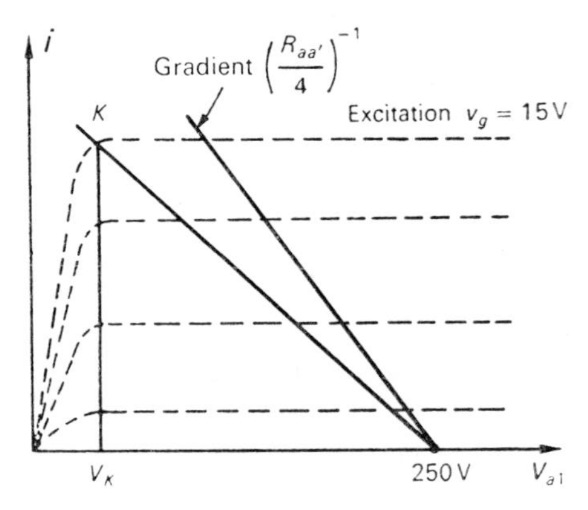

FIG. 4.

We know that the variation v_S of the anode potential is

$$v_S = (i_{a_1} - i_{a_2}) \frac{R_{aa'}}{4}.$$

The variation of V_a with the excitation potential V_g is thus given by drawing a load-line passing through M' and having slope $4/R_{aa'}$.

In the present case the static bias is -15 V so that the maximum signal amplitude is 15 V if the grid is not to go positive. There will be the minimum of distortion for the maximum signal amplitude when the load-line passes through the knee of the curve at K. If the curve is drawn precisely we find

$$V_k = 42 \text{ V}, \quad I_k = 147{\cdot}5 \text{ mA},$$

so that

$$\frac{R_{aa'}}{4} = \frac{250-42}{147{\cdot}5 \times 10^{-3}} \simeq 1400\ \Omega.$$

5. For the first valve we can write

$$i_{a_1} = I_{a0}+A_0+A_1 \cos \omega t+A_2 \cos 2\omega t+A_3 \cos 3\omega t+A_4 \cos 4\omega t,$$

while, for the second, the phase is changed by 180° to give

$$\begin{aligned} i_{a_2} &= I_{a0}+A_0+A_1 \cos (\omega t+\pi)+A_2 \cos 2(\omega t+\pi)+A_3 \cos 3(\omega t+\pi)+A_4 \cos 4(\omega t+\pi) \\ &= I_{a0}+A_0-A_1 \cos \omega t+A_2 \cos 2\omega t-A_3 \cos 3\omega t+A_4 \cos 4\omega t. \end{aligned}$$

The output current is thus

$$i = i_{a_1}-i_{a_2} = 2A_1 \cos \omega t+2A_3 \cos 3\omega t,$$

so that there is only distortion due to the third harmonic.

For $\omega t = 0$ $\qquad i_{max} = 2A_1+2A_3.$

For $\omega t = \dfrac{\pi}{3}$ $\qquad i_{1/2} = A_1-2A_3.$

Then,

$$A_1 = \tfrac{1}{3}(i_{max}+i_{1/2}),$$

$$A_3 = \tfrac{1}{6} i_{max}-\tfrac{1}{3} i_{1/2}$$

and

$$D_3 = 100\left|\frac{A_3}{A_1}\right|.$$

From the composite curves we find

$$i_{max} = 147{\cdot}5 \text{ mA},$$

$$i_{1/2} = 77{\cdot}5 \text{ mA},$$

from which

$$D_3 = 1{\cdot}67\%.$$

The available a.c. power is

$$P = \frac{250 \text{ V}-V_{a\,min}}{\sqrt{2}} \times \frac{i_{max}}{\sqrt{2}} = \frac{250-40}{\sqrt{2}} \times \frac{147{\cdot}5\times 10^{-3}}{\sqrt{2}} = 15{\cdot}5 \text{ W}.$$

The d.c. supply gives a standing current of 67 mA to each valve plus the screen current. Thus, with the 250 V d.c. voltage,

$$P_{DC} = 2\times 250\times(67+3{\cdot}6)\times 10^{-3} = 35{\cdot}3 \text{ W}.$$

The efficiency is thus

$$\eta = \frac{P}{P_{CD}} = \frac{15{\cdot}5}{35{\cdot}3} \simeq 44\%.$$

6. There must be a potential drop of 15 V across R_k. The mean total current producing this is

$$2I_{a0}+2I_{g2} = 2(67+3{\cdot}6)\times 10^{-3}\text{ A}$$

and

$$R_k = \frac{15}{2\times 70{\cdot}6\times 10^{-3}} = 106\ \Omega.$$

The H.T. supply must be increased by 15 V so that

$$\text{H.T.} = 265\text{ V.}$$

7. If there are n_1 turns on one half of the primary, then

$$R_{aa'} = \left(\frac{n_1}{n_2}\right)^2 R_L$$

and with $R_{aa'} = 5600\ \Omega$, $R_L = 7\ \Omega$,

$$\frac{n_1}{n_2} \simeq 28.$$

For a power of 15·5 W, the voltage on the loud-speaker must be

$$V_{\text{rms}} = \sqrt{P\times R_L} = 1{\cdot}04\text{ V,}$$

corresponding to an amplitude

$$V_{\text{max}} = V_{\text{rms}}\sqrt{2} = 14{\cdot}7\text{ V.}$$

8. The gain in voltage of the system is

$$G_v = \frac{14{\cdot}7}{15}$$

and, with a negative feedback of 0·3,

$$1+\beta G_v = 1+0{\cdot}3\times\frac{14{\cdot}7}{15} = 1{\cdot}294,$$

so that the distortion of 1·67% is reduced to

$$D = \frac{D_0}{1+\beta G_v} = 1{\cdot}29\%.$$

Problem No. 2

TRANSISTOR DIFFERENTIAL AMPLIFIER

Most cathode ray tubes require a symmetric supply; i.e. the two plates must receive voltages which have opposite phase. Here it is proposed to study a vertical oscilloscope amplifier which satisfies this condition, the principle of the circuit being shown in Fig. 1. The transistor T_1 acts as a preamplifier which is coupled directly to supply the transistors T_2 and T_3 which, in the common-emitter connection, produce signals automatically of opposite phase. The two deflector plates

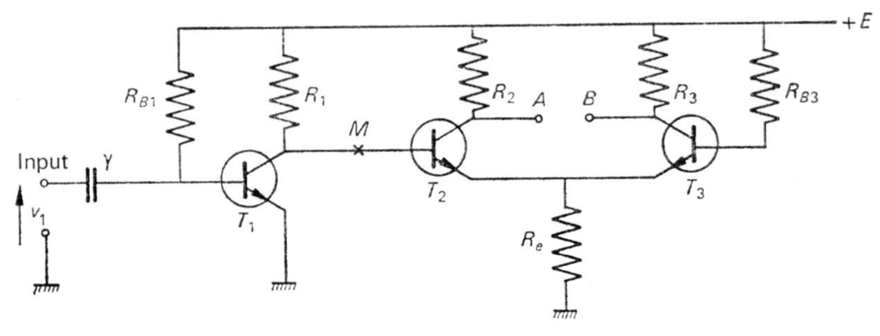

FIG. 1.

are connected to the points A and B. We will consider the conditions for the a.c. potentials at these points to be equal and opposite.

The two fundamental parameters of the first transistor will be taken as β and h_{11}, while those for the identical pair T_2 and T_3 are β' and h'_{11}. Further, any loads are taken small compared to $1/h_{22}$ so that approximate formulae can be used.

1. *First stage alone*

(a) The coupling between the collector of T_1 and the base of T_2 is cut at M so that the first stage is isolated. A low-frequency generator of e.m.f. v_1 (and negligible internal impedance) is connected to the input. An oscilloscope with a high input impedance allows the a.c. output v_2 on the collector to be measured. If a resistance R is inserted between the input terminals and the L.F. generator, v_1 being unchanged, this output is found to become $v_2/(k+1)$. (Assume that, at the frequency considered the condenser γ is effectively a short-circuit.) Calculate β and h_{11} in terms of R, R_1, v_2/v_1 and k.

As a numerical example take: $v_2/v_1 = -100$, $R_1 = 1\ \text{k}\Omega$ and $k = 2$ for $R = 2\ \text{k}\Omega$.

(b) Neglecting the base-emitter voltage of T_1 compared with the supply voltage E, express the collector current I_1 as a function of β, E and R_{B_1}.

(c) When R_{B_1} is varied, it is found that v_2/v_1 and k are inversely proportional to this resistance as long as there is no saturation. Put

$$\frac{v_2}{v_1} = \frac{A}{R_{B_1}} \quad \text{and} \quad k = \frac{B}{R_{B_1}}$$

and deduce the form of the variation of β and h_{11} with I_1. If $R_1 = 1\ \text{k}\Omega$, $R = 2\ \text{k}\Omega$, $A = 2\times10^7$, $B = 4\times10^5$ and $E = 20$ V, what value of R_{B_1} was used in 1(a) above?

2. *Study of the biassing of the output stage.*

In this part we are concerned with the d.c. potentials on the collectors and emitters of the system T_2, T_3. As a first approximation it is supposed that $V_{BE} = 0$ for both T_2 and T_3 which are neither cut-off nor saturated.

(a) Calculate the collector currents I_2 and I_3 in terms of the circuit resistances and the base voltage V_0 of T_2 and sketch I_2 and I_3 graphically as functions of V_0 with $0 \leqslant V_0 \leqslant E$.

What are the limits of V_0 if neither transistor is to be cut-off or saturated? (Assume β' is large enough that $\beta' \simeq \beta'+1$.)

(Note. When one of the transistors is cut-off, its base-emitter voltage is no longer zero but, rather, negative. Account should be taken of this fact in drawing the real variations of I_2 and I_3 for V between 0 and E.)

(b) Find R_{B_3}, in terms of R_2, R_3 and R_E, if we require $V_A = V_B$ for a given value of V_0.

(c) It will be seen from the graph that, if V_0 varies by ΔV_0, the variations ΔI_2 and ΔI_3 of the two collector currents will be of opposite signs and the two voltage variations on the collectors will be of opposite phase. What must be the relation between R_{B_1}, R_2 and R_3 for

$$|\Delta V_A| = |\Delta V_B|\ ?$$

(d) Is it possible to satisfy the two above conditions simultaneously? (Note: calculate R_{B_3}.)

(e) The collector of T_1 and the base of T_2 are reconnected at M and it is assumed that the base current of T_2 is much less than I_1. What numerical relationship must exist between R_{B_3}, R_e and R_3/R_2 if the static voltages at V_A and V_B are to be equal? (Take $\beta' = 200$ and R_{B_1} to have the value found in 1(a).)

3. *Dynamic condition.*

(a) The two transistors T_2 and T_3 will be considered as current generators with gain β' and impedance h_{11} (i.e. h_{12} and h_{22} are taken to be zero). Taking the circuit to be exactly as in Fig. 1 find the two gains

$$G_A = \frac{dV_A}{dV_0}, \quad G_B = \frac{dV_B}{dV_0} \quad \text{and also} \quad p = \frac{G_A}{G_B}.$$

(b) A large condenser is placed between the base of T_3 and earth. Calculate G_A, G_B and p for this case.

(c) If the condition established in 2(e) above is satisfied so that, in the static position, we have, effectively, $V_A = V_B = 15$ V, and knowing that $h'_{11} = 1$ kΩ in the region of the mean bias of T_2 and T_3, calculate R_2, R_3 and R_{B_3} for $R_e = 1$ kΩ.

(d) What is the input impedance of T_2 when the decoupling capacity is placed between the base of T_3 and earth?

Calculate the useful amplifier gain given by

$$G = \frac{\partial(V_A - V_B)}{\partial V_1}.$$

If the cathode ray tube used has a displacement sensitivity of $\frac{1}{3}$ cm per volt potential between the deflector plates, what will be the peak-to-peak amplitude of the sine-wave observed on the screen when a 0·1 mV_{rms} signal is applied at the input?

Solution. 1. (a) For the condition $R_1 \ll 1/h_{22}$, the voltage gain is

$$\frac{v_2}{v_1} = -\frac{\beta R_1}{h_{11}}. \tag{1}$$

When a resistance R is added in series with the input, the potential supplied to the base is reduced to

$$v'_1 = v_1 \frac{h_{11}}{h_{11}+R} \tag{2}$$

and, with the gain unchanged,

$$v'_2 = v_2 \frac{h_{11}}{h_{11}+R} = \frac{v_2}{k+1} \tag{3}$$

with

$$k = \frac{R}{h_{11}} \quad \text{or} \quad h_{11} = \frac{R}{k}. \tag{4}$$

Substitution in (1) gives

$$\beta = -\frac{R}{R_1}\frac{v_2}{v_1}\frac{1}{k} \tag{5}$$

which, with the numerical values given, leads to

$$h_{11} = 1\ \text{k}\Omega \quad \text{and} \quad \beta = 100.$$

(b) The base current is simply $i_B = E/R_{B_1}$ and so

$$I_1 = \frac{\beta E}{R_{B_1}}. \tag{6}$$

(c) We have

$$\frac{v_2}{v_1} = \frac{\beta R_1}{h_{11}} = \frac{A}{R_{B_1}} \tag{7}$$

$$k = \frac{R}{h_{11}} = \frac{B}{R_{B_1}} \tag{8}$$

and, from (6)

$$R_{B_1} = \frac{\beta E}{I_1}. \tag{6}$$

Substituting this value of R_{B_1} in (7) and (8) gives two equations:

$$\beta^2 R_1 E = A h_{11} I \tag{9}$$

$$\beta E R = B h_{11} I. \tag{10}$$

Division of the two sides of (9) by those of (10) gives

$$\beta = \frac{A}{B}\cdot\frac{R}{R_1} \quad \text{(independent of } I_1\text{)} \tag{11}$$

and so

$$h_{11} = \frac{A}{B^2}\cdot\frac{R^2}{R_1}\cdot\frac{E}{I_1}, \tag{12}$$

which is inversely proportional to the collector current.

With the numerical values given

$$\beta = 100 \quad \text{and} \quad h_{11} = \frac{10}{I_1}.$$

In (a) we had found $h_{11} = 10^3\ \Omega$ so that $I_1 = 10$ mA and $R_{B_1} = 200$ kΩ.

2. (a) Since we put $V_{BE} = 0$, the emitter voltage of the two transistors is V_0.

The potential difference across R_{B_3} is thus $(E-V_0)$ so that the current I_3 is

$$I_3 = \beta' \frac{E-V_0}{R_{B_3}}. \tag{13}$$

The resistance R_E carries a current $I_2+I_{B_2}+I_3+I_{B_3}$ so that, neglecting the base current compared to the collector current since $\beta' \gg 1$,

$$V_0 = R_e(I_2+I_3), \tag{14}$$

or

$$I_2 = \frac{V_0}{R_e} - I_3. \tag{15}$$

Together with (13) this gives

$$I_2 = V_0\left[\frac{1}{R_e}+\frac{\beta'}{R_{B_3}}\right] - \frac{\beta' E}{R_{B_3}}. \tag{16}$$

This can be represented graphically as in Fig. 2.

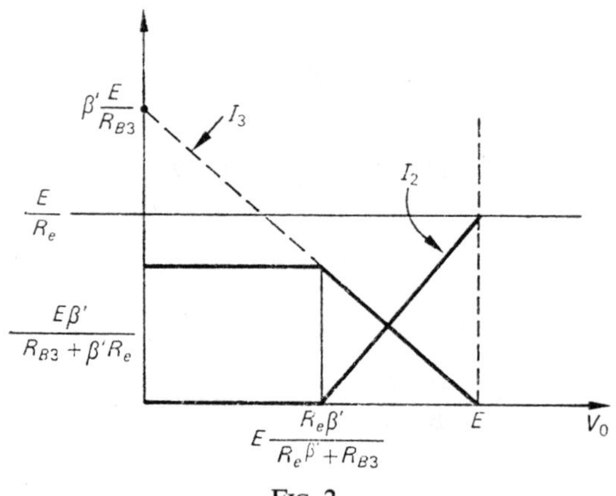

FIG. 2.

For the region

$$V_0 < E\frac{R_e\beta'}{R_e\beta'+R_{B_3}}$$

T_2 will be cut-off: the emitter voltage is no longer V_0 and the calculation must be repeated for this case. The circuit reduces, in this case, to that of Fig. 3. Taking $(\beta'+1) = \beta'$ gives

$$V_E = R_e\beta' I_{B_3} \tag{17}$$

$$I_{B_3} = \frac{E-V_E}{R_{B_3}} = \frac{E-R_e\beta' I_{B_3}}{R_{B_3}}. \tag{18}$$

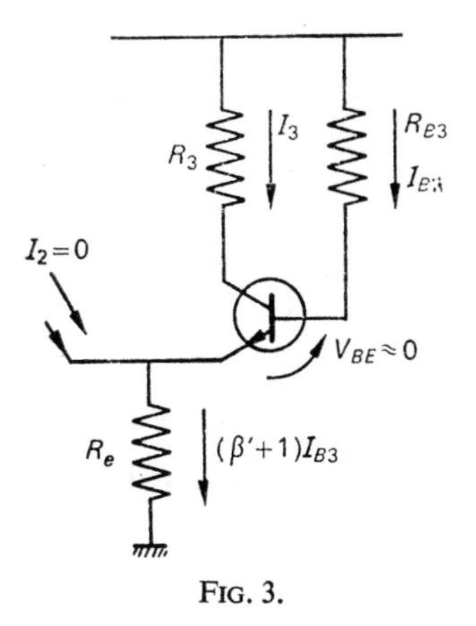

FIG. 3.

Thus

$$I_3 = \beta' \frac{E}{R_{B_3}+\beta' R_e}, \tag{19}$$

which is independent of V_0. In the limit

$$V_0 = E\frac{R_e\beta'}{R_e\beta'+R_{B_3}}, \tag{20}$$

and equation (13) gives (19) directly.

(b) $V_A = E-R_2I_2 = V_B = E-R_3I_3$, which simply requires

$$R_2I_2 = R_3E_3.$$

Thus, from (13) and (16)

$$\frac{1}{R_{B_3}}[\beta'(E-V_0)(R_2+R_3)] = \frac{R_2}{R_e}\cdot V_0,$$

from which

$$R_{B_3} = \beta'\frac{R_e}{V_0}(E-V_0)\left(1+\frac{R_3}{R_2}\right). \tag{21}$$

(c) $|\Delta V_A| = R_2|\Delta I_2| = |\Delta V_B| = R_3|\Delta I_3|$. Here, from (16),

$$\Delta I_2 = \left(\frac{1}{R_e}+\frac{\beta'}{R_{B_3}}\right)\Delta V_0,$$

while, from (13),

$$\Delta I_3 = -\frac{\beta'}{R_{B_3}}\Delta V_0.$$

The required condition is thus

$$R_2\left(\frac{1}{R_e}+\frac{\beta'}{R_{B_3}}\right) = R_3\frac{\beta'}{R_{E_3}},$$

or

$$\frac{R_3}{R_2} = 1+\frac{R_{B_3}}{\beta' R_e}. \tag{22}$$

(d) We must simultaneously have

$$R_{B_3} = \beta' \frac{R_e}{V_0}(E-V_0)\left(1+\frac{R_3}{R_2}\right) = \beta' R_e\left(\frac{R_3}{R_2}-1\right),$$

which gives, on rearranging,

$$-ER_{B_3} = 2\beta' R_e(E-V_0).$$

This would require $R_{B_3} < 0$, which is not possible. The conditions (b) and (c) cannot therefore be satisfied at the same time.

(e) Since for the first stage, I_1 was 10 mA

$$V_0 = E-R_1 I_1 = 10\text{ V}$$

and the condition (21) is, numerically,

$$R_{B_3} = 200\,R_e\left(1+\frac{R_3}{R_2}\right).$$

3. (a) The equivalent circuit for the transistor can be taken as in Fig. 4 and the

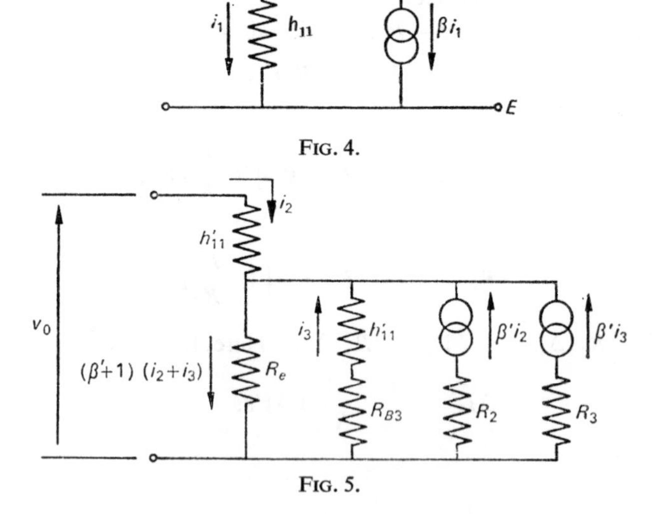

FIG. 4.

FIG. 5.

combined circuit for T_2 and T_3 can be replaced, therefore, by that of Fig. 5, where we can write

$$\begin{cases} v_0 = h'_{11} i_2 + R_e(\beta'+1)(i_2+i_3) = i_2[h'_{11}+R_e(\beta'+1)]+R_e(\beta'+1)i_3 \\ v_0 = h'_{11}\, i_2-(h'_{11}+R_{B_3})i_3, \end{cases}$$

which equations give

$$\begin{cases} i_2 = \dfrac{v_0[h'_{11}+R_{B_3}+R_e(\beta'+1)]}{h'_{11}R_e(\beta'+1)+(R_{B_3}+h'_{11})[h'_{11}+R_e(\beta'+1)]} & (23) \\ i_3 = \dfrac{-v_0[R_e(\beta'+1)]}{h'_{11}R_e(\beta'+1)+(R_{B_3}+h'_{11})[h'_{11}+R_e(\beta'+1)]}. & (24) \end{cases}$$

The output voltages are

$$\begin{cases} v_A = -R_2\beta' i_2 \\ v_B = -R_3\beta' i_3, \end{cases}$$

so that

$$\begin{cases} G_A = \dfrac{-R_2\beta'[h'_{11}+R_{B_3}+R_e(\beta'+1)]}{\Delta} & (25) \\ G_B = \dfrac{R_3\beta'[R_e(\beta'+1)]}{\Delta} & (26) \end{cases}$$

where Δ is the same denominator as in (23) and (24). Then

$$p = -\frac{R_2}{R_3}\,\frac{h'_{11}+R_{B_3}+R_e(\beta'+1)}{R_e(\beta'+1)}. \tag{27}$$

(b) With R_{B_3} decoupled the circuit will behave as if $R_{B_3}=0$ in the equivalent circuit. The preceding equations (25) to (27) then give

$$\begin{cases} G_A = -\dfrac{R_2\beta'[h'_{11}+R_e(\beta'+1)]}{h'_{11}[h'_{11}+2R_e(\beta'+1)]} & (28) \\ G_B = +\dfrac{R_3\beta' R_e(\beta'+1)}{h'_{11}[h'_{11}+2R_e(\beta'+1)]} & (29) \end{cases}$$

$$p = -\frac{R_2}{R_3}\left[1+\frac{h'_{11}}{R_e(\beta'+1)}\right]. \tag{30}$$

(c) The first relation is

$$R_{B_3} = 2\times 10^5\left(1+\frac{R_3}{R_2}\right),$$

while the second is

$$\frac{R_2}{R_3}\left(1+\frac{10^3}{201\times 10^3}\right) = 1,$$

from which we get

$$\frac{R_3}{R_2} = \frac{201}{202} \simeq 1$$

and

$$R_{B_3} = 4\times 10^5\,\Omega.$$

Effectively there are 10 V on the emitter, so that

$$I_3 = 200 \times \frac{10}{4 \times 10^5} = 5 \text{ mA}.$$

Now

$$V_B = 15 \text{ V} = E - R_3 I_3 = 20 - R_3 \times 5 \times 10^{-3},$$

which requires $R_3 = 10^3\ \Omega$. The circuit is thus as in Fig. 6.

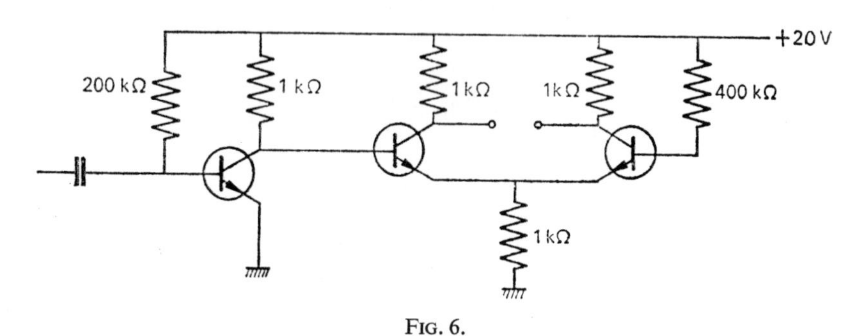

FIG. 6.

(d) From (23)

$$\frac{v_0}{i_2} = \frac{h'_{11} R_e(\beta'+1) + (R_{B_3} + h'_{11})\,[h'_{11} + R_e(\beta'+1)]}{h'_{11} + R_{B_3} + R_e(\beta'+1)}$$

and, putting $R_{B_3} = 0$

$$Z_{i_2} = h_{11} \frac{2R_e(\beta'+1) + h'_{11}}{R_e(\beta'+1) + h'_{11}}.$$

Numerically, this gives $Z_{i_2} = 2000\ \Omega$.

The load of the first stage is $R_1 // Z_{i_2} = 1000 // 2000 = 666\ \Omega$. Its gain is thus

$$\frac{100 \times 666}{1000} = 66{\cdot}6.$$

$$G = \frac{\partial(V_A - V_B)}{\partial V_1} = (1-p)\frac{\partial V_A}{\partial V_1} = (1-p)\frac{\partial V_A}{\partial V_0}\cdot G_1,$$

where G_1, the gain of the first stage, is 66·6 and

$$p = -\frac{R_2}{R_3}\left(1 + \frac{h'_{11}}{R_e(\beta'+1)}\right) \simeq -1.$$

Then

$$G = \frac{\partial V_A}{\partial V_0} \times 2 \times 66{\cdot}6$$

and, since

$$G_A = -\frac{R_2\beta'}{h'_{11}} \cdot \frac{h'_{11}+R_e(\beta'+1)}{h'_{11}+2R_e(\beta'+1)} \simeq -\frac{\beta'}{2} = -100,$$

we have

$$|G| = 13\,320.$$

Then

$$V = 0{\cdot}1 \times 10^{-3} \times 13\,320 = 1{\cdot}33\ \mathrm{V_{rms}},$$

which is equivalent to a peak-to-peak voltage $V_{pp} = 3{\cdot}76$ V and a deflection of 1·25 cm.

Problem No. 3

DIFFERENTIAL AMPLIFIER

Consider the transistor amplifier, shown in Fig. 1, with two identical transistors for which

$$\beta = 70 \quad \text{and} \quad h_{11} = 1500\,\Omega \quad \text{for} \quad I_E = 1\ \text{mA}.$$

1. To find the bias parameters assume that, for a conducting transistor, $V_{BE} = 0$ and that, in the d.c. case, the collector current may be written as

$$I_C = \beta I_B.$$

Calculate the collector currents I_1 and I_2 in terms of the circuit elements. (Take $\beta \gg 1$.) Calculate R_{B_1} and R_{B_3} in terms of β, R and R_2 such that the two transistors both work with the collector-emitter voltage $V_{CE} = E/2$ and find I_1 and I_2 in this case.

As a numerical example take: $R = 1$ kΩ, $R_2 = 10$ kΩ and $E = 20$ V.

2. An a.c. signal is fed to the base of T_1 and the output is taken from the collector of T_2. Calculate the input impedance Z'_i of T_1 and the total input impedance Z_i of the system, as well as the output impedance Z_0. What is the voltage gain, G_v, of the system?

Since the respective loads are small compared to i/h_{22} it is possible to use a simplified equivalent circuit for the transistors with $h_{12} = 0$ and $h_{22} = 0$, as shown in Fig. 2.

Give simple, approximate expressions for Z_i and G_v for $\beta \gg 1$ and $h_{11} \ll \beta R$. With the numerical values given above find Z_i, G_v and Z_0.

3. The complete system may be considered as a four-terminal network $AA'BB'$. Calculate the admittance parameters in terms of Z_i, A_0 and G_v.

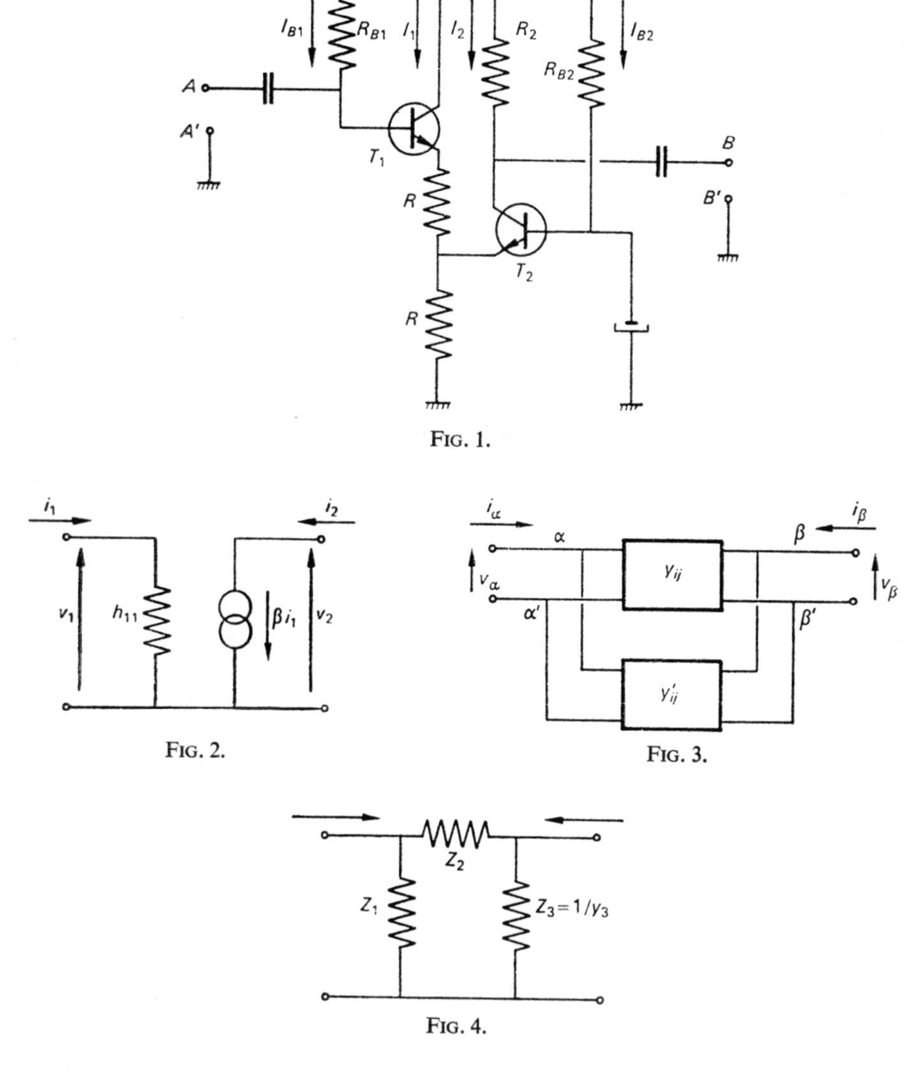

FIG. 1.

FIG. 2.

FIG. 3.

FIG. 4.

4. A passive network with y parameters y'_{11}, y'_{12}, y'_{21}, y'_{22} is placed in parallel with $AA'BB'$ as indicated in Fig. 3. What will be the admittance parameters Y_{ij} of the combination $\alpha\alpha'\beta\beta'$?

Taking i_α, i_β, v_α, v_β to be the input currents and voltages shown, find the relation which must exist between the parameters Y_{ij} such that v_α and v_β can be non-zero for zero values of i_α and i_β. Write this relation in terms of the elements y_{ij} and y'_{ij}.

5. What are the y'_{ij} parameters of the π-network of Fig. 4?

Find these parameters for the particular case:

$$Y_1 = j\omega C_1,$$
$$Y_2 = j\omega C_2,$$
$$Y_3 = \frac{1}{\varrho} + \frac{1}{j\omega L}.$$

6. Consider the oscillator shown in Fig. 5.

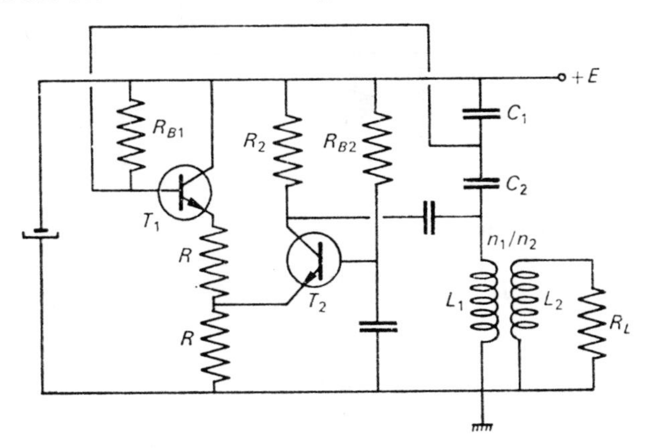

FIG. 5.

Assume that the transformer, of turns ratio $k = n_1/n_2$, is perfect and that the primary has a self-inductance L_1. Also, that the two unmarked capacitors act as coupling or decoupling condensers and do not affect the calculations.

Use the results of the preceding questions to obtain the coupling limit in terms of C_1, C_2, L_1, R_L, k, Z_i and G_v. Use the real part of this expression to give the oscillator frequency.

As a numerical example take $L_1 = 0{\cdot}1$ H, $C_1 = 1$ μF, $C_2 = 0{\cdot}5$ μF, $k = 2$ and $R_L = 10^4\ \Omega$.

(N.B. The first five questions are independent except for the numerical calculations.)

(Partial examination, Paris, 1966)

Solution. 1. $V_{BE} = 0$ so that, following Fig. 6,

$$V_{E_1} = RI_1 + R(I_1 + I_2) = 2RI_1 + RI_2$$
$$V_{E_3} = RI_1 + RI_2$$
$$I_{B_1} = \frac{E - V_{E_1}}{R_{B_1}} = \frac{E - 2RI_1 - RI_2}{R_{B_1}}$$
$$I_{B_1} = \frac{E - V_{E_2}}{R_{B_2}} = \frac{E - RI_1 - RI_2}{R_{B_2}}.$$

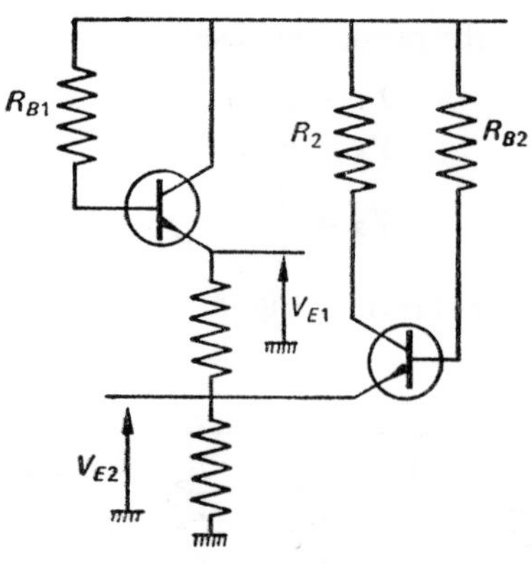

FIG. 6.

$I_c = \beta I_B$ gives two equations

$$I_1 = \beta \frac{E - 2RI_1 - RI_2}{R_{B_1}}$$

$$I_2 = \beta \frac{E - RI_1 - RI_2}{R_{B_2}}.$$

Thus we can write the two relations

$$\begin{cases} I_1(R_{B_1} + 2R\beta) + I_2\beta R = \beta E \\ I_1 R\beta + I_2(R_{B_2} + \beta R) = \beta E, \end{cases}$$

from which

$$I_1 = \frac{\begin{vmatrix} \beta E & \beta R \\ \beta E & (R_{B_2} + \beta R) \end{vmatrix}}{\begin{vmatrix} (R_{B_1} + 2\beta R) & R\beta \\ R\beta & (R_{B_2} + \beta R) \end{vmatrix}} = \frac{\beta E R_{B_2}}{(R_{B_2} + \beta R)(R_{B_1} + 2\beta R) - R^2\beta^2}$$

$$I_2 = \frac{\begin{vmatrix} (R_{B_1} + 2\beta R) & \beta E \\ R\beta & \beta E \end{vmatrix}}{\Delta} = \frac{\beta E(R_{B_1} + \beta R)}{\Delta}.$$

For T_1, $$V_{CE} = E - V_{E_1},$$
for T_2, $$V_{CE} = E - R_2 I_2 - V_{E_2}.$$
Thus we require

$$V_{E_1} = R_2 I_2 + V_{E_2},$$

or

$$RI_1 + R(I_1 + I_2) = R(I_1 + I_2) + R_2 I_2,$$

from which

$$RI_1 = R_2 I_2.$$

This can be written

$$\beta RER_{B_2} = \beta R_2 E(R_{B_1}+\beta R),$$

or

$$RR_{B_2} = R_2(R_{B_1}+\beta R).$$

This relation follows simply from requiring the two V_{CE} to be the same. Taking $V_{CE} = E/2$ gives

$$E-V_{E_1} = E/2, \quad \text{or} \quad V_{E_1} = E/2.$$

Also

$$RI_1+R(I_1+I_2) = R(2I_1+I_2) = E/2 = R\cdot\frac{2\beta ER_{B_2}+E(R_{B_1}+\beta R)}{\Delta},$$

which gives

$$R_{B_2}R_{B_1} = \beta R(2R_{B_2}+R_{B_1}+\beta R),$$

being a second relation between R_{B_1} and R_{B_2}.

From the first relation we get

$$R_{B_2} = \frac{R_2}{R}(R_{B_1}+\beta R),$$

which, on substitution in the second, gives

$$R_{B_1} = \frac{\beta R^2}{R_2}\left(\frac{2R_2}{R}+1\right) = \frac{\beta R(R+2R_2)}{R_2},$$

while

$$R_{B_2} = \beta(3R_2+R).$$

Substituting for β, R_2 and R gives

$$R_{B_2} = 70(3\times 10^4+10^3) = 2{\cdot}17\ \text{M}\Omega$$

$$R_{B_1} = \frac{70\times 10^6}{10^4}\left(\frac{2\times 10^4}{10^3}+1\right) = 7000\times 21 = 147\ \text{k}\Omega.$$

To find I_1 and I_2 it is simplest to start from the condition

$$V_{CE_1} = V_{CE_2} = E/2,$$

so that

$$E-RI_1-R(I_1+I_2) = E-R_2I_2-R(I_1+I_2) = E/2,$$

from which we get straight away

$$I_2 = \frac{E}{2(2R_2+R)} \quad \text{and} \quad I_1 = \frac{ER_2}{2R(2R_2+R)}.$$

Numerically we get

$$I_2 = \frac{10}{21\,000} = 0{\cdot}476 \text{ mA}$$

$$I_1 = \frac{R_2}{R} I_2 = 10 I_2 = 4{\cdot}76 \text{ mA}.$$

2. *The equivalent circuit.*

The currents are as shown in Fig. 7 so that, since the β's are the same for T_1 and T_2 (though not the values of h_{11}),

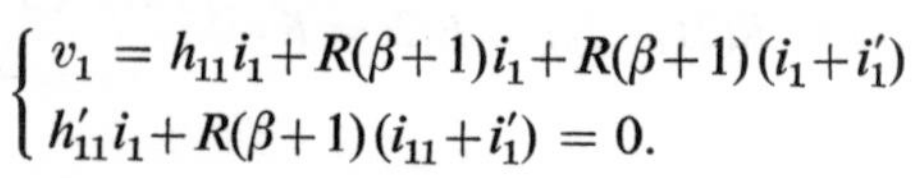

$$\begin{cases} v_1 = h_{11}i_1 + R(\beta+1)i_1 + R(\beta+1)(i_1+i_1') \\ h_{11}'i_1 + R(\beta+1)(i_{11}+i_1') = 0. \end{cases}$$

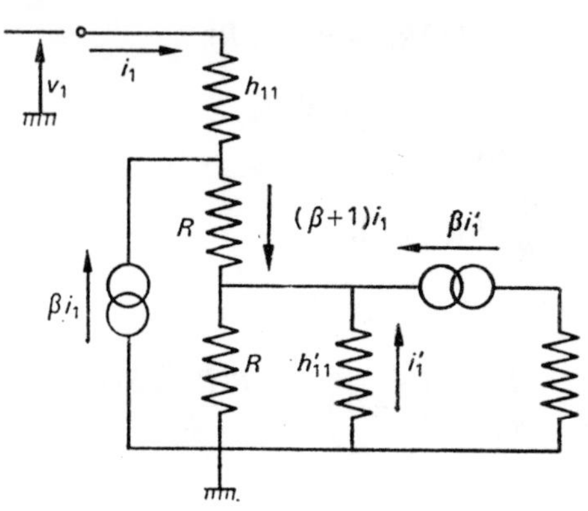

FIG. 7.

From the second expression

$$i_1' = \frac{-(\beta+1)R}{h_{11}'+R(\beta+1)} i_1.$$

Substitution in the first gives

$$v_1 = \left[h_{11}+2R(\beta+1) - \frac{R^2(\beta+1)^2}{h_{11}'+R(\beta+1)}\right] i_1,$$

from which the input impedance of the *first transistor* is

$$Z_i' = h_{11} + R(\beta+1)\left(2 - \frac{R(\beta+1)}{h_{11}'+R(\beta+1)}\right).$$

The current i_1' can be written

$$i_1' = \frac{R(\beta+1)}{h_{11}'+R(\beta+1)} \cdot \frac{v_1}{h_{11}+R(\beta+1)\left(2-\dfrac{R(\beta+1)}{h_{11}'+R(\beta+1)}\right)}$$

from which the voltage gain is

$$G_v = -\frac{R_2 i_1'}{v_1} = \frac{RR_2(\beta+1)}{[h_{11}'+R(\beta+1)]\left[h_{11}+R(\beta+1)\left(2-\dfrac{R(\beta+1)}{h_{11}'+R(\beta+1)}\right)\right]}.$$

If we neglect h_{11} with respect to βR and 1 with respect to β, we have simply

$$Z_i' = \beta R = 70\ \text{k}\Omega.$$

The input impedance of the system is $Z_i = Z_i' \,//\, R_{B_1} = 47{\cdot}5\ \text{k}\Omega \sim 50\ \text{k}\Omega$ and

$$G_v = \frac{R_2}{R} = 10.$$

Taking the approximation $R_2 \ll 1/h_{22}$, the output impedance of the system is R_2.

3. With Z_i and Z_o as the input and output impedances, the system can be represented by the equivalent four-terminal network of Fig. 8.

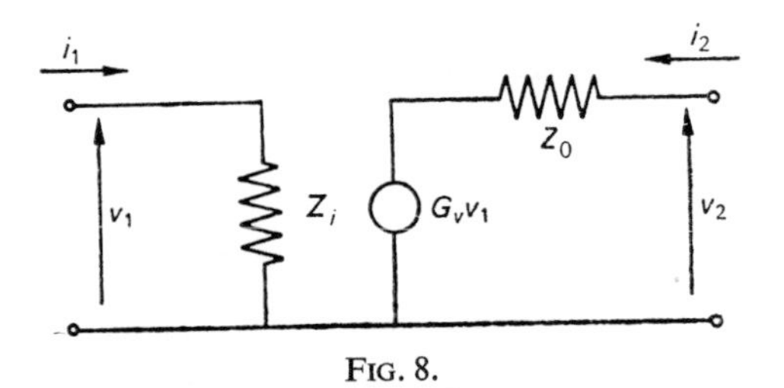

FIG. 8.

Then

$$\begin{cases} v_1 = Z_i i_1 \\ v_2 = G_v v_1 + Z_o i_2, \end{cases}$$

or

$$i_1 = v_1/Z_i$$
$$i_2 = -(G_v/Z_o)v_1 + v_2/Z_o,$$

and so

$$y_{11} = 1/Z_i, \qquad y_{12} = 0,$$
$$y_{21} = -G_v/Z_o, \qquad y_{22} = 1/Z_o.$$

4. $$Y_{11} = y_{11}+y_{11}', \qquad Y_{12} = y_{12}+y_{12}', \text{ etc.}$$

The required condition is that there are allowed non-zero solutions, i.e. $\Delta = 0$. Thus

$$(y_{11}+y_{11}')(y_{22}+y_{22}')-(y_{12}+y_{12}')(y_{21}+y_{21}') = 0.$$

5. From the circuit in Fig. 9, we see that

$$\begin{cases} i_1 = Y_1 v_1 + Y_2(v_1-v_2) \\ i_2 = Y_3 v_2 - Y_2(v_1-v_2) \end{cases}$$

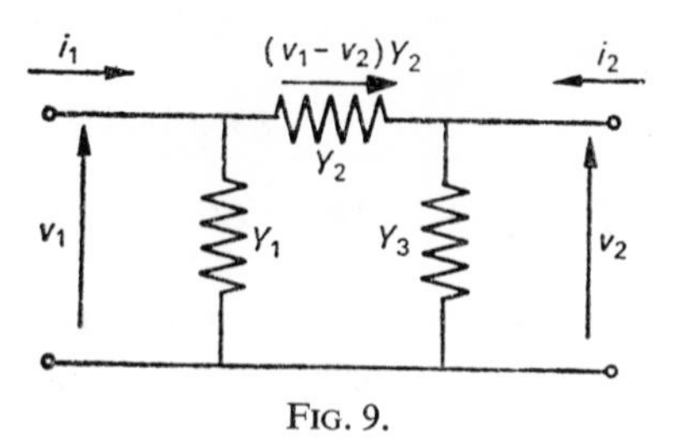

FIG. 9.

or

$$\begin{cases} i_1 = (Y_1+Y_2)v_1-Y_2v_2 \\ i_2 = -Y_2v_1+(Y_2+Y_3)v_2. \end{cases}$$

The admittance matrix is thus

$$(Y) = \begin{pmatrix} Y_1+Y_2 & -Y_2 \\ -Y_2 & Y_2+Y_3 \end{pmatrix}.$$

Substituting the given admittances, this becomes

$$(Y) = \begin{pmatrix} j\omega(C_1+C_2) & -j\omega C_2 \\ -j\omega C_2 & \dfrac{1}{\varrho}+\dfrac{1}{j\omega L}+j\omega C_2 \end{pmatrix}.$$

6. The system of Fig. 5 can be separated into two networks in parallel: the first being the transistor amplifier and the second that shown in Fig. 10. Taking R_L to the transformer primary, this latter is identical with that of the preceding question but with $\varrho = k^2R_L$.

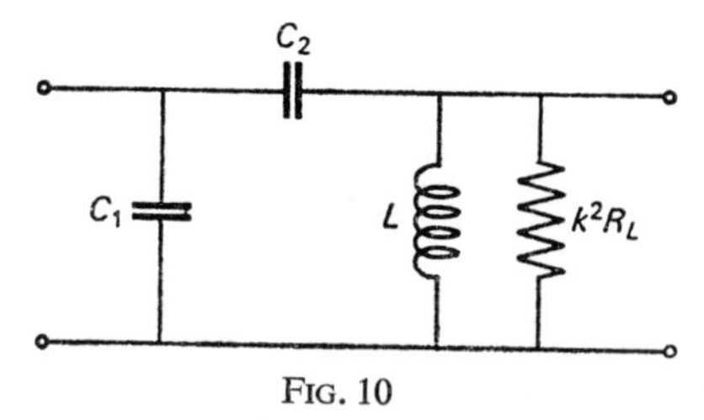

FIG. 10

The condition for the limit of coupling given in solution 4 is

$$\left(j\omega(C_1+C_2)+\frac{1}{Z_i}\right)\left(\frac{1}{\varrho}+\frac{1}{j\omega L}+j\omega C_2+\frac{1}{Z_o}\right)-j\omega C_2\left(j\omega C_2+\frac{G_v}{Z_o}\right)=0.$$

The real part of this expression gives

$$\frac{C_1+C_2}{L}+\frac{1}{Z_i}\left(\frac{1}{\varrho}+\frac{1}{Z_o}\right)-\omega^2C_1C_2=0,$$

while, from the imaginary part,

$$\frac{\omega(C_1+C_2)}{\varrho}+\frac{\omega(C_1+C_2)}{Z_{\bar{O}}}-\frac{1}{\omega L Z_i}+\frac{\omega C_2}{Z_i}-\frac{\omega C_2 G_v}{Z_{\bar{O}}}=0.$$

Multiplying this latter by ω gives

$$\omega^2\left(\frac{C_1+C_2}{\varrho}+\frac{C_1+C_2}{Z_{\bar{O}}}+\frac{C_2}{Z_i}-\frac{C_2 G_v}{Z_{\bar{O}}}\right)=\frac{1}{LZ_i},$$

while the real part gives

$$\omega^2=\frac{C_1+C_2}{LC_1C_2}+\frac{1}{C_1C_2Z_i}\left(\frac{1}{\varrho}+\frac{1}{Z_{\bar{O}}}\right).$$

Putting in the values, we have

$$\varrho = k^2 R_L = 4\times 10^4\,\Omega$$

$$\omega^2 = \frac{1{\cdot}5\times 10^{-6}}{10^{-1}\times 0{\cdot}5\times 10^{-12}}+\frac{1}{0{\cdot}5\times 10^{-12}\times 5\times 10^4}\left(\frac{1}{4\times 10^4}+\frac{1}{10^4}\right)$$

$$= 3\times 10^7+0{\cdot}5\times 10^4.$$

Then

$$\omega = \sqrt{3000{\cdot}5\times 10^4} \sim 5500 \text{ rad/s},$$

corresponding to a frequency of $\sim$875 Hz.

Problem No. 4

TWO-STAGE TRANSISTOR AMPLIFIER

It is proposed to build an amplifier for a low power, portable record player. It will consist of an output stage coupled through a transformer and a preamplifier stage. Only the output stage will be studied here. The two transistors used in the circuit of Fig. 2 are both 2N525 for which the characteristics are given in Fig. 1.

In the following it will be assumed that the working point may be anywhere in the region between the axes and the two lines given by $I_c = 50$ mA and $V_c = 20$ V; i.e. both I_{CE_0} and the voltage V_K at the knee of the characteristics will be neglected.

The supply voltage is 10 volts and the capacitors C_1, C_2 and C_3 are sufficiently large to act as coupling, or decoupling, capacitors without affecting the a.c. gain.

1. If the resistance of the transformer windings is negligible find, in terms of R_6 and E, the current I_c and the d.c. bias voltage V_{CE} that must be used to give a quiescent d.c. power dissipation P in the transistor T_2.

What are these values if $R_6 = 100\,\Omega$, $E = -10$ V, $P = 0{\cdot}1$ W?

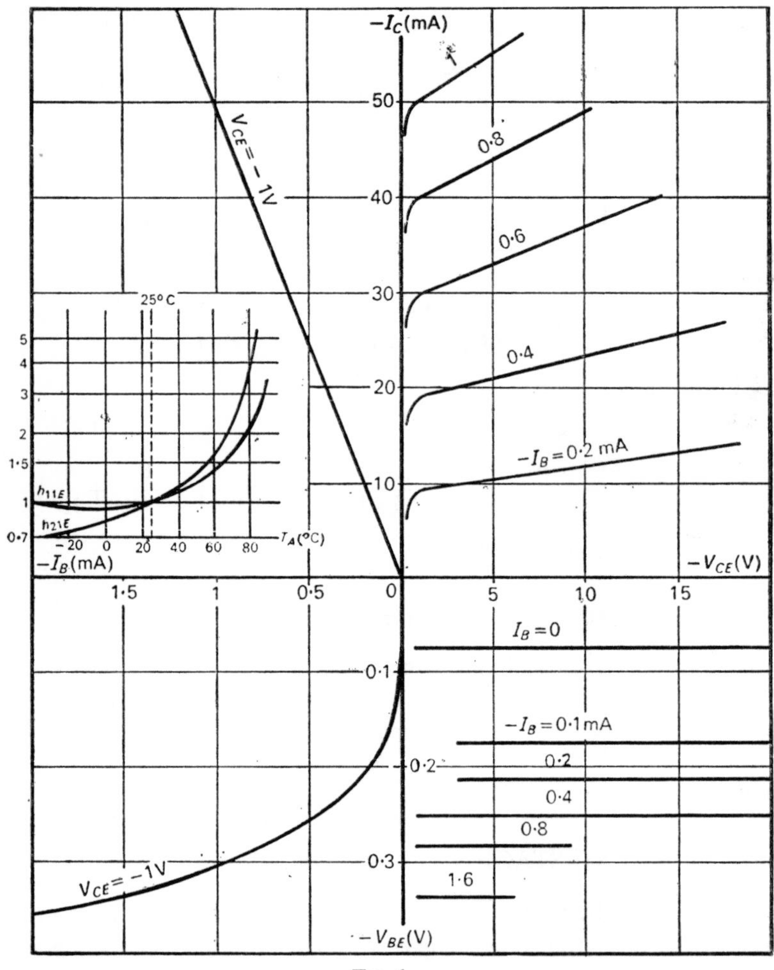

Fig. 1.

Use this particular characteristic to find the corresponding values of I_c, V_{CE}, V_{BE} and I_B.

2. What value of a.c. load on the transformer primary will give the maximum sinusoidal output voltage?

What is the peak-to-peak value of this voltage (measured at the primary) and the corresponding a.c. power?

If r is the equivalent pure resistance of the loud-speaker, what transformer ratio n_1/n_2 is needed to give the optimum load?

Calculate this ratio for $r = 2·5\ \Omega$.

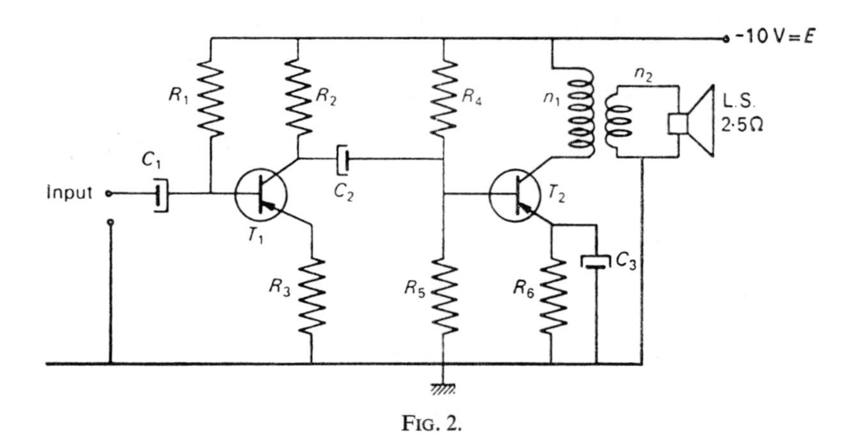

FIG. 2.

3. What are the values of the hybrid parameters for the working point which was determined in 1 above?

In all the following questions, take $h_{12e} = 0$.

4. What is the value of the resistance R_B which, placed on its own between the base and the source in the circuit of Fig. 3, would give the same collector current? (Do not ignore V_{BE} compared with 10 volts.)

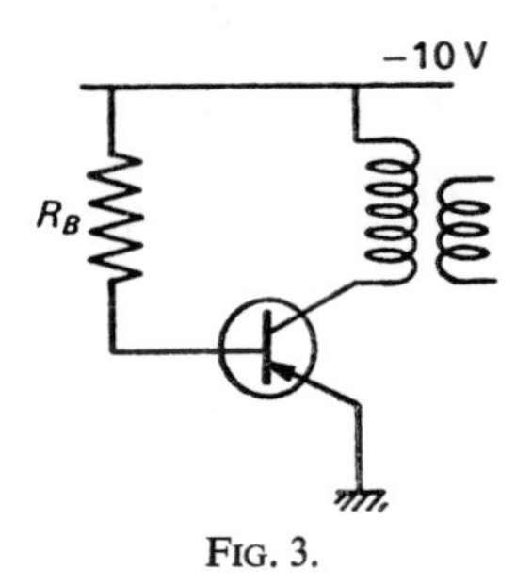

FIG. 3.

5. (a) Find the values of the resistances R_4 and R_5 of Fig. 1 such that a variation ΔI_{CE_0} in the leakage current will give only a fifth of this variation in I_c (i.e. $\Delta I_c/\Delta I_{CE_0} = 1/5$). Find the general formula before finding the numerical values.

(b) If, in Fig. 2, a transistor with a gain β is replaced by one with a gain β' so that I_c changes to I_c', what values of R_4 and R_5 should be chosen so that the change in I_c is reduced to one-fifth?

6. Taking the values of R_4 and R_5 calculated in 5(a) above, calculate the input impedance and the voltage gain of the stage for small changes in V_{BE} about the working point. (Use only the approximation $h_{12e} = 0$.)

7. Assuming that the previous result is valid for large signals, what is the peak-to-peak voltage which must be supplied by the preamplifier if the optimum output power calculated in question 2 above is to be obtained?

(Partial examination, Paris, 1965.)

Solution. 1. The d.c. load-line is only determined by R_6 so that

$$V_{CE} = E - R_6 I_c.$$

The power dissipated is

$$P = VI = I_c(E - R_6 I_c),$$

so that

$$R_6 I_c^2 - E I_c + P = 0$$

and

$$I_c = \frac{E \pm \sqrt{E^2 - 4R_6 P}}{2R_6}$$

$$V_{CE} = \frac{E}{2} \mp \frac{1}{2}\sqrt{E^2 - 4R_6 P}.$$

With the given values we have either

$$I_c = \frac{10 + \sqrt{100 - 40}}{200} \equiv 88{\cdot}75 \text{ mA} > 50 \text{ mA}$$

and must therefore be eliminated, or

$$I_c = \frac{10 - \sqrt{100 - 40}}{200} \equiv 11{\cdot}25 \text{ mA}$$

which is an allowed solution. Thus

$$V_{CE} = 5 + \tfrac{1}{2}\sqrt{60} = 8{\cdot}87 \text{ V}.$$

The other values may be found from the characteristic. It is necessary first of all to draw the load-line passing through, for example,

$$V_{CE} = 10 \text{ V}, \qquad I_c = 0,$$
$$V_{CE} = 5 \text{ V}, \qquad I_c = 50 \text{ mA}.$$

We see from Fig. 4 that the working point lies effectively on the curve for $I_B = 0{\cdot}2$ mA. (In fact I_B is slightly less than 0·2 mA but any value between 0·19 mA and 0·21 mA could be considered sufficiently accurate.)

Examination of the curves in the first quadrant shows that β varies with V_{CE} and we must therefore draw in the second quadrant $I_c = f(I_b)$ for $V_{CE} = 8{\cdot}87$ V. This is easy since it is sufficient to read the values from a vertical line through the working point.

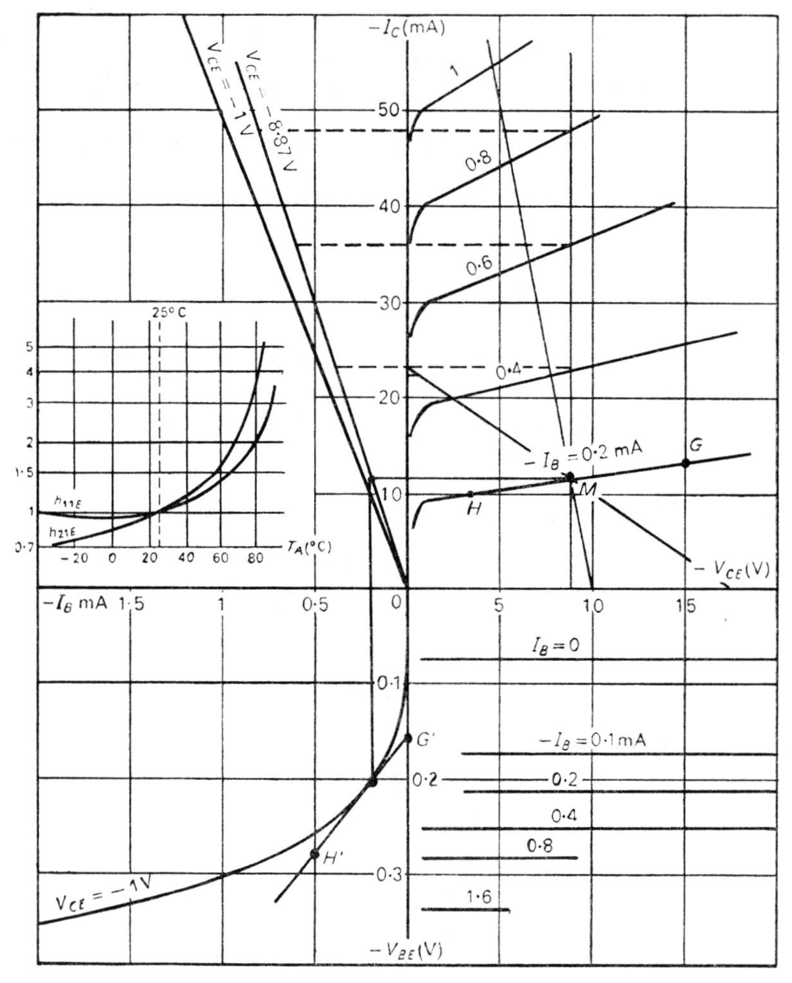

FIG. 4.

Construction from these curves into the third quadrant can be used to find V_{BE} which is seen to be $V_{BE} = 0{\cdot}210$ V.

2. To get the maximum sinusoidal output voltage, the working point must be at the centre of the a.c. load-line. This line must therefore cut the vertical and horizontal axes at $2I_{co}$ and $2V_{co}$ respectively where I_{co} and V_{co} are the co-ordinates of the working point.

The corresponding dynamic load-line is

$$\frac{I_c}{2I_{co}}+\frac{V_{CE}}{2V_{co}}=1,$$

or

$$V_{CE} = 2V_{co} - \frac{V_{co}}{I_{co}} \cdot I_c.$$

Comparison with the normal equation $V = E - ZI$ shows that the dynamic load is

$$Z = \frac{V_{co}}{I_{co}},$$

or

$$Z = \frac{\frac{E}{2} + \frac{1}{2}\sqrt{E^2 - 4R_6P}}{\frac{E - \sqrt{E^2 - 4R_6P}}{2R_6}} = R_6\frac{E + \sqrt{E^2 - 4R_6P}}{E - \sqrt{E^2 - 4R_6P}}.$$

The maximum voltage is

$$V_{c\,\max} = 2V_{co} = E + \sqrt{E^2 - 4R_6P}.$$

The ratio n_1/n_2 is given by the ratio of the impedances

$$\frac{n_2}{n_1} = \sqrt{\frac{r}{Z}}.$$

With the numerical values we have

$$Z = \frac{8 \cdot 87}{11 \cdot 25 \times 10^{-3}} = 788\ \Omega.$$

$$V_{pp\,\max} = 17 \cdot 75\ \text{V}$$

$$P = \frac{V_{pp}^2}{8Z} = 50\ \text{mW}$$

$$\frac{n_2}{n_1} = \sqrt{\frac{2 \cdot 5}{788}} = \frac{1}{17 \cdot 7}.$$

3. $$h_{22} = \left(\frac{\partial I_c}{\partial V_{CE}}\right)_{I_B = \text{constant}} = \frac{(13 \cdot 2 - 10)}{15 - 3 \cdot 7} \times 10^{-3} = 2 \cdot 83 \times 10^{-4}\ \text{S},$$

or

$$\frac{1}{h_{22}} = 3530\ \Omega.$$

(The points used to find the values of I and V are marked as G and H on Fig. 4.)

Note the small value of $1/h_{22}$ which prevents the use of the classic approximation for this stage.

$\beta = h_{21e}$ must be measured from the line $I_c = f(I_B)$ for $V_{CE} = -8{\cdot}87$ V i.e.

$$h_{21e} = 60.$$

For h_{11} we need the tangent to the V_{CE} curve in the third quadrant. In theory we should have to take the curve drawn for $V_{CE} = -8{\cdot}87$ V but because the characteristics are horizontal (i.e. $h_{12} = 0$) they all lie together and, from the points G' and H',

$$h_{11} = \frac{0{\cdot}285 - 0{\cdot}16}{0{\cdot}5 \times 10^{-3}} = 250\,\Omega.$$

4. To get the same collector current it is necessary to generate the same base current, so that

$$R_B = \frac{E - V_{BE}}{I_B} = \frac{10 - 0{\cdot}210}{0{\cdot}2 \times 10^{-3}} = 48\,950\,\Omega.$$

5. (a)

$$\frac{\Delta I_c}{\Delta I_{CE_0}} = \frac{1}{\sigma_0}$$

where we want $\sigma_0 = 5$ and σ_0 follows from Exercise No. 7, Chapter 2.

(b) Taking

$$\sigma_0 = \sigma_\beta = 1 + \frac{\beta R_e}{R_e + \delta},$$

where here $R_e = R_6$, we require

$$1 + \frac{\beta R_6}{R_6 + \delta} = 5$$

with $\delta = R_4 // R_5$. Thus $R_6 = 100\,\Omega$ and $\beta = 60$, this gives

$$\frac{6000}{100 + \delta} = 4,$$

or

$$\delta = 1400\,\Omega.$$

From the relevant part of the circuit which is shown in Fig. 5 we have the equations

$$V_{BE} + V_E = V_B = R_5 i, \tag{1}$$

where V_E is the emitter–earth potential,

$$10 - V_B = R_4(i + I_B) \tag{2}$$

$$R_4 // R_5 = \frac{R_4 R_5}{R_4 + R_5} = \delta = 1400\,\Omega. \tag{3}$$

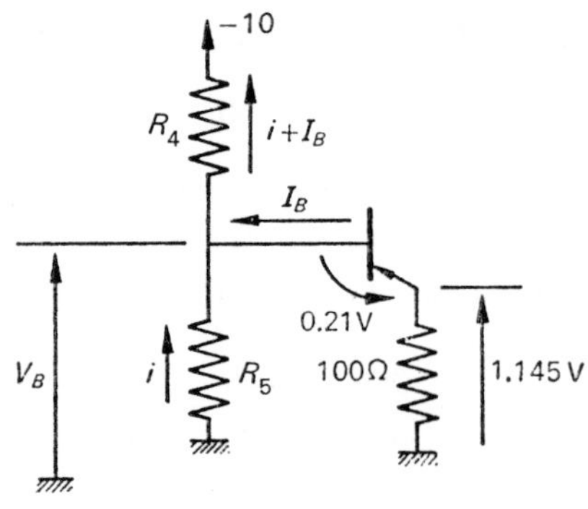

FIG. 5.

From (1) we have

$$i = \frac{V_B}{R_5}.$$

Substituting in (2) we have

$$R_4\left(\frac{V_B}{R_5}+I_B\right) = E-V_B,$$

which, together with (3), gives

$$R_5 = \frac{E\delta}{E-V_B-\delta I_B}$$

$$R_4 = \frac{\delta R_5}{R_5-\delta}, \quad \text{where} \quad V_B = R_E(I_c+I_B)+V_{BE}.$$

The numerical values given lead to

$$V_B = 1{\cdot}145+0{\cdot}210 = 1{\cdot}355 \text{ V}$$
$$R_5 = 1673\ \Omega$$
$$R_4 = 8579\ \Omega.$$

6. The voltage gain has to be calculated using the exact equation. That is,

$$G_v = \frac{\dfrac{h_{21}}{h_{11}}}{h_{22}(S-1)-\dfrac{1}{Z_L}} \quad \text{where, here,} \quad S = \frac{h_{12}h_{21}}{h_{11}h_{22}} = 0, \quad \text{since } h_{12} = 0.$$

Thus

$$G_v = -\frac{60}{250}\left(\frac{1}{2{\cdot}83\times10^{-4}+\frac{1}{788}}\right) = -154.$$

If we had taken the simplified formula with $Z_L \ll 1/h_{22}$ we would have found

$$G_v = -\frac{\beta Z_L}{h_{11}} = -189.$$

The input impedance of the system is obtained with $h_{11} // \delta$, or

$$Z_i = \frac{h_{11}\delta}{h_{11}+\delta} = 212\ \Omega.$$

7. To get a peak-to-peak output of 17·75 V with a gain of -154 requires the preamplifier to give

$$\frac{17 \cdot 75}{154} \equiv 115 \text{ mV peak-to-peak.}$$

Problem No. 5

L.T. TRANSISTOR SUPPLY

It is required to build the circuit of Fig. 1 to supply a current of 1 ampere at a potential of 6 volts to the load R_L.

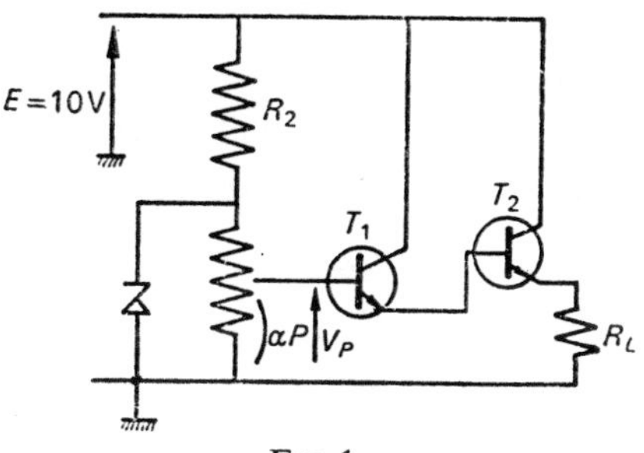

FIG. 1.

Approximate calculation for the circuit elements.

1. What is the approximate working point (I_c, V_{CE}) of the transistor T_2 and the power which it dissipates?

These results suggest the choice of a transistor for which the working point parameters are

$$T_2 \begin{cases} h'_{11e} = 7 \cdot 44\ \Omega & h'_{12e} = 1 \cdot 15 \times 10^{-3} \\ h'_{22e} = 25 \cdot 8 & h'_{22e} = 25 \times 10^{-3}\,\text{S}. \end{cases}$$

2. What is its current gain in the circuit being considered, and what is its base current I_{B_3}?

3. What is the approximate working point and the power dissipated for T_1? These results suggest the choice of T_1 such that

$$T_1 \begin{cases} h_{11e} = 135\ \Omega & h_{12e} = 2{\cdot}4\times 10^{-4} \\ h_{21e} = 67 & h_{22e} = 29\times 10^{-5}\text{S}. \end{cases}$$

4. What is the current gain for T_1 in the chosen circuit and what is its base current I_{B_1}?

5. The voltage gain of the two transistors is very close to 1. Why? The output $V_{\bar{O}}$ will thus follow the input voltage V_p. The reference source producing V_p consists of a Zener diode 14 Z4, a resistance R_2 and a potentiometer P. The Zener diode characteristic is linear between the points

$$(0{\cdot}4\ \text{mA}, 8{\cdot}02\ \text{V}) \quad \text{and} \quad (25\ \text{mA}, 8{\cdot}1\ \text{V}).$$

It can be assumed, therefore, that the Zener voltage is 8 V and, so that possible variations of I_{B_1} shall not change V_p appreciably, the current through P is chosen to be comparatively large, say 5 mA. If the current through the diode is 10 mA, find P and R_2 and the resistance αP of the lower part of the potentiometer when the output voltage is 6 volts.

6. What is the output resistance of the system?

II

1. Calculate the hybrid parameters for the combination of the two transistors in the common-collector connection. (It will be noted that, since the second always works with a higher current than the first, the approximation

$$1/h_{22e} \gg h'_{11e}$$

is always valid.) Deduce the parameters H_{ij} for the combination.

2. Calculate the output impedance of the system from the composite parameters.

3. The voltages V_{EB} of the two transistors are both 0·5 V. Taking account of the effect of the base current of T_1 in the potentiometer but neglecting any internal impedance for the Zener, what is the value of αP to give an output of 6 volts? (Take $V_{\text{Zener}} = 8$ V and R_0 and P to have the values already calculated.)

What is the true value of the output impedance?

(Nantes, October 1963.)

I

Solution. 1. The voltage across the transistor is $10-6 = 4$ V so that the power is

$$p_2 = VI = 4\ \text{W}.$$

2. In the common-collector connection the current gain is

$$G_i' = \frac{h'_{21c}}{1+Z_L h'_{22c}}.$$

Now $h'_{22c} = h'_{22e}$ and $|h'_{21c}| = \beta+1$ while

$$Z_L = 6\,\Omega \quad \text{and} \quad h'_{22} = 25\times 10^{-3},$$

so that

$$G_i = \frac{26{\cdot}8}{1+0{\cdot}15} = 23{\cdot}3,$$

which is close to $h'_{21c} = 26{\cdot}8$.

The base current is then

$$I_{B_2} = \frac{1}{23{\cdot}3} = 42{\cdot}9 \text{ mA}.$$

3. The transistor T_1 must have an emitter current of 42·9 mA and, neglecting V_{EB}, the voltage across its terminals is the same as for T_2 from which the power which it dissipates is

$$p_1 = 4\times 42{\cdot}9\times 10^{-3} = 0{\cdot}171 \text{ W}.$$

4. Like T_2, T_1 is connected with a common-collector with the input impedance of T_2 as its load. This latter impedance is effectively

$$h_{11}+(\beta+1)R_L = 7{\cdot}44+26{\cdot}8\times 6 = 168{\cdot}2\,\Omega,$$

from which its current gain is

$$G_i = \frac{h_{21e}+1}{1+Z_L h_{22e}} = \frac{68}{1+168{\cdot}2\times 29\times 10^{-5}} = 64{\cdot}8$$

so that the base current is

$$\frac{42{\cdot}9}{64{\cdot}8} = 0{\cdot}662 \text{ mA}.$$

5. With the section of the circuit shown in Fig. 2 we see

$$R_2 = \frac{2}{15\times 10^{-3}} = 133\,\Omega$$

$$P = \frac{8}{5\times 10^{-3}} = 1600\,\Omega.$$

When there is a 6 V output there is very closely 6 V across the section αP. Thus

$$\alpha P = \frac{6}{5\times 10^{-3}} = 1200\,\Omega \quad \text{or} \quad \alpha = 0{\cdot}75.$$

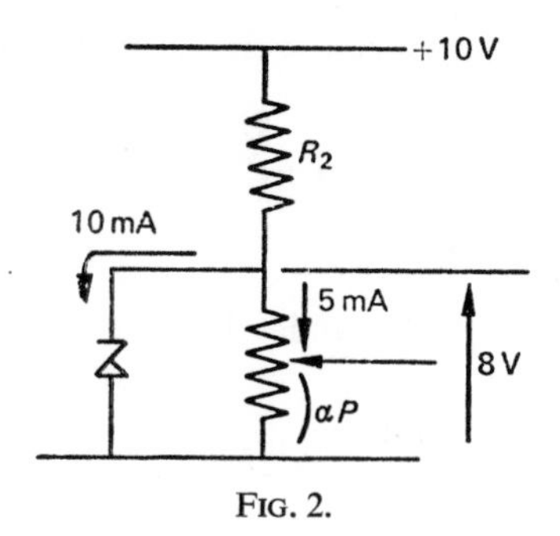

FIG. 2.

6. The combination of the Zener and the associated resistances behave as a source of e.m.f. of 6 V and an internal impedance

$$\frac{\alpha P(1-\alpha)P}{P} = 300\ \Omega$$

where the impedance of the Zener is taken to be negligible.

Now the output of a transistor in the common-collector connection supplied from a source of internal resistance ϱ_g is

$$\frac{\varrho_g + h_{11e}}{\beta+1},$$

so that T_1 has an output impedance

$$Z_{\bar{O}} = \frac{300+135}{68} = 6{\cdot}39\ \Omega.$$

With the same reasoning for T_2

$$Z_{\bar{O}} = \frac{6{\cdot}39+7{\cdot}44}{26{\cdot}8} = 0{\cdot}516\ \Omega.$$

II

1. We have for T_1 (see Fig. 3)

$$\begin{cases} v_1 = h_{11c}i_1 + h_{12c}v_2 \\ i_2 = h_{21c}i_1 + h_{22c}v_2, \end{cases}$$

while, for T_2,

$$\begin{cases} v_2 = h'_{11c}(-i_2) + h'_{12c}v_0 \\ i_0 = h'_{21c}(-i_2) + h'_{22c}v_0. \end{cases}$$

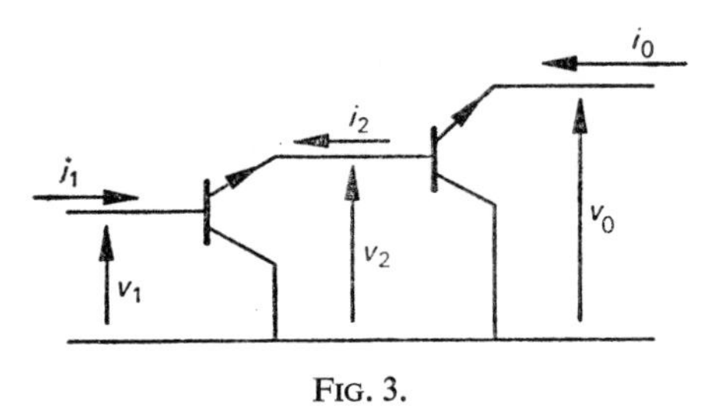

FIG. 3.

From these equations we want to find v_1 and i_0 in terms of i_1 and v_0. Writing them in determinant form we see

$$\begin{cases} v_1 - h_{12c}v_2 + 0 + 0 = h_{11c}i_1 \\ 0 - h_{22c}v_2 + i_2 + 0 = h_{21c}i_1 \\ 0 + v_2 + h'_{11c}i_2 + 0 = h'_{12c}v_0 \\ 0 + 0 + h'_{21c}i_2 + i_0 = h'_{22c}v_0 \end{cases}$$

$$\Delta = \begin{vmatrix} 1 & -h_{12c} & 0 & 0 \\ 0 & -h_{22c} & 1 & 0 \\ 0 & 1 & h'_{11c} & 0 \\ 0 & 0 & h'_{21c} & 1 \end{vmatrix} = -(1 + h_{22c}h'_{11c}).$$

Since $1/h_{22c} = 1/h_{22e}$ is always much greater than the h_{11} for the transistor which follows, we have

$$\Delta \simeq -1.$$

The determinant relating to v_1 is

$$\Delta_{v_1} = \begin{vmatrix} h_{11c}i_1 & -h_{12c} & 0 & 0 \\ h_{21c}i_1 & -h_{22c} & 1 & 0 \\ h'_{12c}v_0 & 1 & h'_{11c} & 0 \\ h'_{22c}v_0 & 0 & h'_{21c} & 1 \end{vmatrix}$$

which is, because of the final column,

$$\Delta_{v_1} = \begin{vmatrix} h_{11c}i_1 & -h_{12c} & 1 \\ h_{21c}i_1 & -h_{22c} & 0 \\ h'_{12c}v_0 & 1 & h'_{11c} \end{vmatrix}$$

or

$$\Delta_{v_1} = -h_{12c}h'_{12c}v_0 - (h_{11c} - h'_{11c}h_{12c}h_{21c} + h'_{11c}h_{11c}h_{22c})i_1.$$

Similarly

$$\Delta_{i_0} = \begin{vmatrix} 1 & -h_{12c} & 0 & i_1 \\ 0 & -h_{22c} & 1 & i_1 \\ 0 & 1 & h'_{11c} & v_0 \\ 0 & 0 & h'_{21c} & v_0 \end{vmatrix}$$

or

$$\Delta_{i_0} = (h_{22c}h'_{21c}h'_{12c} - h_{22c}h'_{11c}h'_{22c} - h'_{22c})v_0 + h'_{21c}h_{21c}i_1.$$

The composite H parameters follow immediately since, e.g., $v_1 = \Delta_{v_1}/\Delta \simeq -\Delta_{v_1}$

$$\begin{aligned} H_{11c} &= h_{11c} + h'_{11c}(h_{11c}h_{22c} - h_{12c}h_{21c}) \\ H_{12c} &= +h_{12c}h'_{12c} \\ H_{21c} &= -h'_{21c}h_{21c} \\ H_{22c} &= h'_{22c} + h_{22c}(h'_{11c}h'_{22c} - h'_{12c}h'_{21c}). \end{aligned}$$

However,

$$h_{11c} = h_{11e}, \qquad h_{21c} = -(h_{21e}+1), \qquad h_{12c} = 1 - h_{12e} \simeq 1 \quad \text{and} \quad h_{22c} = h_{22e}.$$

So, in terms of the common emitter parameters,

$$H_{11c} = H_{11e} = h_{11e} + h'_{11e}(h_{11e}h_{22e} - h_{12e}h_{21e} - h_{12e} + 1 + h_{21e}).$$

But $h_{11e}h_{22e} \ll 1$ and $h_{12e} \ll 1$, so that

$$H_{11c} = H_{11e} = h_{11e} + h'_{11e}(1 + h_{21e}).$$

Similarly

$$H_{12c} = +(1 - h_{12e})(1 - h'_{12e}) = 1 - h_{12e} - h'_{12e} + h_{12e}h'_{12e} = 1 - H_{12e}.$$

Again, since $h_{12e} \ll 1$, we can take

$$\begin{aligned} H_{12e} &= h_{12e} + h'_{12e} \\ H_{21c} &= -(1 + H_{21e}) = -(1 + h_{21e})(1 + h'_{21e}) = -(1 + h_{21e} + h'_{21e} + h_{21e}h'_{21e}) \end{aligned}$$

from which

$$H_{21e} = h_{21e} + h'_{21e} + h_{21e}h'_{21e}.$$

Finally

$$H_{22c} = H_{22e} = h'_{22e} + h_{22e}(h'_{11e}h'_{22e} - h'_{12e}h'_{21e} + 1 + h_{21e})$$

or, very closely,

$$H_{22e} = h'_{22e} + h_{22e}(1 + h_{21e}).$$

2. The "composite transistor" may be characterized by

$$H_{21e} = 25{\cdot}8+67+67\times 25{\cdot}8 = 1821$$
$$H_{11e} = 135+7{\cdot}44(1+67) = 640\ \Omega$$
$$H_{22e} = 25\times 10^{-3}+29\times 10^{-5}(1+67) = 44{\cdot}75\times 10^{-3}\ \text{S}.$$

(Alternatively $1/H_{22e} = 22{\cdot}35\ \Omega$.)

The approximate formula

$$Z_{\bar{O}} = \frac{h_{11}+\varrho_g}{\beta+1}$$

would give a result greatly in error, since $R_L = 6\ \Omega$ is not negligible compared with $1/H_{22e}$.

We use the exact expression

$$Y_0 = H_{22c}\left[1-\frac{S_c H_{11c}}{\varrho_g+H_{11c}}\right]$$

with

$$H_{22c} = H_{22e} = 44{\cdot}75\times 10^{-3}\ \text{S}$$
$$H_{11c} = H_{11e} = 640\ \Omega$$
$$S_c = \frac{H_{12c}H_{21c}}{H_{11c}H_{22c}} \simeq \frac{H_{21c}}{H_{11c}H_{22c}} = -\frac{(H_{21e}+1)}{H_{11e}H_{22e}} = -63{\cdot}7,$$

from which

$$Y_{\bar{O}} = 2{\cdot}1\ \text{S} \quad \text{or} \quad Z_{\bar{O}} = 0{\cdot}478\ \Omega.$$

As will be seen, the rapid calculation carried out in the first part of the question gives a result very close to the exact value of Z_0.

3. The voltage on the slide contact of P must be

$$6+0{\cdot}5+0{\cdot}5 = 7\ \text{V}.$$

Then, as is seen from Fig. 4,

$$I = \frac{7}{P\alpha}$$

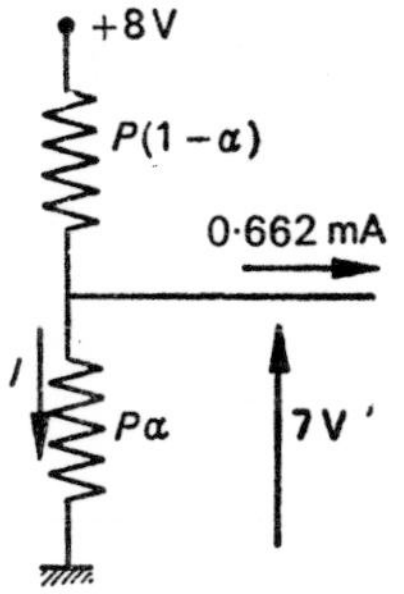

+8V
P(1 − α)
0·662 mA
I
Pα
7V

FIG. 4.

and

$$8-7 = P(1-\alpha)\left(\frac{7}{P\alpha}+0{\cdot}662\times 10^{-3}\right)$$

from which

$$P\alpha = 1424\ \Omega$$
$$P(1-\alpha) = 176\ \Omega.$$

The impedance of the reference source is then

$$\frac{176\times 1424}{176+1424} = 157\Omega.$$

The output admittance of the system is thus

$$Y_{\bar{O}} = 44{\cdot}75\times 10^{-3}\left[1+\frac{63{\cdot}7\times 640}{157+640}\right] = 2{\cdot}33\ \text{S}$$

and

$$Z_{\bar{O}} = 0{\cdot}43\ \Omega.$$

Problem No. 6

H.F. AMPLIFIER

It is required to build a two-stage transistor amplifier as shown in Fig. 1.

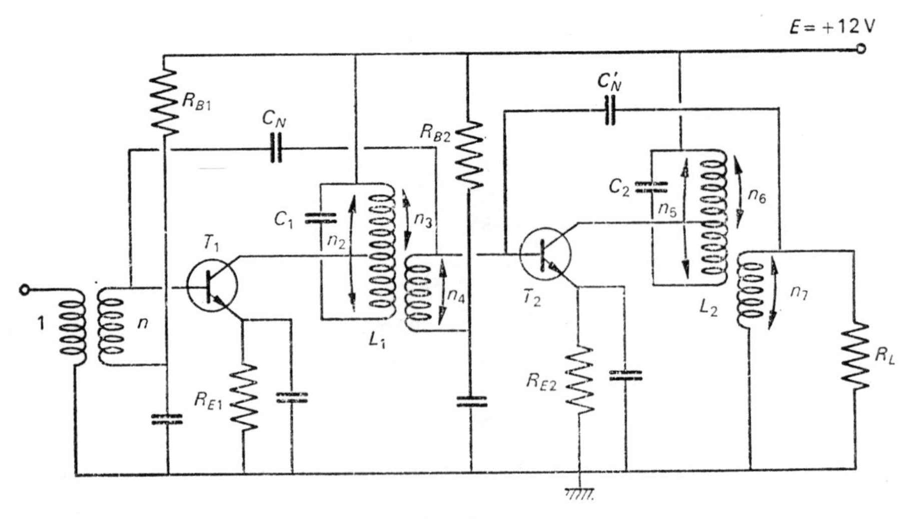

FIG. 1.

1. *Preliminary calculations.*

(a) Calculate the y parameters of a common-emitter connected transistor in terms of the elements of the Giacoletto equivalent circuit shown in Fig. 2.

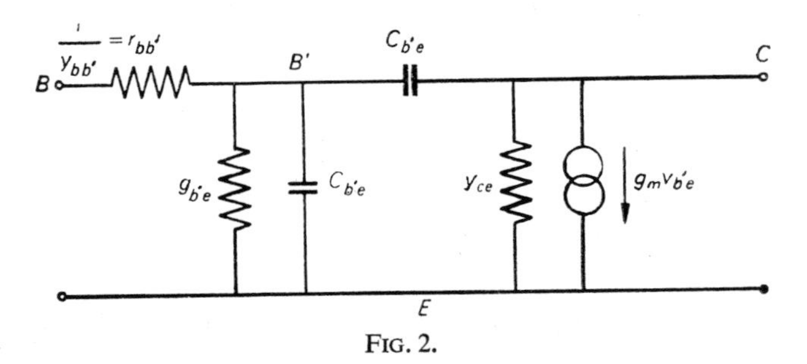

FIG. 2.

It is possible to write

$$y_{b'e} = g_{b'e} + j\omega C_{b'e}$$
$$y_{b'c} = j\omega C_{b'c}.$$

(b) Assuming that the impedance between B' and E is large compared with $r_{bb'}$, i.e.

$$|x| = \left| \frac{g_{b'e} + j\omega C_{b'e}}{y_{bb'}} \right| \ll 1,$$

and taking account of the known orders of magnitude of the other elements, give approximate expressions for the various y_{ij} of the network to first order in x.

(c) In what frequency range is this approximation justified? How does this range change as the emitter frequency increases?

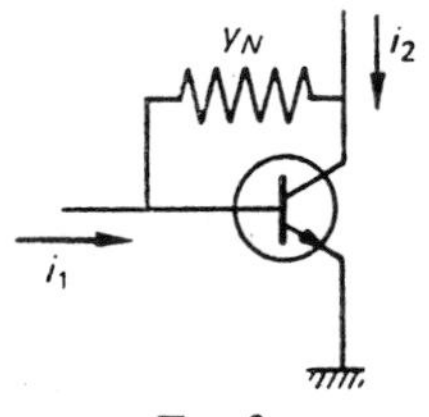

FIG. 3.

(d) An admittance y_N is placed between the collector and base. What should be its value if the transistor input impedance is to be independent of the collector load? Calculate the y parameters of the shunted transistor using the above approximation and deduce the equivalent circuit.

Assume that, in the following, this equivalent circuit is valid for all forms of the shunt y.

2. *Analysis of a single stage.*

The elements of the Giacoletto circuit for the given transistor are

$$\left.\begin{aligned} r_{bb'} &= 100\,\Omega \\ r_{b'e} &= 1500\,\Omega \\ C_{b'e} &= 22\text{ pF} \\ C_{b'c} &= 2\text{ pF} \\ r_{ce} &= 50\text{ k}\Omega \\ g_m &= 35\text{ mA/V} \end{aligned}\right. \quad \text{for} \quad \begin{cases} I_E = 1\text{ mA} \\ V_{EB} = 6\text{ V}. \end{cases}$$

(a) Calculate the cut-off frequency f_β and show whether the preceding approximation is justified.

(b) Assuming that the low-frequency current gain calculated from the equivalent circuit is the same as the d.c. current gain and that I_{CE_0} is zero, calculate the resistances R_{E_1}, R_{E_2} and R_{B_1} and R_{B_2} to give the chosen working point (1 mA, 6 V).

(c) Calculate the output impedance of the amplifier at the resonant frequency as a function of L_2, the inductance of the primary of the second transformer, the ratios n_7/n_5 and n_6/n_5 and the characteristic elements of the transistor. (Take the windings to be lossless.)

(d) Find the pass-band of the output stage when it has a load R_L.

(e) Find the voltage gain of the second stage, with the load R_L, as a function of frequency in the neighbourhood of the resonance.

3. (a) What is the gain of the first stage in the neighbourhood of its matching frequency $\omega_1/2\pi$?

(b) The first stage is matched at ω_1 and the second at ω_2, which is close to ω_1. The primary inductance of the two transformers are, for technical reasons, made to be equal while the ratios of turns are adjusted to give the maximum gain for each stage. What must be the relation between the transformer ratios n_3/n_2 and n_6/n_5 if the two circuits are to have the same loaded Q?

(c) Given the above condition, calculate the overall gain of the amplifier as a function of the frequency in the region of ω_1 and ω_2. Take

$$\omega_0 = \frac{(\omega_2+\omega_1)}{2} \quad \text{and} \quad (\omega_2-\omega_1) = \frac{k\omega_0}{Q},$$

where $k/Q \ll 1$ and take

$$X = 2Qx = 2Q\frac{\omega-\omega_0}{\omega_0}.$$

(d) Show that the maximum of the curve of the absolute gain is very flat for a particular value k_0 and the maximum voltage gain.

(e) If the two frequencies ω_1 and ω_2 are chosen to give $k = k_0$, what is the pass-band of the system at -3 dB?

4. (a) The inductance of the primaries of the transformers is 4 μH. Calculate the unknown elements for an operating frequency of 10 MHz, a pass-band B of 0·1 MHz, a load $R_L = 50\,\Omega$ and the operating conditions given in question 3.

(b) Calculate the capacities required for neutrodyning.

(c) What must be the transformer ratio of the input transformer to give a real part of 50 Ω to the input impedance?

How can the input transformer give a real input impedance?

(d) What overall gain is obtained?

Solution. 1. We use Kirchhoff's laws to give

(a)
$$i_1+(v_{b'e}-v_1)y_{bb'} = 0$$
$$(v_1-v_{b'e})y_{bb'}-y_{b'e}v_{b'e}+(v_2-v_{b'e})y_{b'c} = 0$$
$$(v_{b'e}-v_2)y_{b'e}-v_2 g_{ce}-g_m v_{b'e}+i_2 = 0.$$

Putting $S = y_{bb'}+y_{b'e}+y_{b'c}$ (the sum of the admittances meeting at B') these equations give, on eliminating $v_{b'e}$

$$y_{11} = y_{bb'}\left(1-\frac{y_{bb'}}{S}\right)$$

$$y_{12} = -j\omega\frac{y_{bb'}C_{b'c}}{S}$$

$$y_{21} = y_{bb'}\frac{g_m-j\omega C_{b'c}}{S}$$

$$y_{22} = g_{ce}+j\omega C_{b'c}+j\omega C_{b'c}\left(\frac{g_m-j\omega C_{b'c}}{S}\right).$$

(b) $$S = y_{bb'}+g_{b'e}+j\omega C_{b'e}+j\omega C_{b'c} \simeq y_{bb'}+g_{b'e}+j\omega C_{b'e}$$

(since, in every case, $C_{b'c} \ll C_{b'e}$).

$$\frac{y_{bb'}}{S} = \frac{y_{bb'}}{y_{bc'}+y_{b'n}} = \frac{1}{1+\dfrac{y_{b'e}}{y_{bb'}}} \simeq 1-\frac{y_{b'e}}{y_{bb'}}.$$

Then, to first order in x,

$$\begin{cases} y_{11} = y_{b'e} = g_{b'e} + j\omega C_{b'e} \\ y_{12} = \dfrac{-j\omega C_{b'c}}{1+\dfrac{y_{b'e}}{y_{bb'}}} \simeq -j\omega C_{b'c} \\ y_{21} = g_m - j\omega C_{b'c} \\ y_{22} = g_{ce} + j\omega C_{b'c}\left(1+\dfrac{g_m}{y_{b'e}}\right). \end{cases}$$

(c) We have taken

$$\left|\frac{g_{b'e}+j\omega C_{b'e}}{y_{bb'}}\right| = \left|\frac{g_{b'e}}{y_{bc'}}\left(1+\frac{j\omega C_{b'e}}{g_{b'e}}\right)\right| \ll 1.$$

Now

$$\frac{C_{b'e}}{g_{b'e}} = C_{b'e} r_{b'e} = \frac{1}{\omega_\beta}$$

which is the cut-off frequency for the current gain.

Then

$$\left|\frac{r_{bb'}}{r_{b'e}}\left(1+\frac{j\omega}{\omega_\beta}\right)\right| \ll 1$$

or

$$r_{bb'}\sqrt{1+\left(\frac{\omega}{\omega_\beta}\right)^2} \ll r_{b'e},$$

which requires

$$\frac{\omega}{\omega_\beta} \ll \frac{r_{b'e}}{r_{bb'}}.$$

Here $r_{b'e}$ is the junction resistance corresponding to y_{11} at low frequencies. It decreases as the current increases while $r_{bb'}$ is independent of I_E. Thus the approximation is better at small currents.

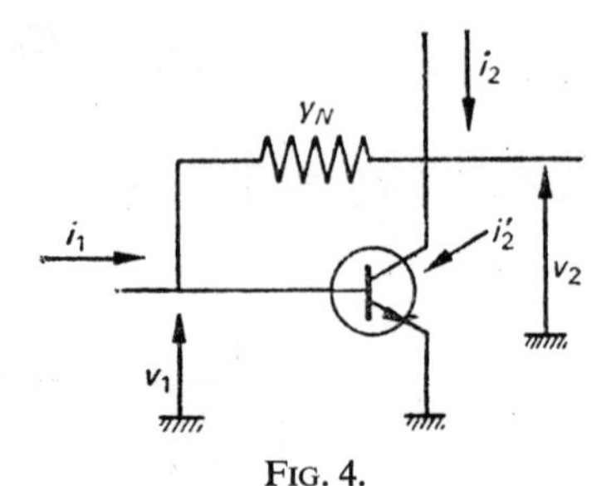

FIG. 4.

(d)

$$\begin{cases} i_1' = i_1+(v_2-v_1)\,y_N \\ i_2' = i_2+(v_1-v_2)\,y_N. \end{cases}$$

Now

$$\begin{cases} i_1' = y_{11}v_1+y_{12}v_2 \\ i_2' = y_{21}v_1+y_{22}v_2. \end{cases}$$

For the complete network this gives

$$\begin{cases} i_1 = (y_{11}+y_N)\,v_1+(y_{12}-y_N)\,v_2 \\ i_2 = (y_{21}-y_N)\,v_1+(y_{22}+y_N)\,v_2, \end{cases}$$

so that the new parameters are

$$\begin{cases} y_{11}' = y_{11}+y_N \quad y_{12}' = y_{12}-y_N \\ y_{21}' = y_{21}+y_N \quad y_{22}' = y_{22}-y_N. \end{cases}$$

If the input impedance is to be independent of the load, we require that there shall be no internal feedback, i.e. $y_{12}' = 0$ or

$$y_N = y_{12}.$$

With the approximation given,

$$y_N = -j\omega C_{b'c}$$

and

$$y_{11}' = g_{b'e}+j\omega C_{b'e}-j\omega C_{b'c} \simeq g_{b'e}+j\omega C_{b'e}$$

$$y_{12}' = 0$$

$$y_{21}' = g_m-j\omega C_{b'c}+j\omega C_{b'c} = g_m$$

$$y_{22}' = g_{ce}+j\omega C_{b'c}\left(1+\frac{g_m}{y_{bb'}}\right) - j\omega C_{b'c} = g_{ce}+j\omega C_{b'c}\frac{g_m}{y_{bb'}}$$

which corresponds to the equivalent circuit of Fig. 5.

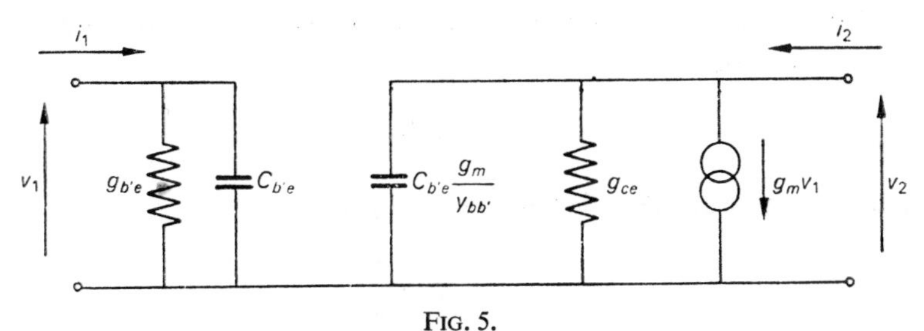

FIG. 5.

2. (a)

$$\omega_\beta = \frac{1}{r_{b'e}C_{b'e}} = \frac{1}{1500\times 22\times 10^{-12}} = \frac{10^9}{33}$$

or

$$f_\beta = \frac{10^9}{2\pi\times 33} = 4\cdot 8 \text{ MHz}.$$

The approximation made in 1 above required that

$$\frac{f}{f_\beta} \ll \frac{r_{b'e}}{r_{bb'}}.$$

Here

$$\frac{f}{f_\beta} \sim 2 \quad \text{and} \quad \frac{r_{b'e}}{r_{bb'}} = 15.$$

There is thus a ratio of 7·5 between the two factors and the resulting error is about 10%.

(b) Clearly

$$R_{E_1} = R_{E_2}, \quad R_{B_1} = R_{B_2}.$$

For $I_c = 1$ mA and $V_c = 6$ V we must lose 6 V in R_E or

$$R_{E_1} = R_{E_2} = \frac{6}{10^{-3}} = 6 \text{ k}\Omega.$$

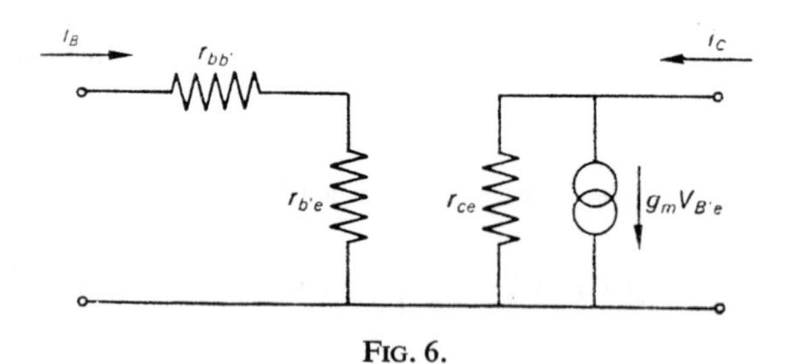

FIG. 6.

The equivalent circuit of Fig. 6 allows the current gain to be calculated and, as stated, this will also be the d.c. gain.

$$V_{b'e} = r_{b'e}I_B.$$

The current generator at the collector is $g_m r_{b'e} I_B$ so that

$$I_c = g_m r_{b'e} I_B + \frac{V_c}{r_{ce}}.$$

Numerically

$$I_c = 52{\cdot}5\, I_B + \frac{V_c}{50\,000}.$$

For $I_c = 1$ mA

$$V_c = 6\, V; \quad I_B = 16{\cdot}8\, \mu\text{A}.$$

Neglecting V_{BE} we have immediately

$$R_{B_1} = R_{B_2} = \frac{12-6}{I_B} = 358\ \text{k}\Omega.$$

(c) Using the results of 1(d), the neutrodyned second stage is as shown in Fig. 7, which can be transformed into Fig. 8 by incorporating all the impedances within the transistor terminals.

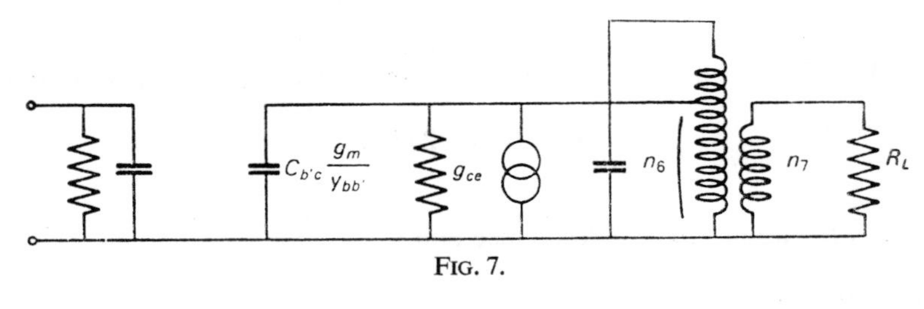

FIG. 7.

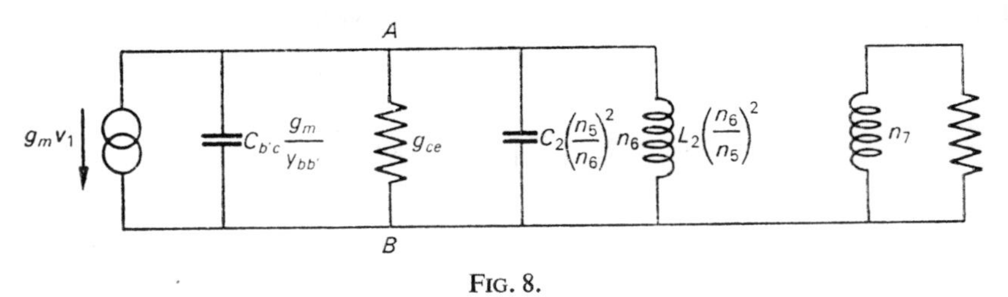

FIG. 8.

At resonance this circuit presents a purely resistive impedance across AB equal to r_{ce}. The output impedance is then

$$R_o = r_{ce}\left(\frac{n_7}{n_6}\right)^2.$$

(d) The complete system can be taken as an oscillating circuit supplied by a current source $g_m v_1$. In a general manner the pass-band is

$$B = f_0/Q.$$

Now

$$Q = \frac{R}{\omega_0 L} = R\sqrt{\frac{C}{L}} \quad \text{and} \quad f_0 = \frac{1}{2\pi\sqrt{LC}}.$$

Thus

$$B = \frac{1}{2\pi RC}.$$

The damping resistance consists of r_{ce} in parallel with the transformed load $R_L(n_6/n_7)^2$. Then

$$B = \frac{1}{2\pi}\,\frac{\dfrac{1}{R_L}\left(\dfrac{n_7}{n_6}\right)^2 + g_{ce}}{\dfrac{g_m}{y_{bb'}}C_{b'e} + C_2\left(\dfrac{n_5}{n_6}\right)^2}.$$

(e) In the neighbourhood of resonance the admittance of an oscillator circuit is

$$Y = G_0(1+2jQx),$$

where G_0 is the admittance at resonance and $x = \dfrac{\Delta\omega}{\omega_2} = \dfrac{\omega-\omega_2}{\omega_2}$, ω_2 being the pulsatance for matching.

From the equivalent circuit of Fig. 9 the collector voltage is seen to be

$$\frac{g_m v_1}{\left[g_{ce} + R_L^{-1}\left(\dfrac{n_6}{n_7}\right)^2\right][1+2jQ_2 x]}.$$

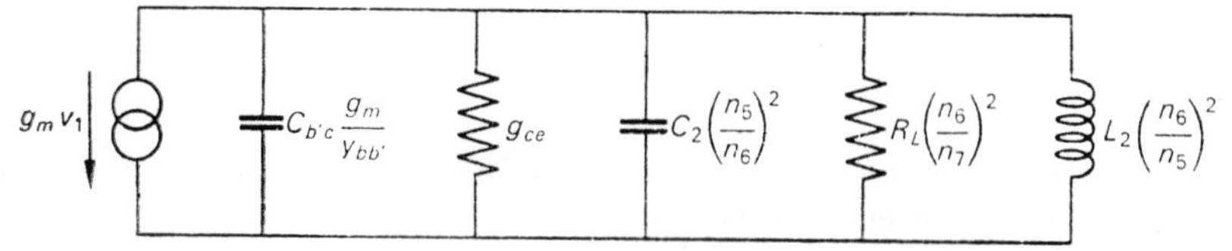

FIG. 9.

The quality factor Q_2 of this circuit is

$$Q_2 = R\sqrt{\frac{C}{L}} = \frac{1}{g_{ce} + R_L^{-1}\left(\dfrac{n_6}{n_7}\right)^2}\sqrt{\frac{C_{b'c}\dfrac{g_m}{y_{bb'}} + C_2\left(\dfrac{n_5}{n_6}\right)^2}{L_2\left(\dfrac{n_6}{n_5}\right)^2}}.$$

From this the voltage gain is

$$G_v = \frac{g_m \left(\frac{n_7}{n_6}\right)}{\left[g_{ce} R_L^{-1} \left(\frac{n_7}{n_6}\right)^2\right] [1+2jQ_0 x]},$$

where account has been taken of the output transformer.

3. (a) For the first stage the treatment is exactly the same. The equivalent circuit obtained by taking the impedances in parallel with the collector is shown in Fig. 10.

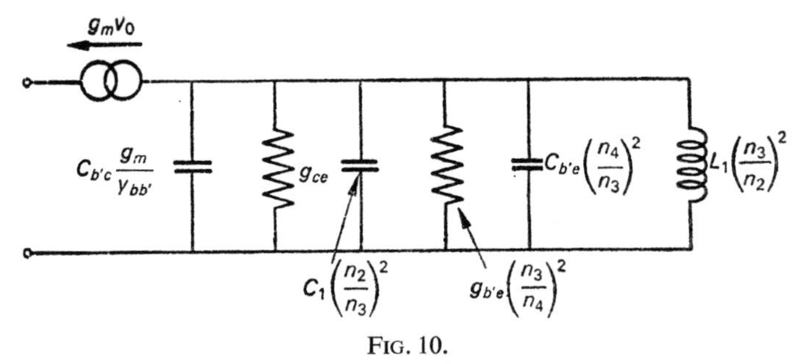

FIG. 10.

Taking account of the transformer ratio n_4/n_3, the voltage gain is

$$G_{V_1} = \frac{g_m \left(\frac{n_4}{n_3}\right)}{\left[g_{ce} + g_{b'e} \left(\frac{n_4}{n_3}\right)^2\right] [1+2jQ_1 x]}, \quad x = \frac{\omega - \omega_1}{\omega_1};$$

with the loaded Q-factor

$$Q_1 = \frac{1}{g_{ce} + g_{b'e} \left(\frac{n_4}{n_3}\right)^2} \sqrt{\frac{C_{b'e} \frac{g_m}{y_{bb'}} + C_1 \left(\frac{n_2}{n_3}\right) + C_{b'e} \left(\frac{n_4}{n_3}\right)^2}{L_1 \left(\frac{n_3}{n_2}\right)^2}}.$$

(b) The maximum power will be obtained at each stage if there is matching of the impedances. This requires that the load transferred to each output is equal to the impedance of each circuit alone. The parallel admittance of the load is then $2g_{ce}$ which requires

$$g_{b'e} \left(\frac{n_4}{n_3}\right)^2 = g_{ce} \quad \text{for the first stage}$$

and

$$R_L\left(\frac{n_6}{n_7}\right)^2 = g_{ce}^{-1} \quad \text{for the second.}$$

The Q-factor is $R/\omega L$ with R as the total parallel resistance. For the first stage

$$Q_1 = \frac{1}{2g_{ce}}\frac{1}{\omega_1}\frac{1}{L}\left(\frac{n_2}{n_3}\right)^2;$$

and for the second

$$Q_2 = \frac{1}{2g_{ce}}\frac{1}{\omega_2}\frac{1}{L}\left(\frac{n_5}{n_6}\right)^2.$$

The relation to be satisfied is then

$$\frac{1}{\omega_1}\left(\frac{n_2}{n_3}\right)^2 = \frac{1}{\omega_2}\left(\frac{n_5}{n_6}\right)^2.$$

(c) In the case of optimum matching and with two equal Q-factors, the gains of the two stages become (ignoring for the moment the input transformer)

$$G_{V_1} = \frac{g(n_4/n_3)}{2g_{ce}(1+2jQx_1)}$$

$$G_{V_2} = \frac{g(n_7/n_6)}{2g_{ce}(1+2jQx_2)}$$

with

$$x_1 = \frac{\omega-\omega_1}{\omega_1}, \quad x_2 = \frac{\omega-\omega_2}{\omega_2}.$$

Now we have

$$2\omega_0 = \omega_1+\omega_2 \quad \text{and} \quad k\frac{\omega_0}{Q} = \omega_2-\omega_1;$$

from which we get

$$\begin{cases} \omega_1 = \omega_0\left(1-\dfrac{k}{2Q}\right), \\ \omega_2 = \omega_0\left(1+\dfrac{k}{2Q}\right), \end{cases}$$

and

$$x_1 = \frac{\omega}{\omega_1}-1 = \frac{\omega}{\omega_0\left(1-\dfrac{k}{2Q}\right)}-1 \approx \frac{\omega}{\omega_0}\left(1+\frac{k}{2Q}\right)-1.$$

Similarly

$$x_2 = \frac{\omega}{\omega_0}\left(1-\frac{k}{2Q}\right)-1.$$

Then

$$G_{V_1} = \frac{g_m n_4}{2g_{ce}n_3}\,\frac{1}{(1+2jQx_1)}$$

$$= \frac{g_m n_4}{2g_{ce}n_3}\,\frac{1}{1+2jQx+jk\dfrac{\omega}{\omega_0}},$$

$$G_{V_2} = \frac{g_m n_7}{2g_{ce}n_6}\,\frac{1}{1+2jQx-jk\dfrac{\omega}{\omega_0}}, \quad \text{with} \quad x = \frac{\omega-\omega_0}{\omega_0}.$$

ω/ω_0 being close to 1, the total gain can be written

$$G = \frac{g_m^2}{4g_{ce}^2}\,\frac{1}{1+2jQx+jk}\,\frac{1}{1+2jQk-jk}\,\frac{n_4}{n_3}\,\frac{n_7}{n_6}.$$

For convenience we put

$$X = 2Qx.$$

Then

$$G = \frac{g_m^2}{4g_{ce}^2}\,\frac{1}{1+j(X+k)}\,\frac{1}{1+j(X-k)}\,\frac{n_4}{n_3}\,\frac{n_7}{n_6}.$$

(d) It is only necessary to consider the modulus of the denominator

$$D = [1+j(X+k)]\,[1+j(X-k)] = 1+k^2-X^2+2jX,$$

$$|D|^2 = (1+k^2-X^2)^2+4X^2.$$

There will be a maximum for $X = 0$, which makes the derivative zero;

$$\frac{d\,|D|^2}{dX} = 4X^3+4X(1-k^2) = 4X[CX^2+1-k^2].$$

The curve will be very flat if the second derivative is also zero. Now

$$\frac{d^2\,|D|^2}{dX} = 12X^2+4-4k^2$$

will be zero for $X = 0$ if $k = \pm 1 = k_0$.

Then

$$G_{v\max} = \frac{g_m^2}{4g_{ce}^2}\,\frac{n_4}{n_3}\,\frac{n_7}{n_6}\left(\frac{1}{1+j}\,\frac{1}{1-j}\right) = \frac{g_m^2}{8g_{ce}^2}\,\frac{n_4}{n_3}\,\frac{n_7}{n_6}.$$

(e) $k = k_0 = 1$. Thus

$$G = \frac{g_m^2}{4g_{ce}^2} \frac{1}{1+j(X+1)} \frac{1}{1+j(X-1)} \frac{n_4}{n_3} \frac{n_7}{n_6}.$$

The limits of the pass-band occur when

$$G = \frac{G_{max}}{\sqrt{2}}$$

or

$$|G| = \frac{g_m^2}{8\sqrt{2}g_{ce}^2} \frac{n_4}{n_3} \frac{n_7}{n_6}.$$

The condition required is then

$$|[1+j(X+1)][1+j(X-1)]| = 2\sqrt{2},$$

or

$$(2-X^2)^2+4X^2 = 8$$

so that

$$X^4+4 = 8, \quad X = \sqrt{2}.$$

The pass-band at 3 dB is thus

$$B = \frac{\sqrt{2}\omega_0}{2\pi Q} = \sqrt{2} \cdot \frac{f_0}{Q}.$$

4. (a) Since the pass-band is given, the common Q of the two circuits may be found immediately as

$$Q = \sqrt{2}\frac{f_0}{B} = \sqrt{2} \times \frac{10}{0{\cdot}1} = 141{\cdot}4.$$

The two resonant frequencies will be

$$f_1, f_2 = f_0\left(1 \mp \frac{k}{2Q}\right), \quad \text{with} \quad k = k_0 = 1;$$

$$f = 10\left(1 \mp \frac{1}{282{\cdot}8}\right),$$

or

$$\begin{cases} f_2 = 10{\cdot}0353 \text{ MHz}, \\ f_1 = 9{\cdot}9647 \text{ MHz}. \end{cases}$$

Above we have written the expressions for Q_1 and Q_2 and we thus get the transformer ratios as

$$\left(\frac{n_6}{n_5}\right)^2 = \frac{1}{Q} \frac{1}{2g_{ce}} \frac{1}{\omega_2} \frac{1}{L} = 0{\cdot}701; \left(\frac{n_6}{n_5}\right) = 0{\cdot}836.$$

With ω_2 and ω_1 nearly equal, n_3/n_2 has almost the same value. The two other transformer ratios are determined by the matching of impedances as

$$\left(\frac{n_6}{n_7}\right)^2 = \frac{1}{R_L g_{ce}} = 1000 \rightarrow \frac{n_6}{n_7} = 31{\cdot}6$$

$$\left(\frac{n_4}{n_3}\right)^2 = \frac{g_{ce}}{g_{b'e}} = \frac{1}{33{\cdot}3} \rightarrow \frac{n_4}{n_3} = \frac{1}{5{\cdot}76}.$$

It remains to find C_1 and C_2 for which we return to the equivalent circuits. For the first stage we have an inductance

$$L_1' = L_1\left(\frac{n_3}{n_2}\right)^2$$

which is tuned by

$$C_{b'c}\frac{g_m}{y_{bb'}} + C_1\left(\frac{n_2}{n_3}\right)^2 + C_{b'e}\left(\frac{n_4}{n_3}\right)^2.$$

Now, for $L_1' = 4\times10^{-6}\times0{\cdot}4 = 2{\cdot}8\ \mu\text{H}$, the total tuning capacity must be

$$C_T = \frac{1}{L_1'\omega_1^2} = 90{\cdot}5\ \text{pF}.$$

Now

$$C_{b'c}\frac{g_m}{y_{bb'}} = 3{\cdot}5\ \text{pF}$$

and

$$C_{b'e}\left(\frac{n_4}{n_3}\right)^2 = 22\times\frac{1}{5{\cdot}76} = 3{\cdot}81\ \text{pF};$$

and, remaining, we have

$$C_1\left(\frac{n_2}{n_3}\right)^2 = 90{\cdot}5-3{\cdot}5-3{\cdot}81 = 83{\cdot}2\ \text{pF},$$

or

$$C_1 = 83{\cdot}2\times0{\cdot}7 = 58{\cdot}3\ \text{pF}.$$

For the second stage, the inductance $L_2' = L(n_6/n_5)^2$ is tuned by

$$C_{b'c}\frac{g_m}{y_{bb'}} + C_2\left(\frac{n_5}{n_7}\right)^2.$$

Since (n_6/n_5) is nearly equal to (n_3/n_2) we get

$$C_2\left(\frac{n_5}{n_6}\right)^2 = 87\ \text{pF}, \quad \text{from which} \quad C_2 = 87\times0{\cdot}7 = 61\ \text{pF}.$$

(b) The transformer ratio between the collector and the output winding is, for the first stage,

$$\frac{n_3}{n_4} = 5{\cdot}76.$$

The condition $y_{12} = y_N$ will be satisfied if

$$C_N = 5{\cdot}76 \times C_{b'c},$$

or

$$C_{N1} = 5{\cdot}76 \text{ pF}.$$

Similarly, for the output stage,

$$C_{N2} = C_{b'c} \times \frac{n_6}{n_7} = 31{\cdot}6 \text{ pF}.$$

(c) The input impedance of the neutrodyned transistor is formed from the resistance $r_{b'e}$ in parallel with the capacity $C_{b'e}$.

At the transformer primary we find

$$\frac{r_{b'e}}{n^2} \quad \text{and} \quad n^2 C_{b'e};$$

$$\frac{1500}{n^2} = 50, \quad \text{from which} \quad n = \sqrt{\frac{1500}{50}} = 5{\cdot}47.$$

For the input impedance to be real we require a heavily damped tuned circuit of resonant frequency ω_0. This will be obtained if the input transformer has a secondary inductance L_i such that

$$\omega_0^2 L_i C_{b'e} = 1$$

or

$$L_i \simeq 11{\cdot}4 \ \mu\text{H}.$$

(d) We have seen that, neglecting the input transformer, the gain of the two stages is

$$G_{v\,\max} = \frac{g_m^2 n_4 n_7}{8 g_{ce}^2 n_3 n_6} = \frac{1}{8}\left(\frac{35 \times 10^{-3}}{0{\cdot}2 \times 10^{-4}}\right)^2 \left(\frac{n_4}{n_3}\,\frac{n_7}{n_6}\right),$$

$$G_{v\,\max} \simeq 2100.$$

Thus, taking this transformer into account

$$G_v = 11\,500.$$

Problem No. 7

DELAY LINE OSCILLATOR

Here we study an L.F. oscillator based on an *RC* phase-change circuit as shown in Fig. 1.

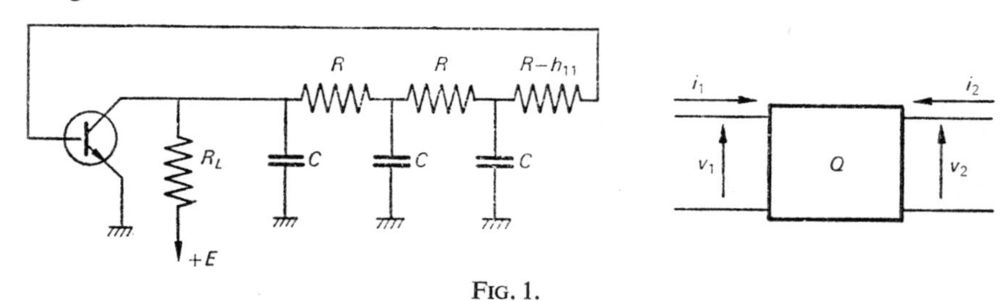

FIG. 1.

1. Draw the equivalent circuit assuming that the internal feedback in the transistor can be ignored. (Take the transistor to be defined by its parameters h_{11e}, $h_{12e} = 0$, $h_{21e} = \beta$, $h_{22e} = 1/r_c$.) This circuit can be put in the form of a current generator feeding into a ladder filter for which the short-circuit output current is the base current of the transistor.

2. What is the expression for the current gain h_{21} for a four-terminal network in terms of the elements of the transfer matrix defined by

$$\begin{pmatrix} v_1 \\ i_1 \end{pmatrix} = (T)\begin{pmatrix} v_2 \\ -i_2 \end{pmatrix}?$$

3. Calculate the h_{12} parameter of the ladder filter of the first question. (Use the substitution $x = \omega CR$.)

4. Find the condition for oscillation and hence the oscillation frequency.

5. What is the smallest current gain of the transistor which will allow oscillation for a given load R_L?

6. Assuming that $h_{22} = 0$ what is the maximum collector voltage in the absence of oscillation which will allow oscillation to commence? Let αE be the collector voltage when the oscillation is cut-off (by, for example, removing the capacitors) and calculate the smallest value of β for oscillation in terms of α.

Solution. 1. With $h_{12} = 0$ the transistor equations become

$$\begin{cases} v_1 = h_{11}i_1 \\ i_2 = \beta i_1 + \dfrac{v_2}{r_c}. \end{cases}$$

The equivalent circuit is thus that of Fig. 2.

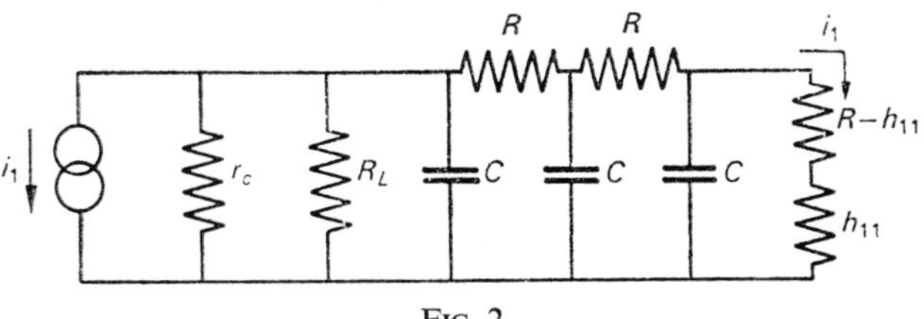

Fig. 2.

2. We have

$$\begin{cases} v_1 = T_{11}v_2 - T_{12}i_2 \\ i_1 = T_{21}v_2 - T_{22}i_2. \end{cases}$$

But $h_{21} = (i_2/i_1)_{v_2=0}$ so that, with $v_2 = 0$, we see

$$h_{21} = \frac{-1}{T_{22}}.$$

3. We wish to calculate the transfer matrix for the network of Fig. 3. For this purpose it is simplest to use the matrix method applied to the sections of the filter taken in series as indicated.

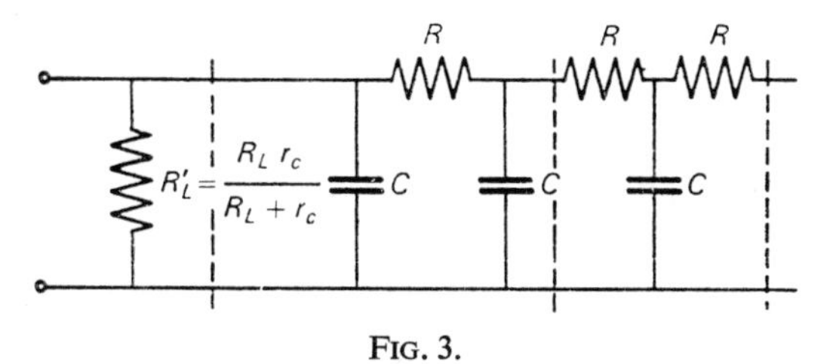

Fig. 3.

For the first element, being R_L in parallel with r_c, as in Fig. 4,

$$v_2 = v_1$$

$$i_1 = \frac{v_2}{R'_L} - i_2,$$

so that

$$(T_1) = \begin{pmatrix} 1 & 0 \\ \dfrac{1}{R'_L} & 1 \end{pmatrix}.$$

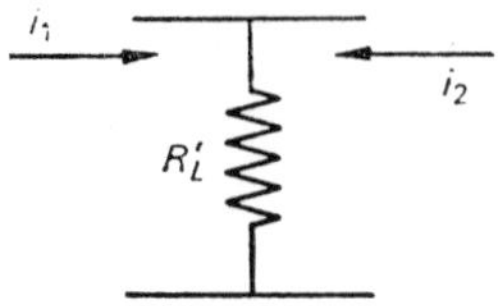

Fig. 4.

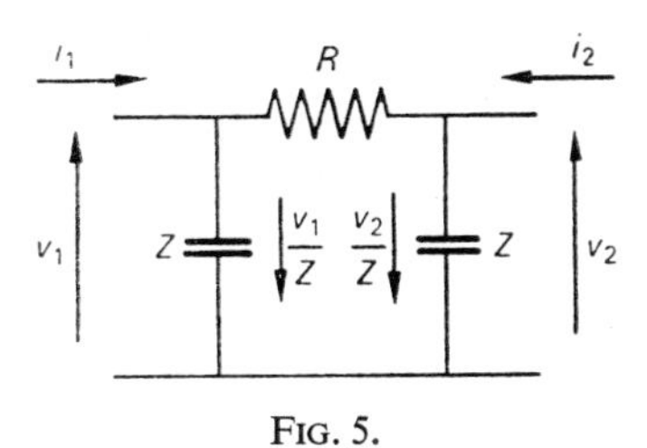

FIG. 5.

For the π-circuit of Fig. 5, Kirchhoff's equations give

$$i_1 - \frac{v_1}{Z} + \frac{v_2 - v_1}{R} = 0$$

$$i_2 - \frac{v_2}{Z} + \frac{v_1 - v_2}{R} = 0.$$

Putting $Z = 1/\text{j}\omega C$ and rearranging gives

$$(T_2) = \begin{pmatrix} 1+\text{j}\omega CR & R \\ \text{j}\omega C(2+\text{j}\omega CR) & 1+\text{j}\omega CR \end{pmatrix}.$$

The same calculation for the T filter gives

$$(T_3) = \begin{pmatrix} 1+\text{j}\omega CR & R(2+\text{j}\omega CR) \\ \text{j}\omega C & 1+\text{j}\omega CR \end{pmatrix}.$$

The total transfer matrix is the product of the three matrices

$$(T) = (T_1)(T_2)(T_3)$$

$$(T_1)(T_2) = \begin{pmatrix} 1 & 0 \\ \frac{1}{R_L'} & 1 \end{pmatrix} \begin{pmatrix} 1+\text{j}\omega CR & R \\ \text{j}\omega C(2+\text{j}\omega CR) & 1+\text{j}\omega CR \end{pmatrix}$$

$$= \begin{pmatrix} 1+\text{j}\omega CR & R \\ \frac{1}{R_L'}(1+\text{j}\omega CR)+\text{j}\omega C(2+\text{j}\omega CR) & \frac{R}{R_L'}+1+\text{j}\omega CR \end{pmatrix}.$$

When we multiply by (T_3) we are only interested in T_{22} which is

$$T_{22} = R\left[\frac{1}{R_L'}(1+\text{j}\omega CR)+\text{j}\omega C(2+\text{j}\omega CR)\right](2+\text{j}\omega CR)$$

$$+\left[\frac{R}{R_L'}+1+\text{j}\omega CR\right](1+\text{j}\omega CR).$$

Putting $x = \omega CR$ gives

$$T_{22} = \left(\frac{3R}{R_L'}+1\right) - x^2\left(\frac{R}{R_L'}+5\right) + \mathrm{j}x\left(6+4\frac{R}{R_L'}-x^2\right) = \frac{-1}{h_{21}}.$$

4. By the definition of the transfer matrix, if $-i_1$ is the output current of the shorted network, the input current is $T_{22}i_1$.

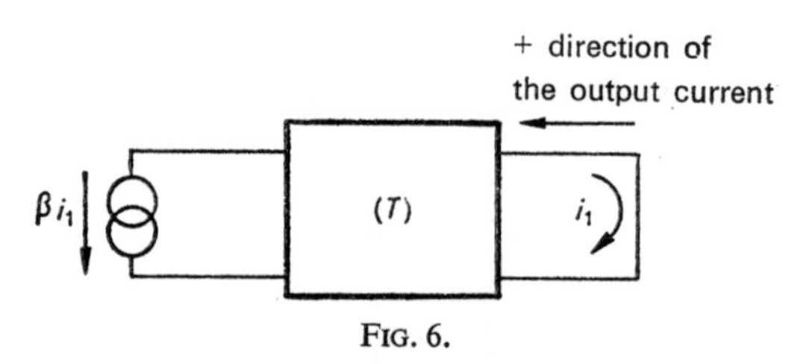

FIG. 6.

The condition for oscillation will be fulfilled if this input current is just equal to the transistor collector current or

$$-\beta i_1 = T_{22}i_1,$$

which requires

$$\beta + \left(\frac{3R}{R_L}+1\right) - x^2\left(\frac{R}{R_L'}+5\right) + \mathrm{j}x\left(6+\frac{4R}{R_L'}-x^2\right) = 0.$$

The imaginary part has to be zero so that

$$6+\frac{4R}{R_L'}-x^2 = 0,$$

or

$$x = \sqrt{6+\frac{4R}{R_L'}} = \omega_0 CR.$$

The oscillation frequency is thus

$$\omega_0 = \frac{1}{CR}\sqrt{6+\frac{4R}{R_L'}}.$$

5. For the real part to be zero at ω_0 requires that

$$\beta + \left(\frac{3R}{R_L'}+1\right) - \left(6+\frac{4R}{R_L'}\right)\left(\frac{R}{R_L'}+5\right) = 0.$$

There is thus a limiting value of β for oscillation and this value is smaller for large values of R_L'. In the limit $R_L' \to \infty$ we have

$$\beta+1-30 = 0$$

$$\beta = 29.$$

If R_L' is finite, the equation requires

$$\beta \geqslant 29+23\frac{R}{R_L'}+4\frac{R^2}{R_L'^2}.$$

6. If E is the collector voltage, the base current is close to

$$\frac{\alpha E}{3R}$$

so that the collector current is

$$I_c \simeq \frac{\beta\alpha E}{3R},$$

which is also

$$\frac{(1-\alpha)E}{R_L}.$$

The condition to be satisfied is thus

$$\frac{R}{R_L} = \frac{\beta}{3}\frac{\alpha}{1-\alpha} = \frac{k\beta}{3}, \quad \text{with} \quad k = \frac{\alpha}{1-\alpha}.$$

For oscillation β must satisfy the inequality of solution 5 which is now

$$\beta \geqslant 29+\frac{23}{3}k\beta+\frac{4}{9}k^2\beta^2,$$

from which

$$\frac{4k^2}{9}\beta^2+\left(\frac{23}{3}k-1\right)\beta+29 \leqslant 0.$$

This inequality will be satisfied if β lies between the roots, provided that there are positive roots for the expression. There will be roots if the discriminant is positive, i.e.

$$65k^2-138k+9 \geqslant 0$$

and this requires k to be outside the interval defined by the roots of this expression,

$$k < \frac{5 \cdot 4}{6 \cdot 5} = 0 \cdot 0846 \quad \text{or} \quad k > \frac{133 \cdot 6}{65} = 2 \cdot 055.$$

Since the two roots for β are the same sign (the product is positive) they will be positive if their sum is positive. That is, if

$$\frac{-23k}{3}+1 > 0$$

or

$$k < \frac{3}{23}.$$

The two conditions are satisfied for $k < 0{\cdot}0846$ which gives

$$\alpha < 0{\cdot}078.$$

For this limiting value, the two roots of β are the same and there exists a single value β which will give oscillation, namely

$$\beta \simeq 55.$$

Problem No. 8

Q-METER FOR MAGNETIC RESONANCE

1. Consider the amplifier represented in Fig. 1 by a four-terminal network supplying a current i into a resistance r where i is related to the input voltage $u = u_0 \cos \omega t$ by $i = au - bu^3$.

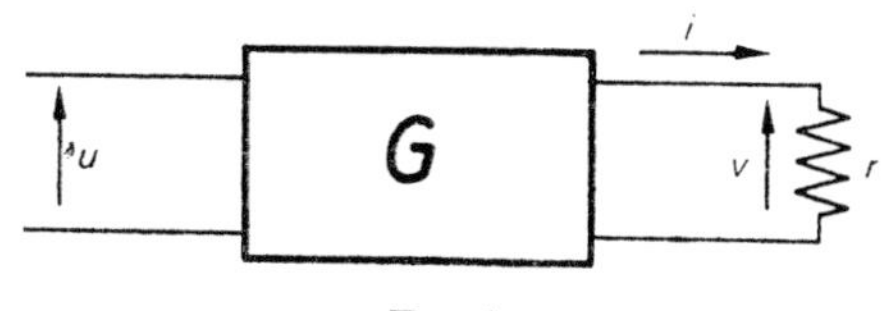

FIG. 1.

(a) If the voltage across r is v, calculate the amplifier gain $G = v/u$, for the fundamental current frequency, in terms of u_0.

(b) Sketch the variation of the gain with input voltage.

2. An oscillator is constructed in the manner of Fig. 2 where Z is the impedance of a parallel resonant circuit as represented in Fig. 3. Assume that the impedance of Z and R in series is very much larger than r.

(a) Calculate the limiting condition for the maintenance of oscillations and their frequency ω_1.

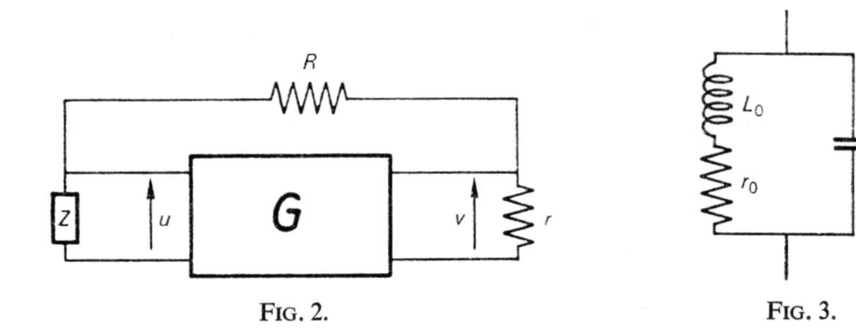

FIG. 2.

FIG. 3.

(b) With this condition satisfied, calculate the amplitude of the oscillations across Z.

(c) Show that the result of 2(b) can be obtained graphically.

3. Consider that, in Fig. 2, the quality factor of the input circuit is changed by ΔQ.

(a) Calculate the fractional amplitude change $\Delta u_0/u_0$ in terms of ar, R, R_0 and $\Delta Q/Q$, R_0 being the impedance of the tuned circuit at resonance.

(b) For a given $\Delta Q/Q$, sketch the variation of $\Delta u_0/u_0$ as a function of $x = R/R_0$, and indicate the region of this curve over which the system will oscillate.

(c) It is required to relate the sensitivity of the oscillation level to the variations of Q in the above circuit, which can be considered as a Q-meter.

For this purpose the L_0C_0 circuit is coupled to a generator (Fig. 4) such that variation ΔQ does not change the total impedance of the load as seen from the terminals A and B. For this purpose, a capacitance C is placed in series with Z so that its impedance is very large at the resonant frequency. The generator can thus be considered to supply a constant current i to the oscillator circuit. Calculate in this case the fractional change $\Delta u_0/u_0$ in terms of $\Delta Q/Q$ and compare this variation with that for the preceding circuit. Deduce the condition for which the circuit of Fig. 4 is more sensitive than the Q-meter.

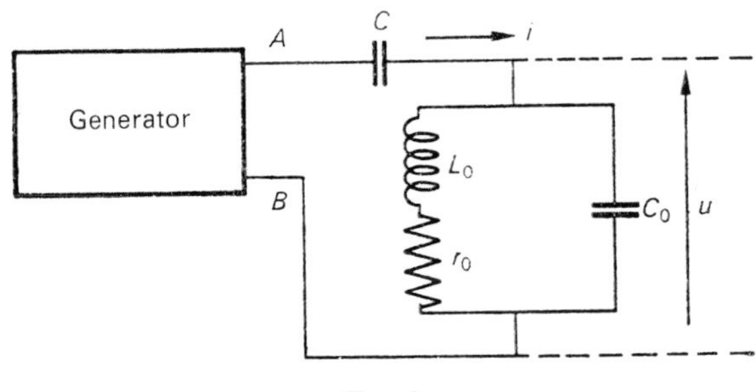

FIG. 4.

4. Considering again Fig. 2, the variation of Q is produced by a variation of L_0 to $L_0(1+\chi)$ where χ is a complex number $(\chi' - j\chi'')$.

(a) Show that this change modifies the elements of the oscillating circuit.

(b) Calculate the change $\Delta Q/Q$ in terms of Q and χ'' assuming $Q \gg 1$ and that χ', which is of the same order as χ'', is very much less than one. Calculate the change $\Delta\omega/\omega_1$ in the oscillator frequency.

(c) An apparatus of this type is used for the detection of a magnetic resonance signal where χ then represents the magnetic susceptibility of a specimen placed inside the coil L_0.

Given that:

$$ar = 1{\cdot}42, \quad R = 4\ \text{k}\Omega, \quad L_0 = 1\ \mu\text{H},$$
$$Q = 100, \quad \omega_0 = 10^8\ \text{rad/s}, \quad \chi'' = 10^{-5},$$

find

$$R_0, \quad x = R/R_0 \quad \text{and} \quad \Delta u_0/u_0.$$

(Besançon, 1965)

Solution. 1. (a) $v = ri = rau - rbu^3$.

Now $\quad u = u_0 \cos \omega t,$

so that

$$u^3 = u_0^3 \cos^3 \omega t = \frac{u_0^2}{4}[3 \cos \omega t + \cos 3\,\omega t]$$

and the gain at the frequency ω is

$$\frac{v(\omega)}{u_0} = r\left[a - \frac{3bu_0^2}{4}\right].$$

(b) This gain has a parabolic variation as shown in Fig. 5.

2. (a) To find the limit conditions we consider that the input circuit is cut as shown in Fig. 6 and require that the voltages at A_1 and A_2 shall be equal.

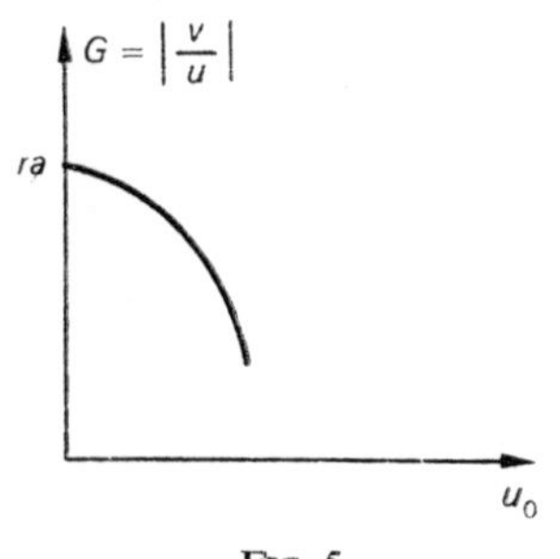

Fig. 5.

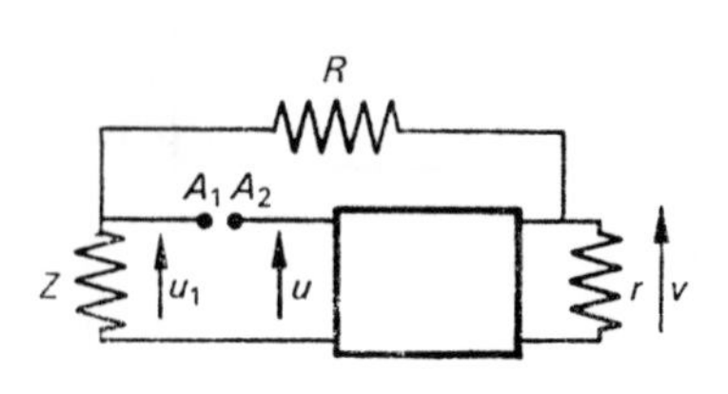

Fig. 6.

At a frequency ω:

$$v = r\left(a - \frac{3bu_0^2}{4}\right)u$$

and

$$u_1 = v\frac{Z}{R+Z} = \frac{v}{\dfrac{R}{Z}+1} = u.$$

This condition can only be satisfied for $(1+R/Z)$ real, i.e. for Z real. This only occurs at resonance. Now

$$Z = \frac{r_0 + j\omega\,[L_0(1-\omega^2 L_0 C_0) - C_0 r_0^2]}{(1-\omega^2 L_0 C_0)^2 + \omega^2 C_0^2 r_0^2}$$

which is real for $\omega = \omega_1$ where, with $\omega_0^2 = (L_0 C_0)^{-1}$,

$$\omega_1^2 = \omega_0^2 - \frac{r_0^2}{L_0^2}.$$

(b) At this frequency

$$Z = R_0 = \frac{1}{\omega_0^2 C_0^2 r_0} = \frac{\omega_0^2 L_0^2}{r_0}.$$

We require

$$r\left(a - \frac{3bu_0^2}{4}\right)u = \left(1 + \frac{R}{R_0}\right)u$$

from which the amplitude of oscillation is

$$u_0 = \sqrt{\frac{4}{3br}\left(ra - 1 - \frac{R}{R_0}\right)}.$$

(c) The graphical construction is especially interesting when the curve $v/u = f(u_0)$ is not algebraically simple. From the relation

$$f(u_0) = 1 + \frac{R}{R_0}$$

for an oscillation frequency close to ω_0 it is sufficient to cut the curve $f(u_0)$ by the horizontal line $(1+R/R_0)$ to give the equilibrium amplitude.

3. If the Q varies, so does R_0 but not ω_0. Putting $Z = Q/\omega_0 C_0$, which is the exact value of the impedance at resonance, we get for u_0,

$$u_0 = \sqrt{\frac{4}{3br}\left(ra - 1 - \frac{\omega_0 C_0 R}{Q}\right)},$$

from which

$$\Delta u_0 = \frac{\frac{2}{3}\frac{\omega_0 C_0 R}{brQ^2}}{\sqrt{\frac{4}{3br}\left(ra-1-\frac{\omega_0 C_0 R}{Q}\right)}}\Delta Q.$$

That is

$$\frac{\Delta u_0}{u_0} = \frac{\omega_0^2 C_0^2 r_0 R}{2(ra-1-\omega_0^2 C_0^2 r_2 R)}\frac{\Delta Q}{Q}.$$

Substituting for R_0 gives

$$\frac{\Delta u_0}{u_0} = \frac{R/R_0}{2(ra-1-R/R_0)}\frac{\Delta Q}{Q}$$

$$= \frac{x}{(ra-1-x)}\frac{\Delta Q}{Q}.$$

(b) For weak oscillations (i.e. near the onset of the oscillations)

$$\frac{v}{u} = ra$$

and the oscillation condition at $\omega = \omega_1$ becomes

$$rau_0\frac{R_0}{R+R_0} = u_0$$

or

$$x = \frac{R}{R_0} = ra-1.$$

If $R > (ra-1)R_0$ there is insufficient feedback oscillation to occur. (This condition follows also from the expression for u_0 where the square root must be real.)

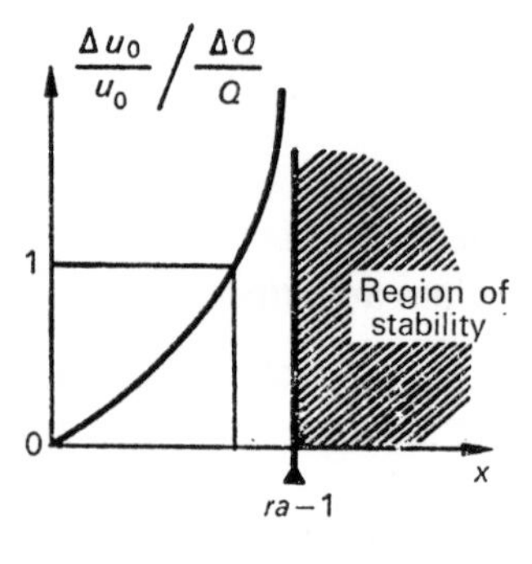

FIG. 7.

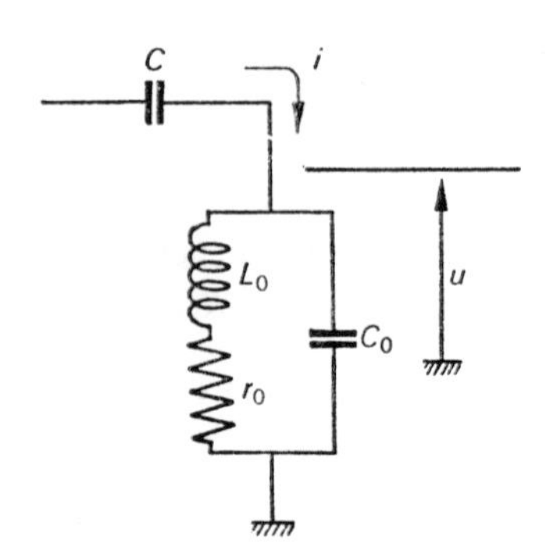

FIG. 8.

(c) If the input frequency is close to ω_0 and the circuit is only slightly damped, we have

$$|u| = |i| \frac{Q}{\omega C_0},$$

from which

$$\Delta u = |i| \frac{\Delta Q}{\omega C_0}.$$

But, effectively, $|i| = u_0/R_0$ and $Q = \omega_0 C_0 R_0 \simeq \omega C_0 R_0$ so that

$$\frac{\Delta u}{u_0} = \frac{\Delta Q}{Q}.$$

The marginal oscillator is more sensitive than the Q-meter if

$$\frac{x}{2(ra-1-x)} > 1,$$

or

$$x > \tfrac{2}{3}(ra-1).$$

4. (a) $$L = L_0(1+\chi'-j\chi'') = L_0(1+\chi')-jL_0\chi''.$$

Thus $j\omega L_0$ becomes

$$j\omega L_0(1+\chi')+\omega L_0\chi''.$$

The circuit behaves as if, in the tuned circuit, C_0 remains constant, L_0 becomes $L_0(1+\chi')$ and r_0 becomes $r_0+\omega L_0\chi''$.

(b) If r is the resistance representing the losses, L the inductance and C' the capacity, then

$$\omega_0^2 LC' = 1$$

and

$$\frac{\Delta L}{L}+\frac{\Delta C'}{C'}+2\frac{\Delta\omega_0}{\omega_0} = 0.$$

Then with C' constant, the resonant frequency obeys the relation

$$\frac{\Delta\omega_0}{\omega_0} = -\frac{1}{2}\frac{\Delta L}{L} = -\frac{1}{2}\chi'.$$

The Q-factor is

$$Q = \frac{1}{\omega_0 C' r}, \quad \text{so that} \quad \frac{\Delta Q}{Q} = -\frac{\Delta r}{r}-\frac{\Delta\omega_0}{\omega_0}.$$

That is

$$\frac{\Delta Q}{Q} = -\frac{\omega L_0 \chi''}{r_0} + \frac{1}{2}\chi'.$$

Now $\frac{\omega L_0}{r_0} \simeq Q$, so that

$$\frac{\Delta Q}{Q} = -Q\chi'' + \frac{1}{2}\chi'.$$

Since Q is large and χ' and χ'' are of the same order we can write

$$\frac{\Delta Q}{Q} = -Q\chi''.$$

The working frequency is given by

$$\omega_1^2 = \omega_0^2 - \frac{r_0^2}{L_0^2} \quad \text{or} \quad \omega_1 \simeq \omega_0\left(1 - \frac{1}{2Q}\right).$$

Since $(1-1/2Q)$ is very nearly unity, we can write

$$\frac{\Delta\omega_1}{\omega_1} \simeq \frac{\Delta\omega_0}{\omega_0} = -\frac{1}{2}\chi'.$$

(c)

$$R_0 = Q\omega_0 L_0 = 10\ \text{k}\Omega$$

$$x = R/R_0 = 0{\cdot}4$$

$$\chi'' = 10^{-5} \quad \text{so that} \quad \frac{\Delta Q}{Q} = -10^{-3}.$$

Then

$$\frac{\Delta u_0}{u_0} = \frac{0{\cdot}4}{2(1{\cdot}42-1-0{\cdot}4)} \times 10^{-3} = 10^{-2}.$$

Problem No. 9

Q-MULTIPLIER CIRCUIT

I

The system of Fig. 1 uses a pentode or some other electronic device having a high input impedance and an a.c. characteristic described by $i = gu$ where i is the a.c. output current and u the input signal. (The biassing circuit has no part in the following discussion and is not included in the diagram.)

1. Calculate the input admittance Y of the system between A and B in the case where $|gZ_2| \gg 1$.

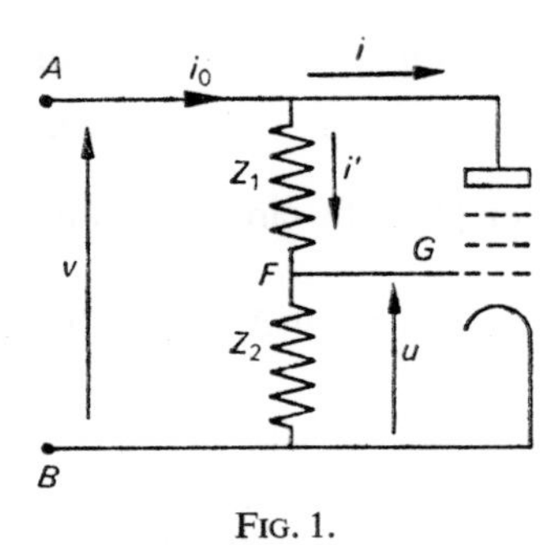

FIG. 1.

2. In the case where Z_1 and Z_2 are pure reactances jX_1 and jX_2, show that the two-terminal network behaves as a pure resistance R_0. Taking Z_1 to be a condenser C and Z_2 to be an inductance L_1 show that there is a frequency range for which R_0 is negative. (Take $\Omega^2LC = 1$ and so define a pulsatance connected to the values chosen for L and C.)

3. For the case where Z_1 is a pure resistance R_1 and Z_2 a pure reactance jX_2, calculate the condition that the network behaves as a pure reactance jX_i.

II

Consider the case of I.2 above. Let a parallel resonant circuit be placed between the terminals A and B as in Fig. 2 and an e.m.f. e of pulsatance ω be applied across this resonant circuit.

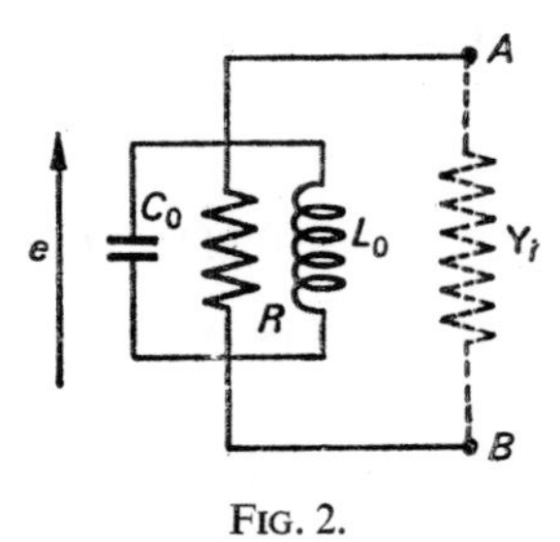

FIG. 2.

1. The L_0, C_0, R circuit resonates at ω_0 with a quality factor Q_0. Show that the value of Ω for the LC circuit (impedances Z_1 and Z_2) can be chosen so that the losses of the resonant circuit are cancelled.

As a numerical example take:

$$C_0 = 0{\cdot}1\ \mu\text{F}; \quad L_0 = 1\ \text{mH}; \quad Q_0 = 20; \quad g = 10\ \text{mA/V}$$

and calculate the required Ω.

2. The L_0C_0 circuit is supplied by a signal of pulsatance

$$\omega = \omega_0 - \Delta\omega \quad (\Delta\omega \ll \omega_0).$$

Calculate the quality factor Q of the circuit for this frequency in terms of Q_0, ω_0 and $\Delta\omega$.

Calculate the numerical value of Q for $\Delta\omega/\omega_0 = 10^{-3}$.

III

Variable Slope Pentode. Frequency Modulation

Consider the case I.3 above with $Z_1 = R_1$ and $Z_2 = (j\omega C)^{-1}$. The circuit of Fig. 3 is placed between A and B.

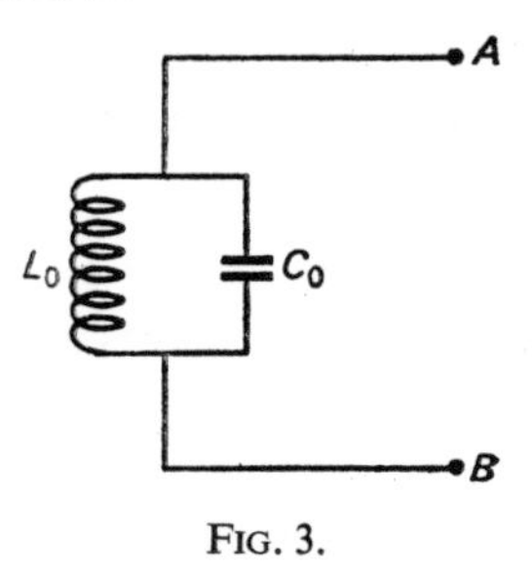

Fig. 3.

1. Supposing that the conditions are fulfilled for Y_i to be a pure susceptance, calculate the frequency of free oscillations of the compound circuit in terms of L_0, C_0, g, R_1, and C.

Calculate this frequency f_1 for

$C_0 = 100$ pF; $L_0 = 0{\cdot}1$ mH; $R_1 = 10$ kΩ; $C = 100$ μF; $g = 10$ mA/V.

2. Consider now the case where the slope or mutual conductance of the valve is a function of the voltage, so that

$$g(u) = g_0 + \alpha u, \quad \text{with} \quad g_0 = 10 \text{ mA/V} \quad \text{and} \quad \alpha = 5\times10^{-4} \text{ A/V}^2.$$

Show that, when u varies, the frequency of the free oscillations of the circuit also varies. Calculate the fractional frequency variation $\Delta f/f_1$ as a function of u in terms of the circuit elements. Calculate the amplitude u_m of the voltage u which leads to a 10% variation in the frequency f_1.

3. It is required now to obtain a linear variation of $\Delta f/f_1$ with u, within about 1%. Calculate the maximum value of u and the corresponding $\Delta f/f_1$ satisfying this restriction.

Calculate the frequency in the case III.2 above. Draw a practical frequency-modulated oscillator based on the system of Fig. 1.

IV

Synchronous Detection

1. Consider again the circuit of Fig. 1 but this time with a voltage source $u' = U \sin \omega_0 t$ and with $Z_1 = jX_1$, $Z_2 = jX_2$, $X_1 \gg X_2$ and $gX_2 \gg 1$. The mutual conductance has again the form $g(u) = g_0 + \alpha u$. Show that in this case the input admittance can be written as

$$Y_i = G_0(1+k \sin \omega_0 t).$$

If Z_1 and Z_2 consist of condensers $C_1 = 10$ pF and $C_2 = 400$ pF, find G_0 and k for $\omega_0 = 10^6$ rad/s, $g_0 = 10$ mA/V, $\alpha = 5 \times 10^{-4}$ A/V^2 and $U = 10$ V.

2. With the network of IV.1 above, the circuit of Fig. 4 is constructed. The e.m.f. of the generator is $e = E \sin \omega t$. Calculate the values of pulsatance and amplitude of the components of v_2 appearing between M and M' in the case where r is very much less than the impedances of the elements R and C of the circuit.

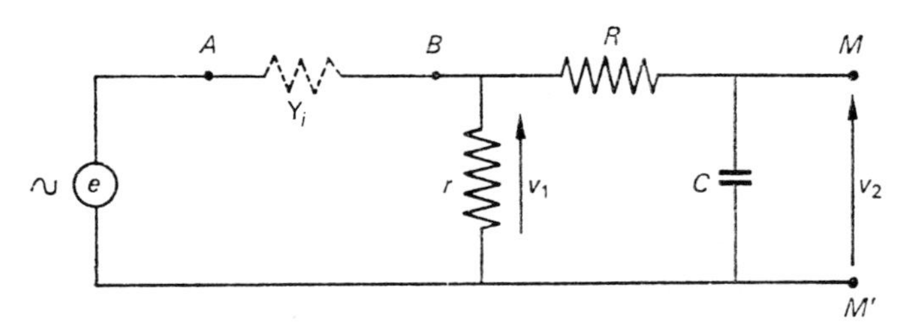

Fig. 4.

With the pulsatance ω nearly equal to ω_0, find the condition imposed on the product RC if only one component of v_2 is to be obtained. Calculate the modulus of this voltage component and find ω to maximize its value. Calculate this maximum voltage for $r = 40\,\Omega$ and $E = 10$ V.

3. The circuit of Fig. 4 can be used as a synchronous detector. We have then

$$Y_i = G_0[1+k \sin (\omega_0 t+\phi)]$$

and the voltage e is amplitude modulated as

$$e = E[1+mf(t)] \sin \omega_0 t.$$

In the conditions of IV.2, show that the signal detected across MM' is proportional to the low-frequency modulation signal. What is the effect of the phase angle ϕ on the signal detected?

4. The circuit can also be used as a harmonic analyser. The voltage e is a periodic function with fundamental pulsatance ω_0 when developed in a Fourier series, and Y_i is of the form

$$Y_i = G_0[1+k \sin(\omega t+\phi)]$$

where, now, ω is to be variable and ϕ can take the values 0 or $\pi/2$.

Show that, if we take $\omega = n\omega_0$, it is possible to find the amplitude and the phase of the nth harmonic of e.

Solution

I.

1. For the pentode we have

$$i = gu = g \cdot \frac{Z_2}{Z_1+Z_2} \cdot v,$$

while, for the arm Z_1Z_2,

$$i' = \frac{v}{Z_1+Z_2}.$$

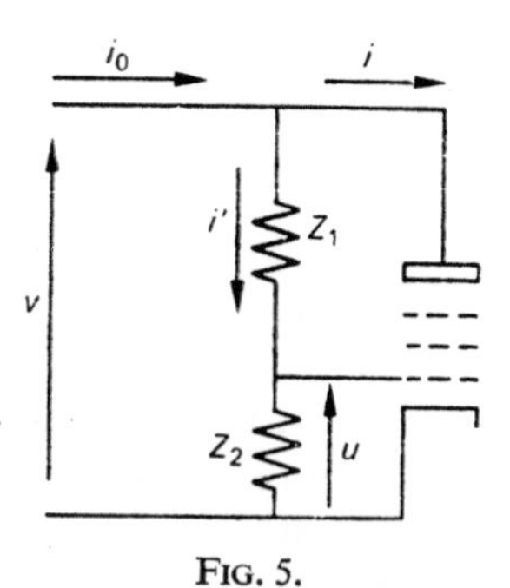

FIG. 5.

The admittance Y_i of the network between A and B becomes

$$Y_i = \frac{i_0}{v} = \frac{i+i'}{v} = \frac{1+gZ_2}{Z_1+Z_2} \simeq \frac{gZ_2}{Z_1+Z_2}$$

or

$$Y_i \simeq \frac{gZ_2}{Z_1+Z_2}.$$

2. $Z_1 = jX_1$ and $Z_2 = jX_2$, so that $Y_i = \dfrac{gX_2}{X_1+X_2} = \dfrac{1}{R_0}$.

In the case where $X_1 = -1/\omega C$ and $X_2 = \omega L$ we have

$$Y_i = \frac{1}{R_0} = \frac{g.\omega L}{\omega L - \dfrac{1}{\omega C}} = \frac{g.\omega^2 LC}{\omega^2 LC - 1} = \frac{g\left(\dfrac{\omega}{\Omega}\right)^2}{\left(\dfrac{\omega}{\Omega}\right)^2 - 1},$$

if we put $\Omega^2 = 1/LC$. The variation with (ω/Ω) is shown in Fig. 6.

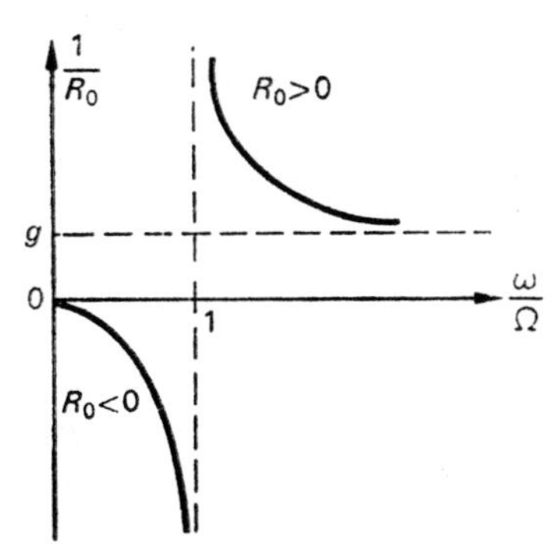

FIG. 6.

For $\omega < \Omega$, we see that R_0 is negative.

3. $Z_1 = R_1$, $Z_2 = jX_2$ gives

$$Y_i = \frac{jgX_2}{R_1 + jX_2} \simeq jg\frac{X_2}{R_1} \quad \text{for} \quad R_1 \gg X_2$$

$$Y_i = \frac{1}{jX_i} \quad \text{where} \quad X_i = \frac{-R_1}{gX_2}.$$

II.

1. Consider the parallel resonant circuit (L_0, C_0, R) across the valve together with the reactance due to L and C. The total admittance of the circuit between A and B is then

$$Y_t = Y_i + \frac{1}{R} + j\left(\omega_0 C_0 - \frac{1}{\omega_0 L_0}\right)$$

which reduces to $(Y_i + 1/R)$ when $\omega_0^2 = 1/LC$.

In order to cancel the losses of the resonant circuit we require

$$Y_i + \frac{1}{R} = 0 \rightarrow \frac{g\left(\dfrac{\omega_0}{\Omega}\right)^2}{\left(\dfrac{\omega_0}{\Omega}\right)^2 - 1} + \frac{1}{R} = 0$$

or

$$\Omega = \omega_0\sqrt{1+gR} \quad \text{and} \quad \omega_0 < \Omega.$$

With the values given

$$\omega_0^2 = \frac{1}{10^{-3}\times 10^{-7}} = 10^{10}, \quad \omega_0 = 10^5 \text{ rad/s},$$

$$f_0 = 16{\cdot}1 \text{ kHz}$$

$$R = 2 \text{ k}\Omega, \quad gR = 10^{-2}\times 2\times 10^3 = 20$$

$$\Omega = \omega_0\sqrt{1+gR} \simeq 4{\cdot}58\omega_0 = 4{\cdot}58\times 10^5 \text{ rad/s}.$$

2. The total conductance of the circuit for $\omega = \omega_0 - \Delta\omega$ is $G = Y_i + 1/R$. The Q-factor is then

$$Q = \frac{1}{\omega L_0 G} \simeq \frac{1}{\left(Y_i + \dfrac{1}{R}\right)\omega_0 L_0} = \frac{R}{\omega_0 L_0}\cdot\frac{1}{(1+RY_i)} = Q_0\cdot\frac{1}{1+RY_i}.$$

Now

$$1+RY_i = 1+\frac{gR\left(\dfrac{\omega}{\Omega}\right)^2}{\left(\dfrac{\omega}{\Omega}\right)^2-1}$$

and

$$\omega = \omega_0 - \Delta\omega = \omega_0\left(1-\frac{\Delta\omega}{\omega_0}\right),$$

so that

$$\left(\frac{\omega}{\Omega}\right)^2 \simeq \left(\frac{\omega_0}{\Omega}\right)^2\left(1-\frac{2\Delta\omega}{\omega_0}\right) = \frac{1}{(1+gR)}\left(1-\frac{2\Delta\omega}{\omega_0}\right).$$

Then

$$1+RY_i \simeq \frac{1+gR\left(\dfrac{1}{1+gR}\right)\left(1-\dfrac{2\Delta\omega}{\omega_0}\right)}{\left(\dfrac{1}{1+gR}\right)\left(1-\dfrac{2\Delta\omega}{\omega_0}\right)-1} = \frac{(1+gR)\cdot\dfrac{2\Delta\omega}{\omega_0}}{\dfrac{2\Delta\omega}{\omega_0}+gR}$$

$$\simeq \frac{2\Delta\omega}{\omega_0}\left(\frac{1+gR}{gR}\right).$$

Since $gR \gg 1$ this leads to

$$Q \simeq Q_0\cdot\frac{\omega_0}{2\Delta\omega}.$$

With $$\frac{\Delta\omega}{\omega_0} = 10^{-3} \quad \text{and} \quad Q_0 = 20, \quad Q = 10^4 \gg Q_0.$$

III

1. $$Z_1 = R_1, \quad Z_2 = \frac{1}{j\omega C} = jX_2 \text{ gives}$$

$$Y_i = jg\frac{X_2}{R_1} = \frac{-jg}{\omega C R_1} \quad \text{provided} \quad R_1 \gg X_2.$$

The total admittance of the circuit is

$$Y_t = Y_i + j\omega C_0 + \frac{1}{j\omega L_0} = j\omega C_0 + \frac{1}{j\omega}\left(\frac{1}{L_0} + \frac{g}{R_1 C}\right).$$

If we put

$$\frac{1}{L} = \frac{1}{L_0} + \frac{g}{R_1 C}$$

then, at resonance,

$$\omega_1^2 L C_0 = 1 \quad \text{and} \quad Y_t = 0.$$

Taking

$$\frac{1}{L} = \frac{1}{L_0} + \frac{g}{R_1 C} = 10^4 + \frac{10^{-2}}{10^4 \times 10^{-10}} = 2\times 10^4 \to L = 50\ \mu\text{H}$$

$$\omega_1^2 = \frac{1}{LC_0} = 2\times 10^4 \times 10^{10}$$

or

$$\omega_1 = 1{\cdot}4\times 10^7 \to f_1 \simeq 2{\cdot}3\ \text{MHz}.$$

(We see that $gX \simeq 7 \gg 1$; $R_1 \gg X_2$ since $10^4\ \Omega \gg 700\ \Omega$.)

2. Now we take $g(u) = g_0 + \alpha u$. The frequency corresponding to the voltage u becomes

$$\omega^2 = \frac{1}{L'C_0} = \frac{1}{C_0}\left[\frac{1}{L_0} + \frac{g(u)}{R_1 C}\right] = \frac{1}{C_0}\left[\frac{1}{L_0} + \frac{g_0}{R_1 C} + \frac{\alpha u}{R_1 C}\right]$$

$$= \frac{1}{C_0}\left[\frac{1}{L} + \frac{\alpha u}{R_1 C}\right].$$

Thus

$$\omega^2 = \omega_1^2\left(1 + \frac{\alpha L}{R_1 C}\cdot u.\right)$$

Putting $\omega = \omega_1+\Delta\omega$ so that $\omega^2 \simeq \omega_1^2\left(1+2\frac{\Delta\omega}{\omega_1}\right)$ we see

$$\frac{\Delta\omega}{\omega_1} \simeq \frac{\alpha L}{2R_1C}\cdot u.$$

If

$$\frac{\Delta\omega}{\omega_1} = \frac{1}{10} = \frac{\alpha L_1}{2R_1C}\cdot u_m,$$

we have

$$u_m = \frac{2R_1C}{10\alpha L} = 8 \text{ V}.$$

3. Rigorously we have

$$\omega^2 = \omega_1^2\left(1+\frac{2\Delta\omega}{\omega_1}+\frac{\Delta\omega^2}{\omega_1^2}\right) = \omega_1^2(1+Au) \quad \text{where} \quad A = \frac{\alpha L}{R_1C}.$$

The only allowed solution is

$$\frac{\Delta\omega}{\omega_1} = -1+\sqrt{1+Au}.$$

Thus, if $Au \ll 1$, we have

$$\sqrt{1+Au} \simeq 1+\frac{Au}{2}-\frac{A^2u^2}{8}$$

and

$$\frac{\Delta\omega}{\omega_1} = \frac{Au}{2}-\frac{A^2u^2}{8}.$$

For the frequency change to be linear within 1% it is necessary that

$$\frac{1}{8}A^2u^2 \leqslant \frac{1}{100}\cdot\frac{Au}{2}$$

or

$$Au \leqslant \frac{1}{25}.$$

The limit for us is

$$u_M = \frac{1}{25A} = \frac{5u_m}{25} \Rightarrow u_M = \frac{u_m}{5} = 1{\cdot}6 \text{ V}.$$

It follows that

$$\frac{\Delta\omega}{\omega_1} = \frac{Au}{2} \leqslant \frac{1}{50}.$$

In the preceding case we had a frequency change $\Delta f \simeq 200$ kHz. To get the same frequency change linear with u to 1% we must choose ω_1 such that

$$\omega_1 = 50\,\Delta\omega \Rightarrow f_1 = 50\,\Delta f = 11 \text{ MHz}.$$

A practical circuit must have, besides the elements already considered, a generator G giving a modulating voltage u and an oscillator triode T_1 supplying the tuned circuit. A large condenser C_1 is used to isolate the arm R_1, C from the H.T. supply.

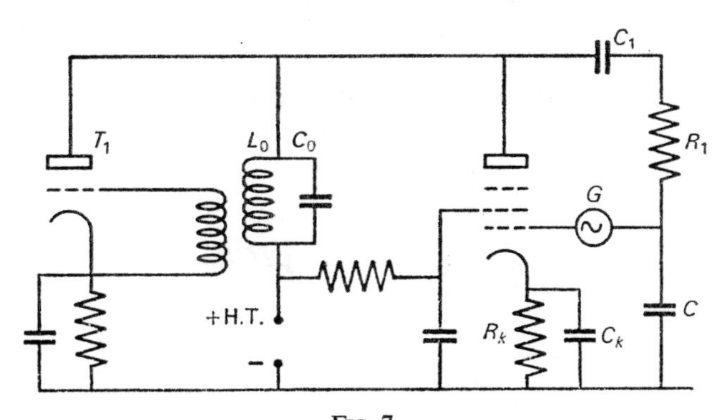

FIG. 7.

IV

1. The voltage u in Fig. 1 is

$$u = u' + \frac{Z_2}{Z_1+Z_2} v.$$

If one takes $Z_1 \gg Z_2$ (i.e. $X_1 \gg X_2$) we will have $u \sim u'$. Then

$$Y_i = \frac{gX_2}{X_1+X_2} = \frac{X_2(g_0+\alpha u)}{X_1+X_2} = \frac{g_0 X_2}{X_1+X_2}\left(1+\frac{\alpha}{g_0} U \sin \omega_0 t\right),$$

i.e.

$$Y_i = G_0 (1+k \sin \omega_0 t)$$

with

$$G_0 = \frac{g_0 X_2}{X_1+X_2} \simeq \frac{g_0 X_2}{X_1} \quad \text{and} \quad k = \frac{\alpha U}{g_0}.$$

Substituting the values given,

$$X_1 = \frac{1}{\omega C_1} = 10^5\,\Omega \gg X_2 = \frac{1}{\omega C_2} = 5200\,\Omega$$

and

$$g_0 X_2 = \frac{g_0}{\omega C_2} = 25 \gg 1$$

$$G_0 \simeq g_0 \frac{X_2}{X_1} = g_0 \frac{C_1}{C_2} = 10^{-2} \times \frac{1}{40} = 2{\cdot}5 \times 10^{-4}\ \text{ohm}^{-1}$$

or

$$G_0 = 2{\cdot}5\times10^{-4}\ \mathrm{S}$$

$$k = \frac{\alpha U}{g_0} = \frac{5\times10^{-4}\times10}{10^{-2}} = 0{\cdot}5.$$

2. We take a general value for Ω and calculate v_2 from the potential divider, so that

$$v_2 = \frac{\dfrac{1}{\mathrm{j}\Omega C}}{R+\dfrac{1}{\mathrm{j}\Omega C}}\cdot v_1 = \frac{v_1}{1+\mathrm{j}\Omega CR}.$$

Also

$$v_1 = \frac{\dfrac{1}{Y'}}{\dfrac{1}{Y_i}+\dfrac{1}{Y'}}\cdot e = \frac{Y_i}{Y_i+Y'}e \simeq Y_i re,$$

since

$$Y' = \frac{1}{r}+\frac{\mathrm{j}\Omega C}{1+\mathrm{j}\Omega CR} \simeq \frac{1}{r}.$$

It follows that

$$v_2 = \frac{rY_i e}{1+\mathrm{j}\Omega CR}$$

where

$$Y_i e = G_0(1+k\sin\omega_0 t)E\sin\omega t = G_0E[\sin\omega t+k\sin\omega_0 t\sin\omega t]$$

$$= G_0E\left[\sin\omega t+\frac{k}{2}\cos(\omega_0-\omega)t-\frac{k}{2}\cos(\omega_0+\omega)t\right].$$

There are thus three voltages across MM' with pulsatances ω, $\omega_0-\omega$ and $\omega_0+\omega$ respectively and amplitudes

$$\left|\frac{rG_0E}{1+\mathrm{j}\omega CR}\right|, \quad \left|\frac{krG_0E}{2[1+\mathrm{j}(\omega_0-\omega)CR]}\right| \quad \text{and} \quad \left|\frac{krG_0E}{2[1+\mathrm{j}(\omega_0+\omega)CR]}\right|.$$

If we take $\omega CR \gg 1$ and $\omega \sim \omega_0$, the first and third amplitudes will be of the same order and will be small compared with the second. The voltage across MM' will thus be dominated by the voltage with pulsatance $\omega_0-\omega$ and amplitude

$$|V_2| = \frac{krG_0E}{2|1+\mathrm{j}(\omega_0-\omega)CR|} = \frac{krG_0E}{2\sqrt{1+(\omega_0-\omega)^2C^2R^2}}.$$

This voltage is greatest for $\omega = \omega_0$ when

$$V_{2\,\max} = \frac{krG_0E}{2} = \frac{0{\cdot}5\times 40\times 2{\cdot}5\times 10^{-4}\times 10}{2} = 25\text{ mV}.$$

3. In this case

$$eY_i = G_0E[1+mf(t)][1+k\sin(\omega_0t+\phi)]\sin\omega_0t$$

$$= G_0E[1+mf(t)]\left[\sin\omega_0t+\frac{k}{2}\cos\phi-\frac{k}{2}\cos(2\omega_0t+\phi)\right].$$

The filter composed of R and C will eliminate the components ω_0 and $2\omega_0$ so that the voltage across MM' is

$$v_2 = \tfrac{1}{2}krG_0E[1+mf(t)]\cos\phi.$$

The d.c. component can be easily eliminated while the a.c. component is proportional to the low-frequency modulation signal $f(t)$. With the signal multiplied by $\cos\phi$ the best detection is clearly for $\phi = 0$ or π, while $\phi = \pm\pi/2$ gives no signal to detect.

4. The voltage e is of the form

$$e = E_0+\sum_1^\infty E_n\cos(n\omega_0t-\phi_n) = E_0+\sum_1^\infty E'_n\cos n\omega_0t+\sum_1^\infty E''_n\sin n\omega_0t.$$

The product Y_ie is then

$$Y_ie = G_0\left[E_0+\sum_1^\infty E'_n\cos n\omega_0t+\sum_1^\infty E''_n\sin n\omega_0t\right]$$

$$+kG_0\left[E_0\sin(\omega t+\phi)+\sum_1^\infty E'_n\cos n\omega_0t\sin(\omega t+\phi)\right.$$

$$\left.+\sum_1^\infty E''_n\sin n\omega_0t\sin(\omega t+\phi)\right]$$

$$= G_0e+kG_0\left\{E_0\sin(\omega t+\phi)+\tfrac{1}{2}\sum_1^\infty E'_n\sin[(n\omega_0+\omega)t+\phi]\right.$$

$$-\tfrac{1}{2}\sum_1^\infty E'_n\sin[(n\omega_0-\omega)t-\phi]+\tfrac{1}{2}\sum_1^\infty E''_n\cos[(n\omega_0+\omega)t+\phi]$$

$$\left.-\tfrac{1}{2}\sum_1^\infty E''_n\cos[(n\omega_0-\phi)t-\phi]\right\}.$$

When $\omega = n\omega_0$ the signal obtained from $Y_ie.r$, after filtering with the RC network, is

$$v_2 = rG_0E_0+\frac{rkG_0}{2}(E'_n\sin\phi-E''_n\cos\phi).$$

After rG_0E_0 has been found from a d.c. measurement (i.e. $\omega = 0$) E'_n and E''_n can be measured by taking, in turn, $\phi = 0$ and $\phi = \pi/2$. Then

$$\tan \phi_n = \frac{E''_n}{E'_n} \quad \text{and} \quad E_n = \frac{E'_n}{\cos \phi}.$$

Problem No. 10

THE GYRATOR

A gyrator is a four-terminal network Q for which the admittance matrix has the form

$$|Y_0| = \begin{vmatrix} 0 & -\dfrac{\alpha}{A} \\ \alpha A & 0 \end{vmatrix}$$

with α and A real and positive, being the gyrator constant and the gain respectively. (The currents are taken to be positive when they *enter* the gyrator.)

I

Properties of the Gyrator

1. What is the input impedance of a gyrator defined by the parameters (α, A) when a capacitance C is connected across its output?

2. Two gyrators (α, A) and (α', A') are placed in series. What is the hybrid matrix [H] of the combination? Under what conditions is this combination equivalent to a perfect transformer of ratio n? How is n related to α and α'?

3. Consider the circuit of Fig. 1 formed by two current sources (1) and (2) in series with two identical conductances, g. These sources are driven by the voltages $v'_2 = v_{M'} - v_{N'}$ and $v'_1 = v_M - v_N$. Calculate the elements of the admittance matrix between AB and $A'B'$ in terms of g, g_1 and g_2.

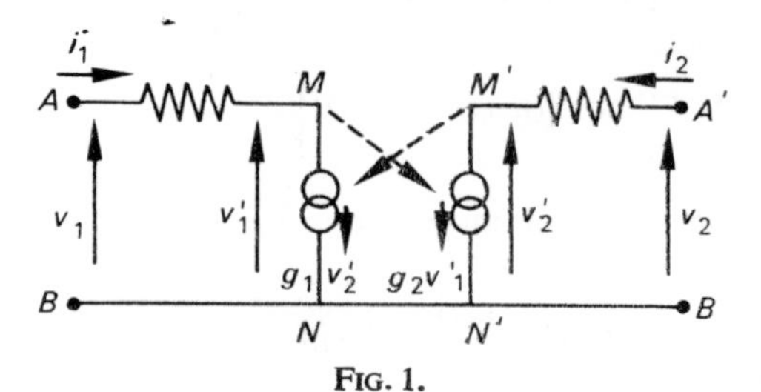

FIG. 1.

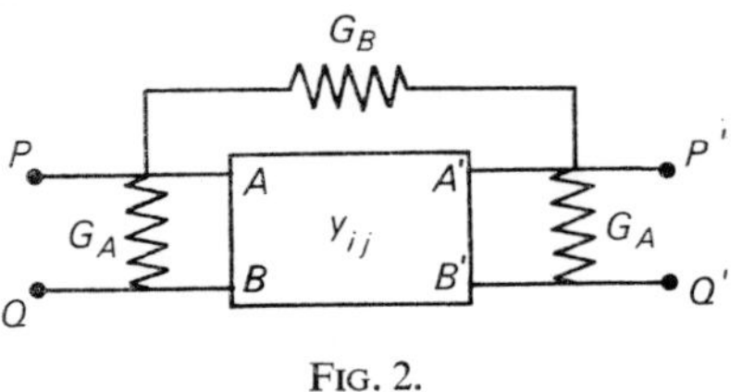

FIG. 2.

4. Calculate the elements of the admittance matrix of the four-terminal network Q' formed by placing the above network Q in parallel with a symmetric π-network formed by the conductances G_A and G_B as shown in Fig. 2.

What relation must exist between the elements y_{ij}, G_A, G_B, α and A if the network is to form a perfect gyrator defined by α and A? (Express the y_{ij} in terms of G_A, G_B, α and A.)

5. Express the conductance g_1 in terms of the elements y_{ij}. Then, knowing that g_1 is necessarily positive, show that A cannot take all values and that, in particular, it is impossible to produce a passive gyrator with $A = 1$ with the proposed circuit. (Use the y_{ij} obtained in question 4.)

6. With A, α and G_A given and $g_1 > 0$, what is the minimum value for G_B?

7. As a numerical example consider that, in constructing the circuit of Fig. 2, the output is connected to a capacity of 2 μF while an input impedance is required equivalent to an inductance of 10 henries. If $G_A = 2\times 10^{-4}$ S and $A = 5$ what is the minimum value of G_B?

With $G_B = 5\times 10^{-4}$ S, find g, g_1 and g_2.

II

PRACTICAL CIRCUIT OF A GYRATOR

To give the required current sources one has recourse to transistors.

1. Show that, to a first approximation, the circuit of Fig. 3 will satisfactorily give a current source $i = g_1 u$ controlled by the voltage u. Give an expression for g_1 by taking β and h_{11} as the appropriate hybrid parameters of the transistor with common emitter while $h_{12} = h_{22} = 0$.

2. The use of this type of source leads to the type of circuit shown in Fig. 4. Give the values of the resistances R_2, R_3, R_2', R_3' and R_4 in terms of the conductances G_A, G_B and g of Figs. 1 and 2.

Is it possible to find convenient values for the resistances such that the transistors T_1 and T_2 shall be biassed normally? Explain.

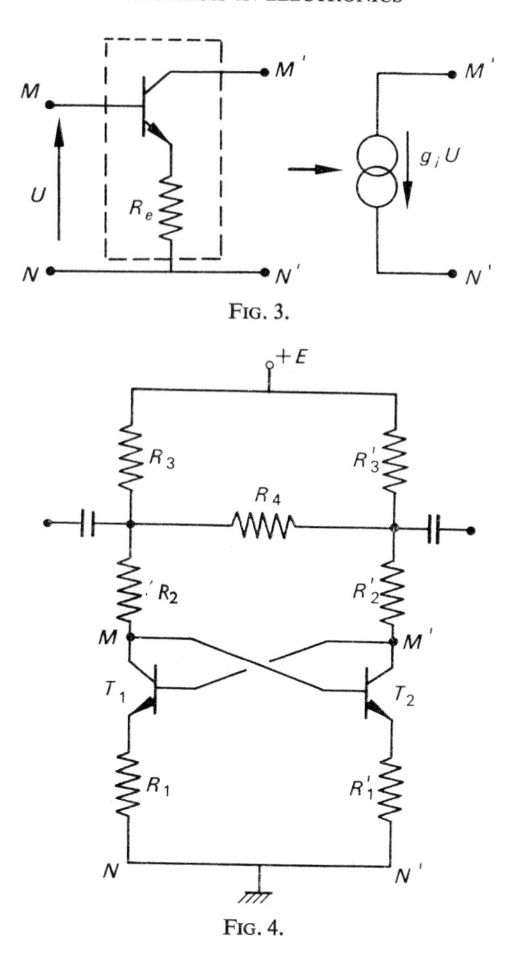

FIG. 3.

FIG. 4.

(Make any necessary assumptions about the relative values of β the potentials at M and M' and assume that the base-emitter voltages on the transistors are negligible.)

3. The circuit used in practice is that of Fig. 5. Take the d.c. values of β for the two transistors to be identical as well as the base emitter voltages which are both V_0. It is required that the points P and P' shall be at the same d.c. potential V_p while the resistances R_2, R_3, R'_2, R'_3 and R_4 have the values found above. Calculate R_{B_1} and R_{B_2}, in terms of E, V_p, V_0, R, R_2, R_3, R'_1, R'_2 and R'_3.

4. What is the minimum allowable value for V_p? (Take as R the larger of the two resistances R_1 and R'_1.)

5. With the same conditions as in I.7 and with identical transistors characterized by $\beta = 200$ (for both d.c. and a.c.) and $h_{11} = 1\,\text{k}\Omega$, calculate the resistances

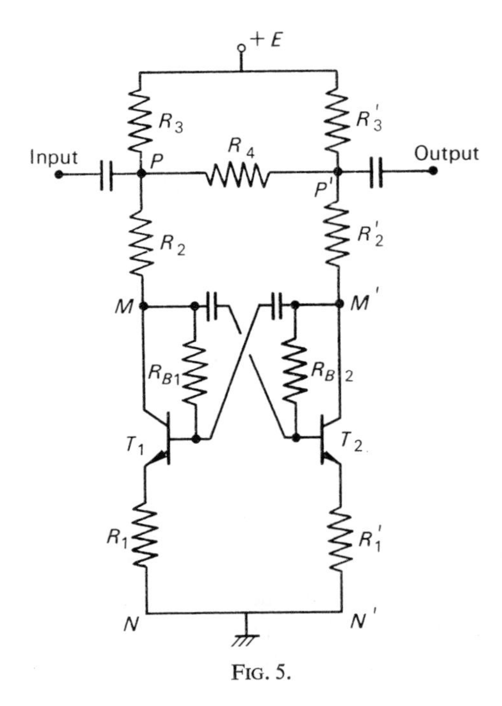

FIG. 5.

R_1, R'_1, R_2, R'_2, R_3, R'_3 and R_4. (Assume that the decoupling condensers act perfectly at the frequencies employed for this circuit.)

If $E = 25$ volts, what is the minimum value of V_p?

If V_p is taken as 20 V and $V_0 = 0{\cdot}5$ V what are R_{B_1} and R_{B_2}?

III

INFLUENCE OF IMPERFECTIONS IN THE CURRENT SOURCES

The current sources produced as indicated in Fig. 3 differ from ideal current sources in that:

their output impedance is not infinite;
their control circuit (i.e. the base) has a non-infinite input impedance;
there is an internal feedback between input and output.

1. In a general fashion the real current source provided by a transistor can be represented by a network with admittance parameters a, b, c and d as indicated in Fig. 6. Establish the equivalent circuit for the system of Fig. 1 in which each current source is formed from a transistor having the equivalent circuit of Fig. 6. What are the new elements y_{ij} of the admittance matrix for the system of Fig. 1

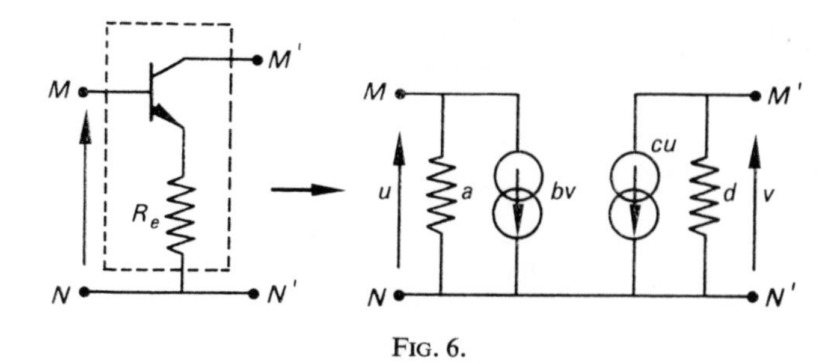

FIG. 6.

in this case? (Take the elements of the admittance matrices of the two transistors to be a, b, c, d, a', b', c', d' and put

$$g_1' = c'+b \qquad G_1 = d'+a$$
$$g_2' = b'+c \qquad G_2 = a'+d.)$$

2. Explain why, under these conditions, it is no longer possible to form a perfect gyrator using the principle of Fig. 2. It is possible, however, if two different conductances G_A are used. Let G_A be the conductance placed between the points P and Q, and G_A' that placed between P' and Q'. Calculate the difference $G_A - G_A'$ which will give a perfect gyrator, expressing the value in terms of g, g_1', g_2', G_1' and G_2'.

3. Calculate the elements a, b, c and d in terms of the parameters y_{ije} and the emitter resistance R_e where y_{ije} are the admittance parameters of the transistor with common emitter.

4. The transistors used are identical with

$$y_{11e} = 10^{-3}\ \text{S}, \quad y_{12e} = -10^{-7}\ \text{S}$$
$$y_{21e} = 0{\cdot}05\ \text{S}, \quad y_{22e} = 2\times 10^{-5}\ \text{S}.$$

Calculate a, b, c, d, a', b', c', d' for the current sources in Fig. 5, together with the value of the difference $G_A - G_A'$ which will establish equilibrium in the system.

(Paris, June 1967)

Solution

I

1. The form of the matrix Y leads to the equations

$$i_1 = -\frac{\alpha}{A} v_2$$

$$i_2 = \alpha A v_1$$

while, at the load, $i_2 = -j\omega C v_2$.

Thus,

$$Z_i = \frac{v_1}{i_1} = \frac{j\omega C}{\alpha^2} = j\omega L,$$

giving an input impedance equivalent to an inductance $L = C/\alpha_2$.

2. Taking i_1, i_2, v_1, v_2 to be the currents and voltages for the first gyrator and i_1', i_2', v_1', v_2' for the second, we see

$$v_2 = v_1' \quad \text{and} \quad i_2 = -i_1',$$

while

$$i_1 = -\frac{\alpha}{A} v_2, \qquad i_1' = -\frac{\alpha'}{A'} v_2'$$

$$i_2 = \alpha Au, \qquad i_2' = \alpha' A' v_1'.$$

We want to write

$$v_1 = h_{11} i_1 + h_{12} v_2'$$

$$i_2' = h_{21} i_1 + h_{22} v_2'$$

and by identifying terms we find

$$h_{11} = 0, \qquad h_{12} = \frac{\alpha'}{\alpha} \cdot \frac{1}{AA}$$

$$h_{21} = -\frac{\alpha'}{\alpha} AA' \quad h_{22} = 0.$$

Now, for a perfect transformer,

$$h_{11} = 0 \qquad h_{12} = n$$

$$h_{21} = -n \quad h_{22} = 0$$

and the system may be taken equivalent to a perfect transformer if

$$AA' = \frac{1}{AA'} \quad (\text{i.e. } AA' = 1)$$

when

$$n = \alpha'/\alpha.$$

3. We can write

$$i_1 = g_1 v_2' \quad v_1' = v_1 - \frac{i_1}{g}$$

$$i_2 = g_2 v_1' \quad v_2' = v_2 - \frac{i_2}{g}.$$

Thus

$$i_1\left[1-\frac{g_1g_2}{g^2}\right] = g_1v_2-\frac{g_1g_2}{g}v_1$$

$$i_2\left[1-\frac{g_1g_2}{g^2}\right] = g_2v_1-\frac{g_1g_2}{g}v_2,$$

which is equivalent to writing

$$y_{11} = y_{22} = \frac{-g_1g_2g}{g^2-g_1g_2}$$

$$y_{12} = \frac{g_1g^2}{g^2-g_1g_2} \quad \text{and} \quad y_{21} = \frac{g_2g^2}{g^2-g_1g_2}.$$

4. The y parameters of the two networks in parallel just add together. It is therefore necessary to calculate the y parameters of the π-network formed by the

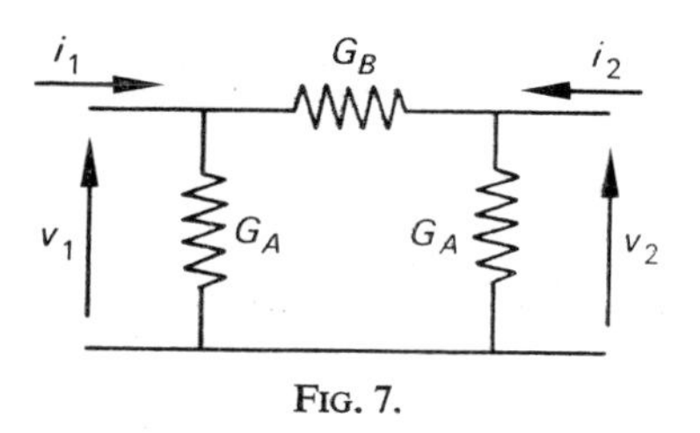

FIG. 7.

three conductances G shown in Fig. 7. If the output is short-circuited $v_2 = 0$ and

$$y'_{11} = \frac{i_1}{v_1} = G_A+G_B$$

$$y'_{21} = \frac{i_2}{v_1} = -G_B.$$

By symmetry $y'_{22} = y'_{11}$ and $y'_{12} = y'_{21}$. Thus the y parameters of the combined networks are

$$Y_{11} = y_{11}+G_A+G_B = Y_{22}$$

$$Y_{12} = y_{12}-G_B, \quad Y_{21} = y_{21}-G_B.$$

The complete network will be a gyrator if $Y_{11} = Y_{22} = 0$, which will require that $y_{22} = y_{11} = -(G_A+G_B)$. Also

$$Y_{12} = y_{12}-G_B = -\frac{\alpha}{A}, \quad Y_{21} = y_{21}-G_B = \alpha A,$$

so that

$$y_{21} = G_B + \alpha A. \qquad y_{12} = G_B - \frac{\alpha}{A}.$$

5. It is simplest to calculate

$$y_{12}y_{21} - y_{11}y_{22} = \frac{g_1 g_2 g^4 - g_1^2 g_2^2 g^2}{(g^2 - g_1 g_2)^2} = \frac{g_1 g_2 g^2}{g^2 - g_1 g_2} = g_1 y_{21},$$

so that

$$g_1 = \frac{y_{12}y_{21} - y_{11}^2}{y_{21}} > 0.$$

Now

$$y_{21} = G_B + \alpha A > 0,$$

so that we must have

$$y_{12}y_{21} - y_{11}^2 = \left(G_B - \frac{\alpha}{A}\right)(G_B + \alpha A) - (G_A + G_B)^2 > 0,$$

or

$$\alpha G_B\left(A - \frac{1}{A}\right) > \alpha^2 + G_A^2 + 2G_A G_B.$$

The right hand term is positive, which implies

$$A - \frac{1}{A} > 0,$$

or

$$A > 1.$$

6. The preceding condition gives $G_B > \dfrac{\alpha^2 + G_A^2}{\alpha\left[A - \dfrac{1}{A}\right] - 2G_A}$.

7. Here

$$\alpha = \sqrt{C/L} = 4{\cdot}48 \times 10^{-4}\,\Omega$$

$$G_{B\,\min} = 1{\cdot}37 \times 10^{-4}\,\text{S}$$

$$y_{11} = -7 \times 10^{-4}\,\text{S}, \quad y_{12} = 4{\cdot}1 \times 10^{-4}\,\text{S}, \quad y_{21} = 2{\cdot}74 \times 10^{-3}\,\text{S}$$

$$g_1 = 2{\cdot}31 \times 10^{-4}\,\text{S}, \quad g_2 = 1{\cdot}54 \times 10^{-3}\,\text{S}, \quad g = 9{\cdot}05 \times 10^{-4}\,\text{S}.$$

II

1. With the values allowed for the h_{ij} of the transistor, the equivalent circuit becomes that shown in Fig. 8. Hence

$$v = (\beta+1)R_E i_B$$

$$u = [h_{11}+(\beta+1)\,R_E]\,i_B$$

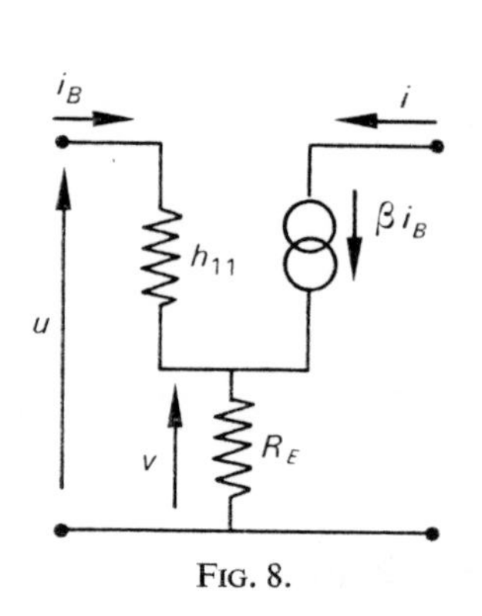

FIG. 8.

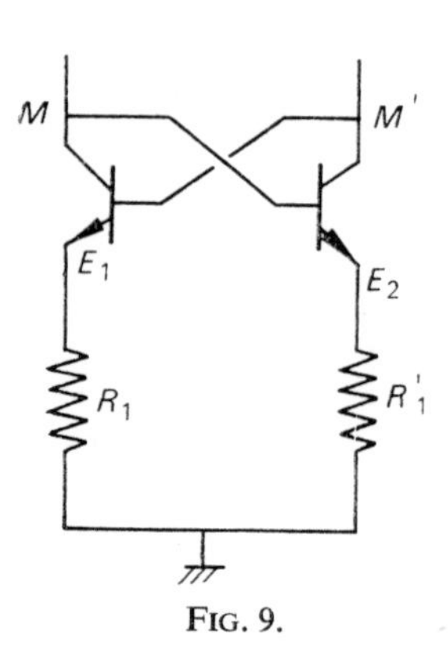

FIG. 9.

and

$$i = \beta i_B = \frac{\beta u}{h_{11}+(\beta+1)R_E} \simeq \frac{u}{R_E},$$

since, normally, $h_{11} \ll (\beta+1)R$ and $\beta \gg 1$.

Thus

$$g_1 = \frac{\beta}{h_{11}+(\beta+1)R_E} \simeq \frac{1}{R_E}.$$

2. It is obvious that

$$R_2 = R_2' = \frac{1}{g}, \quad R_3 = R_3' = \frac{1}{G_A} \quad \text{and} \quad R_4 = \frac{1}{G_B}.$$

The collectors are normally at a higher potential than the base so that, in Fig. 9,

$$V_M > V_{E_1}$$
$$V_{M'} > V_{E_2},$$

while, if the base voltage V_{BE} is small,

$$V_M \simeq V_{E_2}$$
$$V_{M'} \simeq V_{E_1}.$$

The conditions stated become

$$\left.\begin{aligned} V_M > V_{E_1} \Rightarrow V_M > V'_M \\ V'_M > V_{E_2} \Rightarrow V'_M > V_M \end{aligned}\right\} \text{ impossible!}$$

It is thus not possible to bias the transistors in a suitable manner.

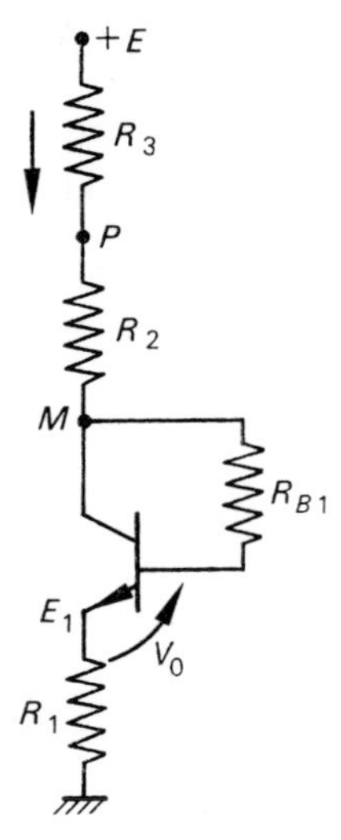

FIG. 10.

3. If $V_p = V_{p'}$, there is no current through R_4. Then, from Fig. 10

$$I = (\beta+1)I_B$$

$$E - V_p = R_3 I$$

$$R_1 I + V_0 + R_{B_1} I + R_2 I = V_p,$$

from which can be found

$$R_{B_1} = \left[R_3 \frac{V_p - V_0}{E - V_p} - (R_1 + R_2)\right](\beta+1)$$

$$R_{B_2} = \left[R'_3 \frac{V_p - V_0}{E - V_p} - (R'_1 + R_2)\right](\beta+1).$$

4. It is necessary that $V_M > V_{E_1}$, so that

$$V_p - R_2 I > R_1 I_1$$

from which

$$V_p > E \frac{R + R_2}{R + R_2 + R_3}.$$

5. It has been seen that
$$g_1 = \frac{\beta}{(\beta+1)R_E+h_{11}}$$

from which
$$R_1+\frac{h_{11}}{\beta} = \frac{1}{g_1} = 4{\cdot}34\ \text{k}\Omega \simeq R_1$$

$$R_1'+\frac{h_{11}}{\beta} = \frac{1}{g_2} = 650\ \Omega \simeq R_1'$$

$$R_2 = R_2' = \frac{1}{g} = 1{\cdot}1\ \text{k}\Omega$$

$$R_3 = R_3' = \frac{1}{G_A} = 5\ \text{k}\Omega$$

$$R_4 = \frac{1}{G_B} = 2\ \text{k}\Omega$$

$$V_{PM} = 13\ \text{V}$$

$$R_{B_1} = 2{\cdot}8\ \text{M}\Omega, \quad R_{B_2} = 3{\cdot}6\ \text{M}\Omega.$$

III

1. The required equivalent circuit is given in Fig. 11,

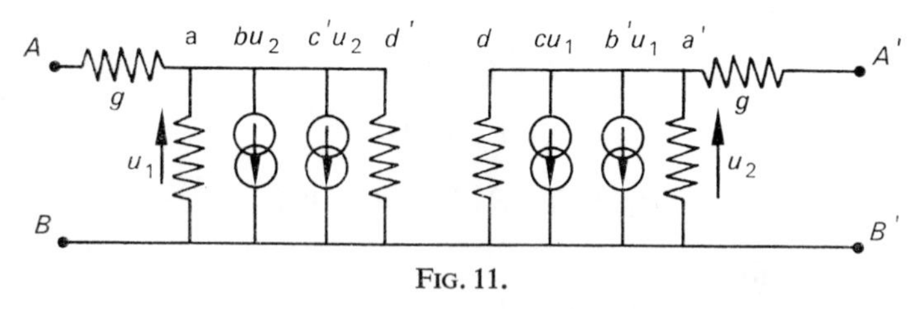

FIG. 11.

which can be reduced to that of Fig. 12.

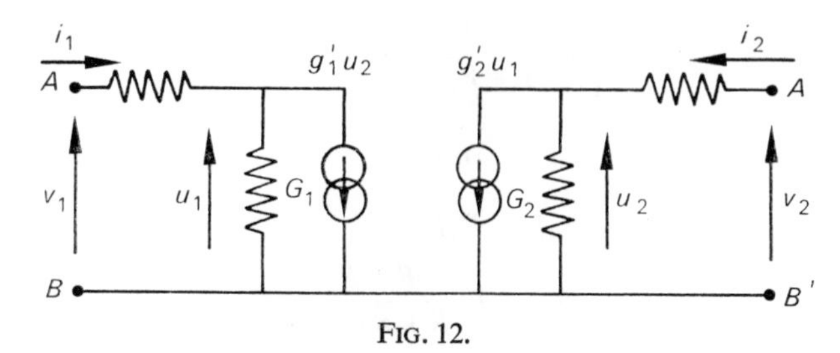

FIG. 12.

It is possible to write

$$\begin{cases} u_1 = v_1 - \dfrac{i_1}{g} & i_1 = G_1u_1 + g_1'u_1 = G_1\left(V_1 - \dfrac{i_1}{g}\right) + g_1'\left(V_2 - \dfrac{i_2}{g}\right) \\ u_2 = v_2 - \dfrac{i_2}{g} & i_2 = G_2u_2 + g_2'u_2 = G_2\left(V_2 - \dfrac{i_2}{g}\right) + g_2'\left(V_1 - \dfrac{i_2}{g}\right). \end{cases}$$

From these equations it follows that

$$\begin{cases} i_1\left(1 + \dfrac{G_1}{g}\right) + i_2\left(\dfrac{g_1'}{g}\right) = G_1v_1 + g_1'v_2 \\ i_1\left(\dfrac{g_2'}{g}\right) + i_2\left(1 + \dfrac{G_2}{g}\right) = g_2'v_1 + G_2v_2. \end{cases}$$

Resolving these equations into the form

$$\begin{cases} i_1 = Y_{11}v_1 + Y_{12}v_2 \\ i_2 = Y_{21}v_1 + Y_{22}v_2 \end{cases}$$

gives

$$Y_{11} = \frac{gG_1(g+G_1) - g_1'g_2'g}{D}, \quad Y_{22} = \frac{gG_2(g+G_1) - g_1'g_2g}{D},$$

$$Y_{12} = \frac{g_1'g^2}{D}, \quad Y_2 = \frac{g_2'g^2}{D},$$

where

$$D = (g+G_1)(g+G_2) - g_1'g_2'.$$

2. It is not possible to have a perfect gyrator for $Y_{11} \neq Y_{22}$. With the addition of the external elements G_A, $G_{A'}$, and G_B a gyrator will be formed if

$$Y_{11} + G_A + G_B = 0$$
$$Y_{22} + G_{A'} + G_B = 0,$$

from which

$$G_A - G_{A'} = Y_{22} - Y_{11} = \frac{g^2(G_2 - G_1)}{D}.$$

3. From Fig. 13 it is possible to write

$$i_1 = y_{11e}u_1 + y_{12e}u_2$$
$$i_2 = y_{21e}u_1 + y_{22e}u_2$$
$$v_1 = u_1 + R_E(i_1 + i_2)$$
$$v_1 = u_2 + R_E(i_1 + i_2)$$

which can be solved for i_1 and i_2 in terms of v_1 and v_2.

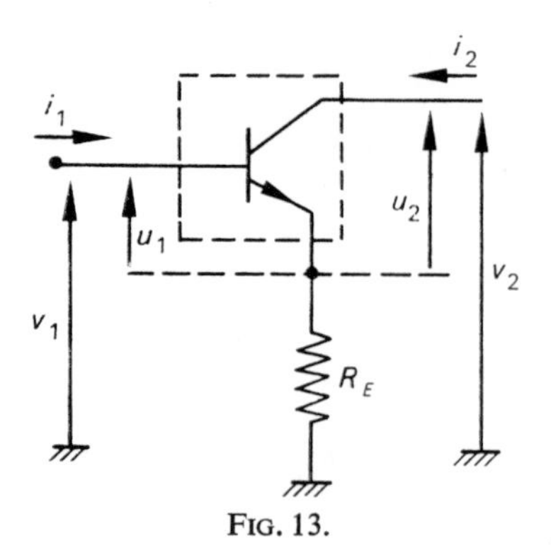

FIG. 13.

Putting
$$\delta = y_{11e}y_{22e} - y_{12e}y_{21e}$$
and
$$\Sigma = y_{11e} + y_{12e} + y_{21e} + y_{22e}$$
the solution leads to

$$a = \frac{y_{11e} + R_E\delta}{1 + R_E\Sigma}, \quad b = \frac{y_{12e} - R_E\delta}{1 + R_E\Sigma},$$

$$c = \frac{y_{21e} - R_E\delta}{1 + R_E\Sigma}, \quad d = \frac{y_{22e} + R_E\delta}{1 + R_E\Sigma}.$$

4. With the values given, $\delta = 2{\cdot}5 \times 10^{-8}$, $\Sigma = 5 \times 10^{-2}$

$$R_1\Sigma = 217 \qquad R'_1\Sigma = 32{\cdot}5$$
$$R_1\delta = 10{\cdot}8 \times 10^{-5} \qquad R'_1\delta = 16{\cdot}25 \times 10^{-6}.$$

Then $a = 4{\cdot}7 \times 10^{-6}$ S, $b \simeq -0{\cdot}5 \times 10^{-6}$ S, $c \simeq y_{21e}/R_1\Sigma = 2{\cdot}36 \times 10^{-4}$ S, $d \simeq (y_{21e} + R_1\delta)/R_1\Sigma = 5{\cdot}6 \times 10^{-7}$ S. Similarly $a' = 3{\cdot}08 \times 10^{-5}$ S, $b' = -0{\cdot}5 \times 10^{-6}$ S, $c' = 1{\cdot}54 \times 10^{-3}$ S, $d' = 1{\cdot}1 \times 10^{-6}$ S.

$$g'_1 = 1{\cdot}54 \times 10^{-3}\text{ S}, \quad g'_2 = 2{\cdot}36 \times 10^{-4}\text{ S}, \quad g = 9{\cdot}05 \times 10^{-7}\text{ S}$$
$$G_1 = 5{\cdot}8 \times 10^{-6}\text{ S}, \quad G_2 = 3{\cdot}08 \times 10^{-5}\text{ S}$$
$$G_A - G'_A = 4{\cdot}2 \times 10^{-5}\text{ S}, \quad G'_A = 0{\cdot}79G_A = 1{\cdot}58 \times 10^{-5}\text{ S and } 1 - G'_A/G_A = 0{\cdot}21.$$

Problem No. 11

PHASE SENSITIVE DETECTOR

The input signal v_1 in the circuit of Fig. 1 below is sinusoidal of the form
$$v_1 = V_1 \cos(\omega' t + \phi).$$

v_2 is a square wave voltage which alternately cuts off and saturates the transistor. Assume that, when the transistor is saturated, $V_A = 0$.

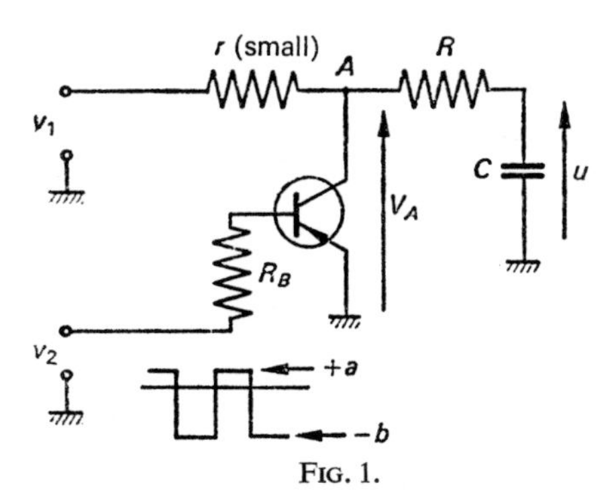

FIG. 1.

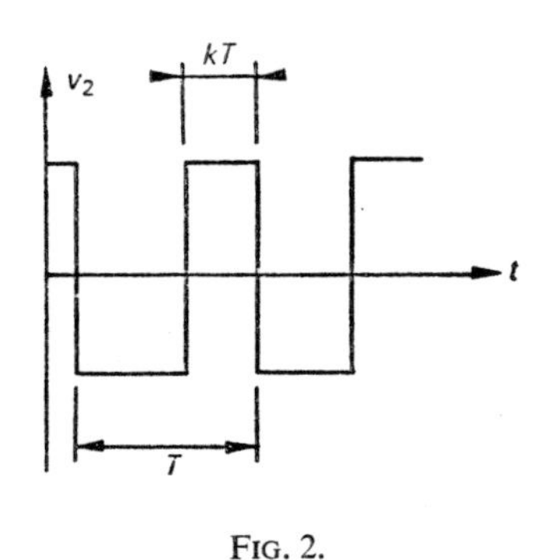

FIG. 2.

1. In the absence of the *RC* circuit, what inequalities must be satisfied by V_1, the input amplitude, and the values of a and b of v_2 if the transistor is to act as a perfect commutator? Take β_f and β_r as the forward and reverse current gains of the transistor.

2. The integrator network being replaced, calculate $u(t)$ in the general case where the square wave is asymmetric by the ratio k shown in Fig. 2, and has frequency $\omega/2\pi$. (Assume $r \ll R$.)

3. In the particular case where $\omega = \omega'$ express the d.c. component in terms of the different amplitudes and of the phase ϕ.

4. In the general case $\omega \neq \omega'$, for what frequencies is there a d.c. component of u? What is the behaviour of the output in the neighbourhood of these frequencies and, in particular, how does the amplitude of the low-frequency component of u vary as a function of the difference $(\omega - \omega')$? What is this behaviour for the case $k = \frac{1}{2}$?

Solution. 1. The chopper in the circuit blocks the current when it is open and has zero voltage across its terminals when it is closed. It is required to function irrespective of the sense of the voltages which are applied to it.

A transistor can approximately satisfy these conditions within certain limits. No current passes when $I_B = 0$ (if we neglect I_{CE_0}) while it behaves as a very low resistance when it is saturated. We must examine four cases.

(i) $$v_1 < 0, \quad v_2 < 0.$$

The transistor behaves in a normal manner, its base current being about v_2/R_B, its collector current v_1/r for saturation, which requires that

$$\frac{v_1}{r} < \beta_f \frac{v_2}{R_B},$$

or

$$v_1 < \beta_f \frac{r}{R_B} \cdot v_2.$$

In this case, when v_2 has the maximum negative value, b, this becomes

$$v_1 < \beta_f \frac{r}{R_B} \cdot b.$$

(ii) $$v_1 < 0, \quad v_2 > 0.$$

The base-emitter junction is cut-off and the second condition is thus that v_2 shall not exceed the Zener voltage, i.e.

$$a < V_Z.$$

(iii) $$v_1 > 0, \quad v_2 < 0.$$

The control base current is in the forward direction but the collector voltage is reversed. It is as if the transistor is reverse biassed, the collector acting as emitter and vice versa. We have the saturation condition as in (i) but with β_r replacing the much larger β_f. The condition thus obtained is

$$v_1 < \beta_r \frac{r}{R_B} \cdot b$$

which supersedes the condition in (i).

(iv) $$v_1 > 0, \quad v_2 > 0.$$

With v_1 positive, the input junction is biassed so that the transistor is cut-off and no current flows. It is necessary, however, that $|v_1| < |v_2|$ if the cut-off condition is to hold. This requires here that

$$v_1 < a.$$

2. Since $r \ll R$ when the transistor is cut-off, the voltage at A is almost equal to v_1 while, when the transistor is saturated, this voltage is zero. Then, effectively

$$v_A(t) = v_1(t) \times g(t)$$

where $g(t)$ is a square wave function, being either 0 or 1; 0 when the transistor is saturated, 1 when it is cut-off. $g(t)$ can be expressed as a Fourier series

$$g(t) = a_0 + \sum_{n=1}^{\infty} a_n \cos n\omega t = \sum_{n=0}^{\infty} a_n \cos n\omega t.$$

The circuit of interest is then that of Fig. 3,

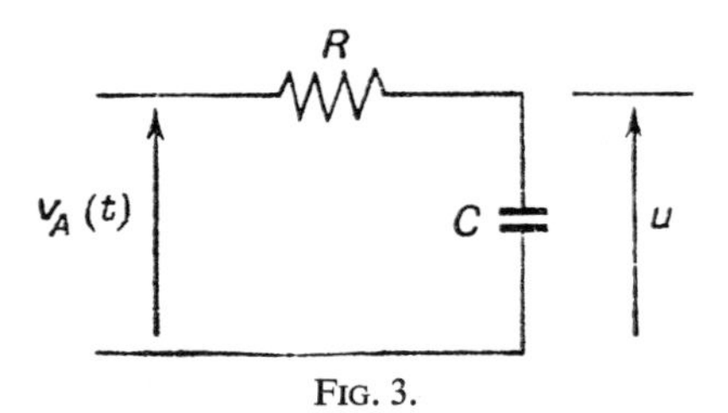

FIG. 3.

For this circuit there are two conditions:

$$u(t) = v_A(t) - Ri(t)$$

$$u(t) = \frac{1}{C}\int_0^t i(t)\,\mathrm{d}t.$$

Eliminating i gives the equation to be solved, namely,

$$RC\frac{\mathrm{d}u}{\mathrm{d}t} + u = v_1(t)\,g(t),$$

which is equivalent to

$$RC\frac{\mathrm{d}u}{\mathrm{d}t} + u = V_1 \cos(\omega' t + \phi)\sum_0^\infty a_n \cos n\omega t.$$

This is a first order differential equation with complementary function

$$u = K\mathrm{e}^{-t/RC}$$

which is a transient solution only.

Taking the complete solution as $n = f(t)\mathrm{e}^{-t/RC}$ gives

$$u(t) = \frac{V_1}{2RC}\sum_{n=0}^\infty a_n \left\{ \frac{\frac{1}{RC}}{\frac{1}{R^2C^2} + (\omega' + n\omega)^2}\cos[(\omega' + n\omega)t + \phi] \right.$$

$$+ \frac{\omega' + n\omega}{\frac{1}{R^2C^2} + (\omega' + n\omega)^2}\sin[(\omega' + n\omega)t + \phi]$$

$$+ \frac{\frac{1}{RC}}{\frac{1}{R^2C^2} + (\omega' - n\omega)^2}\cos[(\omega' - n\omega)t + \phi]$$

$$\left. + \frac{\omega' - n\omega}{\frac{1}{R^2C^2} + (\omega' - n\omega)^2}\sin[(\omega' - n\omega)t + \phi] \right\}$$

which may be reduced to

$$u(t) = \frac{V_1}{2RC}\sum_0^\infty a_n\left\{\frac{1}{\sqrt{\frac{1}{R^2C^2}+(\omega'+n\omega)^2}}\cos[(\omega'+n\omega)t+\phi+\psi_+]\right.$$

$$\left.+\frac{1}{\sqrt{\frac{1}{R^2C^2}+(\omega'-n\omega)^2}}\cos[(\omega'-n\omega)t+\phi+\psi_-]\right\}$$

with

$$\tan\psi_\pm = (\omega'\pm n\omega)RC.$$

The coefficients a_n can be calculated in terms of the ratio k since

$$g(t) = k+\frac{2}{\pi}\sum_n\frac{\sin kn\pi\cos n\omega t}{n},$$

from which

$$a_0 = k$$

and

$$a_n = \frac{2}{\pi}\,\frac{\sin kn\pi}{n}.$$

3. $\omega = \omega'$. Then, with $1/RC = \omega_0$,

$$u = \frac{V_1\omega_0}{2}\sum_0^\infty a_n\left\{\frac{1}{\sqrt{\omega_0^2+(n+1)^2\omega^2}}\cos[(n+1)\omega t+\phi+\psi_+]\right.$$

$$\left.+\frac{1}{\sqrt{\omega_0^2+(n-1)^2\omega^2}}\cos[(n-1)\omega t+\phi+\psi_-]\right\}.$$

The d.c. component corresponds to $n = 1$, $(n-1) = 0$, or

$$u_0 = \frac{V_1\omega_0}{2}a_1\frac{1}{\omega_0}\cos(\phi+\psi_-)$$

with $\cos\psi_- = 1$, so that $\psi_- = 0$ or 2π, etc.

$$u_0 = \frac{V_1a_1}{2}\cos\phi = \frac{V_1\sin k\pi}{\pi}\cdot\cos\phi.$$

In the particular case $k = \frac{1}{2}$, the symmetric square wave,

$$u_0 = \frac{V_1}{\pi}\cos\phi.$$

This d.c. term is superimposed on the a.c. terms of which those of lowest frequency correspond to $n = 0$ and $n = 2$, giving

$$u_1 = \frac{V_1\omega_0}{2}\frac{1}{\sqrt{\omega_0^2+\omega^2}}[a_0\cos(\omega t+\phi+\psi_+)+a_2\cos(\omega t+\phi+\psi_-)]$$

which has an amplitude less than

$$\frac{V_1}{2}\cdot\frac{1}{\sqrt{1+\omega^2/\omega_0^2}}(a_0+a_2).$$

If $\omega \gg \omega_0$

$$|u_\omega| \leqslant \frac{V_1}{2}\frac{\omega_0}{\omega}(a_0+a_2)$$

depending on ψ_+ and ψ_-.

4. There will be a d.c. component for each case $(\omega'-n\omega) = 0$, i.e. for $\omega' = n\omega$, the magnitude being

$$u_{0n} = \frac{V_1 a_n}{2}.$$

In the neighbourhood of these particular points which give a d.c. component there will be a low-frequency contribution

$$u_{\text{L.F.}} = \frac{V_1\omega_0}{2}\frac{a_n}{\sqrt{\omega_0^2+\delta_n^2}}\cos(\delta_n t+\phi+\psi_-)$$

with $\delta_n = \omega'-n\omega$.

The amplitude can be written with $x_n = \delta_n/\omega_0$

$$u_{\text{L.F.}} = \frac{V_1}{2}a_n\frac{1}{\sqrt{1+x_n^2}}.$$

This decreases by 3 dB when the frequency is $\pm\omega_0$ from $n\omega$ as indicated in Fig. 4.

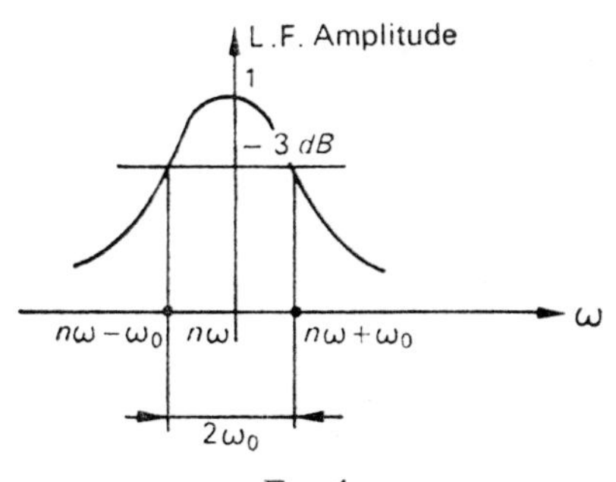

FIG. 4.

Problem No. 12

THE DIODE CIRCUIT

1. In the circuit of Fig. 1 the diode is assumed perfect, with an infinite reverse resistance and negligible forward resistance.

A voltage generator supplies a symmetric square wave $(+a)$ to $(-a)$ across the input AB.

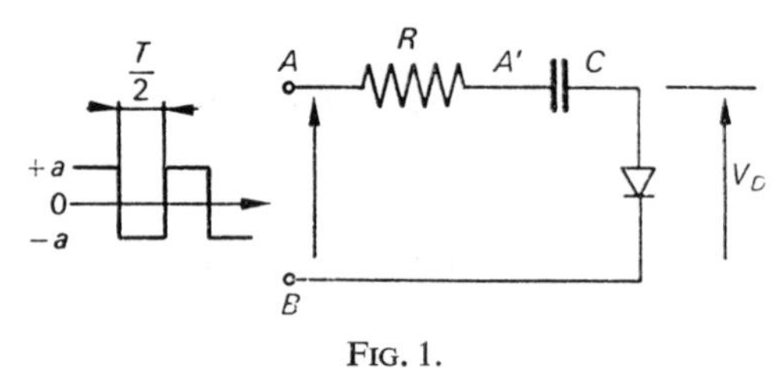

FIG. 1.

If the period of the square wave is long compared with RC, what is the form of the output voltage $V_D(t)$ across the diode?

2. What is the form of $V_D(t)$ if a resistance R_1 is placed between the anode of the diode and a positive voltage E as in Fig. 2?

Sketch $V_D(t)$ and $V_{A'}(t)$ in the case where $R = 10\ \text{k}\Omega$, $C = 0{\cdot}1\ \mu\text{F}$, $R_1 = 100\ \text{k}\Omega$, $E = 10$ V, $a = 4$ V, $T = 2 \times 10^{-2}$ s. What would be the forms of $V_D(t)$ and $V_{A'}(t)$ if $RC \ll T$ and $(R+R_1)C \gg T$?

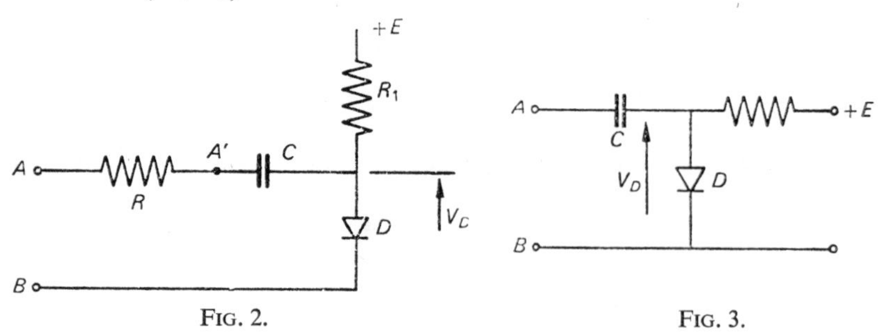

FIG. 2. FIG. 3.

3. The circuit of Fig. 2 is now replaced by that of Fig. 3 with $R = 0$ and a sinusoidal voltage $V_1 = a \sin \omega t$ applied across AB at $t = 0$. Calculate the charging current of the capacitor and find the point t_1 at which the diode cuts-off.

4. Taking the time constant R_1C to be very much larger than the period $2\pi/\omega$ what will be the form of the voltage across the diode? (Consider how the charge on C varies.)

5. In the general case, R_1C is not large with respect to the period. Write the differential equation for the variation of v, the voltage across C, after the instant t_1 calculated in 3. Find the form of $v(t)$ without calculating the integration constant explicitly. How can the calculation be developed to give the instant t_2 at which the diode begins to conduct again?

6. Find t_2 and $V_D(t)$ in the case where

$$\frac{\omega}{\omega_0} = \frac{4}{\sqrt{2}}, \quad \frac{E}{a} = 2, \quad E = 10\text{ V}.$$

Use the following table to give a graphical solution.

$\exp\left(-\frac{\sqrt{2}}{4}\cdot\frac{\pi}{4}\right) = 0{\cdot}758$	$\exp\left(-\frac{\sqrt{2}}{4}\cdot\pi\right) = 0{\cdot}33$
$\exp\left(-\frac{\sqrt{2}}{4}\cdot\frac{\pi}{2}\right) = 0{\cdot}574$	$\exp\left(-\frac{\sqrt{2}}{4}\cdot\frac{3\pi}{2}\right) = 0{\cdot}189$
$\exp\left(-\frac{\sqrt{2}}{4}\cdot\frac{3\pi}{4}\right) = 0{\cdot}435$	$\exp\left(-\frac{\sqrt{2}}{4}\cdot 2\pi\right) = 0{\cdot}11$

Solution. 1. Suppose that we start from $V_A = -a$ with the equilibrium potentials indicated in Fig. 4. The diode is cut-off and V_A will change abruptly from $-a$ to $+a$. As long as the diode does not conduct the current in the circuit will be

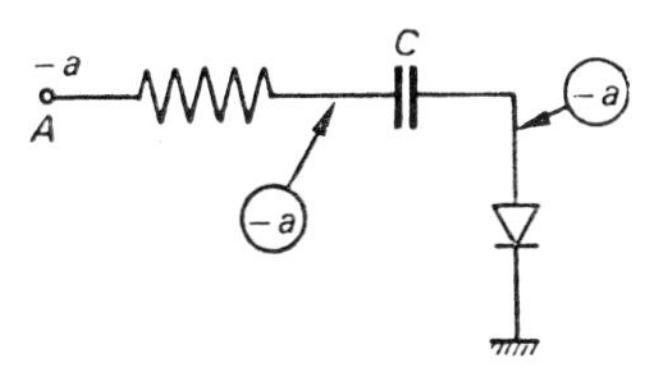

FIG. 4.

zero and the voltage V_0 will follow any variation of V_A. When the change in V_A occurs instantaneously the diode will conduct, connecting the right hand plate of C to earth. C will then begin to charge up. We have supposed $T \gg RC$, so that the potential at A' will follow Fig. 5 and, just before the potential V_A drops to $-a$, the potentials will be as shown in Fig. 6.

When V_A falls to $-a$, the other potentials will follow and the diode cuts-off. The capacity C cannot change its charge instantaneously, so that when V_A drops to

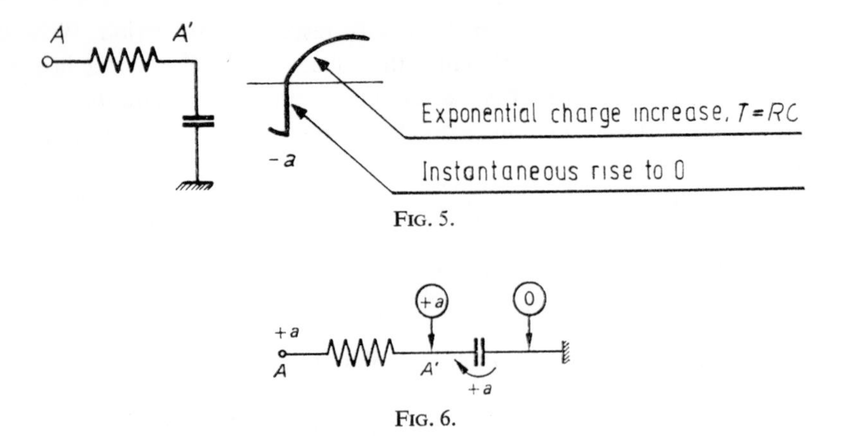

FIG. 5.

FIG. 6.

$-a$, V_D drops to $-2a$, so that a voltage a remains across C, giving the voltage distribution of Fig. 7. This state is stable since no current can flow through the diode and C cannot discharge. When V_A goes again to $+a$, the other potentials follow, V_D goes back to zero without the diode conducting and the state of Fig. 6 is repeated. The potential $V_D(t)$ is thus a square wave between zero and $-2a$ in Fig. 8.

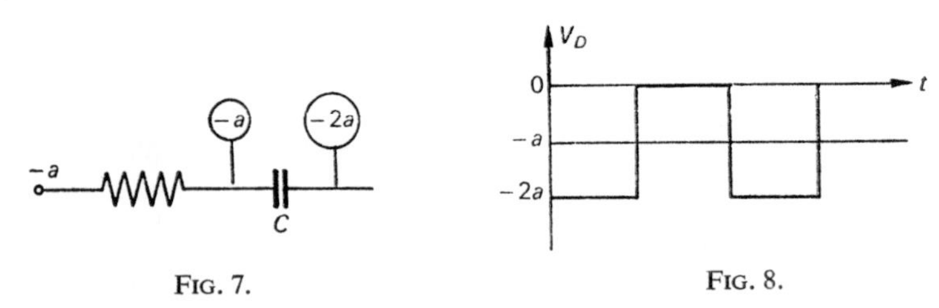

FIG. 7. FIG. 8.

2. In this case, if we start again from $V_A = -a$, equilibrium will be established as follows. The diode conducts through R_1 and $V_D = 0$, $V_{A'} = -a$. V_A changes abruptly towards $+a$. Neither $V_{A'}$ nor the charge on C can change instantaneously, V_D remaining zero while the diode conducts. $V_{A'}$ climbs exponentially as far as $+a$ with the time constant RC.

Just before the potential at A drops again we have

$$V_A = +a, \quad V_{A'} = +a, \quad V_D = 0.$$

The capacity is charged by the voltage a. V_A now falls to $-a$. A negative pulse is sent to the diode which cuts-off. Immediately after this change, the circuit will be in the state of Fig. 9.

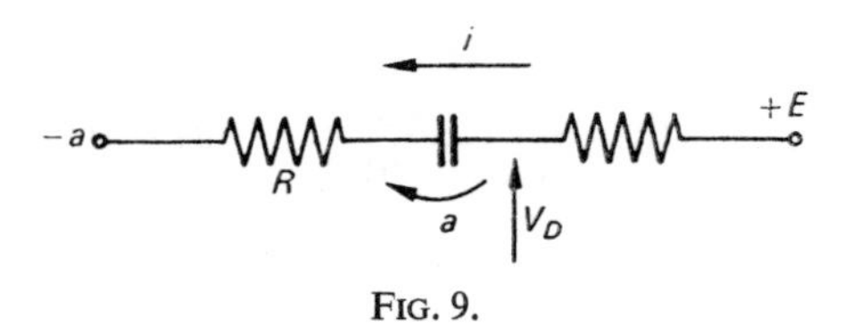

FIG. 9.

The capacity C behaves instantaneously as a battery, the current through the circuit being

$$i = \frac{E+2a}{R+R_1},$$

so that

$$V_D = E - R_1 \frac{E+2a}{R+R_1} = V_{D_0}$$

$$V_{A'} = E + a - R_1 \frac{E+2a}{E+R_1} = V_{A'_0}.$$

The voltages now change, with the time constant $(R+R_1)C$, towards an equilibrium state with $V_{A'} = -a$, $V_D = +E$.

However, when V_D reaches 0 the diode will begin to conduct. Immediately before this point the state of the circuit is that of Fig. 10, while immediately afterwards it becomes that of Fig. 11 with the diode earthing the right hand plate of C. $V_{A'}$ continues to change towards $-a$ but with the new time constant RC Immediately before V_A drops to $-a$ we thus find the initial state $V_{A'} = -a$, $V_D = 0$ and the cycle begins again.

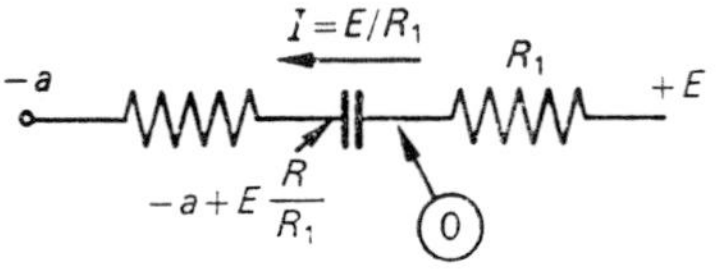

FIG. 10.

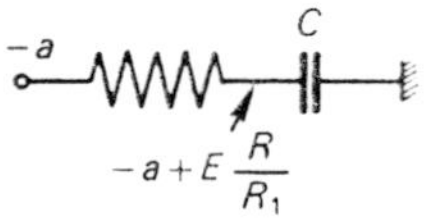

FIG. 11.

Putting in the numerical values gives

$$V_{D_0} = -6{\cdot}35, \quad V_{A'_0} = 2{\cdot}35, \quad RC = 10^{-3}\text{ s}, \quad (R+R_1)C = 11\times 10^{-3}\text{ s}.$$

The cut-off period for the diode is given by

$$T_c = (R+R_1)C \ln\left(1 - \frac{V_{\min}}{E}\right),$$

which is here

$$T_c = 11\times10^{-3}\ln\left(1+\frac{6\cdot35}{10}\right) = 5\cdot41\times10^{-3}\ \text{s}.$$

$V_{A'}(t)$ and $V_D(t)$ are then as shown in Fig. 12.

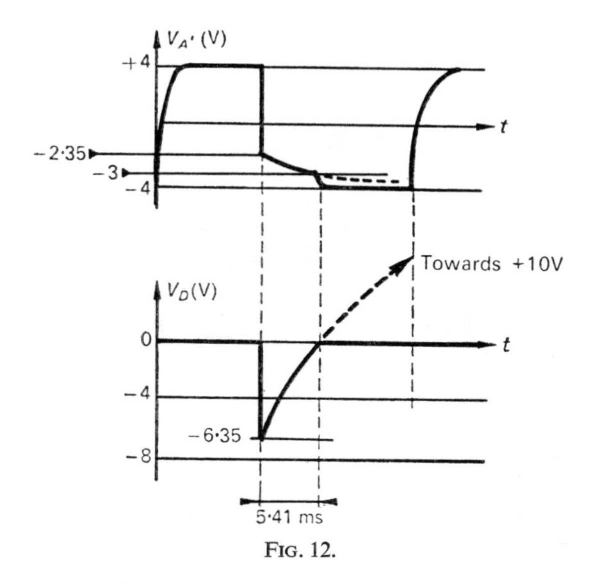

FIG. 12.

If $(R+R_1)C \gg T$ the increase in V_D is negligible during a half-cycle and this voltage follows very closely a square wave (Fig. 13).

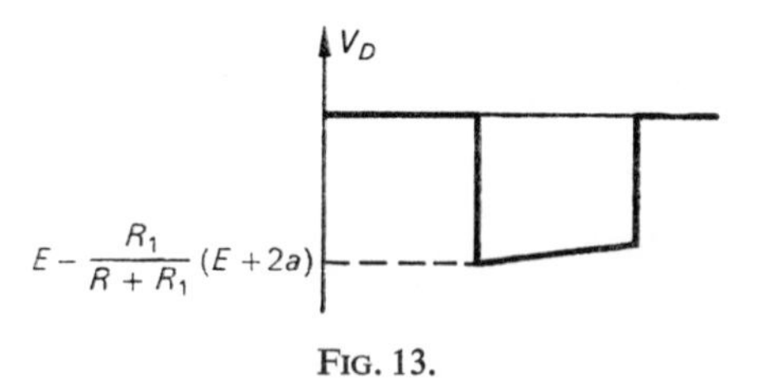

FIG. 13.

3. At first, however much V_A increases, the diode remains conducting and $V_D = 0$. The voltage v across C is thus V_A and the charging current is

$$i = -C\frac{dv}{dt} = -aC\omega \cos \omega t$$

(the negative sign is needed since i must be negative if $dv/dt > 0$). This current necessarily flows through R_1 while the current in the diode is (see Fig. 14)

$$I_D = \frac{E}{R_1} - i.$$

The diode will cut-off when no current is supplied, i.e. when

$$\frac{E}{R_1} + aC\omega \cos \omega t = 0,$$

or

$$\cos \omega t_1 = -\frac{E}{R_1 a C\omega} = -\frac{E}{a}\frac{\omega_0}{\omega}$$

where $\omega_0 = 1/RC$.

4. When the diode is cut-off, the circuit is equivalent to Fig. 15.

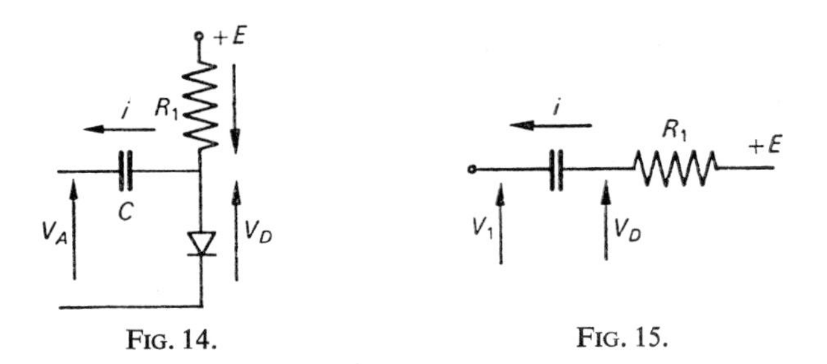

FIG. 14. FIG. 15.

With a large time constant R_1C, the charge on the condenser and the voltage across it cannot vary appreciably over one period. Now, at the moment of cut-off, $V_D = 0$ and

$$V_1 = a \sin \omega t_1 = a\sqrt{1 - \frac{E^2}{a^2} \cdot \frac{\omega_0^2}{\omega^2}} = V_{10}.$$

Since the voltage across C does not vary, V_D will follow the variations of V_1 approximately as far as V_{10}. This will continue up to the point where

$$V_D = V_1(t) - V_{10}$$

becomes positive, when the diode conducts (time t_2) and V_D remains zero. The time constant of the charge on C becomes zero and the situation is exactly analogous to that which existed for $t < t_1$. The voltage V_D has then the form of a limited sine wave. (Note that the reasoning would be far more complex if the forward resistance of the diode were not neglected, since then the voltage across C would not be able to follow the variations of V_1 exactly and, at the end of one period, the charge on C would not be exactly the same as for the preceding period.

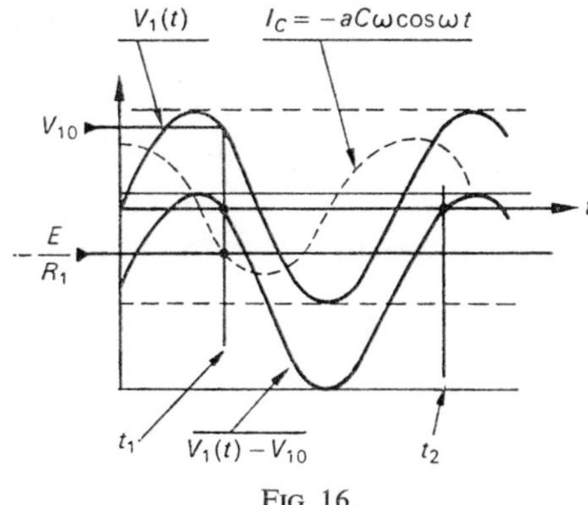

FIG. 16.

A permanent pattern would not be established theoretically until an infinite time had elapsed, in practice before several complete periods. Here, however, the stable pattern is obtained in a single period, which greatly simplifies the calculation.)

Variation of $V_1(t)$ is shown in Fig. 16 and the required form of V_D derived from it is in Fig. 17.

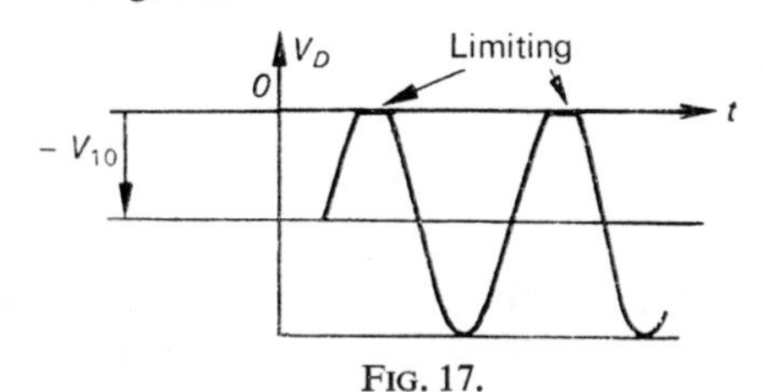

FIG. 17.

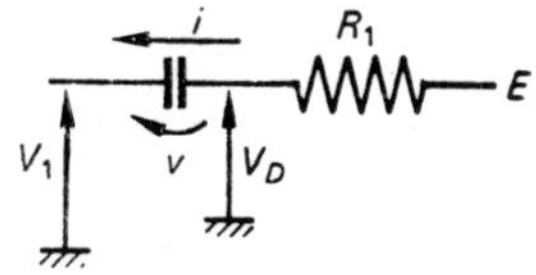

FIG. 18.

5. At the moment of cut-off $V_D = 0$; the charging current is then $i = E/R_1$ and $V_1 = V_{10}$. The charging current is also (Fig. 18)

$$i = -C\frac{\mathrm{d}v}{\mathrm{d}t}.$$

Also we can write

$$E - R_1 i + v = V_1$$

so that

$$R_1 C\frac{\mathrm{d}v}{\mathrm{d}t} + v = a \sin \omega t - E.$$

Writing the solution as $v = K\mathrm{e}^{-t/R_1C} = K\mathrm{e}^{-\omega_0 t}$ gives for K the equation

$$\frac{\mathrm{d}K}{\mathrm{d}t} = \frac{1}{R_1 C}[a \sin \omega t - E]\,\mathrm{e}^{t/R_1C} = \omega_0[a \sin \omega t - E]\,\mathrm{e}^{+\omega_0 t},$$

or

$$K = -E\,e^{\omega_0 t} + \int \omega_0 a\, e^{\omega_0 t} \sin \omega t\, dt.$$

Integration by parts [or writing $\sin \omega t = (e^{i\omega t} - e^{-i\omega t})/2i$] gives

$$K = -E\,e^{\omega_0 t} + \frac{\omega_0 a}{\omega_0^2 + \omega^2} (\omega_0 \sin \omega t - \omega \cos \omega t)\, e^{\omega_0 t} + K'$$

and, substituting in the expression for v, we have

$$v = -E + \frac{a\omega_0^2}{\omega_0^2+\omega^2} \sin \omega t - a \frac{\omega\omega_0}{\omega_0^2+\omega^2} \cos \omega t + K'\, e^{-\omega_0 t}.$$

The constant K' is determined from the boundary conditions that, for $t = t_1$,

$$v = V_{10} = a\sqrt{1 - \frac{E^2\omega_0^2}{a^2\omega^2}}, \quad \cos \omega t_1 = -\frac{E\omega_0}{a\omega}, \quad \sin \omega t_1 = \sqrt{1 - \frac{E^2}{a^2}\frac{\omega_0^2}{\omega^2}}$$

which, on substitution, gives

$$K' = \left(1 - \frac{\omega_0^2}{\omega_0^2+\omega^2}\right)\left(E + a\sqrt{1 - \frac{E^2}{a^2}\frac{\omega_0^2}{\omega_0}}\right) e^{\omega_0 t_1}.$$

The curve can be traced, point by point, from the following table:

ωt	$\sin \omega t$	$\cos \omega t$	$\frac{5}{9} \sin \omega t$	$-\frac{20}{9\sqrt{2}} \cos \omega t$	$27{\cdot}7 \exp\left(-\frac{\sqrt{2}}{4}\omega t\right)$	v	v_1
0	0	1	0	−1·572	27·7	16·13	0
$\frac{\pi}{4}$	$\sqrt{2}/2$	$\sqrt{2}/2$	0·393	−1·111	21	10·28	3·54
$\frac{2\pi}{4}$	1	0	0·555	0	15·9	6·45	5
$\frac{3\pi}{4}$	$\sqrt{2}/2$	$\sqrt{2}/2$	+0·393	+1·111	12·05	3·54	3·54
$\frac{4\pi}{4}$	0	−1	0	+1·572	9·15	0·722	0
$\frac{6\pi}{4}$	−1	0	−0·555	0	5·24	−5·31	−5
$\frac{8\pi}{4}$	0	1	0	−1·572	3·05	−8·52	0

The voltage across the diode is

$$V_D = V_1 - v = a \sin \omega t - v = E + a\left(1 - \frac{\omega_0^2}{\omega_0^2+\omega^2}\right) \sin \omega t + \frac{a\omega\omega_0}{\omega_0^2+\omega^2} \cos \omega t - K' e^{-\omega_0 t}.$$

The time t_2 will be found on taking the solutions for $V_D = 0$.

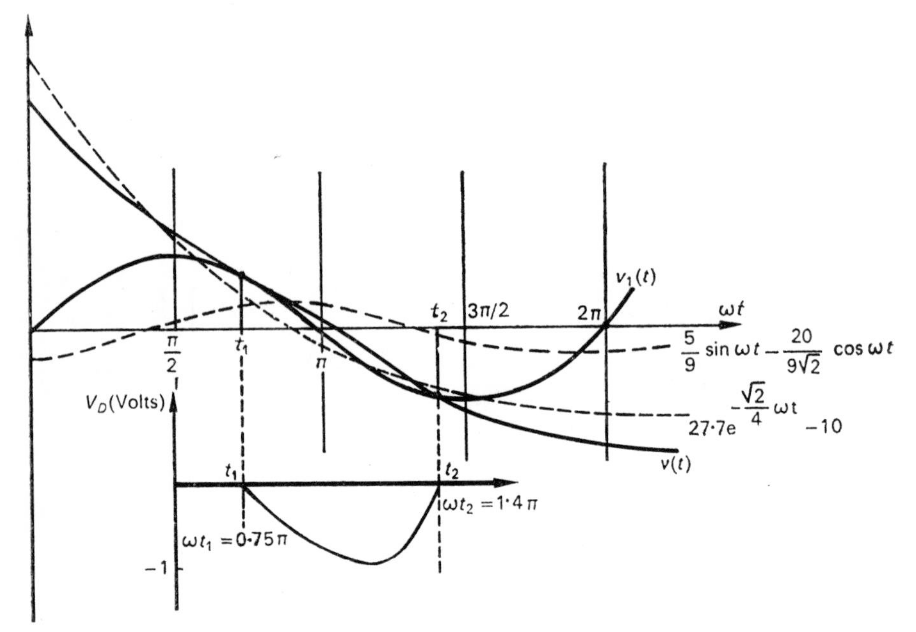

FIG. 19.

6. The numerical values given allow the coefficients of the above equation to be calculated easily (see Fig. 19).

$$\cos \omega t_1 = -2\frac{\sqrt{2}}{4} = \frac{-\sqrt{2}}{2}, \quad \text{from which} \quad \omega t_1 = \frac{3\pi}{4};$$

$$\omega_0 t_1 = \frac{\omega_0}{\omega} \times \omega t_1 = \frac{\sqrt{2}}{4} \cdot \frac{3\pi}{4} = \frac{3\sqrt{2}\pi}{16}$$

$$\exp\left(-\frac{3\sqrt{2}\pi}{16}\right) = 0{\cdot}435$$

$$K' = 27{\cdot}7$$

$$v = -10 + \frac{5}{9}\sin \omega t - \frac{20}{9\sqrt{2}}\cos \omega t + 27{\cdot}7 \exp\left(-\frac{\sqrt{2}}{4}\omega t\right).$$

The intersection of $v(t)$ with $v_1(t)$ gives t_2; here

$$\omega t_2 \simeq 1{\cdot}4\pi.$$

Moreover, during cut-off,

$$V_D = v_1 - v,$$

from which we have the curve of $V_D(t)$.

Problem No. 13

IMPROVEMENT OF SIGNAL-TO-NOISE RATIO

1. Firstly, the modulator system. Consider the circuit of Fig. 1 where the diodes are supposed perfect, with zero impedance in the forward direction. The transformers T_1 and T_2, of turns ratio 1 : 1, are also perfect with no magnetic leakage and have infinite self-inductance for the windings. Show that, if $u(t) = 0$, the gain of the system is zero with $v_2(t) = 0$.

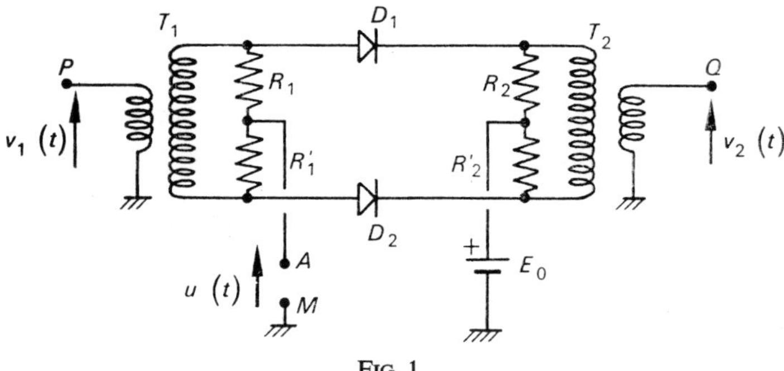

FIG. 1.

With no signal applied (i.e. the input "floating") what is the form of $v_2(t)$ when a square-wave pulse of length ε and amplitude $+2E_0$ is applied to the control input AM? What must be the values of the resistances if $v_2(t)$ is to remain zero?

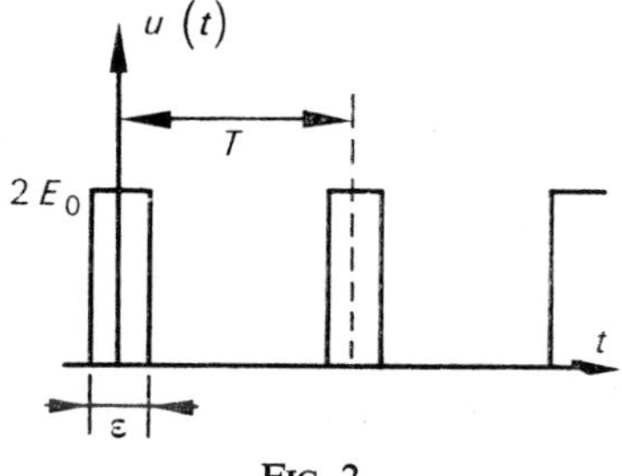

FIG. 2.

With the above conditions retained, how would the gain $G(t)$ given by

$$G(t) = \frac{v_2(t)}{v_1(t)}$$

vary when a series of square pulses of length ε, amplitude $2E_0$ and recurrence period T are applied at AM?

Assuming that the gain can be represented as a Fourier series

$$G(t) = \sum_{n=0}^{\infty} c_n \exp jn \frac{2\pi t}{T}$$

calculate the coefficients c_n.

2. The variable gain circuit of Fig. 1 is placed between two filters having transfer functions $A_i(\omega)$ and $A_0(\omega)$ respectively as in Fig. 3. The Fourier transforms of the

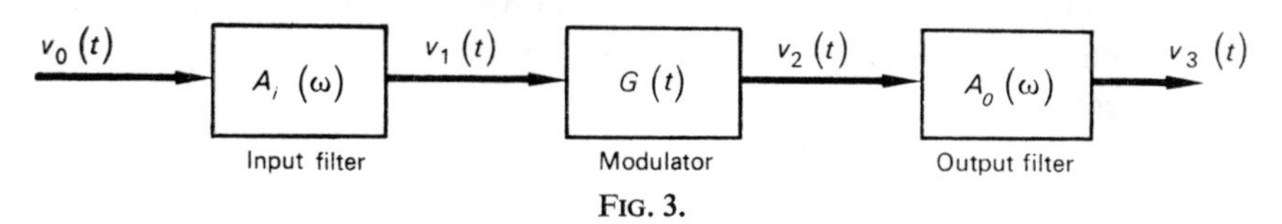

FIG. 3.

signals will be indicated by $V_1(j\omega)$, $V_2(j\omega)$, $V_3(j\omega)$, their spectral energy density by $S_1(\omega)$, $S_2(\omega)$, $S_3(\omega)$ and their powers by P_1, P_2, P_3. In the most general case, $v_1(t)$ will be composed of a useful signal $v_{1S}(t)$ which may be swamped in a noise signal $v_{1N}(t)$. Consider first the transmission of the noise through the system and then the signal:

(a) State, first of all, the definitions of S and P for a non-periodic signal.

(b) The noise.

Calculate $S_{3N}(\omega)$, the spectral distribution of the noise at the output, in terms of S_{1N}, $A_0(\omega)$ and the coefficients c_n.

In the case where there is white noise filtered by $A_i(\omega)$ at the input, calculate the power P_{3N} of the output noise in terms of P_{1N}. Let N^2 be the spectral density of the white noise before the input filter and put

$$\frac{1}{K_i} = \int_{-\infty}^{\infty} |A_i(\omega)|^2 \, d\omega.$$

The filters $A_i(\omega)$ and $A_0(\omega)$ are low-pass RC networks with time constants T_i and T_0 given by

$$T_i = R_i C_i = 10 \ \mu\text{s}$$
$$T_0 = R_0 C_0 = 1 \text{ s}.$$

To simplify the calculations consider these low-pass filters to act perfectly, i.e.

$$A_i(\omega) = 1 \quad \text{for} \quad |\omega| < \frac{1}{T_i}.$$

$$A_i(\omega) = 0 \quad \text{for} \quad |\omega| > \frac{1}{T_i},$$

$$A_0(\omega) = 1 \quad \text{for} \quad |\omega| < \frac{1}{T_0},$$

$$A_0(\omega) = 0 \quad \text{for} \quad |\omega| > \frac{1}{T_0}.$$

Further, take

$$\Omega = \frac{2\pi}{T} = 500 \text{ rad/s}$$

and put

$$z = \frac{\varepsilon}{T_i}.$$

Under these conditions find P_{3N} in terms of P_{1N}, T_i, T_0, T and z, replacing the summation over n by an integration. (Discuss the justification of this approximation in terms of the magnitude of z, taking

$$Si(z) = \int_0^z \frac{\sin u}{u} du.,$$

(c) The signal.

We have seen the transformation, by the sampling system, of white noise of power P_{1N} to the reduced power P_{3N} after it had passed the modulator and low-pass output filter. We now study the effect of the system on a periodic signal of pulsatance Ω_S close to Ω. At the output of A_i

$$v_{1S} = \sum_k s_k \exp(jk\,\Omega_S t)$$

and, after the modulator,

$$v_{2S}(t) = G(t)\,v_{1S}(t).$$

If the output filter A_0 will only passfrequencies which are small compared with the signal ($T \ll T_0$) what is the form of $v_{3S}(t)$? Note that if the c_n, the Fourier coefficients for $G(t)$, are independent of n for those values of n where the coefficients s_n are non-zero, then $v_{3S}(t)$ will exactly reproduce $v_{1S}(t)$, but at a much lower frequency.

Assume that the coefficients s_k are only appreciable for $k \leqslant 10$. What is the physical significance of such an assumption? What limit does it imply for the value of z? Under this condition, give an expression for P_{3S}/P_{1S} in terms of z, T_i and T.

3. Signal-to-noise ratio.

The noise improvement factor of the system may be defined as

$$F = \frac{\dfrac{P_{3S}}{P_{3N}}}{\dfrac{P_{1S}}{P_{1N}}}.$$

This factor F is a measure of the improvement in the signal-to-noise ratio given by the system. With the restriction imposed above on the c_n and the limitation which this imposes on z, find the value of F. Remember that, for z small,

$$S_i = z - \frac{z^3}{3\times 3!}.$$

Solution. 1. The diodes D_1 and D_2 are reverse biassed by the battery E_0 and have infinite impedance. Then, for a signal generated across the secondary of the transformer T_1 the diodes are mounted back-to-back and $v_2(t)$ will be zero irrespective of $v_1(t)$.

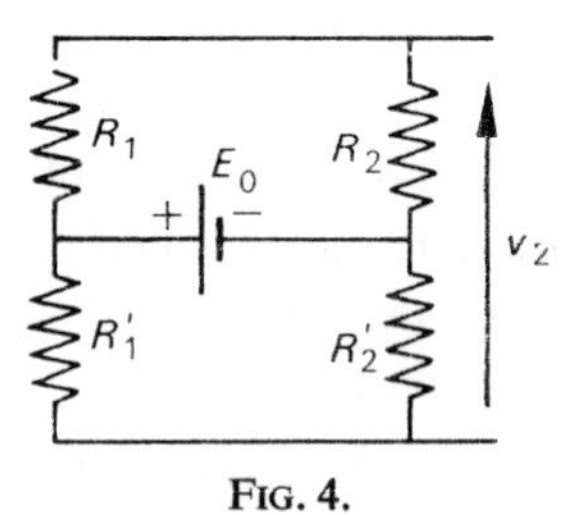

FIG. 4.

For the duration of the pulse $2E_0$ the equivalent circuit of the system is that of Fig. 4 from which

$$v_2 = E_0 \left(\frac{R_2}{R_1 + R_2} - \frac{R_2'}{R_1' + R_2'} \right)$$

which will be zero if $R_2/R_1 = R_2'/R_1'$. (In practice $R_1 = R_1'$, $R_2 = R_2'$.) If $u(t) = +2E_0$ the diodes conduct and $v_2 = v_1$. The system gain thus varies between

0 and 1 with the frequency of the input pulses (Fig. 5). Putting $\Omega = 2\pi/T$

$$c_n = \frac{1}{T}\int_{-T/2}^{+T/2} G(t)\exp(-\mathrm{j}n\,\Omega t)\,\mathrm{d}t = \frac{1}{T}\int_{-\varepsilon/2}^{+\varepsilon/2} \exp(-\mathrm{j}n\,\Omega t)\,\mathrm{d}t.$$

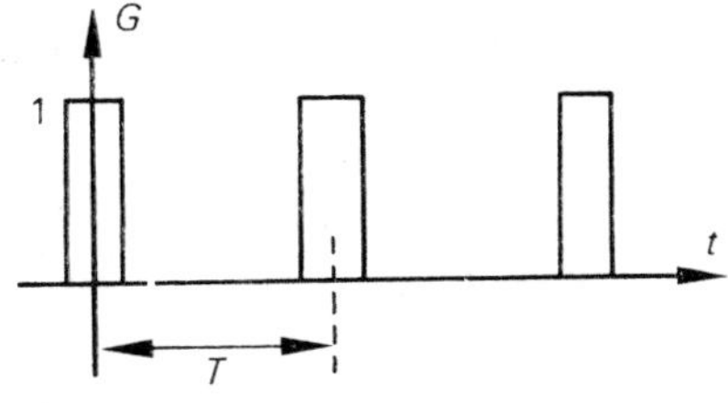

FIG. 5.

With $\mathrm{e}^{\mathrm{j}\alpha} = \cos\alpha + \mathrm{j}\sin\alpha$ this gives immediately

$$c_n = \frac{\sin n\Omega\dfrac{\varepsilon}{2}}{n\pi} = \frac{\varepsilon}{T}\,\frac{\sin n\dfrac{\varepsilon}{2}}{n\dfrac{\varepsilon}{2}}.$$

2. (a) For a periodic signal the mean power may be written from Parceval's theorem in the form

$$\frac{1}{T}\int_{T/2}^{T/2} f^2(t)\,\mathrm{d}t = \sum_n |c_n|^2,$$

the term $|c_n|^2$ representing the power carried by the nth harmonic.

For a non-periodic signal it is not possible to define a mean power on the basis of a finite interval of time since the result will depend on the position of this interval. For a completely general signal $f(t)$ the function $f_T(t)$ is introduced such that

$$f_T(t) = f(t) \quad \text{for} \quad -\frac{T}{2} < t < +\frac{T}{2}$$

$$f_T(t) = 0 \quad \text{for} \quad |t| > \frac{T}{2}.$$

This function has a Fourier transform

$$F_T(\mathrm{j}\omega) = \int_{-T/2}^{T/2} f_T(t)\,\mathrm{e}^{-\mathrm{j}\omega t}\,\mathrm{d}t.$$

The mean energy delivered during the time T is then of the form

$$P_T = \frac{1}{T} \int_{-T/2}^{T/2} f_T^2(t)\,\mathrm{d}t = \frac{1}{T} \int_{-\infty}^{+\infty} |F_T(\mathrm{j}\omega)|^2 \frac{\mathrm{d}\omega}{2\pi},$$

which is a generalization of Parceval's theorem. Going to the limit the mean power may be defined as

$$P = \lim_{T\to\infty} \frac{1}{T} \int_{-\infty}^{+\infty} |F_T(\mathrm{j}\omega)|^2 \frac{\mathrm{d}\omega}{2\pi},$$

which is usually written in terms of the spectral density as

$$P = \int_{-\infty}^{\infty} S(\omega) \frac{\mathrm{d}\omega}{2\pi}$$

with

$$S(\omega) = \lim_{T\to\infty} \frac{|F_T(\mathrm{j}\omega)|^2}{T}.$$

(b) We have

$$v_{1N} = \int_{-\infty}^{\infty} v_{1N}(\mathrm{j}\omega)\,\mathrm{e}^{\mathrm{j}\omega t} \frac{\mathrm{d}\omega}{2\pi}$$

and

$$v_{2N}(t) = G(t) v_{1N}(t) \quad \text{with} \quad G(t) = \sum_n c_n \exp(\mathrm{j}n\Omega t).$$

Then

$$V_{2N}(\mathrm{j}\omega) = \int_{-\infty} v_{2N}(t)\,\mathrm{e}^{-\mathrm{j}\omega t}\,\mathrm{d}t = \sum_n c_n \int_{-\infty} v_{1N}(t)\,\mathrm{e}^{-\mathrm{j}(\omega - n\Omega)t}\,\mathrm{d}t.$$

That is,

$$V_{2N}(\mathrm{j}\omega) = \sum_n c_n V_{1N}[\mathrm{j}(\omega - n\Omega)].$$

On passing the output filter this becomes

$$V_{3N}(\mathrm{j}\omega) = A_0(\omega) V_{2N}(\mathrm{j}\omega) = A_0(\omega) \sum_n c_n V_{1N}[\mathrm{j}(\omega - n\Omega)].$$

However, it must be noted that the noise $V_{3N}(\mathrm{j}\omega)\,\mathrm{e}^{\mathrm{j}\omega t}$ is not a single sine wave of amplitude $A_0(\omega)\sum_n c_n V_{1N}[\mathrm{j}(\omega - n\Omega)]$ for which the power would be proportional to

$$\left| A_0(\omega) \sum_n c_n V_{1N}[\mathrm{j}(\omega - n\Omega)] \right|^2.$$

We have here a sum of sine waves all of pulsatance ω but with completely random phases. The instantaneous power is thus

$$|A_0(\omega)|^2 \sum_n |c_n|^2 V_{1N}[j(\omega - n\Omega)]^2.$$

The spectral distribution is then

$$S_{3N}(\omega) = |A_0(\omega)|^2 \sum_n |c_n|^2 S_{1N}(\omega - n\Omega).$$

The power output from the first filter is

$$P_{1N} = \int_{-\infty}^{\infty} N^2 |A_i(\omega)|^2 \frac{d\omega}{d\pi} = \frac{N^2}{2\pi K_i},$$

or, again,

$$S_{1N}(\omega) = N^2 |A_i(\omega)|^2$$

and

$$S_{1N}(\omega - n\Omega) = N^2 |A_i(\omega - n\Omega)|^2.$$

Substituting in the expression for S_{3N} this gives

$$S_{3N}(\omega) = |A_0(\omega)|^2 N^2 \sum_n |c_n|^2 |A_i(\omega - n\Omega)|^2$$

and the final noise power is

$$P_{3N} = \int_{-\infty}^{\infty} S_{3N}(\omega) \frac{d\omega}{d\pi} = \frac{N^2}{2\pi} \int_{-\infty}^{\infty} |A_0(\omega)|^2 \sum_n |c_n|^2 |A_i(\omega - n\Omega)|^2 \, d\omega$$

$$= K_i P_{1N} \sum_n |c_n|^2 \int_{-\infty}^{\infty} |A_0(\omega)|^2 |A_i(\omega - n\Omega)|^2 \, d\omega,$$

where we see the convolute introduced to give

$$P_{3N} = P_{1N} K_i \sum_n |c_n|^2 [|A_0(\omega)|^2_* |A_i(\omega)|^2]_{n\Omega}$$

With $T_i \ll T \ll T_0$, we calculate first of all the convolute

$$\int_{-\infty}^{\infty} |A_0(\omega)|^2 |A_i(\omega - n\Omega)|^2 \, d\omega.$$

The integrand will be different from zero only as long as

$$|\omega| \leqslant \frac{1}{T_0} = 1 \text{ rad/s}, \quad \text{for which} \quad |A_0(\omega)|^2 = 1$$

and

$$|\omega - n\Omega| \leqslant \frac{1}{T_i} = 10^5 \text{ rad/s},$$

or, taking note that $|\omega| < 1$,

$$|n| < \frac{1}{T_i\Omega} = \frac{T}{2\pi T_i}.$$

For a value of n which satisfies this condition, the convolute reduces to

$$\int_{-1/T_0}^{1/T_0} d\omega = \frac{2}{T_0},$$

when

$$P_{3N} = P_{1N}K_i \sum_{|n|<1/T_i\Omega} |c_n|^2 \frac{2}{T_0}$$

and

$$\frac{1}{K_i} = \int_{-\infty}^{\infty} |A_i(\omega)|^2 \, d\omega = \int_{-1/T_i}^{1/T_i} d\omega = \frac{2}{T_i}.$$

The summation over n gives

$$\sum_{|n|<1/T_i\Omega} |c_n|^2 = \sum_{|n|<1/T_i\Omega} \frac{\sin^2 n\Omega \dfrac{\varepsilon}{2}}{n^2\pi^2} = \sum_{|n|<1/T_i\Omega} \left(\frac{zT_i\Omega}{2\pi}\right)^2 \frac{\sin^2\left(\dfrac{nzT_i\Omega}{2}\right)}{\left(\dfrac{nzT_i\Omega}{2}\right)^2}.$$

If the term $nzT_i\,\Omega/2$ varies very slowly when n varies (i.e. $zT_i\,\Omega/2 \ll \pi$ or $z \ll T/T_i$) the summation over n can be replaced by an integral. If we put $x = nzT_i\,\Omega/2$ the sum becomes $\sum(\sin^2 x/x^2)$ which represents the sum of segments of which the abscissae are multiples of $(zT_i\,\Omega/2)$ and the heights are $(\sin^2 x/x^2)$. Now the area of the elementary rectangles of Fig. 6 are equal to $(\sin^2 x/x^2)\times(zT_i\,\Omega/2)$ and, on taking the limit when $(zT_i\,\Omega/2)$ is small,

$$\sum \frac{\sin^2 x}{x^2} \to \frac{1}{\left(\dfrac{zTi\Omega}{2}\right)} \int \frac{\sin^2 x}{x^2}\,dx$$

where the limits $|n| < 1/T_i\Omega$ require $|x| < z/2$, so that, finally,

$$\sum |c_n|^2 = \frac{zT_i\Omega}{2\pi^2} \int_{x=-z/2}^{x=z/2} \frac{\sin^2 x}{x^2}\,dx,$$

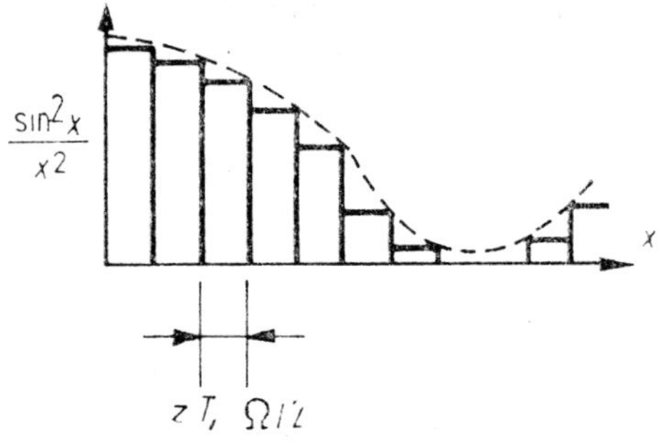

FIG. 6.

or, since the function is even,

$$\sum |c_n|^2 = \frac{zT_i\Omega}{\pi^2}\int_0^{z/2}\frac{\sin^2 x}{x^2}\,dx.$$

Integration by parts gives

$$\int_0^{z/2}\frac{\sin^2 x}{x^2}\,dx = \left(\frac{-\sin^2 x}{x}\right)_0^{z/2} + \int_0^{z/2}\frac{\sin 2x}{x}\,dx$$

and, putting $2x = u$,

$$\sum |c_n|^2 = \frac{zT_i\Omega}{\pi^2}\left(Si(z) - \frac{\sin^2\frac{z}{2}}{\frac{z}{2}}\right).$$

Substituting in the expression for P_{3N} and taking account of the relative magnitudes of the time constants gives

$$P_{3N} = P_{1N}\frac{2T_i^2 z}{\pi T T_0}\left(Si(z) - \frac{2\sin^2\left(\frac{z}{2}\right)}{z}\right).$$

If the sampling pulses are narrow so that z is small

$$P_{3N} \simeq P_{1N}\frac{T_i^2 z^2}{\pi T T_0} \quad \text{to second order in } z.$$

(c) The signal.

$$v_{1S}(t) = \sum_k s_k \exp(jk\Omega_S t),$$

so that

$$v_2S(t) = G(t)v_{1S}(t) = \sum_k\sum_n s_k c_n \exp j(k\Omega_S + n\Omega)t.$$

Since the output filter will only pass frequencies which are small compared with the signal frequency, only those terms having $n+k = 0$ will be transmitted and the output signal will be of the form

$$v_{3S}(t) = \sum_k s_k c_{-k} \exp \mathrm{j}(k\,\Omega_S - k\,\Omega)t,$$

or, since $c_{-k} = c_k$ from symmetry,

$$v_{3S}(t) = \sum_k s_k c_k \exp \mathrm{j}k(\Omega_S - \Omega)t.$$

If the values of c_k are independent of k for those cases where $s_k \neq 0$, then $v_{3S}(t)$ will exactly reproduce $v_{1S}(t)$ but at a frequency $(\Omega_S - \Omega)$ instead of Ω_S. The limitation $k \leqslant 10$ implies that no harmonic greater than 10 takes part in the signal, which will thus have relatively low-frequency contributions.

The condition that c_k is independent of k for $k \leqslant 10$ can be written as

$$\frac{\sin k\,\Omega\varepsilon/2}{k\pi} = c_0 \quad \text{for} \quad k \leqslant 10,$$

which implies that $10 \times \Omega\ \varepsilon/2 \ll \pi$ or $z \ll T/10\ T_i$. (In this case the condition that the summation over n can be replaced by an integral is, *a fortiori*, satisfied.)

Then

$$|c|_{k\leqslant 10} \simeq \frac{zT_i}{T},$$

$$P_{1S} = \sum_k |s_k|^2,$$

$$P_{3S} = \sum_k |s_k|^2 |c_k|^2 = |c_k|^2_{k\leqslant 10} \sum_k |s_k|^2$$

and

$$\frac{P_{3S}}{P_{1S}} = |c_k|^2_{k\leqslant 10} = \frac{z^2 T_i^2}{T^2}.$$

3. From the preceding results the noise improvement is

$$F = \frac{\dfrac{P_{3S}}{P_{3N}}}{\dfrac{P_{1S}}{P_{1N}}} = \frac{\dfrac{P_{3S}}{P_{1S}}}{\dfrac{P_{3N}}{P_{1N}}} = \frac{z^2T_i^2}{T^2} \cdot \frac{\pi T T_0}{2T_i^2 z} \, \frac{1}{\left[Si(z) - 2\dfrac{\sin^2 \dfrac{z}{2}}{z}\right]}$$

$$= \frac{\pi z T_0}{2T} \cdot \frac{1}{\left[Si(z) - 2\dfrac{\sin^2 \dfrac{z}{2}}{z}\right]}.$$

For small values of z,

$$Si(z) - 2\frac{\sin^2 z/2}{z} \sim z/2$$

and

$$F \simeq \frac{\pi T_0}{T}.$$

For $T_0 = 1$ s and $T = 2\pi/\Omega = 2\pi/500$, $F = 250$.

Problem No. 14

MATCHING OF A TRANSMISSION LINE TO A GENERATOR

(This problem requires the use of a Smith chart.)

A lossless line of characteristic impedance $Z_0 = 50\ \Omega$ and length $L = 24{\cdot}25$ m is terminated in a load impedance $Z_L = (20+\mathrm{j}\ 10)\ \Omega$. The phase velocity in the line is $c_0 = 3\times10^8$ m/s.

The source gives a peak voltage of 5 V at 150 MHz and has an internal impedance of $Z_g = (30+\mathrm{j}\ 5)\ \Omega$.

1. What is the power given to the load Z_L?

2. What would be the input impedance for there to be matching between the line and the generator? What would be the power supplied to the load in this case?

3. To obtain the matching required in question 2 another 50 Ω line is placed in parallel with the first (a stub) and is short-circuited at a distance l from the connection. Calculate the length l and the distance s which the stub must be placed from the load for there to be matching.

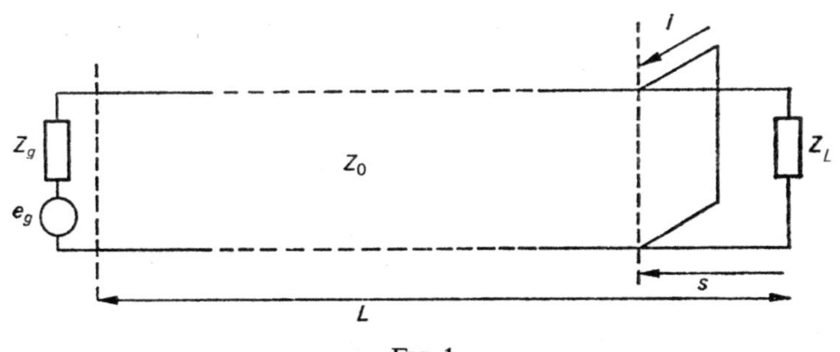

FIG. 1.

Solution. 1. At the frequency $f = 150$ MHz, the wavelength is

$$\lambda = \frac{c_0}{f} = 2\,\mathrm{m} \quad \text{and} \quad \frac{L}{\lambda} = 12{\cdot}125.$$

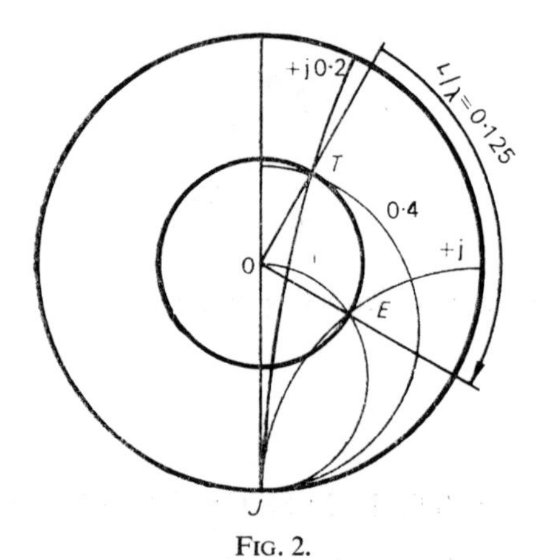

FIG. 2.

The reduced load is $z_L = Z_L/Z_0 = 0{\cdot}4 + \mathrm{j}\,0{\cdot}2$ and is represented on the chart of Fig. 2 by the point T.

The reduced input impedance z_i is found by turning through L/λ around the circle of constant V.S.W.R. through T. Since one turn of the chart corresponds to a distance $\lambda/2$ we must go round 24·25 times and so arrive at E. This gives $z_i = 1+\mathrm{j}$ and $Z_i = (50+\mathrm{j}\,50)\ \Omega$.

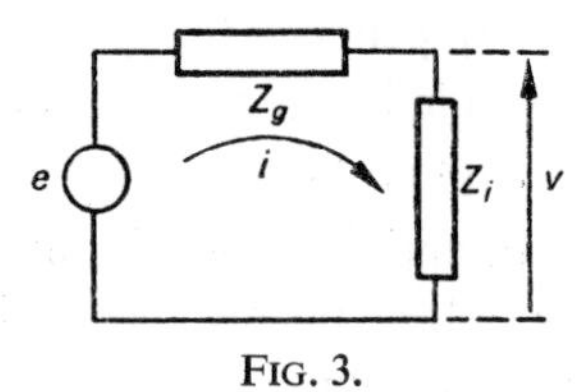

FIG. 3.

The equivalent input circuit is shown in Fig. 3. The power P supplied to the line is the real part of the product

$$\Pi = \frac{1}{2}\,vi^* = \frac{1}{2}\,e\,\frac{Z_i}{Z_g+Z_i}\cdot\frac{e^*}{(Z_g+Z_i)^*} = \frac{|e|^2}{2}\,\frac{Z_i}{|Z_g+Z_i|^2}\,.$$

$$P = \frac{|e|^2}{2}\,\frac{R_i}{|Z_g+Z_i|^2} = \frac{25}{2}\,\frac{50}{|80+\mathrm{j}\,55|^2} = 66{\cdot}3\ \text{mW}.$$

Since the line is lossless the whole of this power of 66·3 mW is transmitted to the load.

2. There will be matching if $Z_i = Z_g^* = (30-\mathrm{j}\,5)\ \Omega$. In that case

$$P' = \frac{|e|^2}{8R_i} = 104\ \text{mW}.$$

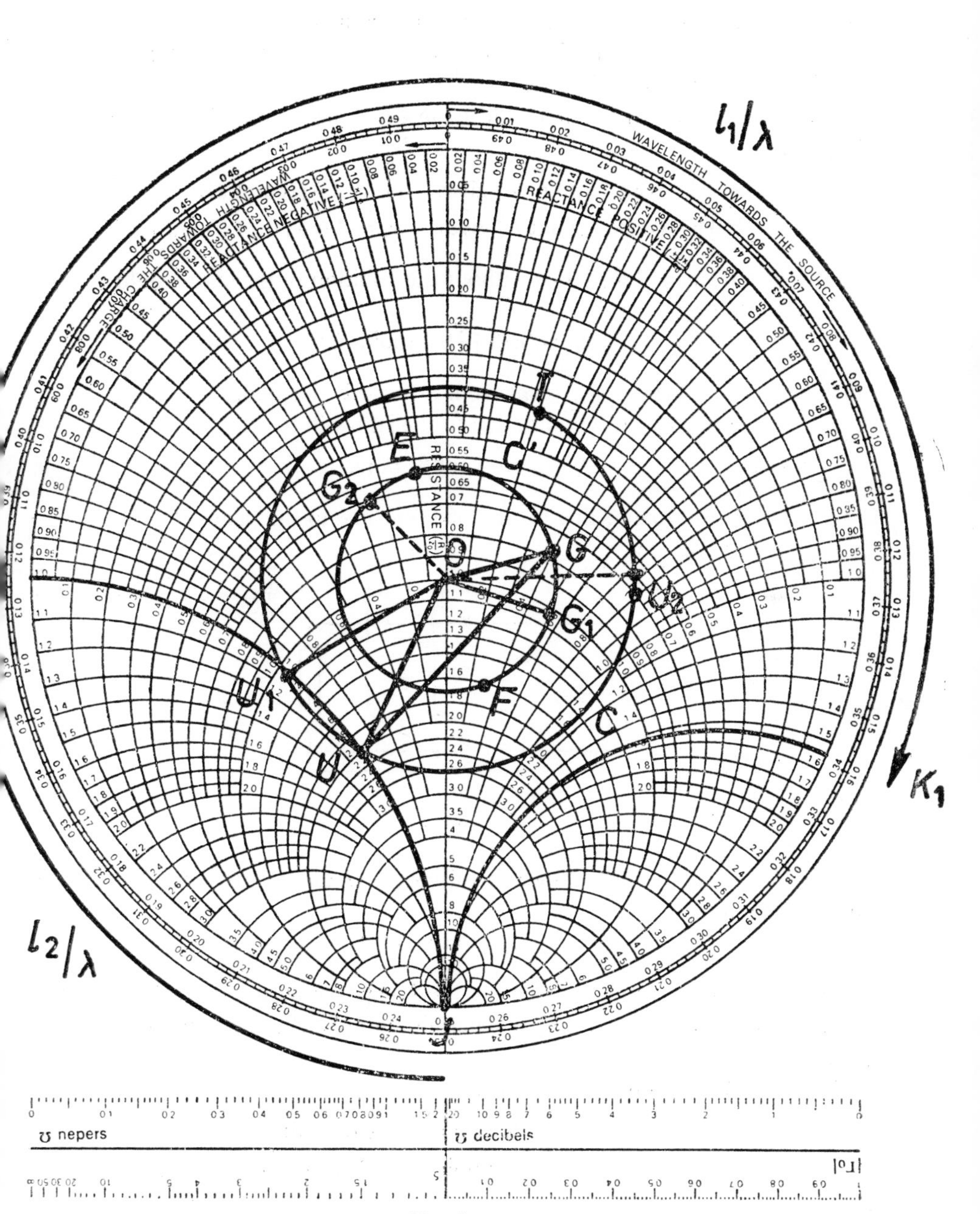

l_1/λ
l_2/λ
K_1
WAVELENGTH TOWARDS THE SOURCE
WAVELENGTH TOWARDS THE CHARGE
REACTANCE POSITIVE
REACTANCE NEGATIVE
RESISTANCE
E
C'
G
G₁
G₂
O
U
U₁
U₂
F
C
I
nepers
decibels

FIG. 4.

3. Since the load impedance Z_L and the required input impedance, $Z_i = (30-j5)\,\Omega$, are known, the position and length of the stub can be calculated. The procedure uses the chart of Fig. 4 as follows:

(a) Since the stub is in parallel with the line it is most convenient to work with admittances:

$$z_L = 0{\cdot}4+j\,0{\cdot}2 \quad \text{(point } T\text{) gives} \quad y_L = 2-j \text{ (point } U\text{)}$$
$$z_i = 0{\cdot}6-j\,0{\cdot}1 \quad \text{(point } E\text{) gives} \quad y_i = 1{\cdot}6+j\,0{\cdot}28 \text{ (point } F\text{)}.$$

In order to have $y_i = 1{\cdot}6+j\,0{\cdot}28$ without the addition of the stub would require a load admittance

$$y'_L = 0{\cdot}75+j\,0{\cdot}4 \text{ (point } G\text{)}$$

which is found by taking $24\frac{1}{4}$ turns from F towards the load.

(b) Now the reduced admittance due to the stub will be

$$y_s = -j \cot \beta l = jb.$$

Let $y(s)$ be the impedance due to the load and line in the plane Π at which the stub is inserted (Fig. 5).

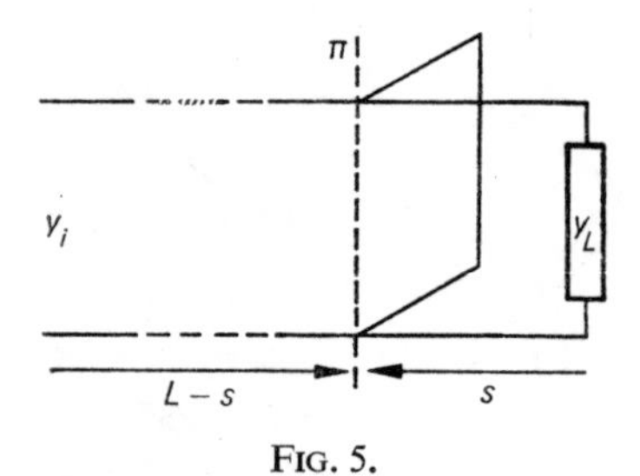

FIG. 5.

Let $y'(L-s)$ be the admittance at this plane Π due to the input impedance. Then we require

$$y'(L-s) = y(s)+jb.$$

Since the stub can only compensate the imaginary parts, the real parts of y' and y must be identical. We must determine graphically the plane Π at which $g' = g$.

If we start from the plane of the load

$$y_L \quad \text{is the point} \quad U(s = 0),$$
$$y'_L \quad \text{is the point} \quad G(L-s = L).$$

We now move towards the source (Fig. 4) when

U moves around the circle C (V.S.W.R. $= 2{\cdot}6$)
G moves around the circle C' (V.S.W.R. $= 1{\cdot}7$).

The triangle UOG rotates about 0 without changing shape so that we have $g = g'$ when the triangle is at U_1OG_1 or U_2OG_2.

First case. U_1OG_1; $g = g' = 1{\cdot}02$,

at U_1 $\quad y_1 = 1{\cdot}02 - \mathrm{j}\,1{\cdot}01$,

at G_1 $\quad y_1' = 1{\cdot}02 + \mathrm{j}\,0{\cdot}55$.

The position of the stub is given by the angle $(\boldsymbol{OU}, \boldsymbol{OU}_1) \equiv 0{\cdot}048$. Thus

$$s_1 = 0{\cdot}048\lambda + \frac{n\lambda}{2}.$$

The first plane Π is thus situated at

$$s_1 = 9{\cdot}6 \text{ cm}.$$

Since the stub equalizes the imaginary parts, we require

$$0{\cdot}55 = -1{\cdot}01 + b_1, \quad \text{or} \quad b_1 = 1{\cdot}56.$$

Second case. U_2OG_2; $g = g' = 0{\cdot}65$,

at U_2 $\quad y_2 = 0{\cdot}65 + \mathrm{j}\,0{\cdot}73$,

at G_2 $\quad y_2 = 0{\cdot}65 - \mathrm{j}\,0{\cdot}27$.

We then find

$$s_2 = 0{\cdot}336\lambda + \frac{n\lambda}{2}$$

with the first plane Π at $s_2 = 67{\cdot}2$ cm and the susceptance due to the stub $b_2 = -0{\cdot}27 - 0{\cdot}73 = 1$.

(c) The stub length is calculated from the distance from the short-circuit point J to the susceptance point b on the circle.

Thus for $b_1 = 1{\cdot}56$ we have the point K_1 and

$$\frac{l_1}{\lambda} = 0{\cdot}158 + 0{\cdot}25 = 0{\cdot}408,$$

so that

$$l_1 = 0{\cdot}408\lambda + \frac{n\lambda}{2}.$$

For $b_2 = -1$ (point K_2)

$$l_2 = 0{\cdot}125\,\lambda + \frac{n\lambda}{2}.$$

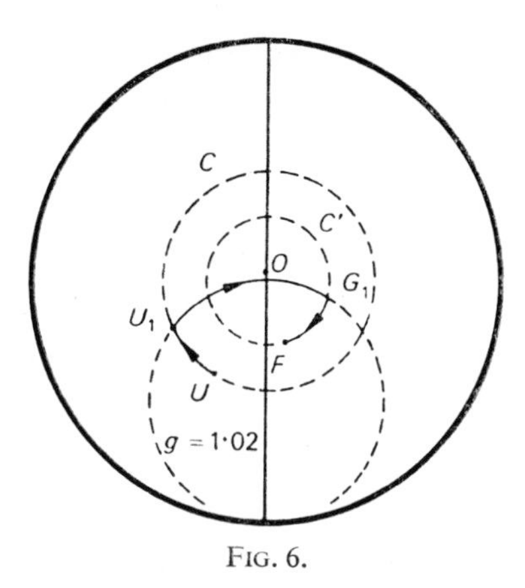

FIG. 6.

The chart of Fig. 6summarizes the method, commencing at the load, in the first case. U is displaced to U_1 on the circle C with S.W.R. of 2·6. In the plane Π the stub is added, which corresponds to a displacement around the circle g = constant = 1·02 to G_1. From G_1 we move around the circle of S.W.R. 1·7 by an angle $(L-s)/\lambda$ towards the source and this brings us to the point F.

Problem No. 15

MATCHING WITH TWO STUBS

This problem can be completely solved by means of the Smith chart.

A lossless 50 Ω line is terminated by an unknown load Z_L. To measure this impedance, Z_L is replaced by a short circuit and the distance between two consecutive minima is determined. A separation of 10 cm is found and the position of a specific minimum is noted.

When the impedance Z_L is placed at the end of the line the V.S.W.R. is found to be $S = 3$ while the previously noted minimum is displaced by 1·4 cm, towards the load.

1. What is the value of Z_L?

2. In order to match the load two stubs, composed of short-circuited lengths of lossless 50 Ω line, are placed in parallel with the main line at 6·6 cm and 9·6 cm from the load respectively (Fig. 1). Calculate the lengths of the stubs which will give matching. (Consider all the possible cases.)

3. Calculate the lengths of the two stubs which will match a load Z'_L for which $S = 3$ and the displacement of the minimum towards the load is 2·28 cm.

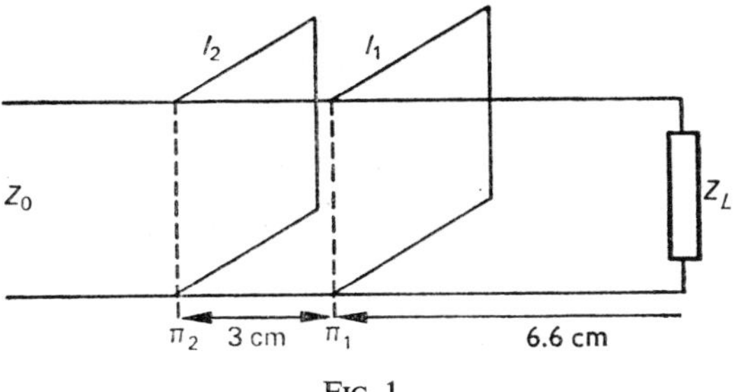

FIG. 1.

4. What impedances cannot be matched by this system?

Solution. 1. To find Z_L. Since the separation of two consecutive minima is $\lambda/2$ we have $\lambda = 20$ cm.

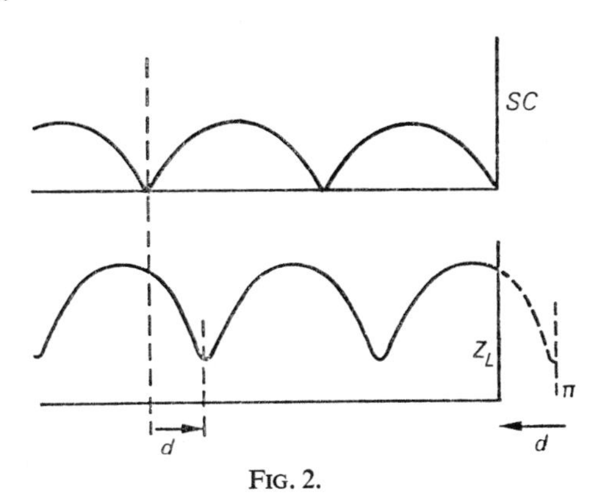

FIG. 2.

With the minimum displaced by d towards Z_L as shown in Fig. 2, the load is found from the chart as follows:

(a) The circle C of S.W.R. 3 is traced as in Fig. 3. The point m, the intersection of C with OI, represents all the points on the line where the impedance is a minimum and, in particular, the plane π where there is a virtual minimum behind the load.

(b) The distance between the load and the plane π is d. Starting from π, i.e. from m on the chart, we go round the circle C by $d/\lambda = 1{\cdot}4/20 = 0{\cdot}07$ towards the source. We arrive at T representing z_L such that

$$z_L = 0{\cdot}4 + \mathrm{j}\,0{\cdot}4$$

and

$$Z_L = (20 + \mathrm{j}\,20)\ \Omega.$$

2. Since the stubs are in parallel, we calculate the admittances.

The reduced admittance y_L is represented by the point U, symmetric with T through 0, i.e.

$$y_L = 1{\cdot}25 - \mathrm{j}\,1{\cdot}25.$$

The first stub is 6·6 cm from the load (plane π_1) which gives an angle

$$\frac{s_1}{\lambda} = \frac{6{\cdot}6}{20} = 0{\cdot}33.$$

The second stub is 3 cm from the first (plane π_2) and

$$\frac{s_2}{\lambda} = 0{\cdot}15.$$

(a) What is the admittance in the plane π_1?

Movement along the line is represented, in this case, by movement from U around the circle C. The admittance at π_1 due to the admittance y_t is y_t' and is represented by the point U' where $(\boldsymbol{OU}, \boldsymbol{OU'}) = 0{\cdot}33$. Thus $y_t' = 0{\cdot}8+\mathrm{j}$.

If l_1 is the length of the first stub, the input admittance of this stub is

$$y_{i_1} = -\mathrm{j}\cot\beta l_1 = \mathrm{j}b_{i_1}(l_1).$$

The total admittance at π_1 is

$$y_{\pi_1} = 0{\cdot}8+\mathrm{j}\,[1+b_{i_1}(l_1)].$$

As the length l_1 varies, the locus of the points y_{π_1} is the circle C_1 with $g_1 = 0{\cdot}8 =$ constant. l_1 is not fixed here nor is the impedance here taken to the plane π_2.

(b) What is the admittance at π_2?

We know that a movement along the line can be represented on the chart by movement around a circle of constant S centred on 0. The movement from π_1 to π_2 corresponds to $s_2/\lambda = 0{\cdot}15$. Thus each point on C_1 is displaced by $s_2/\lambda = 0{\cdot}15$. Thus J becomes J″ following a displacement around the circle $S = \infty$, U' becomes U'' (movement round circle $S = 3$) and similarly for all points on C_1. In effect C_1 is displaced to become the circle C_2.

Thus C_2 is the locus of all admittances brought to the plane π_2 when l_1 varies and varies y_{π_1}'. The total admittance in the plane π_2 is

$$y_{\pi_2} = y_{\pi_1}'+y_{i_2},$$

where $y_{i_2} = \mathrm{j}b_{i_2}(l_2)$ is the input admittance of the second stub. Thus

$$y_{\pi_2} = g_{\pi_1}'(l_1)+\mathrm{j}[b_{\pi_1}'(l_1)+b_{i_2}(l_2)].$$

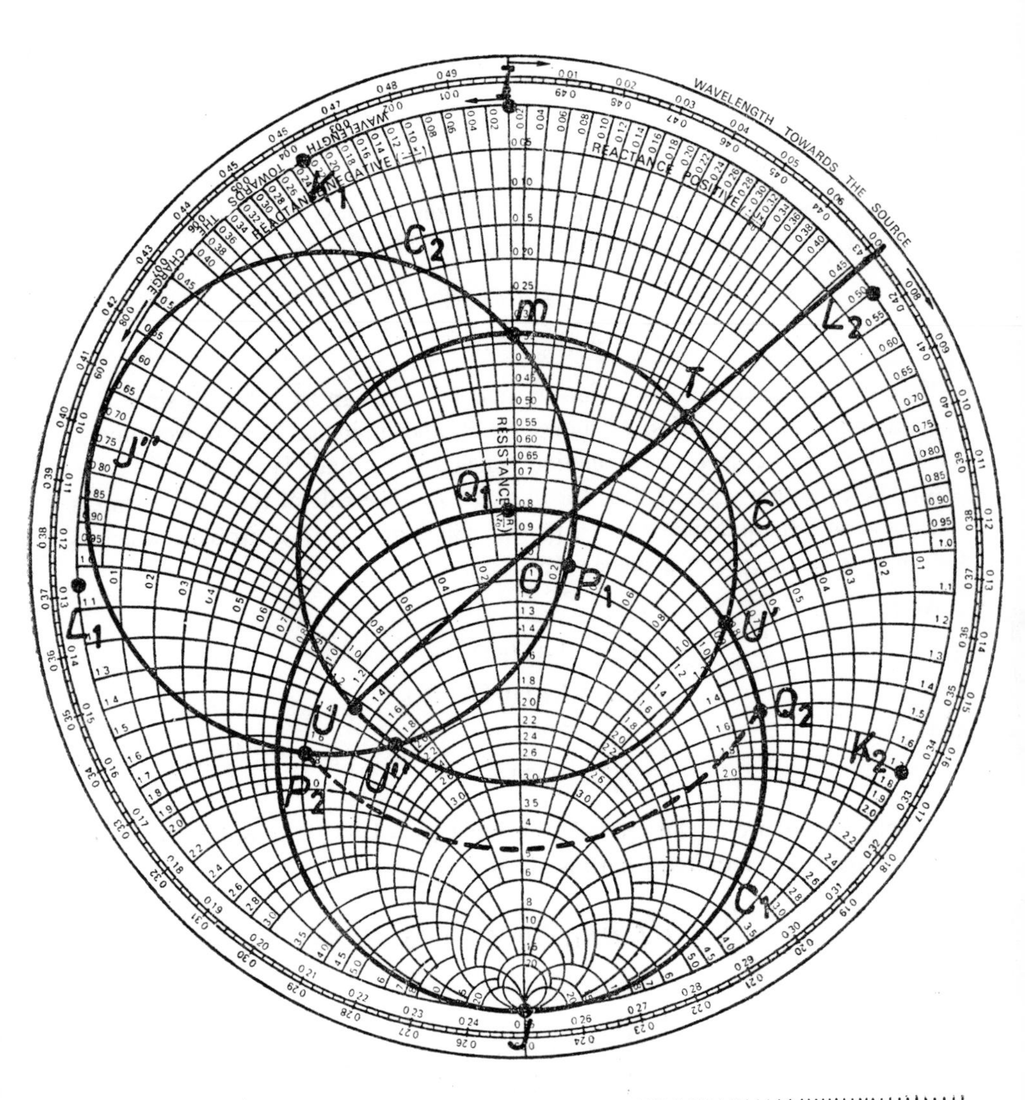
WAVELENGTH TOWARDS THE SOURCE
WAVELENGTH TOWARDS THE CHARGE
REACTANCE POSITIVE
REACTANCE NEGATIVE
RESISTANCE
C
C_1
C_2
K_1
K_2
L_1
L_2
m
T
O
P_1
P_2
Q_1
Q_2
U
U'
U''
J
J'''
ʊ nepers
ʊ decibels
$|\Gamma_0|$

FIG. 3.

The stub is seen to give a pure susceptance in the plane π_2 and hence has no effect on g'_{π_1}. For matching in the plane π_2 we require

$$y_{\pi_2} = 1+\mathrm{j}\,0,$$

which implies

$$g'_{\pi_1} = 1.$$

The two points P_1 and P_2 for which $g'_{\pi_1} = 1$ are situated at the intersection of the circle C and the circle $g = 1$. These are

P_1 $$y_{\pi_2} = 1+\mathrm{j}\,0{\cdot}22$$

P_2 $$y_{\pi_2} = 1-\mathrm{j}\,1{\cdot}7.$$

How can we determine the points P_1 and P_2? We can fix P_1 or P_2 since l_2 is not yet fixed. When l_1 varies, U' moves on C_1 and its image in π_2 moves on C_2. To fix P_1 or P_2 we fix l_1, a calculation which will be made later.

(c) The condition $g'_{\pi_1} = 1$ fixes for us P_1 or P_2. The role of the second stub is to compensate the imaginary parts, namely j 0·22 or −j 1·7, a problem which is the same as in the case of a single stub.

At P_1

$$y'_{\pi_1} = 1+\mathrm{j}\,0{\cdot}22$$

$$y_{\pi_2} = 1+\mathrm{j}\,(0{\cdot}22+b_{i_2})$$

and

$$b_{i_2} = -0{\cdot}22.$$

The input impedance of the second stub must be −j 0·22 (the point K_1 on the outside circle). The required length is given since, from K_1,

$$\frac{l_2}{\lambda} = (\boldsymbol{OJ}, \boldsymbol{OK}_1) = 0{\cdot}466-0{\cdot}25 = 0{\cdot}216.$$

At P_2

$$y'_{\pi_1} = 1-\mathrm{j}\,1{\cdot}7$$

$$y'_{\pi_2} = 1+\mathrm{j}\,(b_{i_2}-1{\cdot}7)$$

and

$$b'_{i_2} = 1{\cdot}7 \quad \text{(point } K_2\text{)}.$$

On the circle the length l'_2 is given since

$$\frac{l'_2}{\lambda} = (\boldsymbol{OJ}, \boldsymbol{OK}_2) = 0{\cdot}165+0{\cdot}25 = 0{\cdot}415.$$

(d) Having resolved the problem at π_2 we return to π_1. When C_2 becomes C_1, $P_1 \rightarrow Q_1$ and $P_2 \rightarrow Q_2$. Q_1 and Q_2 are the images of P_1 and P_2 in π_1 and they are found by rotating through $s_2/\lambda = 0{\cdot}15$ towards the load.

$$Q_1 \text{ corresponds to } \quad y_{\pi_1} = 0{\cdot}8 - \mathrm{j}\,0{\cdot}04.$$
$$Q_2 \text{ corresponds to } \quad y_{\pi_2} = 0{\cdot}8 + \mathrm{j}\,1{\cdot}5.$$

However, we know that

$$y_{\pi_1} = 0{\cdot}8 + \mathrm{j}[1 + b_{i_1}(l_1)].$$

For the condition Q_1, the length l_1 must give

$$1 + b_{i_1} = -0{\cdot}04 \quad \text{or} \quad b_{i_1} = -1{\cdot}04.$$

For the condition Q_2, there will be a length l_1' to give

$$1 + b_{i_1}' = 1{\cdot}5 \quad \text{or} \quad b_{i_1}' = 0{\cdot}5.$$

The length of the stub is calculated from the input impedance b_{i_1} as

$$b_{i_1} = -1{\cdot}04; \qquad \frac{l_1}{\lambda} = (\boldsymbol{OJ}, \boldsymbol{OL}_1) = 0{\cdot}372 - 0{\cdot}25 = 0{\cdot}122;$$

$$b_{i_1}' = 0{\cdot}5; \qquad \frac{l_1'}{\lambda} = (\boldsymbol{OJ}, \boldsymbol{OL}_2) = 0{\cdot}25 + 0{\cdot}074 = 0{\cdot}324.$$

The two pairs of solutions are then

$$l_1 = 2{\cdot}44 \text{ cm} + n \times 10 \text{ cm}, \qquad l_2 = 4{\cdot}32 \text{ cm} + n' \times 10 \text{ cm},$$
$$l_1' = 6{\cdot}48 \text{ cm} + n \times 10 \text{ cm}, \qquad l_2' = 8{\cdot}3 \text{ cm} + n' \times 10 \text{ cm}.$$

3. (Chart in Fig. 4.) For the second load, the minimum is displaced by $d = 22{\cdot}8$ mm so that $d/\lambda = 0{\cdot}114$.

The movement by d/λ around the circle C ($S = 3$) towards the source gives

$$z_L = 0{\cdot}55 + \mathrm{j}\,0{\cdot}72 \quad \text{(point } T\text{)}.$$

The corresponding admittance is $y_L = 0{\cdot}67 - \mathrm{j}\,0{\cdot}9$ (point U).
This admittance, transferred to the plane π_1, is

$$y_L' = 1{\cdot}60 + \mathrm{j}\,1{\cdot}32 \quad \text{(point } U'\text{)}.$$

The total admittance in the plane π_1 is thus

$$y_{\pi_1} = 1{\cdot}60 + \mathrm{j}[1{\cdot}32 + b_{i_1}(l_1)].$$

The circle C_1, which is the locus of y_{π_1} when l_1 varies, is the circle for $g = 1{\cdot}60$; in the plane π_2, C_1 becomes C_2, a circle tangential to the circle $g = 1$ at P, such that

$$y_{\pi_1}' = 1 - \mathrm{j}\,0{\cdot}72.$$

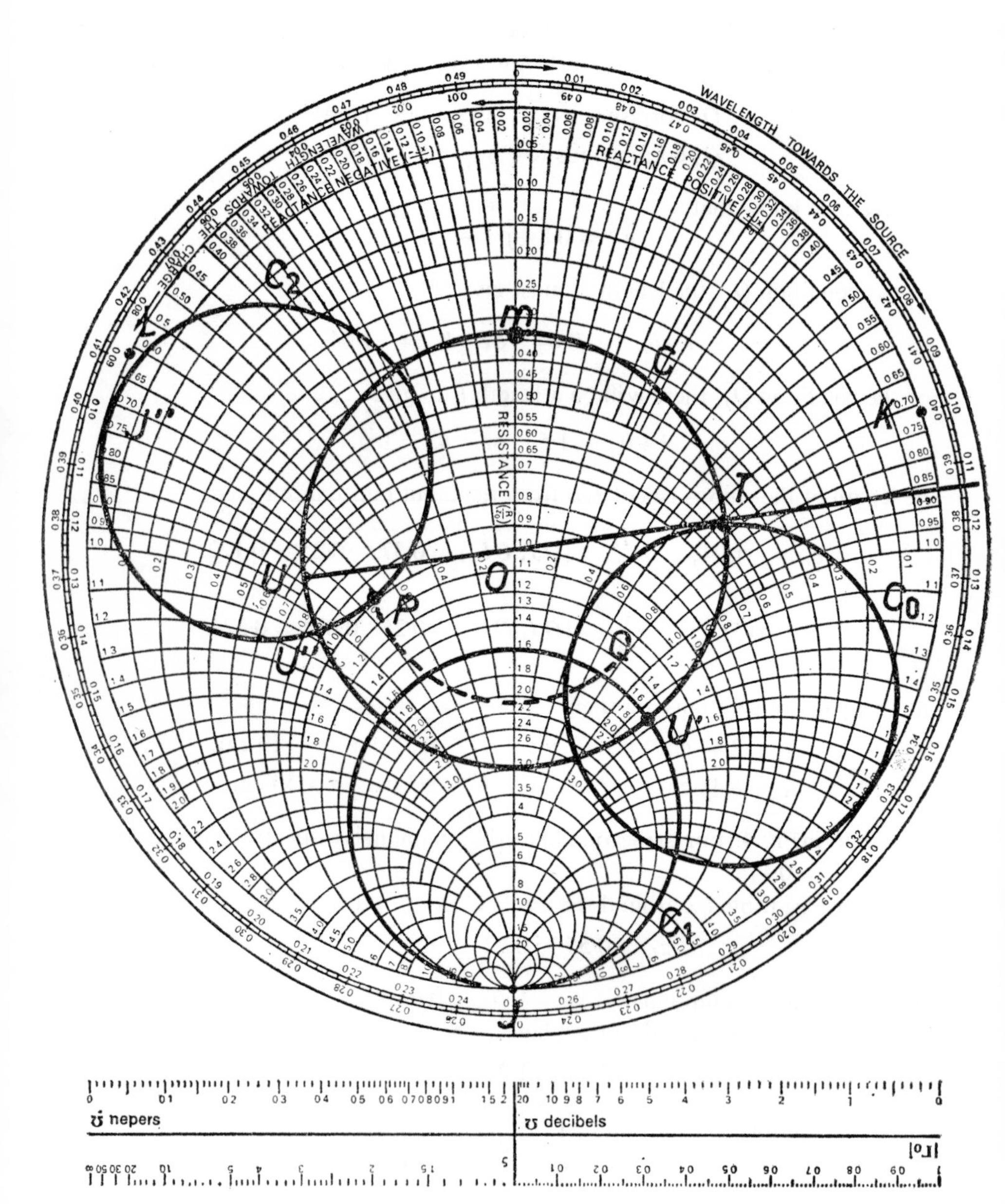
WAVELENGTH TOWARDS THE SOURCE
REACTANCE POSITIVE
REACTANCE NEGATIVE
RESISTANCE
C2
C
C0
C1
m
T
K
O
P
Q
U
nepers
decibels

FIG. 4.

Thus we must have $b_{i_2} = 0{\cdot}72$ or

$$\frac{l_2}{\lambda} = (\boldsymbol{OJ}, \boldsymbol{OK}) = 0{\cdot}098+0{\cdot}25 = 0{\cdot}348.$$

When P is transferred into π_1 to give Q we have

$$y_{\pi_1} = 1{\cdot}60+\mathrm{j}\,0{\cdot}7.$$

Then

$$1{\cdot}60+\mathrm{j}\,0{\cdot}7 = 1{\cdot}60+\mathrm{j}\,1{\cdot}32+\mathrm{j}\,b_{i_1}(l_1)$$

or

$$b_{i_1}(l_1) = 0{\cdot}7-1{\cdot}32 = -0{\cdot}62.$$

The stub length is then given by

$$\frac{l_1}{\lambda} = (\boldsymbol{OJ}, \boldsymbol{OL}) = 0{\cdot}412-0{\cdot}25 = 0{\cdot}162.$$

The only solution for this case is thus

$$l_1 = 3{\cdot}24 \text{ cm}+n\times 10 \text{ cm},$$
$$l_2 = 6{\cdot}96 \text{ cm}+n'\times 10 \text{ cm}.$$

4. We have just seen that the two solutions become the same when the circle C_2 is tangential to $g = 1$. Also, we see that a circle centred on OJ'' and not cutting $g = 1$ will not give a solution.

The circle C_2 on the second chart is the limiting circle of admittances which cannot be matched in the plane π_2. It gives C_1 in the plane π_1 and, in the plane of the load, it gives C' deduced from C_1 by a rotation of $s_1/\lambda = 0{\cdot}33$ towards the load.

The locus of non-matchable impedances is the circle C_0 symmetric with C' and deduced from C_1 by a rotation of $0{\cdot}33-0{\cdot}25 = 0{\cdot}08$ towards the load. No impedance within C_0 can be matched and it is obvious that T is actually on C_0.

Problem No. 16

MATCHING BAND-WIDTH

It is required to match a resistance R_L to a line L of characteristic impedance R_i with the aid of a quarter-wave line of characteristic impedance Z_0.

Let S be the V.S.W.R. in the quarter-wave line and S' the V.S.W.R. in L, π the input plane of the quarter-wave line which is quarter-wave at a frequency f_0.

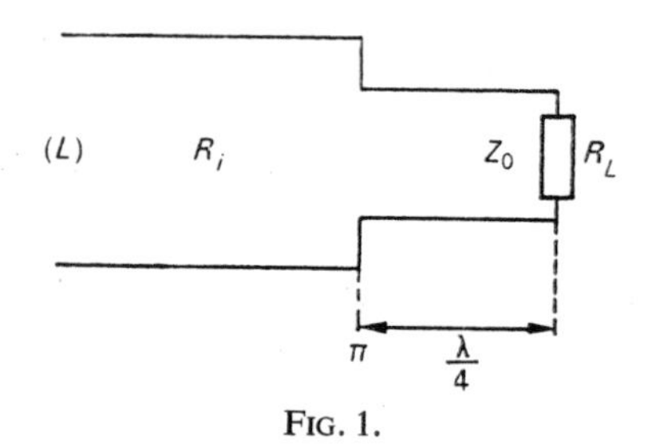

FIG. 1.

I

1. Calculate Z_0 in terms of R_i and R_L and give its value for $R_i = 1$ kΩ, $R_L = 40$ Ω.

2. Calculate S in terms of R_i and R_L and show that, depending on the ratio of R_i and R_L, there will be two distinct cases.

3. Calculate S'.

4. Using the Smith chart, represent the locus S = constant as the frequency is varied. (Take Z_0 as the reference impedance.) What are the intercepts of this locus with the real axis? Calculate the numerical values for $R_i = 1$ kΩ, $R_L = 40$ Ω.

II

If the frequency is changed the system will no longer be matched. With the aid of the Smith chart the bandwidth for matching, defined as the range of frequency for which S' is less than 1·5, can be found.

5. Calculate the points of intersection of the locus $S' = 1{\cdot}5$ with the real axis for $R_i = 1$ kΩ, $R_L = 40$ Ω. Hence deduce their representation (points A and B) on the Smith chart, the reference impedance being again Z_0.

6. The locus $S' = 1{\cdot}5$ is a circle of diameter AB which cuts the locus S = constant in two points which represent the extremities of the band for matching. Hence find $\Delta f/f_0$ for R_i and R_L given in 5.

7. What is the corresponding result if R_i is changed to 160 Ω, and what conclusion do you draw?

III

It is now required to find $\Delta f/f_0$ for $S' = 1{\cdot}5$ by calculation in the two preceding cases.

8. Calculate the impedance Z_i' transferred to the input of the quarter-wave line for a frequency f in the neighbourhood of f_0.

Take $\delta f = f - f_0$, $\Delta f = 2\,\delta f$ with $x = (\pi/2)\,(\delta f/f_0)$ such that $\tan x \simeq x$.

9. Hence deduce the reflection coefficient Γ from the line L in the plane π.

10. Express $|\Gamma|$ in terms of S and x.

11. Hence deduce S' in the form $S' = 1+\alpha$ with $\alpha = \Phi(x, S)$.

12. Calculate $\Delta f/f_0$ for $S' = 1{\cdot}5$ in the two previous cases ($R_i = 1\ \text{k}\Omega$, $R_L = 40\ \Omega$; $R_i = 160\ \Omega$, $R_L = 40\ \Omega$).

(Partial examination, Paris 1966)

Solution

I

1. The impedance transferred to the plane π is given by the normal transformation equation

$$Z_\pi = Z(s_0) = Z_0 \frac{R_L + jZ_0 \tan \beta s_0}{Z_0 + jR_L \tan \beta s_0}, \qquad \beta = \frac{2\pi}{\lambda}.$$

With $s_0 = \lambda/4$, $\beta s_0 = \pi/2$ and $\tan \beta s_0 = \infty$ so that

$$Z_\pi = \frac{Z_0^2}{R_L}.$$

There will be matching if

$$R_i = Z_\pi = \frac{Z_0^2}{R_L},$$

or

$$Z_0 = \sqrt{R_i R_L}.$$

For $R_L = 40\ \Omega$ and $R_i = 1000\ \Omega$, $Z_0 = 200\ \Omega$.

2. $$S = \frac{1+|\Gamma|}{1-|\Gamma|} \quad \text{and} \quad \Gamma = \frac{R_L - Z_0}{R_L + Z_0}.$$

First case:

$$R_L > R_i. \quad \text{Since} \quad Z_0 = \sqrt{R_i R_L}$$
$$R_L > Z_0 > R_i$$

$$|\Gamma| = \frac{R_L - Z_0}{R_L + Z_0},$$

$$S = \frac{1 + \dfrac{R_L - Z_0}{R_L + Z_0}}{1 - \dfrac{R_L - Z_0}{R_L + Z_0}} = \frac{R_L}{Z_0}, \quad \text{or} \quad R_L = SZ_0.$$

Then

$$S = \frac{R_L}{\sqrt{R_i R_L}} = \sqrt{\frac{R_L}{R_i}}.$$

Second case:

$$R_i > R_L, \quad R_i > Z_0 = \sqrt{R_i R_L} > R_L$$

$$|\Gamma| = \frac{Z_0 - R_L}{Z_0 + R_L},$$

$$S = \frac{Z_0}{R_L} \quad \text{or} \quad R_L = \frac{Z_0}{S}.$$

Then

$$S = \frac{\sqrt{R_i R_L}}{R_L} = \sqrt{\frac{R_L}{R_i}}.$$

In the case quoted

$$S = \sqrt{\frac{1000}{40}} = 5.$$

3. $S' = 1$ since we have specified a matched system.

4. The locus $S =$ constant will represent all the values of impedance along the line. This is the circle C_0 centred on 0 and passing through the point T which represents $R_L = Z_0/S$. In reduced impedances this is $r_L = 1/S = 0{\cdot}2$.

This circle cuts the real axis at T for the minimum impedance, and for the maximum impedance at $Z_m = SZ_0 = R_i$, which in reduced impedance is $r_i = S = 5$.

II

5. The V.S.W.R. S' is the S.W.R. in the line L of characteristic impedance R_i. The locus $S' =$ constant will cut the real axis at A and B with $Z = R$ and $Z' = R'$.

$$S' = \frac{1+|\Gamma|}{1-|\Gamma|}, \qquad \Gamma = \frac{R-R_i}{R+R_i}.$$

First case: $R' > R_i$.

$$|\Gamma| = \frac{R-R_i}{R+R_i}, \quad S' = \frac{R}{R_i} \quad \text{which gives} \quad R = S'R_i.$$

Second case: $R' < R_i$.

$$|\Gamma| = \frac{R_i-R'}{R_i+R'}, \quad S' = \frac{R_i}{R'} \quad \text{and} \quad R' = \frac{R_i}{S'}.$$

On the chart with reference Z_0 the points A and B are given such that

$$r = \frac{R}{Z_0} = \frac{S'R_i}{Z_0} = S'S = 5S'$$

$$r' = \frac{R'}{Z_0} = \frac{1}{S'}\frac{R_i}{Z_0} = \frac{S}{S'} = \frac{5}{S'}.$$

If $S' = 1{\cdot}5$, $r = 7{\cdot}5$ and $r' = 3{\cdot}33$ – points A and B on the chart of Fig. 2.

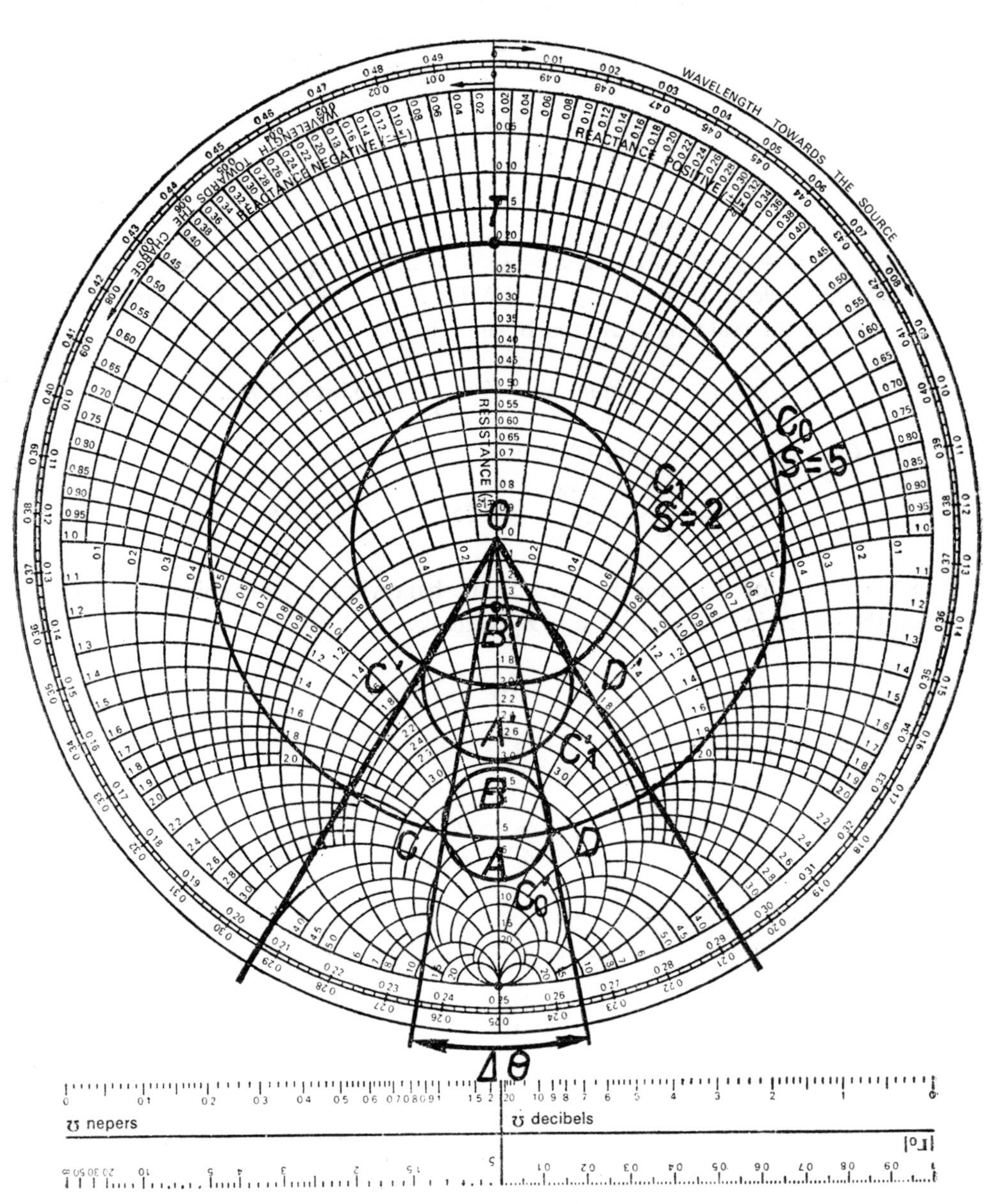
WAVELENGTH TOWARDS THE SOURCE
WAVELENGTH TOWARDS THE CHARGE
REACTANCE NEGATIVE
REACTANCE POSITIVE
RESISTANCE
T
O
C_0 S=5
C_1 S=2
B'
A'
B
A
C'
D'
C
D
C_1'
C_0'
Δθ
ʊ nepers
ʊ decibels
|Γ₀|

FIG. 2.

6. Now trace the circle C_0' on the diameter AB. C_0' cuts C_0 at two points, C and D, which give the limits of the band. Then

$$\frac{\Delta\theta}{\theta_0} = \frac{\Delta(1/\lambda_0)}{1/\lambda_0} = \frac{\Delta f}{f_0} = \frac{0{\cdot}028}{0{\cdot}25} = 0{\cdot}11,$$

or

$$\frac{\Delta f}{f_0} = 11\%.$$

7. For $R_L = 40\,\Omega$, $R_i = 160\,\Omega$, $S = \sqrt{\frac{160}{40}} = 2$, $Z_0 = \sqrt{R_L R_i} = 80$.

Trace the circle C_1 representing the S.W.R., $S = 2$. The circle C_1' representing $S' = 1{\cdot}5$ cuts the real axis at A' and B' such that

$$r = S'S = 3 \text{ (point } A'), \quad r' = \frac{S}{S'} = 1{\cdot}33 \text{ (point } B').$$

C_1' cuts C_1 in the points C' and D' with

$$\frac{\Delta\theta'}{\theta_0} = \frac{\Delta f'}{f_0} = \frac{0{\cdot}088}{0{\cdot}25} = 0{\cdot}35 \quad \text{or} \quad \frac{\Delta f'}{f_0} = 35\%.$$

It is seen that, for the same S' (i.e. 1·5), the bandwidth for matching is greater when the separation between R_L and R_i is smaller.

III

8. $f \neq f_0$, $\delta f = f - f_0$.

The length of the quarter-wave line is fixed as

$$l = \frac{\lambda_0}{4} = \frac{v}{4f_0}$$

where v is the phase velocity of the wave.

The propagation constant becomes $\beta = \beta_0 + \delta\beta$ and the angle $\beta_0 l$ becomes

$$\beta l = \beta_0 l + \delta\beta l_0 = \frac{\pi}{2} + 2\pi\frac{\delta f_0}{v}\cdot\frac{v}{4f_0} = \frac{\pi}{2} + \frac{\pi}{2}\frac{\delta f}{f_0} = \frac{\pi}{2} + x.$$

Then

$$\tan\beta l = \tan(\beta_0 l + \delta\beta l) = \frac{\tan\pi/2 + \tan x}{1 - \tan\pi/2\,\tan x} = -\frac{1}{\tan x} \simeq -\frac{1}{x}.$$

The impedance transferred to the input of the quarter-wave line becomes

$$Z_i' = Z_0\frac{R_L + jZ_0\tan\beta l}{Z_0 + jR_L\tan\beta l} = Z_0\frac{R_L - jZ_0/x}{Z_0 - jR_L/x},$$

or

$$Z_i' = Z_0 \frac{R_L x - jZ_0}{Z_0 x - jR_L}.$$

(For $x = 0$ this is $Z_i' = Z_0^2/R_L$.)

9.

$$\Gamma = \frac{Z_i' - R_i}{Z_i' + R_i} = \frac{Z_0 \dfrac{R_L x - jZ_0}{Z_0 x - jR_L} - R_i}{Z_0 \dfrac{R_L x - jZ_0}{Z_0 x - jR_L} + R_i}$$

$$= \frac{Z_0 R_L x - jZ_0^2 - R_i Z_0 x + jR_i R_L}{Z_0 R_L x - jZ_0^2 + R_i Z_0 x - jR_i R_L}$$

$$= \frac{Z_0 x(R_L - R_i) + j(R_L R_i - Z_0^2)}{Z_0 x(R_L + R_i) - j(R_L R_i + Z_0^2)}.$$

We know that, from the first question, $Z_0 = \sqrt{R_L R_i}$, so that

$$\Gamma = \frac{Z_0 x(R_L - R_i)}{Z_0 x(R_L + R_I) - j2Z_0^2} = \frac{x(R_L - R_i)}{x(R_L + R_i) - 2jZ_0}.$$

10. (a) $R_L > Z_0 > R_i$, $R_L = SZ_0$, $R_i = Z_0/S$.
(b) $R_L < Z_0 < R_i$, $R_i = SZ_0$, $R_L = Z_0/S$.

In both cases,

$$|\Gamma| = \frac{x(S - 1/S)Z_0}{[x^2(S + 1/S)^2 Z_0^2 + 4Z_0^2]^{1/2}} = \frac{x(S - 1/S)}{[x^2(S + 1/S)^2 + 4]^{1/2}}.$$

11.

$$S' = \frac{1 + |\Gamma|}{1 - |\Gamma|} = 1 + \frac{2|\Gamma|}{1 - |\Gamma|} = 1 + \alpha,$$

$$\alpha = \frac{\dfrac{2x(S - 1/S)}{\left[x^2\left(S + \dfrac{1}{S}\right)^2 + 4\right]^{1/2}}}{1 - \dfrac{x(S - 1/S)}{\left[x^2\left(S + \dfrac{1}{S}\right) + 4\right]^{1/2}}}.$$

$$\alpha = \frac{2x(S - 1/S)}{\left[x^2\left(S + \dfrac{1}{S}\right)^2 + 4\right]^{1/2} - x\left(S - \dfrac{1}{S}\right)}.$$

12. From the above equations we find

$$\alpha^2\left[x^2\left(S+\frac{1}{S}\right)^2+4\right] = x^2\left(S-\frac{1}{S}\right)^2(2+\alpha)^2,$$

$$x^2\left[\alpha^2\left(S+\frac{1}{S}\right)^2-(2+\alpha)^2\left(S-\frac{1}{S}\right)^2\right] = -4\alpha^2,$$

$$x^2 = \frac{4\alpha^2}{(2+\alpha)^2\left(S-\frac{1}{S}\right)^2-\alpha^2\left(S+\frac{1}{S}\right)^2};$$

from which

$$x = \pm\frac{2\alpha}{\left[(2+\alpha)^2\left(S-\frac{1}{S}\right)^2-\left(S+\frac{1}{S}\right)^2\alpha^2\right]^{1/2}} = \frac{\pi}{2}\frac{\delta f}{f_0}$$

$$\frac{\Delta f}{f_0} = \frac{4}{\pi}\frac{2\alpha}{\left[(2+\alpha)^2\left(S-\frac{1}{S}\right)^2-\left(S+\frac{1}{S}\right)^2\alpha^2\right]^{1/2}}.$$

$$\alpha = 0{\cdot}5\begin{cases} S = 5 \rightarrow \dfrac{\Delta f}{f_0} = 10{\cdot}8\%; \\ S = 2 \rightarrow \dfrac{\Delta f}{f_0} = 36\%. \end{cases}$$

Problem No. 17

RESONANT LOAD ON A LINE

Consider a lossless line of characteristic impedance Z_0 for which the phase velocity v may be taken to be independent of the frequency (i.e. a dispersionless line). The line is supplied by a matched source (i.e. of internal impedance Z_0). At a distance s_0 from the end of the line a series resonant circuit of negligible dimensions, consisting of an inductance L, capacitance C and resistance R, is placed ries with the line as in Fig. 1.

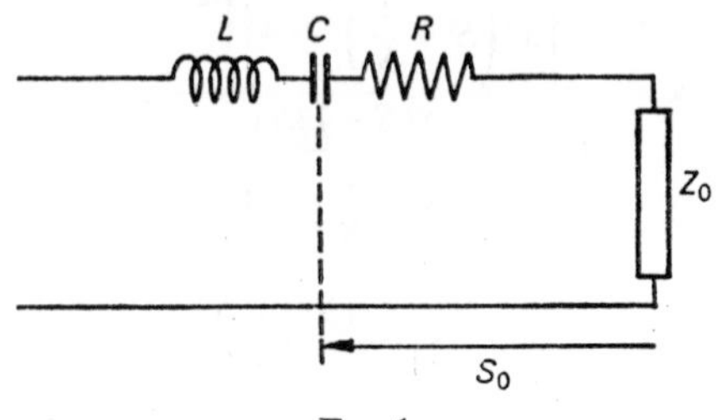

FIG. 1.

1. If the line is terminated by an impedance Z_0, what is the locus on the Smith chart of the points representing the coefficient of reflection $\Gamma(s_0)$, calculated at s_0, when the frequency of the source oscillator is varied? For what frequency $\omega_0/2\pi$ is the reflection coefficient a minimum? To what does this frequency correspond? What is the reflection coefficient at $s = s_0$ for this frequency $\omega_0/2\pi$?

2. The impedance Z_0 is replaced by a short-circuit at the end of the line. What is the locus of the points representing the reflection coefficient $\Gamma'(s_0)$ as the frequency varies? If it is supposed that s_0 is very much smaller than the wavelength λ, so that $2\pi s_0/\lambda \ll 1$ and the coefficient of reflection is found to be minimum for a frequency $\omega_1/2\pi$, calculate L and C in terms of ω_0, ω_1 and s_0 and the constants of the line.

3. As a numerical example, consider that the line is bifilar and placed in air (which may be taken to have the same constants as free space) and that $Z_0 = 140\,\Omega$, $s_0 = 5$ cm, $\omega_0/2\pi = 100{\cdot}09$ MHz and $\omega_1/2\pi = 99{\cdot}48$ MHz.

4. A quarter-wave line of characteristic impedance Z_0 is inserted between the resonant circuit and the section of line s_0 terminated by the short-circuit. For what frequency $\omega_2/2\pi$ will the coefficient of reflection be minimal in the plane of the resonance circuit?

5. As a numerical example calculate $\omega_2/2\pi$ with the data of question 3.
For the rest of the question the system is taken as in question 1 and the line is terminated in Z_0.

6. Show that the system behaves as a resonant circuit and calculate the Q-factor in terms of ω_0, ω_1, the line constants and the ratio $a = R/Z_0$.

7. As a numerical example take the data of question 3 and $a = 0{\cdot}67$.

8. Let P_0 be the maximum power available at the terminals of the generator. Express in terms of the magnitudes of P_0, $\Gamma(s_0)$ and a only

(a) the incident power, P_i

(b) the reflected power, P_r

(c) the power dissipated in the impedance Z of the resonant circuit, P_Z

(d) the power transmitted to Z_0, P_{Z_0}.

9. For what frequency is P_{Z_0} a maximum? Hence express this maximum power in terms of P_0 and a only.

10. At what distance s from the impedance Z, in the direction of the source, must a stub be placed in parallel with the line in order to match the system at the frequency $\omega_0/2\pi$?

11. What must be the length of the stub if it is formed from a short-circuited $50\,\Omega$ line? As a numerical example take $a = 0{\cdot}67$.

Solution. 1. The impedance of the resonant circuit is $Z = R+\mathrm{j}\left(\omega L-\dfrac{1}{\omega C}\right)$. Since the line is terminated in its characteristic impedance, the impedance transferred to s_0 is Z_0 and

$$Z(s_0) = Z+Z_0 = R+\mathrm{j}\left(\omega L-\frac{1}{\omega C}\right)+Z_0,$$

or, in terms of reduced impedances

$$z(s_0) = \frac{R+Z_0}{Z_0}+\mathrm{j}\frac{1}{Z_0}\left(\omega L-\frac{1}{\omega C}\right).$$

The locus representing $z(s_0)$ on the Smith chart is $r =$ constant with $r = 1+R/Z_0$ It is a circle outside the circle $r = 1$, say the circle C_0 on the chart of Fig. 2.

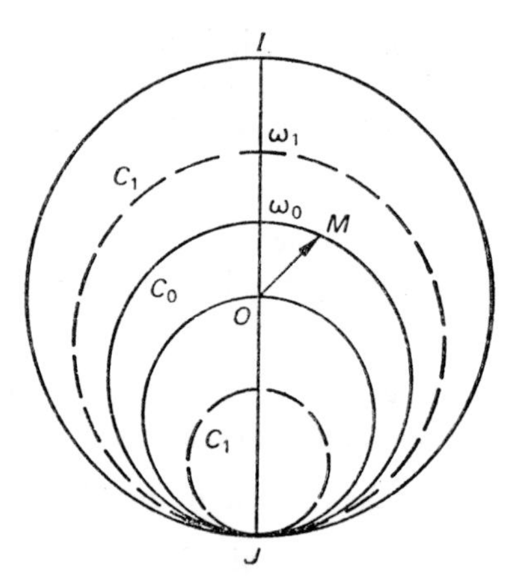

FIG. 2.

On the chart $\Gamma(s_0)$ is the radius vector $\boldsymbol{OM}$ and the locus of $\Gamma(s_0)$ is thus the circle C_0.

$$\text{If } \omega \to 0, \qquad z(s_0) \to r-\mathrm{j}\infty.$$

$$\text{If } \omega \to \infty, \qquad z(s_0) \to r+\mathrm{j}\infty.$$

For $\omega = \omega_0 = 1/\sqrt{LC}$, $z(s_0) = r$ and it is seen from the chart that $\Gamma(s_0)$ is a minimum for $\omega = \omega_0$. In this case

$$\Gamma = \frac{z-1}{z+1} = \frac{R}{Z_0}\frac{1}{2+\dfrac{R}{Z_0}}.$$

2. With a short-circuit at the end of the line the reduced impedance transferred to s_0 is

$$z = \mathrm{j} \tan \beta s_0,$$

with

$$\beta s_0 = \frac{2\pi}{\lambda_0} s_0 \ll 1,$$

so that

$$z = \mathrm{j}\frac{2\pi s_0}{\lambda} = \mathrm{j}\frac{\omega s_0}{v}.$$

The total reduced impedance at s_0 is then

$$\begin{aligned} z'(s_0) &= \frac{R}{Z_0}+\mathrm{j}\left[\frac{1}{Z_0}\left(\omega L-\frac{1}{\omega C}\right)+\frac{\omega s_0}{v}\right] \\ &= r'+\mathrm{j}x'. \end{aligned}$$

The locus representing $z'(s_0)$ on the chart is a circle C with $r = R/Z_0$ which may be inside or outside the circle $r = 1$. The modulus of the reflection coefficient $\Gamma'(s_0)$ is a minimum at the frequency $\omega_1/2\pi$ where $x'(\omega_1) = 0$. That is, where

$$\frac{1}{Z_0}\left(\omega_1 L-\frac{1}{\omega_1 C}\right)+\frac{\omega_1 s_0}{v} = 0$$

or

$$\frac{1}{Z_0}(LC\omega_1^2-1)+\frac{C\omega_1^2 s_0}{v} = 0$$

which becomes, on putting $\omega_0^2 LC = 1$

$$\frac{1}{Z_0}\left[\left(\frac{\omega_1}{\omega_0}\right)^2-1\right]+\frac{C\omega_1^2 s_0}{v} = 0,$$

which requires

$$C = \frac{-v}{Z_0 s_0 \omega_1^2}\left[\left(\frac{\omega_1}{\omega_0}\right)^2-1\right],$$

for which there is only a solution if $\omega_1 < \omega_0$. The complete solution is

$$C = \frac{1}{Z_0}\frac{v}{s_0}\frac{1}{\omega_1^2}\left[1-\left(\frac{\omega_1}{\omega_0}\right)^2\right],$$

$$L = \frac{s_0}{v} Z_0 \frac{1}{\left(\frac{\omega_0}{\omega_1}\right)^2-1}.$$

3. From the values given, $Z_0 = 140\,\Omega$, $s_0 = 5$ cm, $v = c_0 = 3\times 10^{10}$ cm/s, $v/s_0 = 6\times 10^9\ \text{s}^{-1}$.

$$\frac{\omega_0}{2\pi} = 100{\cdot}09 \text{ MHz}, \qquad \frac{\omega_1}{2\pi} = 99{\cdot}48 \text{ MHz}.$$

$$\left(\frac{\omega_1}{\omega_0}\right) = 0{\cdot}994, \qquad \left(\frac{\omega_1}{\omega_0}\right)^2 = 0{\cdot}988, \qquad \left(\frac{\omega_0}{\omega_1}\right)^2 = 1{\cdot}012;$$

from which

$$L = 1{\cdot}94\times 10^{-6}\text{ H}, \qquad C = 1{\cdot}28 \text{ pF}.$$

4. The impedance transferred to the plane of the impedance Z is now

$$z_2 = \frac{1}{z_1} = -\mathrm{j}\frac{v}{\omega s_0}.$$

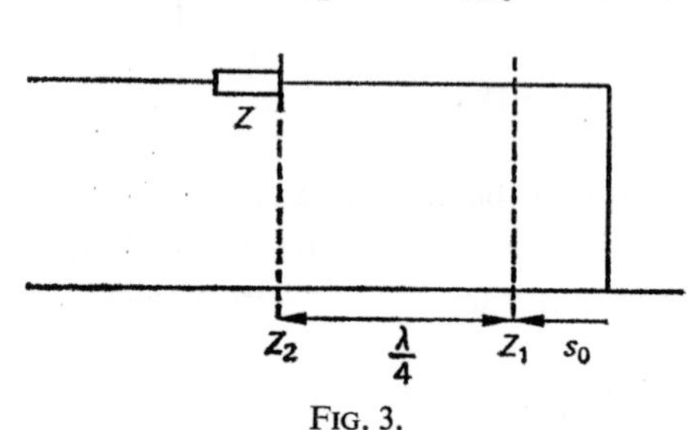

FIG. 3.

The total impedance at $s_0+\lambda/4$ is

$$z'' = \frac{R}{Z_0} + \mathrm{j}\left[\frac{1}{Z_0}\left(\omega L - \frac{1}{\omega C}\right) - \frac{v}{\omega s_0}\right] = r'' + \mathrm{j}x''.$$

Γ'' is a minimum for the frequency $\omega_2/2\pi$ where $x''(\omega_2) = 0$. Then

$$\frac{1}{Z_0}\left(\omega_2 L - \frac{1}{\omega_2 C}\right) - \frac{v}{\omega_2 s_0} = 0$$

$$\frac{1}{Z_0}\left[\left(\frac{\omega_2}{\omega_0}\right)^2 - 1\right] = \frac{Cv}{s_0},$$

so that

$$\left(\frac{\omega_2}{\omega_0}\right)^2 = \frac{CvZ_0}{s_0} + 1,$$

$$\omega_2 = \omega_0\sqrt{1 + \frac{CvZ_0}{s_0}}.$$

5. Numerical example:

$$\omega_2 = \omega_0\sqrt{2{\cdot}075} = 1{\cdot}44\omega_0,$$
$$f_2 = 144{\cdot}12 \text{ MHz}.$$

6.
$$Q = \frac{\omega_0 L}{R_{\text{total}}},$$

$R_t = R+2Z_0$, since the generator also has an impedance Z_0. For the impedance Z alone

$$Q_0 = \frac{\omega_0 L}{R};$$

$$Q = Q_0 \frac{a}{2+a}, \quad \text{with} \quad \frac{R}{Z_0} = a;$$

$$Q_0 = \omega_0 \frac{Z_0}{R} \frac{s_0}{v} \frac{1}{\left(\frac{\omega_0}{\omega_1}\right)^2 - 1}.$$

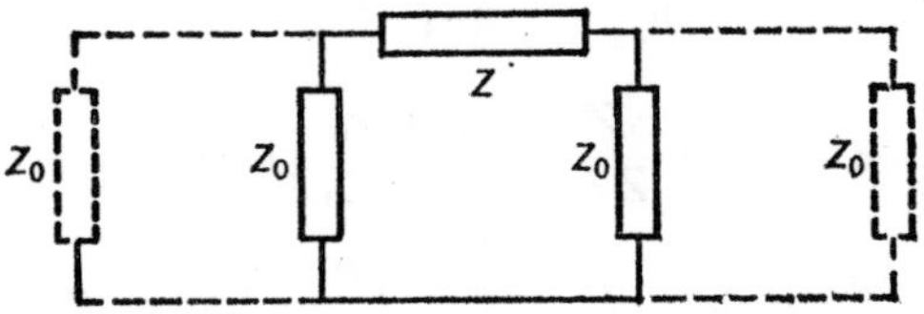

FIG. 4.

Thus

$$Q = \omega_0 \frac{s_0}{v} \frac{1}{\left(\frac{\omega_0}{\omega_1}\right)^2 - 1} \frac{1}{2+a}.$$

7. Taking the numerical values, $Q = 3{\cdot}2$.

8. (a) $P_i = P_0$.

(b) $P_r = P_0 \Gamma\Gamma^* = P_0 |\Gamma|^2$.

(c) Refer to Fig. 5.

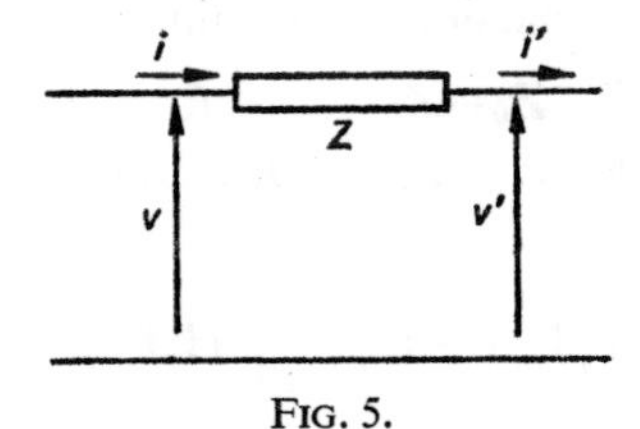

FIG. 5.

We have

$$v = v_0(1+\Gamma), \qquad v' = v_0(1+\Gamma) - \frac{v_0 Z}{Z_0}(1-\Gamma),$$

$$i = \frac{v_0}{Z_0}(1-\Gamma), \qquad i' = \frac{v_0}{Z_0}(1-\Gamma).$$

The power dissipated in Z and Z_0 is

$$P_t = \tfrac{1}{2}\,\mathrm{Re}\,(vi^*),$$

$$\frac{1}{2}\,vi^* = \frac{v_0 v_0^*}{2Z_0}(1+\Gamma)(1-\Gamma^*)$$

$$= \frac{|v_0|^2}{2Z_0}(1-|\Gamma|^2-\Gamma^*+\Gamma),$$

which, taking the real part, is

$$P_t = \frac{|v_0|^2}{2Z_0}(1-|\Gamma|^2).$$

The power dissipated in Z_0 alone is

$$P_{Z_0} = \tfrac{1}{2}\,\mathrm{Re}\,(v'i'^*),$$

$$\frac{1}{2}\,v'i'^* = \frac{1}{2}\left[v_0(1+\Gamma)-\frac{v_0 Z}{Z_0}(1-\Gamma)\right]\frac{v_0^*}{Z_0}(1-\Gamma^*)$$

$$= \frac{1}{2}\,\frac{|v_0|^2}{Z_0}\left[1+\Gamma-\Gamma^*-|\Gamma|^2-\frac{R+\mathrm{j}X}{Z_0}(1+|\Gamma|^2-\Gamma-\Gamma^*)\right];$$

from which

$$P_{Z_0} = P_0\left[1-|\Gamma|^2-\frac{R}{Z_0}(1+|\Gamma|^2-\Gamma-\Gamma^*)\right]$$

$$= P_0\left[1-\frac{R}{Z_0}-|\Gamma|^2\left(1+\frac{R}{Z_0}\right)+\frac{R}{Z_0}(\Gamma+\Gamma^*)\right.$$

$$= P_0[1-a-|\Gamma|^2(1+a)+a(\Gamma+\Gamma^*)].$$

Consequently, the power dissipated in Z is

$$P_Z = P_0 a[1+|\Gamma|^2-(\Gamma+\Gamma^*)].$$

9. The transmitted power is maximum for minimum Γ, i.e. for $\omega = \omega_0$. Thus

$$\Gamma = \frac{R+Z_0-Z_0}{R+Z_0+Z_0} = \frac{R}{R+2Z_0} = \frac{a}{2+a}.$$

Under these conditions

$$P_{Z_0} = P_0\left[1-a-\left(\frac{a}{2+a}\right)^2(1+a)+\frac{2a^2}{2+a}\right],$$

$$P_{Z_0} = \frac{4P_0}{(2+a)^2},$$

and the power transmitted to Z and Z_0 is

$$P_t = P_0\,\frac{4(1+a)}{(2+a)^2}.$$

10. For $\omega = \omega_0$

$$Z = R + Z_0,$$

$$z = 1 + a = 1{\cdot}67,$$

so that

$$y \simeq 0{\cdot}6.$$

From the chart the solutions are

$$s_1/\lambda = 0{\cdot}146 \qquad (\text{point } P_1,\ 1+\text{j}\,0{\cdot}54);$$
$$s_2/\lambda = 0{\cdot}354 \qquad (\text{point } P_2,\ 1-\text{j}\,0{\cdot}54).$$

Now $\lambda = c_0/f_0 = 2{\cdot}99$ m and

$$s_1 = 43{\cdot}6 \text{ cm}, \qquad s_2 = 105 \text{ cm}.$$

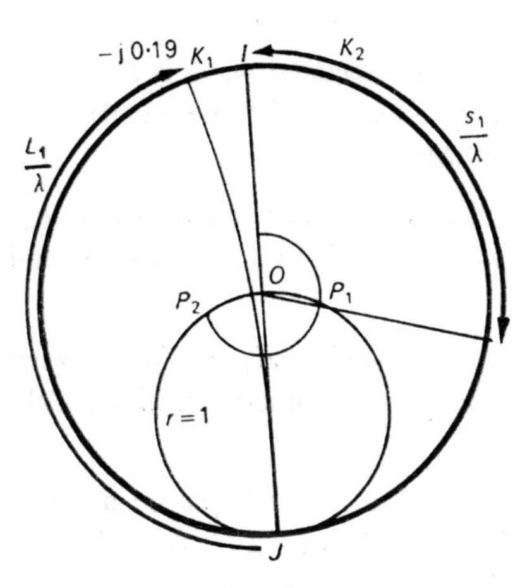

FIG. 6.

11. The admittance due to the stub is $Y_s = -\text{j}Y_0' \cot \beta l$, Y_0' being the characteristic admittance of the stub. Taking $Y_0 = 1/Z_0$ as the reference admittance gives

$$y_s = -\text{j}\frac{Y_0'}{Y_0} \cot \beta l = -\text{j}\frac{Z_0}{Z_0'} \cot \beta l,$$

$$= \text{j}b \cdot \frac{140}{50} = \text{j}\,2{\cdot}8b \quad \text{with} \quad b = -\cot \beta l.$$

The imaginary part of the admittance brought to the point P_1, i.e. $+\text{j}\,0{\cdot}54$, must be compensated by $y_s = -\text{j}\,0{\cdot}54 = \text{j}\,2{\cdot}8b$. Thus

$$b_1 = -0{\cdot}19.$$

For the point P_2, $b_2 = +0{\cdot}19$ and we have

$$\frac{l_1}{\lambda} = 0{\cdot}22, \qquad l_1 = 65{\cdot}7 \text{ cm},$$

$$\frac{l_2}{\lambda} = 0{\cdot}28, \qquad l_2 = 83{\cdot}7 \text{ cm}.$$

Problem No. 18

MATCHING IN A WAVE GUIDE

A U.H.F. wave is propagated in the TE_{01} mode in a lossless rectangular guide of dimensions $a = 1{\cdot}016$ cm, $b = 2{\cdot}286$ cm. The guide is air-filled.

1. What is the cut-off wavelength λ_c and cut-off frequency f_c?

2. If the frequency of the propagated wave is $f > f_c$, express the guide wavelength λ_g as a function of λ_0 and λ_c, λ_0 being the wavelength in air for the frequency f. What is λ_g for $f = 9800$ MHz?

3. The guide is now filled with a dielectric with $\mu = \mu_0$, $\varepsilon = \varepsilon_r\varepsilon_0$. What is the form of the new guide wavelength λ_g' in terms of λ_0, λ_c and ε_r? What is its value for $f = 9800$ MHz, $\varepsilon_r = 2{\cdot}25$?

4. The wave impedance is defined as $Z_0 = |E_T/H_T|$ where E_T and H_T are the transverse components of the electromagnetic field. Calculate the wave impedances Z_0 for $\varepsilon = \varepsilon_0$ and Z_0' for $\varepsilon = \varepsilon_r\varepsilon_0$.

5. A plate of dielectric of thickness t is placed across the guide as shown in Fig. 1 and is followed by a matched load C. What is the input impedance Z_1 in the plane (2, 2′) of the dielectric?

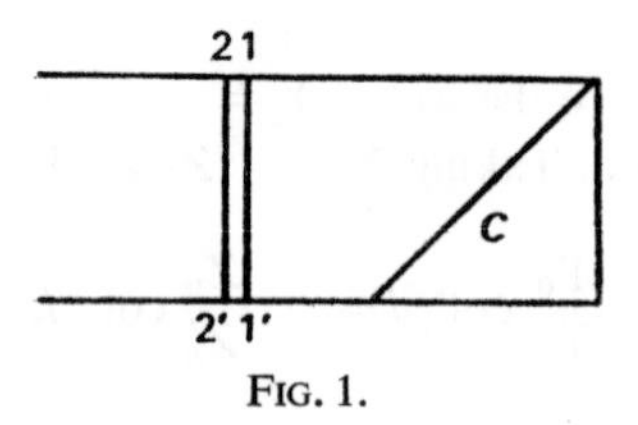

FIG. 1.

6. Given that t is small, so that $2\pi t/\lambda_g' \ll 1$, show that the reduced impedance $z_1 = Z_1/Z_0$ can be written in the form

$$z_1 = 1 + jx_1$$

and express x_1 as a function of λ, λ_g' and $K = \lambda_g'/\lambda_g$.

7. Give an expression for x in terms of λ_g, λ_0, t and ε_r and show that the dielectric plate behaves as a capacitance in parallel with the matched load. What is the value of this capacity for $f = 9800$ MHz, $\varepsilon_r = 2{\cdot}25$, $t = 0{\cdot}7$ mm?

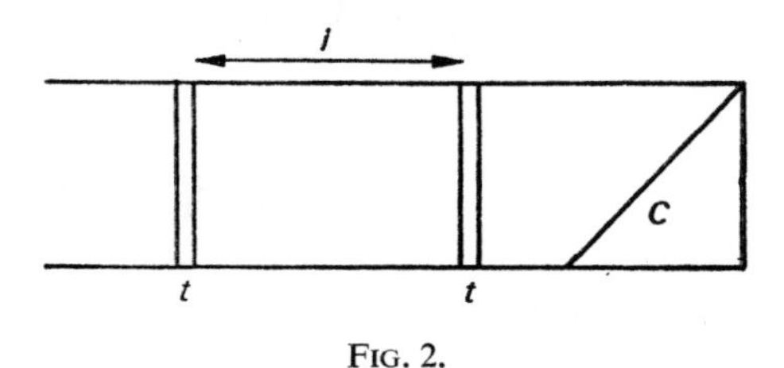

FIG. 2.

8. An enclosure is formed in the guide (see Fig. 2) by two identical plates of thickness t and separated by a distance l. Calculate this separation l for the system to be matched at 9800 MHz.

Solution. 1.

1. $$\lambda_c = 2b = 4{\cdot}572 \text{ cm}, \qquad f_c = \frac{c_0}{\lambda_c} = 6550 \text{ MHz}.$$

2. $$\frac{1}{\lambda_g^2} = \frac{1}{\lambda_0^2} - \frac{1}{\lambda_c^2}.$$

Multiplying through by λ_0^2 gives

$$\frac{\lambda_0^2}{\lambda_g^2} = 1 - \frac{\lambda_0^2}{\lambda_c^2},$$

and

$$\lambda_g = \frac{\lambda_0}{\sqrt{1-\left(\dfrac{\lambda_0}{\lambda_c}\right)^2}}.$$

For $f = 9800$ MHz, $\lambda_0 = 3{\cdot}06$ cm, $\lambda_g = 4{\cdot}12$ cm.

3. If the guide is filled with dielectric, λ_0 must be replaced by $\lambda = \lambda_0/\sqrt{\varepsilon_r}$ so that

$$\frac{1}{\lambda_g'^2} = \frac{1}{\lambda^2} - \frac{1}{\lambda_c^2},$$

from which

$$\frac{1}{\lambda_g'^2} = \frac{\varepsilon_r}{\lambda_0^2} - \frac{1}{\lambda_c^2},$$

and

$$\lambda'_g = \frac{\lambda_0}{\sqrt{\varepsilon_r - \left(\frac{\lambda_0}{\lambda_c}\right)^2}}.$$

For $f = 9800$ MHz

$$\varepsilon_r = 2{\cdot}25, \quad \lambda = \frac{\lambda_0}{1{\cdot}5} = 2{\cdot}04 \text{ cm}, \quad \lambda'_g = 2{\cdot}28 \text{ cm}.$$

4. The wave impedance is given by

$$Z_0 = \frac{\omega\mu}{\beta_g} = \eta_0 \frac{\lambda_g}{\lambda_0},$$

where η_0 is the impedance of free space

$$\eta_0 = \sqrt{\mu_0/\varepsilon_0} = 377\ \Omega,$$

$$Z'_0 = \frac{\omega\mu}{\beta'_g} = \eta_0 \frac{\lambda'_g}{\lambda_0},$$

$$Z_0 = 507\ \Omega, \quad Z'_0 = 281\ \Omega.$$

5. Since the load C is matched, the impedance in the plane (1, 1′) is Z_0, which gives an impedance Z_1 in the plane (2, 2′) as

$$Z_1 = Z'_0 \frac{Z_0 + jZ'_0 \tan \beta'_g t}{Z'_0 + jZ_0 \tan \beta'_g t}$$

with $\beta'_g = 2\pi/\lambda'_g$.

6. With $z_1 = Z_1/Z_0$ and $K = Z'_0/Z_0 = \lambda'_g/\lambda_g$ we have

$$z_1 = K \frac{1 + jK \tan \beta'_g t}{K + j \tan \beta'_g t} \simeq K \frac{1 + jK\beta'_g t}{K + j\beta'_g t}$$

$$= \frac{1 + jK\beta'_g t}{1 + j\frac{\beta'_g t}{K}} \simeq (1 + jK\beta'_g t)\left(1 - \frac{j\beta'_g t}{K}\right)$$

or

$$z_1 = 1 + \beta'^2_g t^2 + j\beta'_g t\left(K - \frac{1}{K}\right).$$

Since $\beta'_g t \ll 1$ this is, to good approximation,

$$z_1 = 1 + j\beta'_g t\left(K - \frac{1}{K}\right).$$

7.

$$z_1 = 1+\mathrm{j}\frac{2\pi}{\lambda'_g}t\left(\frac{\lambda'_g}{\lambda_g}-\frac{\lambda_g}{\lambda'_g}\right)$$

$$= 1+\mathrm{j}2\pi t\left(\frac{1}{\lambda_g}-\frac{\lambda_g}{\lambda'^2_g}\right).$$

Thus

$$x_1 = \frac{2\pi t}{\lambda_g}\left(1-\frac{\lambda_g^2}{\lambda'^2_g}\right)$$

$$\left(\frac{\lambda_g}{\lambda'_g}\right)^2 = \frac{\varepsilon_r-(\lambda_0/\lambda_c)^2}{1-(\lambda_0/\lambda_c)^2}$$

$$1-\left(\frac{\lambda_g}{\lambda'_g}\right)^2 = \frac{1-\varepsilon_r}{1-\left(\frac{\lambda_0}{\lambda_c}\right)^2}.$$

Then

$$x_1 = -2\pi t\left(\frac{\lambda_g}{\lambda_0^2}\right)(\varepsilon_r-1).$$

The impedance in the plane (2, 2′) can be put in the form $z_1 = 1+\mathrm{j}x_1$, or $Z_1 = Z_0(1+\mathrm{j}x_1)$ with $x_1 < 0$. It is equivalent toa capacitance in series with the matched load. The plate of dielectric alone behaves like a capacitance given by

$$Z_0x_1 = \frac{1}{\omega C_0},$$

or

$$C_0 = \frac{1}{Z_0x_1\omega} \simeq 0{\cdot}13 \text{ pF}.$$

8. The equivalent circuit of the system in terms of a transmission line can be taken as in Fig. 3. The impedance in the plane of the first plate is

$$z = 1-\mathrm{j}|x_1|.$$

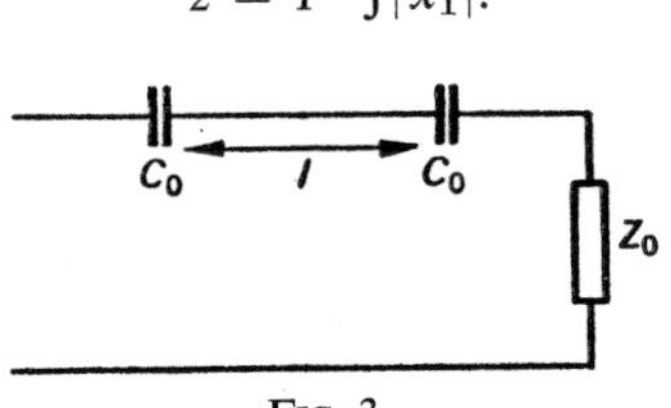

FIG. 3.

At a distance l the transferred impedance can be written, with $\theta = \tan\beta_g l$,

$$z' = \frac{z+\mathrm{j}\theta}{1+\mathrm{j}\theta z}.$$

Taking into account the capacitance due to the second plate, the input impedance becomes

$$z_i = -\mathrm{j}\,|x_1| + z'.$$

There will be matching if $z_i = 1$ or

$$1 = -\mathrm{j}\,|x_1| + \frac{z+\mathrm{j}\theta}{1+\mathrm{j}\theta z}$$

$$(1+\mathrm{j}\,|x_1|)(1+\mathrm{j}\theta z) = z+\mathrm{j}\theta.$$

Thus

$$1+\mathrm{j}\,|x_1|+\mathrm{j}\theta(1-\mathrm{j}\,|x_1|)(1+\mathrm{j}\,|x_1|) = 1-\mathrm{j}\,|x_1|+\mathrm{j}\theta$$

$$|x_1|+(1+|x_1|^2)\theta = -|x_1|+\theta.$$

This is satisfied if, beside the solution $|x_1| = 0$ for $\varepsilon_r = 1$, we have

$$2+|x_1|\,\theta = 0, \quad \text{or} \quad \theta = -\frac{2}{|x_1|}$$

and, since

$$|x_1| = 2\pi t\frac{\lambda_g}{\lambda_0^2}(\varepsilon_r-1) = 0{\cdot}24,$$

we have

$$\theta = -8{\cdot}33 = \tan\beta_g l.$$

Then

$$\beta_g l = \pi - 1{\cdot}45 + n\pi,$$

from which

$$\frac{l}{\lambda_g} = \frac{1{\cdot}69}{2\pi}+\frac{n}{2},$$

$$l = 0{\cdot}269\lambda_g + \frac{n\lambda_g}{2}$$

$$= 1{\cdot}108 \text{ cm} + n\times 2{\cdot}06 \text{ cm}.$$

Problem No. 19

MATCHING WITH QUARTER-WAVE STEPS

A rectangular waveguide is supplied with power by a klystron at a frequency $f_0 = 9450$ MHz and is terminated by a load Z_L. The dimensions of the guide section are $a = 1{\cdot}016$ cm, $b = 2{\cdot}286$ cm.

I

In order to match the load, use is made of a quarter-wave adaptor (at the frequency f_0) placed a distance d from Z_L. This adaptor consists of a parallelepiped of dielectric of length L and with the same cross-section as the guide. The relative permittivity of the dielectric is $\varepsilon_r = 1{\cdot}81$ and the complete system is shown in Fig. 1.

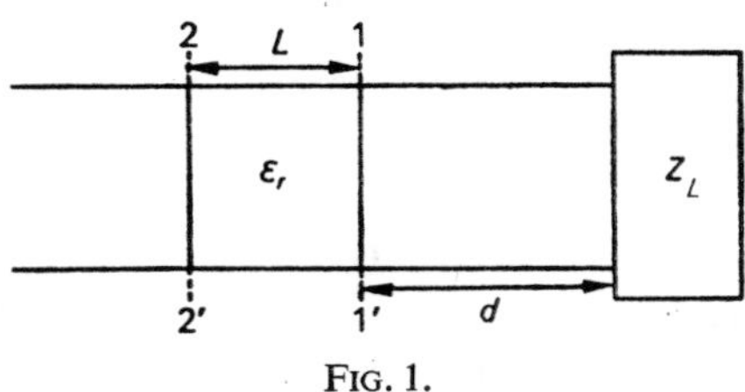

FIG. 1.

1. Calculate the guide wavelength λ_g and the wavelength λ'_g in the dielectric assuming that a TE_{01} mode is propagated. Hence deduce L.

2. The wave impedance is defined as $Z_0 = |E_T/H_T|$ where the suffix T indicates the transverse components of the fields. Calculate the wave impedances Z_0 and Z'_0 for the guide empty and with the dielectric respectively.

3. What must be the value of Z_d, the impedance transferred to the plane (1, 1′), if there is to be matching in the plane (2, 2′)?

4. If the adaptor can be moved along the guide between the positions $d = d_0 =$ 2 cm and $d = d_0 + \lambda_g/2$, trace on the Smith chart the locus of the reduced impedances $z_L = Z_L/Z_0$ that can be matched by this system.

5. At what position must the adaptor be placed to match $z_L = 1{\cdot}8 - j$?

II

To improve the system, two identical quarter-wave adaptors are used separated by the distance l as shown in Fig. 2 where this distance l is variable.

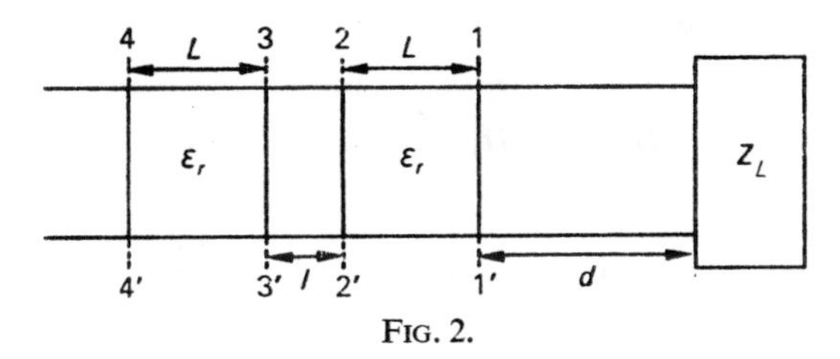

FIG. 2.

(For each value of l, the adaptor can be moved between d_0 and $d_0 + \lambda_g/2$.)

1. Calculate the impedance transferred to the input of the second adaptor (plane 3, 3′) in terms of Z_d, Z_0, Z_0' and $\theta = \tan \beta_g l$ where $\beta_g = 2\pi/\lambda_g$.
2. Hence deduce the reduced impedance z_d for matching in the plane (4, 4′).
3. Trace on the Smith chart the locus of z_d when l varies from zero to $\lambda_g/4$.
4. Find the locus of impedances which it is impossible to match with d between d_0 and $d_0+\lambda_g/2$.
5. Calculate d and l for matching with $z_L = 1{\cdot}4-\text{j}\,0{\cdot}6$.

Solution

I

1. In the guide we have

$$\frac{1}{\lambda_g^2} = \frac{1}{\lambda_0^2} - \frac{1}{\lambda_c^2}.$$

For the TE_{01} mode

$$\lambda_c = 2b = 4{\cdot}572 \text{ cm}, \quad \lambda_0 = c_0/f_0 = 3{\cdot}172 \text{ cm},$$

so that

$$\lambda_g = \frac{\lambda_0}{\sqrt{1-\left(\frac{\lambda_0}{\lambda_c}\right)^2}} = 4{\cdot}4 \text{ cm}.$$

In the dielectric filled guide

$$\frac{1}{\lambda_g'^2} = \frac{\varepsilon_r}{\lambda_0^2} - \frac{1}{\lambda_c^2},$$

so that

$$\lambda_g' = \frac{\lambda_0}{\sqrt{\varepsilon_r-\left(\frac{\lambda_0}{\lambda_c}\right)^2}} = \frac{\lambda_0}{1{\cdot}154} = 2{\cdot}75 \text{ cm},$$

from which

$$L = \frac{\lambda_g'}{4} = 0{\cdot}687 \text{ cm}.$$

2. The wave impedance for the TE_{01} mode is

$$\left|\frac{E_T}{H_T}\right| = \left|\frac{E_x}{H_y}\right| = \frac{\omega\mu}{\beta_g} = \eta_0 \frac{\lambda_g}{\lambda_0},$$

where

$$\eta_0 = \sqrt{\frac{\mu_0}{\varepsilon_0}} = 377\ \Omega, \quad \beta_g = \frac{2\pi}{\lambda_g}.$$

Thus, in the empty guide, $Z_0 = 523\ \Omega$ and in the dielectric filled guide $Z_0' = \eta_0\beta_g'/\lambda_0 = 327\ \Omega$.

3. If Z_d is the impedance transferred to the plane (1, 1′) we have, in the plane (2, 2′),

$$z = Z_0' \frac{Z_d + jZ_0' \tan \beta_g' L}{Z_0' + jZ_d \tan \beta_g' L},$$

with

$$\beta L_g' = \frac{2\pi}{\lambda_g'} \frac{\lambda_g'}{4} = \frac{\pi}{2},$$

which gives

$$\tan \beta_g' L = \infty, \quad Z = \frac{Z_0'^2}{Z_d}.$$

There will be matching if

$$Z = Z_0 = \frac{(Z_0')^2}{Z_d}, \quad \text{or} \quad Z_d = 204\ \Omega.$$

4. We have

$$z_d = \frac{Z_d}{Z_0} = \left(\frac{Z_0'}{Z_0}\right)^2 = \left(\frac{\lambda_g'}{\lambda_g}\right)^2 = 0{\cdot}39,$$

which is represented by the point A on the chart of Fig. 3. If d can vary from d_0 to $d_0 + \lambda_g/2$, all the impedances on the circle of radius OA can be matched.

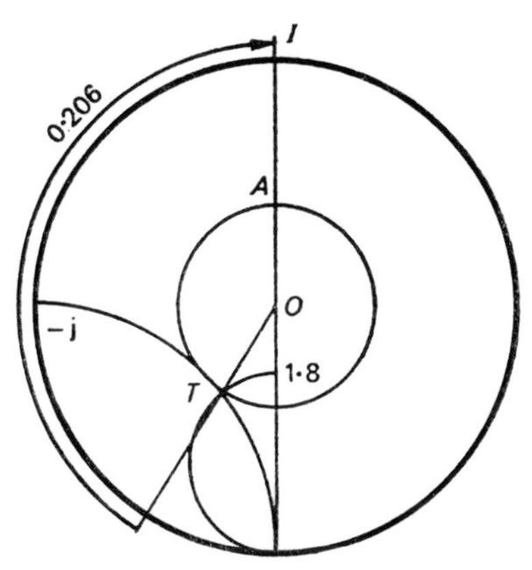

FIG. 3.

5. $z_L = 1{\cdot}8 - j$ is represented by the point T on the chart. The distance d from the load to the adaptor is given by the angle ($\boldsymbol{OT}$, $\boldsymbol{OI}$) as

$$\frac{d}{\lambda_g} = 0{\cdot}206,$$

or

$$d = 0{\cdot}906 \text{ cm} + \frac{n\lambda_g}{2}.$$

As

$$d_0 < d < d_0 + \frac{\lambda_g}{2}$$

we take as the solution

$$d = 0{\cdot}906 + 2{\cdot}2 = 3{\cdot}106 \text{ cm.}$$

II

1. We know that the impedance transferred to the plane (2, 2′) is

$$Z_2 = \frac{(Z_0')^2}{Z_d}.$$

In the plane (3, 3′) this transferred impedance is given by the usual formula

$$Z_3 = Z_0 \frac{Z_2 + jZ_0 \tan \beta_g l}{Z_0 + jZ_2 \tan \beta_g l} = Z_0 \frac{Z_2 + jZ_0\theta}{Z_0 + jZ_2\theta}.$$

Replacing Z_2 by its value calculated above

$$Z_3 = Z_0 \frac{\dfrac{(Z_0')^2}{Z_d} + jZ_0\theta}{Z_0 + j\dfrac{(Z_0')^2}{Z_d}\theta}. \tag{1}$$

2. If we require that $Z = Z_0$ in the plane (4, 4′) then we must have

$$Z = Z_0 = \frac{Z_0'^2}{Z_3}, \quad \text{or} \quad Z_3 = \frac{Z_0'^2}{Z_0},$$

which, with (1), becomes

$$\frac{Z_0'^2}{Z_0} = Z_0 \frac{\dfrac{Z_0'^2}{Z_d} + jZ_0\theta}{Z_0 + j\dfrac{Z_0'^2\theta}{Z_d}}.$$

Putting

$$k = \left(\frac{Z_0'}{Z_0}\right)^2 = \left(\frac{\lambda_g'}{\lambda_g}\right)^2 = 0{\cdot}39$$

we obtain

$$k\left(Z_0 + j\frac{Z_0'^2\theta}{Z_d}\right) = \frac{Z_0'^2}{Z_d} + jZ_0\theta.$$

On passing to reduced impedances with $Z_d/Z_0 = z_d$, this becomes

$$k\left(1+\mathrm{j}\frac{k\theta}{z_d}\right) = \frac{k}{z_d}+\mathrm{j}\theta\,,$$

$$k-\mathrm{j}\theta = \frac{k}{z_d}(1-jk\theta)$$

and

$$z_d = \frac{k(1-\mathrm{j}k\theta)}{k-\mathrm{j}\theta}$$

$$= \frac{k^2(1+\theta^2)}{k^2+\theta^2}+\mathrm{j}\frac{k\theta(1-k^2)}{k^2+\theta^2} = r_d+\mathrm{j}x_d\,.$$

3. l varies from 0 to $\lambda_g/4$. Thus:

for $l = 0$ $\quad \theta = 0 \quad r_d = 1 \quad x_d = 0 \quad (D_1)$

$l = \frac{\lambda_g}{16} \quad \theta = 0{\cdot}414 \quad r_d = 0{\cdot}54 \quad x_d = 0{\cdot}42 \quad (D_2)$

$l = \frac{\lambda_g}{8} \quad \theta = 1 \quad r_d = 0{\cdot}256 \quad x_d = 0{\cdot}285 \quad (D_3)$

$l = \frac{3\lambda_g}{16} \quad \theta = 2{\cdot}413 \quad r_d = 0{\cdot}17 \quad x_d = 0{\cdot}132 \quad (D_4)$

$l = \frac{\lambda_g}{4} \quad \theta = \infty \quad r_d = k^2 = 0{\cdot}152 \quad x_d = 0 \quad (D_5)$

where the points $D_1 \ldots D_5$ are shown on the chart in Fig. 4.

The locus is the semicircle C of diameter D_1D_4 passing through the five points given.

4. If the adaptor is moved along the guide by $\lambda_g/2$, the semicircle will move as a whole and make a complete circuit of the chart. All the impedances within the circle C_0 can therefore be matched.

Only impedances outside C_0 cannot be matched.

5. $z_L = 1{\cdot}4-\mathrm{j}\,0{\cdot}6$ is represented by the point T, while the impedance in the plane (1, 1′) must be on the circle C for there to be matching. Thus $z_d = 0{\cdot}7+\mathrm{j}\,0{\cdot}4$ (point D), which is obtained on taking the intercept of C_0', the circle through T_1 and C. Then

$$\frac{d}{\lambda_g} = 0{\cdot}192+0{\cdot}092 = 0{\cdot}284$$

and

$$d = 1{\cdot}249+\frac{\lambda_g}{2} = 3{\cdot}449 \text{ cm}.$$

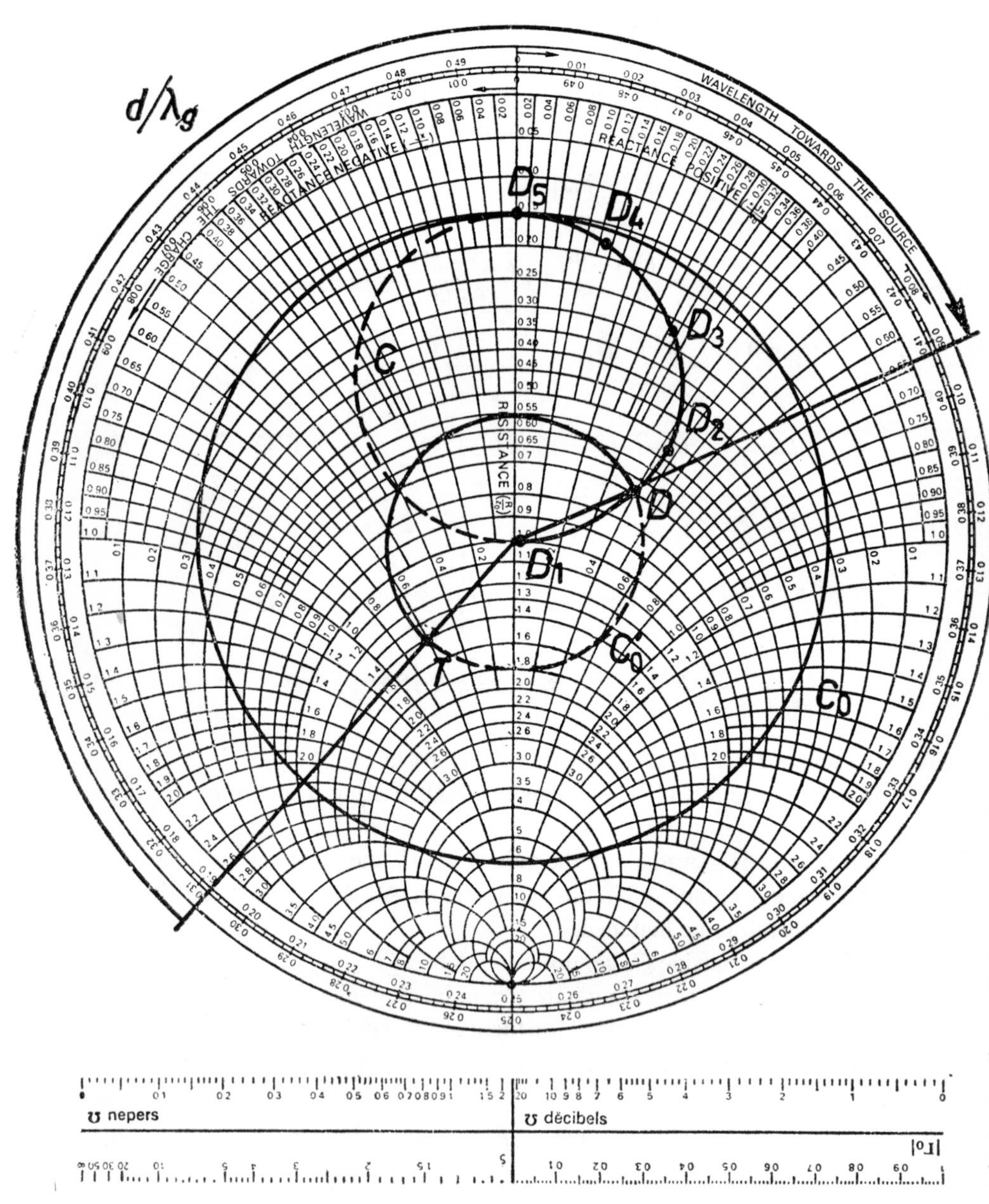
d/λg
WAVELENGTH TOWARDS THE SOURCE
WAVELENGTH TOWARDS THE CHARGE
REACTANCE POSITIVE
REACTANCE NEGATIVE
RESISTANCE
D5
D4
D3
D2
D
D1
C
C0
C0
υ nepers
υ décibels
|Γ0|

FIG. 4.

As $r_d = 0{\cdot}7$ we can write

$$\frac{k^2(1+\theta^2)}{k^2+\theta^2} = 0{\cdot}7,$$

which gives l, since $0 < l < \lambda_g/4$.

We find

$$\theta^2 = \frac{0{\cdot}3k^2}{0{\cdot}7-k^2} = 0{\cdot}081,$$

$$\theta = 0{\cdot}285, \quad \beta_g l = 0{\cdot}278,$$

$$l = 0{\cdot}044\lambda_g = 0{\cdot}19 \text{ cm}.$$

Problem No. 20

RESONANT SYSTEMS

I

1. Consider a series L, C, R circuit supplied by a sine wave of pulsatance ω. What are the values of the complex impedance Z, the resonant pulsatance ω_0 and the Q-factor, Q_0? Express the complex impedance in terms of R, Q_0, ω_0 and ω.

2. Show that the Q-factor can be expressed by a generalized relationship between energies, the stored energy being taken as W.

3. The circuit is supplied by a constant current source. Compare the amplitude of the voltage across the inductance with that across the resistance and comment on this ratio at resonance.

4. Supposing that the supply is of the constant-voltage type and Q_0 is large, calculate $\omega_2-\omega_1$ where ω_1 and ω_2 are the pulsatances for which the power dissipated in the resistance R is reduced to one-half of its maximum value.

5. The circuit supply, which is at the resonant frequency, is cut at the instant t_0. Find the time $t_0+\tau$ at which the stored energy is reduced to a fraction $1/e$ of its initial value, where e is the base of the natural logarithms and Q_0 is again assumed large.

II

Consider two metallic conductors P_1 and P_2 each of which is limited by two plane, parallel surfaces. They are placed perpendicular to the z-axis and they are taken to be very much thicker than the skin depth. The air gap between the conductors (taken equivalent to free space) is of thickness d along the z-axis and their opposing surfaces, each of area S, are of dimensions very much greater than the wavelengths of any electromagnetic radiation considered. (This last condition will permit the integration of Maxwell's equations as if the planes were infinite and without edge effects.)

6. Show that uniform plane waves with plane polarization cannot propagate between P_1 and P_2 along the z-axis except for well-defined frequencies. (For this part of the question and the following assume that the conductivity is infinite.)

7. Calculate the field components of E and H in terms of the energy W stored between the plates P_1 and P_2 within the cylinder of base area S and of the number of electric field nodes N along the z-axis.

8. In this part account is taken of the finite conductivity σ of the metal of the plates, while the field distribution is taken to be the same as in the preceding question. Calculate the current density in the plates P_1 and P_2 as a function of z. What is the power dissipated in P_1 and P_2?

9. Calculate the Q-factor Q_0 of the resonator consisting of the two plates P_1 and P_2. Construct the curve of Q_0 as a function of N. Can this result be foreseen from energy considerations?

10. As a numerical example take the plates P_1 and P_2 to be of copper and the frequency to be 10 000 MHz. ε and μ for the metal are taken to be ε_0 and μ_0 respectively and

$$\left.\begin{aligned} \varepsilon_0 &= \frac{1}{36\pi}\times 10^{-9} \\ \mu_0 &= 4\pi\times 10^{-7} \\ \sigma &= 5{\cdot}8\times 10^{7} \end{aligned}\right\} \text{S.I.}$$

III

The plate P_1 is considered to be pierced with holes so that electromagnetic energy can be radiated from this plate. An electromagnetic horn collects this energy and supplies it to a rectangular waveguide propagating in the H_{01} mode with characteristic impedance Z_0. (Suppose that the horn is lossless and remember that no knowledge of the theory of this instrument is necessary for this question.)

The system of resonator and horn may be represented as a series resonant circuit which, through the intermediary of the transformer of ratio n, closes one end of a lossless line A of characteristic impedance Z_0 (see Fig. 1).

11. Calculate in terms of L, C, R, n and Z_0 the Q-factor Q_L. of the resonator coupled to the waveguide. Calculate also the ratio α between the power radiated by the holes and the power dissipated by the walls when the waveguide is terminated by its characteristic impedance. Establish a relationship between this Q_L and Q_0 and express L, C and R in terms of Q_0, α, n, Z_0 and ω_0, the resonant pulsatance.

12. Consider now that it is proposed to find, by experiment, the fundamental

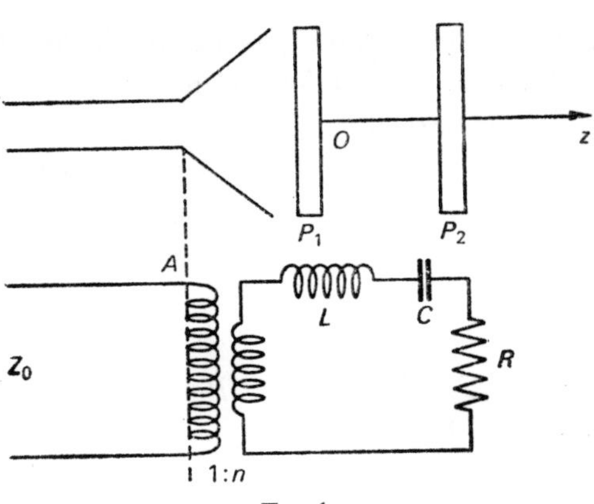

FIG. 1.

parameters of the resonator composed of P_1 and P_2. The waveguide is supplied by a matched klystron giving a power P_0 at a fixed pulsatance ω. A standing-wave detector is placed between the klystron and the horn.

Calculate the complex reflection coefficient at the point A and the standing wave ratio in the guide in terms of α, Q_0, ω, N and the separation d of P_1 and P_2.

13. Suppose now that the plate P_2 can be moved through a measurable distance relative to the fixed plate P_1. Show that it is reasonable to expect that α will vary little during such movement. What is the locus of the impedance of the resonator on the Smith chart as d varies?

14. The variation of d is taken to be around the value d_0 at which the resonance occurs at the pulsatance ω for $N = 2$. How can one determine experimentally the quality factors Q_0 and Q_L? For what value of α is the stored energy in the resonator a maximum?

IV

It is proposed to study in some detail the result of the incidence of a uniform plane wave on an infinite surface composed of a metal for which conductivity σ and for which $\varepsilon = \varepsilon_0$ and $\mu = \mu_0$. Take the wave to be propagated along the z-axis such that the metal surface is in the plane $z = 0$ with metal in the region $z > 0$, vacuum in the region $z < 0$.

15. Take the electric field in the form

$$E_x = E_0\, e^{-j\beta z} + BE_0\, e^{j\beta z}$$

and, by writing the equations of the fields at the surface of the metal, determine the value of B and the amplitude of the electromagnetic field inside the metal. (It is convenient to write $K = \sqrt{(\varepsilon/\mu)}/\sigma\delta$ where δ is the skin depth.)

Show that, with the normal magnitudes for the parameters, an approximation may be made in the calculations. Hence show that $|B|^2 = 1-4K$.

16. Show that, from this last result, it is possible to calculate the energy lost from the electromagnetic wave after M reflections. Hence deduce, with the help of question 5, the factor Q_0 for the resonator.

(Paris-Orsay, June 1963)

Solution

I.

1. The complex impedance is given by $Z = R+j(\omega L-1\omega/\omega C)$. At the resonant pulsatance $Z = R$ and $\omega_0 = 1/\sqrt{LC}$.

The Q-factor is

$$Q_0 = \frac{\omega_0 L}{R}.$$

The impedance can be written

$$Z = R\left[1+j\frac{1}{R}\left(\omega L-\frac{1}{\omega C}\right)\right]$$

$$= R\left[1+j\frac{\omega_0 L}{R}\left(\frac{\omega}{\omega_0}-\frac{1}{LC\omega\omega_0}\right)\right]$$

$$= R\left[1+jQ_0\left(\frac{\omega}{\omega_0}-\frac{\omega_0}{\omega}\right)\right].$$

2. There is energy stored both in the inductance and the capacitance and there will be an exchange of energy between the two. At an instant t, when the current has its maximum value I, the energy stored in the inductance will be

$$W = \tfrac{1}{2}LI^2.$$

The energy lost over a complete cycle in the resistance is

$$W_J = \tfrac{1}{2}RI^2T.$$

Thus

$$\frac{W}{W_J} = \frac{L}{RT} = \frac{L}{R\cdot\dfrac{2\pi}{\omega_0}} = \frac{\omega_0 L}{2\pi R}$$

and hence

$$Q_0 = 2\pi\frac{W}{W_J}.$$

3. The voltage across R has amplitude $|V_R| = RI$ while, across L, $|V_L| = \omega LI$.

$$\left|\frac{V_L}{V_R}\right| = \frac{\omega L}{R}.$$

At resonance

$$\frac{|V_L|}{|V_R|} = \frac{\omega_0 L}{R} = Q_0.$$

4. The power dissipated in the resistance at the frequency ω is

$$P = \frac{1}{2} Rii^* = \frac{1}{2} R \frac{vv^*}{ZZ^*},$$

$$P = \frac{1}{2} \frac{V^2}{R} \frac{1}{1+Q_0^2\left(\frac{\omega}{\omega_0}-\frac{\omega_0}{\omega}\right)^2}.$$

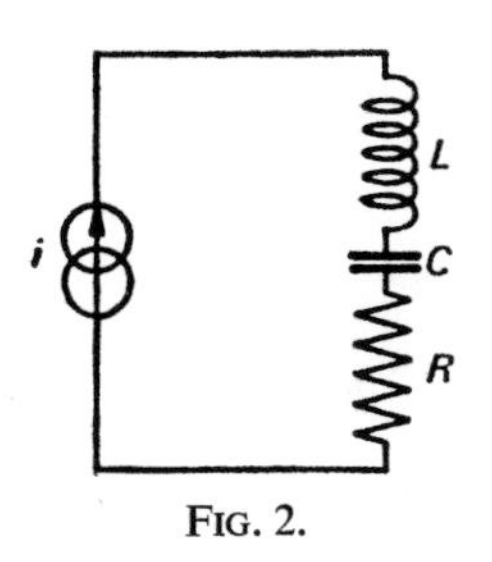

FIG. 2.

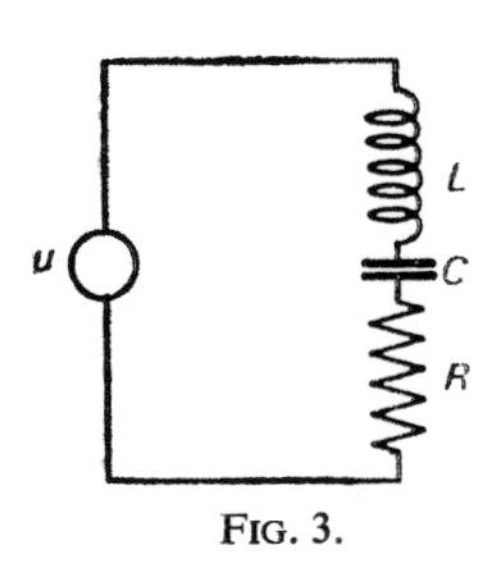

FIG. 3.

This power is a maximum for $\omega = \omega_0$:

$$P = P_0 = \frac{V^2}{2R}.$$

Also, $P = P_0/2$ for

$$Q_0^2\left(\frac{\omega}{\omega_0}-\frac{\omega_0}{\omega}\right)^2 = 1,$$

which corresponds to

$$\frac{\omega_1}{\omega_0}-\frac{\omega_0}{\omega_1} = \frac{1}{Q_0},$$

$$\frac{\omega_2}{\omega_0}-\frac{\omega_0}{\omega_2} = -\frac{1}{Q_0}.$$

We then get

$$\frac{\omega_1-\omega_2}{\omega_0} = \frac{1}{Q_0},$$

from which

$$Q_0 = \frac{\omega_0}{\omega_1-\omega_2}.$$

5. The stored energy is proportional to I^2. If the supply is cut-off we have

$$Ri+L\frac{di}{dt}+\frac{1}{C}\int i\,dt = 0,$$

or

$$LC\frac{d^2i}{dt^2}+RC\frac{di}{dt}+i = 0,$$

for which the characteristic equation can be written

$$p^2+\frac{R}{L}p+\frac{1}{LC} = 0,$$

$$p^2+\frac{\omega_0}{Q_0}p+\omega = 0,$$

so that

$$p = -\frac{\omega_0}{2Q_0}\pm j\omega_0\sqrt{1-\frac{1}{4Q_0^2}} \simeq -\frac{\omega_0}{2Q_0}\pm j\omega_0 .$$

The modulus of the current is proportional to $\exp[-(\omega_0/2Q_0)t]$ and the energy is thus proportional to $\exp[-(\omega_0/Q_0)t]$.

The stored energy is thus reduced to $1/e$ of its initial value when

$$\frac{\omega_0}{Q_0}\tau = 1,$$

so that

$$Q_0 = \omega_0\tau = 2\pi\frac{\tau}{T}.$$

II

6. In air, Maxwell's equations may be written in complex form as

$$\begin{aligned}\text{curl}\,\boldsymbol{E} &= -j\omega\mu_0\boldsymbol{H}\\ \text{curl}\,\boldsymbol{H} &= j\omega\varepsilon_0\boldsymbol{E}\\ \text{div}\,\boldsymbol{E} &= 0\\ \text{div}\,\boldsymbol{H} &= 0.\end{aligned}$$

For a plane, uniform wave propagating along the z-axis we know that

(a) the properties of the wave do not depend on x and y;

(b) the transverse components of the electromagnetic field are perpendicular to each other.

From (a) we have

$$\operatorname{div} \boldsymbol{E} = \frac{\partial E_z}{\partial z} = 0,$$

$$\operatorname{div} \boldsymbol{H} = \frac{\partial H_z}{\partial z} = 0.$$

Since we are only interested in variable fields this requires that

$$E_z = H_z = 0.$$

Taking, from (b), the components $E = E_x$ and $H = H_y$ we get

$$\operatorname{curl} \boldsymbol{E} = -\mathrm{j}\omega\mu_0 \boldsymbol{H}, \quad \text{so that} \quad \frac{\partial E_x}{\partial z} = -\mathrm{j}\omega\mu_0 H_y;$$

$$\operatorname{curl} \boldsymbol{H} = \mathrm{j}\omega\varepsilon_0 \boldsymbol{E}, \quad \text{so that} \quad \frac{-\partial H_y}{\partial z} = \mathrm{j}\omega\varepsilon_0 E_x.$$

The propagation equations are thus

$$\frac{\partial^2 E_x}{\partial z^2} = -\beta_0^2 E_x, \quad \beta_0^2 = \omega^2 \varepsilon_0 \mu_0;$$

$$\frac{\partial^2 H_y}{\partial z^2} = -\beta_0^2 H_y,$$

$$H_y = a\, \mathrm{e}^{-\mathrm{j}\beta_0 z} + b\, \mathrm{e}^{+\mathrm{j}\beta_0 z},$$

$$E_x = -\frac{1}{\mathrm{j}\omega\varepsilon_0} \frac{\partial H_y}{\partial z} = \eta_0 [a\, \mathrm{e}^{-\mathrm{j}\beta_0 z} - b\, \mathrm{e}^{+\mathrm{j}\beta_0 z}]$$

where $\eta_0 = \sqrt{\mu_0/\varepsilon_0}$.

The two plates of infinite conductivity (perfectly conducting metals) impose the conditions that:

E is perpendicular to the walls or is zero,

H is tangential to the walls or is zero.

Thus:

$$\text{at} \quad z = 0, \quad E_x(0) = 0 \quad \text{from which} \quad b = a;$$

$$\text{at} \quad z = d \quad E_x(d) = 0 \quad \text{from which} \quad \sin \beta_0 d = 0,$$

which gives

$$\beta_0 d = p\pi,$$

with p an integer and

$$\beta_0 = p\frac{\pi}{d} = \omega\sqrt{\varepsilon_0 \mu_0}.$$

Then

$$f = \frac{p}{2d\sqrt{\varepsilon_0\mu_0}}.$$

This condition is really just

$$\beta_0 = \frac{2\pi}{\lambda_0} = p\frac{\pi}{d}, \quad \text{or} \quad d = p\frac{\lambda_0}{2}.$$

7.

$$H_y = 2a \cos p\frac{\pi z}{d}, \quad H_x = H_z = 0,$$

$$E_x = -2\text{j}a\eta_0 \sin p\frac{\pi z}{d}, \quad E_y = E_z = 0.$$

The stored energy is

$$W = \frac{1}{2}\int_v \left(\frac{\varepsilon_0 EE^*}{2} + \frac{\mu_0 HH^*}{2}\right) \mathrm{d}v.$$

Since the stored energy is equally divided between the electric and magnetic fields we can write

$$W = \frac{1}{2}\int_v \mu_0 HH^* \,\mathrm{d}v = \frac{1}{2}\int_v \mu_0 4a^2 \cos^2 p\frac{\pi z}{d}\,\mathrm{d}v = 2a^2 S\mu_0 \int_0^d \cos^2 p\frac{\pi z}{d}\,\mathrm{d}z,$$

$$W = a^2 S\mu_0 d,$$

from which

$$a = \sqrt{\frac{W}{\mu_0 Sd}}$$

if a is taken to be real. The number of nodes is related to the number of half-wavelengths contained between the planes P_1 and P_2 by the equation

$$N = p+1.$$

The complex fields can thus be written

$$H_y = 2\sqrt{\frac{W}{\mu_0 Sd}} \cos (N-1)\pi\frac{z}{d},$$

$$E_x = -2\text{j}\sqrt{\frac{W}{\varepsilon_0 Sd}} \sin (N-1)\pi\frac{z}{d},$$

or, in real notation,

$$\mathcal{H}_y = 2\sqrt{\frac{W}{\mu_0 Sd}} \cos (N-1)\pi\frac{z}{d} \cos \omega t$$

$$\mathcal{E}_x = 2\sqrt{\frac{W}{\varepsilon_0 Sd}} \sin (N-1)\pi\frac{z}{d} \sin \omega t.$$

8. When the metal is no longer a perfect conductor, Maxwell's equations must be written with

$$\operatorname{curl} \boldsymbol{E} = -j\omega\mu\boldsymbol{H}$$
$$\operatorname{curl} \boldsymbol{H} = \sigma\boldsymbol{E} + j\omega\varepsilon\boldsymbol{E}.$$

With $\sigma \gg j\omega\varepsilon$, the equation of propagation inside the metal is approximately

$$\frac{\partial^2 H}{\partial z^2} = j\omega\mu\sigma H = \gamma^2 H,$$

γ being the propagation constant in the metal. This gives

$$\gamma = \frac{1+j}{\sqrt{2}}\sqrt{\omega\mu\sigma},$$

or

$$\gamma = \frac{1+j}{\delta},$$

where

$$\delta = \sqrt{\frac{2}{\omega\mu\sigma}} \quad \text{is the skin depth.}$$

The magnetic field inside the metal is thus of the form

$$H = c\,e^{-\gamma z} + d\,e^{+\gamma z}.$$

In the plane P_1 the wave is propagated in the negative z-direction and in the plane P_2 in the positive z-direction. Thus

$$H = d\,e^{+\gamma z} \quad (P_1), \qquad H = c\,e^{-\gamma z} \quad (P_2).$$

At the surface between any two media the tangential component of H is continuous, so that

$$H_{t\,(\text{metal})} = H_{t\,(\text{air})} \quad \text{for both} \quad z = 0 \quad \text{and} \quad z = d.$$

If we take the fields between the two planes to be the same as in the case of the perfectly conducting plates we have

(a)
$$d = 2\sqrt{\frac{W}{\mu_0 S d}},$$

(b)
$$c = \pm 2\sqrt{\frac{W}{\mu_0 S d}}\,e^{\gamma d}$$

with
+ for $(N-1)$ even,
− for $(N-1)$ odd.

In the plate P_1

$$H = H_y = 2\sqrt{\frac{W}{\mu_0 Sd}} \exp\left(\frac{1+j}{\delta}z\right)$$

while in the plate P_2

$$H = H_y = \pm 2\sqrt{\frac{W}{\mu_0 Sd}} \exp\left(-\frac{1+j}{\delta}(z-d)\right).$$

Also, in the metal,

$$i = \sigma E,$$

with σ the conductivity of the metal and

$$E = E_x = -\frac{1}{\sigma}\frac{\partial H_y}{\partial z}.$$

In the plate P_1

$$i_x = -\frac{\partial H_y}{\partial z} = -2\sqrt{\frac{W}{\mu_0 Sd}}\,\frac{1+j}{\delta} \exp\left(\frac{1+j}{\delta}z\right)$$

while in the plate P_2

$$i_x = -\frac{\partial H_y}{\partial z} = \pm 2\sqrt{\frac{W}{\mu_0 Sd}}\,\frac{1+j}{\delta} \exp\left(-\frac{1+j}{\delta}(z-d)\right).$$

The power dissipated is given by the real part of the complex Poynting vector across the area S. Thus, for P_1, this power is

$$P = -\text{Re}\,\frac{1}{2}\int_S (\boldsymbol{E} \wedge \boldsymbol{H}^*)\, \mathrm{d}\boldsymbol{S} = -\text{Re}\,\frac{1}{2}\int_S E_x H_y^*\, \mathrm{d}S,$$

$$\frac{1}{2}\int_S E_x H_y^*\, \mathrm{d}S = \frac{1}{2}\int_S -\frac{2}{\sigma}\sqrt{\frac{W}{\mu_0 Sd}}\,\frac{1+j}{\delta}\cdot 2\sqrt{\frac{W}{\mu_0 Sd}}\, \mathrm{d}S;$$

from which

$$P = \frac{1}{2}\,\frac{4W}{\mu_0 d}\,\frac{1}{\sigma\delta}.$$

The power dissipated in P_2 is exactly the same as for P and so

$$P_{\text{total}} = \frac{4W}{\mu_0 d}\,\frac{1}{\sigma\delta}.$$

9. The quality factor of the resonator can be deduced as

$$Q_0 = \omega_0 \frac{W}{P} = \omega_0 \frac{W}{4W}\mu_0 d\sigma\delta,$$

$$Q_0 = \frac{\omega_0 \mu_0 d\sigma\delta}{4} = \frac{d}{2\delta}.$$

The number of nodes may be varied in two different ways.

(a) The frequency is maintained constant while d is varied. Then

$$d = \frac{p\pi}{\omega_0\sqrt{\varepsilon_0\mu_0}} = \frac{(N-1)\pi}{\omega_0\sqrt{\varepsilon_0\mu_0}}$$

from which

$$Q_0 = \frac{(N-1)\pi\sigma\delta}{4}\sqrt{\frac{\mu_0}{\varepsilon_0}}$$

$$= (N-1)\frac{\pi}{4}\frac{\eta_0}{R_S},$$

$R_s = 1/\sigma\delta$ being the surface resistance,
$\eta_0 = \sqrt{\mu_0/\varepsilon_0}$ being the wave impedance in free space.

As N is increased the stored energy increases while the losses remain constant. Thus Q_0 increases with W and $W \propto d$ (Fig. 4).

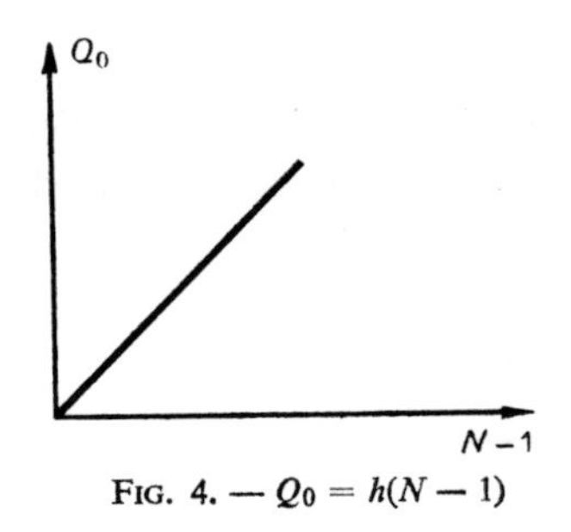

FIG. 4. — $Q_0 = h(N-1)$

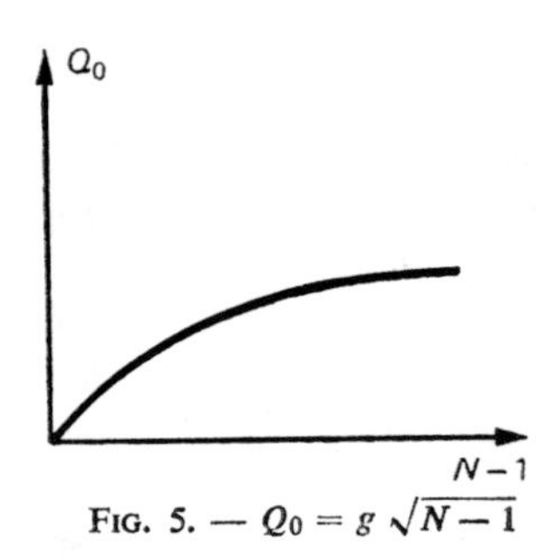

FIG. 5. — $Q_0 = g\sqrt{N-1}$

(b) The separation d is maintained constant while the frequency is varied. (The skin depth will then vary also.) Thus

$$Q_0 = \frac{d}{2\delta} = \frac{d}{2\sqrt{2}}\sqrt{\mu_0\sigma}\sqrt{\omega_0},$$

with

$$\omega_0 = \frac{\pi}{d\sqrt{\varepsilon_0\mu_0}}(N-1).$$

Thus, with d fixed, the number of nodes increases directly as the frequency while the losses increase as $f^{+1/2}$. The Q-factor increases as $f^{+1/2}$ as shown in Fig. 5.

10. Numerical example:

$$f = 10^{10}\text{ Hz},$$

$$R_S = \frac{1}{\sigma\delta} = \sqrt{\frac{\mu\omega}{2\sigma}}.$$

As $\mu = \mu_0$

$$R_s = 2{\cdot}61 \times 10^{-2}\ \Omega$$
$$\eta_0 = 377\ \Omega$$
$$Q_0 = (N-1) \times 11\,300, \quad \text{with} \quad N > 2.$$

III

11. The equivalent circuit of the system is shown in Fig. 6.

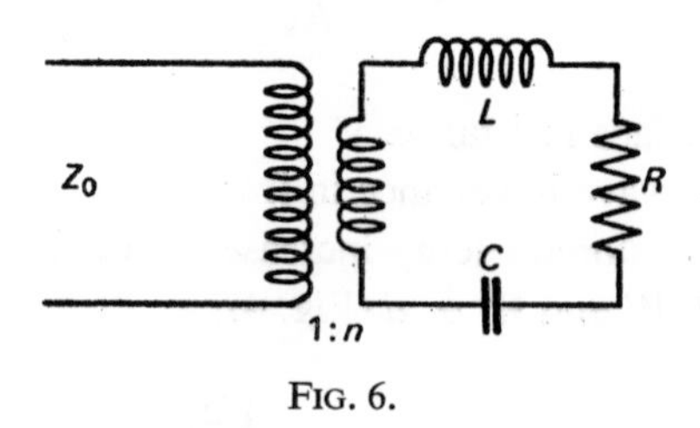

FIG. 6.

The impedance due to the source is Z_0. The cavity coupled to the guide can thus be represented by the circuit of Fig. 7 and the Q-factor of the loaded cavity is

$$Q_L = \frac{L\omega_0}{R+n^2Z_0} = \frac{\sqrt{L/C}}{R+n^2Z_0}.$$

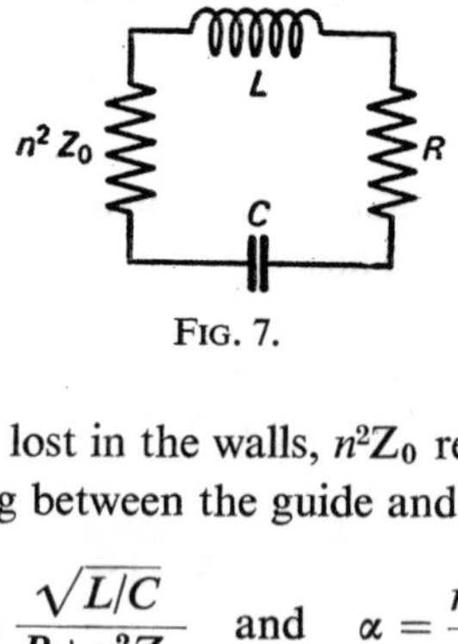

FIG. 7.

Here R represents the power lost in the walls, n^2Z_0 represents the energy radiated by the holes (i.e. the coupling between the guide and resonator).

$$Q_L = \frac{\sqrt{L/C}}{R+n^2Z_0} \quad \text{and} \quad \alpha = \frac{n^2Z_0}{R}.$$

Since $Q_0 = \omega_0 L/R$ we have

$$\frac{1}{Q_L} = \frac{1}{Q_0} + \frac{1}{Q_E},$$

with the external Q given by

$$\frac{1}{Q_E} = \frac{n^2Z_0}{\omega_0 L} = \frac{n^2Z_0}{R}\,\frac{R}{\omega_0 L} = \frac{\alpha}{Q_0}.$$

Finally

$$Q_L = \frac{Q_0}{1+\alpha}.$$

Now $\alpha = n^2 Z_0/R$ gives

$$R = \frac{n^2 Z_0}{\alpha}$$

and $Q_0 = \omega_0 L/R$ gives

$$L = \frac{Q_0 R}{\omega_0} = \frac{Q_0}{\omega_0}\frac{n^2 Z_0}{\alpha},$$

$$C = \frac{1}{L\omega_0^2} = \frac{\alpha}{Q_0 n^2 Z_0 \omega_0}.$$

12. The impedance for a series resonant circuit at a frequency different from resonance is

$$Z = R + \mathrm{j}\left(\omega L - \frac{1}{\omega C}\right) = R\left(1 + \mathrm{j}Q_0\left(\frac{\omega}{\omega_0} - \frac{\omega_0}{\omega}\right)\right).$$

The impedance transferred to the input plane of the transformer is

$$Z_A = \frac{Z}{n^2} = \frac{R}{n^2}\left[1 + \mathrm{j}Q_0\left(\frac{\omega}{\omega_0} - \frac{\omega_0}{\omega}\right)\right].$$

In reduced impedances

$$z_A = \frac{Z_A}{Z_0} = \frac{R}{n^2 Z_0}\left[1 + \mathrm{j}Q_0\left(\frac{\omega}{\omega_0} - \frac{\omega_0}{\omega}\right)\right],$$

where

$$\frac{R}{n^2 Z_0} = \frac{1}{\alpha};$$

from which

$$\Gamma = \frac{z_A - 1}{z_A + 1} = \frac{\dfrac{1}{\alpha}\left[1 + \mathrm{j}Q_0\left(\dfrac{\omega}{\omega_0} - \dfrac{\omega_0}{\omega}\right)\right] - 1}{\dfrac{1}{\alpha}\left[1 + \mathrm{j}Q_0\left(\dfrac{\omega}{\omega_0} - \dfrac{\omega_0}{\omega}\right)\right] + 1},$$

or

$$\Gamma = \frac{1 - \alpha + \mathrm{j}Q_0\left(\dfrac{\omega}{\omega_0} - \dfrac{\omega_0}{\omega}\right)}{1 + \alpha + \mathrm{j}Q_0\left(\dfrac{\omega}{\omega_0} - \dfrac{\omega_0}{\omega}\right)}.$$

The standing wave ratio is

$$S = \frac{1 + |\Gamma|}{1 - |\Gamma|}$$

and if we put

$$u = Q_0\left(\frac{\omega}{\omega_0} - \frac{\omega_0}{\omega}\right)$$

so that

$$|\Gamma| = \left[\frac{(1-\alpha)^2+u^2}{(1+\alpha)^2+u^2}\right]^{1/2}$$

we obtain

$$S = \frac{\sqrt{(1+\alpha)^2+u^2}+\sqrt{(1-\alpha)^2+u^2}}{\sqrt{(1+\alpha)^2+u^2}+\sqrt{(1-\alpha)^2+u^2}}.$$

Also

$$\omega_0 = (N-1)\frac{\pi}{d\sqrt{\varepsilon_0\mu_0}},$$

from which

$$u = Q_0\left[\frac{\omega d\sqrt{\varepsilon_0\mu_0}}{(N-1)\pi} - \frac{(N-1)\pi}{\omega d\sqrt{\varepsilon_0\mu_0}}\right].$$

13. Since α represents the ratio between the power dissipated by the radiation from the holes and that dissipated in the walls and these two powers vary in the same manner when d is changed, it may be considered that α is a constant. Then

$$z_A = \frac{1}{\alpha}\left[1 + jQ_0\left(\frac{\omega d\sqrt{\varepsilon_0\mu_0}}{(N-1)\pi} - \frac{(N-1)\pi}{d\sqrt{\varepsilon_0\mu_0}\,\omega}\right)\right].$$

When d changes, the real part of z_A remains constant and the locus of z_A on the Smith chart is the circle $1/\alpha$ = constant.

14. For $N = 2$

$$d = d_0, \quad \omega = \frac{\pi}{d_0\sqrt{\varepsilon_0\mu_0}},$$

from which

$$z_A = \frac{1}{\alpha}\left[1 + jQ_0\left(\frac{d}{d_0} - \frac{d_0}{d}\right)\right].$$

To obtain Q_L and Q_0, it is possible to trace the locus of $z_A(d)$ on the Smith chart. For $d = d_0$, $z_A = 1/\alpha$, from which α can be found. For $d = d_1$ such that $Q_0(d_1/d_0 - d_0/d_1) = 1$ we have

$$z_A = \frac{1+j}{\alpha}.$$

We look for the intersection of the circle $1/\alpha$ = constant and the circle $X = 1$. From this Q_0 and Q_L can be deduced. (The Q-factor can be found from the curve of $S(d)$ on the chart as in Problem 24.)

The stored energy in the resonator will be a maximum for $d = d_0$ and for $\alpha = 1$ since the condition $z_A = 1$ is equivalent to the absence of reflected power.

IV

15. In vacuum

$$E_x = E_0 \, e^{-j\beta_0 z} + BE_0 \, e^{+j\beta_0 z}$$

$$H_y = -\frac{1}{j\omega\mu_0} \frac{\partial E_x}{\partial z} = \sqrt{\frac{\varepsilon_0}{\mu_0}} \, [E_0 \, e^{-j\beta_0 z} - BE_0 \, e^{j\beta_0 z}].$$

In the metal, there will be propagation along the z-axis for $z > 0$ for which

$$H_y = H \exp\left(-\frac{1+j}{\delta} z\right),$$

$$E_x = -\frac{1}{\sigma} \frac{\partial Hy}{\partial z} = \frac{1+j}{\sigma\delta} H \exp\left(-\frac{1+j}{\delta} z\right).$$

The boundary conditions at $z = 0$ will require continuity of the tangential components of the electromagnetic field. Thus

$$E_0 + BE_0 = \frac{1+j}{\sigma\delta} H,$$

$$\frac{\varepsilon_0}{\mu_0} (E_0 - BE_0) = H.$$

Eliminating E_0 and H gives

$$\sqrt{\frac{\mu_0}{\varepsilon_0}} \frac{1+B}{1-B} = \frac{1+j}{\sigma\delta},$$

so that

$$\frac{1+B}{1-B} = 1(+j)K;$$

from which

$$B = \frac{(1+j)K-1}{(1+j)K+1} = \frac{K-1+jK}{K+1+jK}.$$

Now $K = R_s/\eta_0 = (2{\cdot}61 \times 10^{-2})/377 \simeq 0{\cdot}7 \times 10^{-4}$ so that

$$|B|^2 = \frac{(K-1)^2 + K^2}{(K+1)^2 + K^2} \simeq \frac{1-2K}{1+2K} \simeq (1-4K),$$

$$|B| = 1-2K.$$

As

$$B = \frac{K-1+jK}{K+1+jK} = -\frac{(1-K)(1+j \tan \phi_1)}{(1+K)(1+j \tan \phi_2)},$$

with

$$\tan\phi_1 = \frac{-K}{1-K} \simeq -K \quad \text{and} \quad \tan\phi_2 = \frac{K}{1+K} \simeq K,$$

then

$$B = -(1-2K)\,\mathrm{e}^{-2jK}.$$

Substituting for H gives

$$H = \sqrt{\frac{\varepsilon_0}{\mu_0}}\,E_0[1-(1-2K)\,\mathrm{e}^{-2jK}].$$

16. The stored energy is proportional to $|E_0|^2$. After each reflection $|E_0|^2$ is reduced by a fraction $|B|^2$ so that, after M reflections,

$$W = W_0(1-4K)^M.$$

These M reflections will occur in a time τ where

$$\tau = \frac{Md}{c_0}, \qquad c_0 = \frac{1}{\sqrt{\varepsilon_0\mu_0}},$$

so we can write

$$W = W_0(1-4K)^{c\tau/d}.$$

The energy is reduced by a factor e in the time τ_0 where

$$\frac{W}{W_0} = \frac{1}{e} = (1-4K)^{c\tau_0/d},$$

so that

$$-1 = \frac{c\tau_0}{d}\ln(1-4)K \simeq -\frac{4K\tau_0 c}{d},$$

from which

$$\tau_0 = \frac{d}{4Kc}.$$

For an oscillatory circuit, the rate of decrease of the energy is as $\mathrm{e}^{-(R/L)t}$ with

$$\frac{R}{L} = \frac{\omega_0}{Q_0}.$$

The energy will be reduced by a factor e in time τ_0 with

$$\frac{\omega_0\tau_0}{Q_0} = 1,$$

from which

$$Q_0 = \omega_0\tau_0 = \frac{\omega_0 d}{4Kc} = \frac{(N-1)\pi}{4K},$$

which is the expression found in question 9.

Problem No. 21

Q-FACTOR OF A RECTANGULAR CAVITY

We consider a rectangular waveguide, such as that shown in Fig. 1, with a wave propagating in the TE_{01} mode.

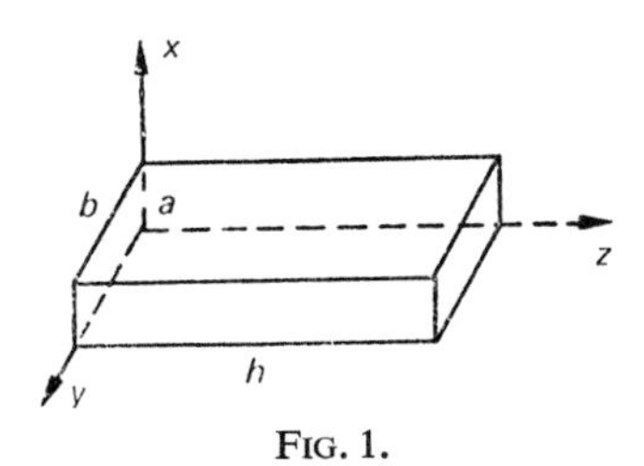

FIG. 1.

1. Write down the complex field components of the electromagnetic field in the guide.

2. Consider that a cavity is formed from this waveguide by placing two conductors at $z = 0$ and $z = h$. Neglecting the coupling hole, find the field components inside the cavity for the TE_{01p} mode. What are the resonant frequencies of this cavity?

3. Calculate the quality factor Q_{01p} for the empty cavity.

4. In practice only the conductor placed at $z = 0$ is fixed while the other at $z = h$ can be moved along the z-axis.

Calculate the different lengths h_p for which there will be resonance.

5. State the law which governs the variation of the Q-factor with h_p. Sketch the curve for this variation for

$$f = 10\,000 \text{ MHz}, \quad a = 1{\cdot}016 \text{ cm}, \quad b = 2{\cdot}286 \text{ cm}$$

for a cavity of copper, of conductivity $5{\cdot}8 \times 10^7$ S.I. and permeability $\mu = \mu_0$.

6. If the length h_p is made equal to b what must be the source frequency? What will be the form factor of the cavity?

7. With $h = h_1 = b$, the source frequency is varied. What are the frequencies at which the cavity will resonate? To what modes do these frequencies correspond? Can they be observed?

What is the law for the variation of the Q-factor with frequency?

Solution. 1. For a wave propagating in the z-direction towards $z > 0$

$$\left.\begin{aligned} H_z &= H_0 \cos \frac{\pi y}{b}\, e^{-j\beta_g z} \\ E_x &= \frac{j\omega\mu_0}{\beta_c} H_0 \sin \frac{\pi y}{b}\, e^{-j\beta_g z} \\ H_y &= \frac{j\beta_g}{\beta_c} H_0 \sin \frac{\pi y}{b}\, e^{-j\beta_g z} \end{aligned}\right\}$$

where

$$\beta_c = \frac{2\pi}{\lambda_c}, \qquad \beta_g = \frac{2\pi}{\lambda_g}.$$

For the TE_{01} mode $\lambda_c = 2b$ and

$$\beta_c^2 + \beta_g^2 = \beta_0^2 = \omega^2 \varepsilon_0 \mu_0 .$$

2. The total field in the cavity is the sum of an incident field (indicated by $+$) propagating in the $+z$-direction and a reflected field (indicated by $-$) propagating in the $-z$-direction. Then

$$H_z^+ = H_0 \cos \frac{\pi y}{b}\, e^{-j\beta_g z},$$

$$E_x^+ = \frac{j\omega\mu_0}{\beta_c} H_0 \sin \frac{\pi y}{b}\, e^{-j\beta_g z},$$

$$H_y^+ = \frac{j\beta_g}{\beta_c} H_0 \sin \frac{\pi y}{b}\, e^{-j\beta_g z},$$

$$H_z^- = H_0' \cos \frac{\pi y}{b}\, e^{+j\beta_g z},$$

$$E_x^- = \frac{j\omega\mu_0}{\beta_c} H_0' \sin \frac{\pi y}{b}\, e^{+j\beta_g z},$$

$$H_y^- = -\frac{j\beta_g}{\beta_c} H_0' \sin \frac{\pi y}{b}\, e^{+j\beta_g z},$$

from which the total field for the component H_z is

$$H_z = H_z^+ + H_z^- = H_0 \cos \frac{\pi y}{b}\, e^{-j\beta_g z} + H_0' \cos \frac{\pi y}{b}\, e^{+j\beta_g z}.$$

Taking the boundary conditions:
at $z = 0$, $H_z = 0$, so that $H_0 + H_0' = 0$,

$$H_z = -2jH_0 \cos \frac{\pi y}{b} \sin \beta_g z;$$

at $z = h$, $H_z = 0$ which requires $\sin \beta_g h = 0$ or $\beta_g h = p\pi$. This gives the well-known result

$$h = p\frac{\lambda_g}{2}.$$

Hence we can write

$$H_z = -2jH_0 \cos \frac{\pi y}{b} \sin p\frac{\pi z}{h},$$

$$E_x = \frac{j\omega\mu_0}{\beta_c} H_0 \sin \frac{\pi y}{b} (e^{-j\beta_g z} - e^{+j\beta_g z})$$

$$= \frac{2\omega\mu_0}{\beta_c} H_0 \sin \frac{\pi y}{b} \sin p\frac{\pi z}{h},$$

$$H_y = j\frac{\beta_g}{\beta_c} H_0 \sin \frac{\pi y}{b} (e^{-j\beta_g z} + e^{+j\beta_g z})$$

$$= \frac{2j\beta_g}{\beta_c} H_0 \sin \frac{\pi y}{b} \cos p\frac{\pi z}{h}.$$

The conditions $\beta_g = \pi p/h$ and $\beta_g^2 + \beta_c^2 = \beta_0^2$ give

$$\left(p\frac{\pi}{h}\right)^2 + \left(\frac{\pi}{b}\right)^2 = \omega^2 \varepsilon_0 \mu_0,$$

from which

$$f_{01p} = \frac{1}{2\pi\sqrt{\varepsilon_0\mu_0}} \sqrt{\left(p\frac{\pi}{h}\right)^2 + \left(\frac{\pi}{b}\right)},$$

which can be written in the form

$$f_{01p} = \frac{1}{2b\sqrt{\varepsilon_0\mu_0}} \sqrt{1 + p^2\frac{b^2}{h^2}}.$$

3. The Q-factor is given by

$$Q = \frac{2}{\delta} \frac{\int_V |H|^2 \, dv}{\int_S |H_t|^2 \, dS},$$

where δ is the skin depth and H_t the tangential magnetic field at the walls, the integration being taken over the whole of the surface of the cavity.

Because there is equipartition of energy we can write

$$\varepsilon_0 \int_V |E|^2 \, dv = \mu_0 \int_V |H|^2 \, dv$$

where

$$|H|^2 = |H_y|^2 + |H_z|^2, \qquad |E|^2 = |E_x|^2.$$

In practice it is more convenient to calculate $\int_V |E|^2 \, dv$ to give

$$\int_V |H|^2 \, dv = \frac{\varepsilon_0}{\mu_0} \int_V |E|^2 \, dv$$

$$= \frac{\varepsilon_0}{\mu_0} \frac{4\omega^2 \mu_0^2}{\beta_c^2} H_0^2 \int_0^a dx \int_0^b \sin^2 \frac{\pi y}{b} \, dy \int_0^h \sin^2 \frac{p\pi z}{h} \, dz,$$

from which

$$\int_V |H|^2 \, dv = \left(\frac{\beta_0}{\beta_c}\right)^2 H_0^2 V,$$

with $V = abh$. Now we calculate $\int_S |H_t|^2 \, dS$.

(a) First face; $x = 0$. Here $|H_t|^2 = |H_z|^2 + |H_y|^2$ and

$$\int_S |H_z|^2 \, dS = 4H_0^2 \int_0^b \cos^2 \frac{\pi y}{b} \, dy \int_0^h \sin^2 \frac{p\pi z}{h} \, dz = H_0^2 bh,$$

$$\int_S |H_y|^2 \, dS = \frac{4\beta_g^2}{\beta_c^2} H_0^2 \int_0^h \cos^2 \frac{p\pi z}{h} \, dz = H_0^2 \left(\frac{\beta_g}{\beta_c}\right)^2 bh.$$

For the second face, $x = a$, the result is exactly the same and

$$\int_{1,2} |H_t|^2 \, dS = 2H_0^2 bh \left(1 + \frac{\beta_g^2}{\beta_c^2}\right).$$

(b) For the faces 3 and 4, $y = 0$ and $y = b$, $H_t = H_z$ and

$$\int_3 |H_z|^2 \, dS = 4H_0^2 \int_0^a dx \int_0^h \sin^2 \frac{p\pi z}{h} \, dz = 4H_0^2 \frac{ah}{2};$$

from which

$$\int_{3,4} |H_t|^2 \, dS = 4H_0^2 ah.$$

(c) For faces 5 and 6, $z = 0$ and $z = h$, $H_t = H_y$ and

$$\int_S |H_y|^2 \, dS = 4H_0^2 \left(\frac{\beta_g}{\beta_c}\right)^2 \int_0^a dx \int_0^b \sin^2 \frac{\pi y}{b} \, dy = 2H_0^2 \left(\frac{\beta_g}{\beta_c}\right)^2 ab,$$

from which

$$\int_{5,6} |H_t|^2 \, \mathrm{d}S = 4H_0^2 ab\left(\frac{\beta_g}{\beta_c}\right)^2.$$

The total integral is thus

$$\int_S |H_t|^2 \, \mathrm{d}S = 2H_0^2 bh\left(1+\frac{\beta_g^2}{\beta_c^2}\right)+4H_0^2 ah+4H_0^2 ab\left(\frac{\beta_g}{\beta_c}\right)^2.$$

Now

$$1+\left(\frac{\beta_g}{\beta_c}\right)^2 = 1+\left(\frac{p\pi/h}{\pi/b}\right)^2 = 1+p^2\frac{b^2}{h^2},$$

$$\begin{aligned}\int_S |H_t|^2 \, \mathrm{d}S &= 2H_0^2\left[bh\left(1+p^2\frac{b^2}{h^2}\right)+2ah+2p^2\frac{b^2}{h^2}\,ab\right]\\ &= 2H_0^2\left[bh+2ah+p^2\frac{b^2}{h^2}(bh+2ab)\right];\end{aligned}$$

from which

$$Q_{01p} = \frac{1}{\delta}\,\frac{\left(1+p^2\dfrac{b^2}{h^2}\right)V}{bh+2ah+p^2\dfrac{b^2}{h^2}(bh+2ab)}.$$

4. When f is constant and h varies we have the condition

$$\beta_g^2+\beta_c^2 = \beta_0^2,$$

and

$$\left(p\frac{\pi}{h}\right)^2+\left(\frac{\pi}{b}\right)^2 = \omega^2\varepsilon_0\mu_0 = 4\pi^2 f^2\varepsilon_0\mu_0,$$

so that

$$\left(\frac{p}{h}\right)^2 = 4f^2\varepsilon_0\mu_0-\frac{1}{b^2} = \text{constant},$$

from which

$$\boxed{\frac{p}{h} = \text{constant}}$$

a condition which could have been obtained from the relation

$$h = p\frac{\lambda_g}{2},$$

since λ_g is constant.

$$h = \frac{p}{4f^2\varepsilon_0\mu_0 - \dfrac{1}{b^2}}; \qquad p = 1 : h = h_1 = \frac{1}{4f^2\varepsilon_0\mu_0 - \dfrac{1}{b^2}} \tag{1}$$

with here $p > 1$ and p/h_p = constant = $1/h_1$, from which $h_p = ph_1$, where h_1 is given by (1).

There will be a discrete series of lengths h_p for which resonance can occur.

5. The Q-factor can be put in the form

$$Q_{01p} = \frac{1}{\delta} \frac{1+(p/h)^2 b^2}{bh_p + 2ah_p + (p/h_p)^2 b^2(bh_p + 2ab)} abh_p$$

with

$$\frac{p}{h_p} = \text{constant} = \frac{1}{h_1},$$

from which

$$Q_{01p} = \frac{1+(b/h_1)^2}{\delta} \frac{abh_p}{2ab(b/h_1)^2 + h_p[b+2a+b(b/h_1)^2]}.$$

We have then

$$Q_{01p} = A\frac{h_p}{B+h_p},$$

with

$$A = \frac{1+(b/h_1)^2}{b+2a+b(b/h_1)^2} \frac{ab}{\delta}$$

$$B = \frac{2ab(b/h_1)^2}{b+2a+b(b/h_1)^2}.$$

For f = 10 000 MHz, a = 1·016 cm and b = 2·286 cm

$$h_1 = \frac{1}{\sqrt{\left(\dfrac{2}{3}\right)^2 - \left(\dfrac{1}{2{\cdot}286}\right)^2}} = \frac{1}{\sqrt{0{\cdot}44 - 0{\cdot}19}} = \frac{1}{\sqrt{0{\cdot}25}} = 2 \text{ cm},$$

$$\left(\frac{b}{h_1}\right)^2 = 1{\cdot}3, \qquad ab = 2{\cdot}32 \times 10^{-4}\ \text{cm}^2,$$

$$\delta = \sqrt{\frac{2}{\omega\mu\sigma}} = \left[\frac{1}{\pi \times 10^{10} \times 4\pi \times 10^{-7} \times 5{\cdot}8 \times 10^7}\right]^{1/2} = \frac{1}{10^6\sqrt{2{\cdot}32}}$$

$$\simeq 0{\cdot}655 \times 10^{-6}\ \text{m}.$$

$$A = \frac{2{\cdot}3 \times 2{\cdot}32 \times 10^{-4}}{7{\cdot}2 \times 10^{-2} \times 0{\cdot}655 \times 10^{-6}} = 1{\cdot}13 \times 10^4,$$

$$B = \frac{6{\cdot}03\times10^{-4}}{7{\cdot}2\times10^{-2}} = 0{\cdot}83\times10^{-2},$$

$$Q_{01p} = 1{\cdot}13\times10^{4}\,\frac{h_p}{0{\cdot}83+h_p},$$

where h_p is in cm.

Now

$$h_p = ph_1 = p\times2\ \text{cm},$$

from which

$$Q_{01p} = 11\,300\,\frac{p}{0{\cdot}415+p} = Q_{01\infty}\,\frac{p}{0{\cdot}415+p}.$$

The variation of Q with h_p is shown in Fig. 2,

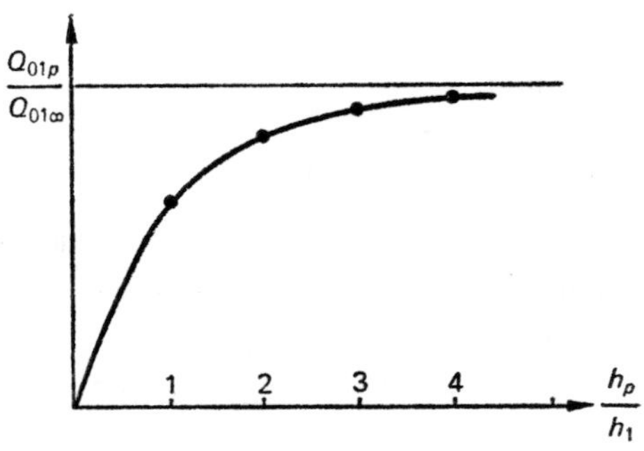

FIG. 2.

with

$$Q_{011} \simeq 8000,$$
$$Q_{012} \simeq 9350,$$
$$Q_{013} \simeq 9925,$$
$$Q_{014} \simeq 10\,240,$$
$$\frac{Q_{01\infty}-Q_{014}}{Q_{01\infty}} \simeq 10\,\%.$$

6. If we chose $h_1 = b$ then

$$\lambda_g = 2h_1 = \lambda_c = 2b$$

and

$$\frac{1}{\lambda^2} = \frac{1}{\lambda_g^2}+\frac{1}{\lambda_c^2} = \frac{2}{\lambda_g^2},$$

so that

$$\lambda = \frac{\lambda_g}{\sqrt{2}} = b\sqrt{2} = 3{\cdot}232\ \text{cm}, \qquad f = 9280\ \text{MHz}.$$

The Q-factor of the cavity then becomes

$$Q_{01p} = \frac{ab}{\delta}\,\frac{1}{a+b}\,\frac{1}{\dfrac{ab}{a+b}+h_p} = \frac{2}{\delta}\,\frac{abh_p}{2(ab+h_pa+h_pb)} = \frac{2}{\delta}\,\frac{V}{S}.$$

Since the form factor is given by

$$Q = \frac{2}{\delta}F\frac{V}{S},$$

we have $F = 1$ independent of the cavity mode.

In effect, this result is the same as for a uniform field distribution. Then we would have

$$Q_0 = \frac{2}{\delta}\,\frac{\int_V |H|^2\,dv}{\int_S |H_t|^2\,ds}$$

and, if $|H| = |H_t| = $ constant,

$$Q_0 = \frac{2}{\delta}\,\frac{V}{S}.$$

7. Here the piston is fixed at $h = h_1 = b$. As the frequency varies, λ_g varies and there will only be resonance if $b = p\lambda_g/2$. This requires

$$\frac{1}{\lambda^2} = \frac{1}{\lambda_c^2}+\frac{1}{\lambda_g^2} = \frac{1}{(2b)^2}+\frac{p^2}{(2b)^2},$$

from which

$$\lambda = \frac{2b}{\sqrt{1+p^2}}.$$

$$p = 1, \quad \lambda = b\sqrt{2} = 2{\cdot}232 \text{ cm}, \quad f = f_{011} = 9\,280 \text{ MHz};$$

$$p = 2, \quad \lambda = \frac{2b}{\sqrt{5}} = 2{\cdot}04 \text{ cm}, \quad f = f_{012} = 14\,700 \text{ MHz}$$

with only the first resonance being observable.

The Q-factor is

$$Q_{01p} = \frac{1}{\delta}\,\frac{\left(1+p^2\dfrac{b^2}{h^2}\right)V}{bh+2ah+p^2\dfrac{b^2}{h^2}(bh+2ab)},$$

which, for $b = h$, gives

$$Q_{01p} = \frac{ab^2}{\delta}\,\frac{1+p^2}{b^2+2ab+p^2(b^2+2ab)} = \frac{1}{\delta}\,\frac{ab^2}{b^2+2ab} \propto \sqrt{f},$$

since δ is proportional to $1/\sqrt{f}$.

Problem No. 22

MEASUREMENT OF DIELECTRIC CONSTANT

I

An electromagnetic wave is propagated in the TE_{22} mode along a waveguide with dimension a in the x-direction and dimension b in the y-direction with $a < b$. The guide walls are considered to be perfect conductors.

1. If the guide is filled with a dielectric of permittivity $\varepsilon = \varepsilon' \varepsilon_0$

(a) what is the cut-off wavelength λ_c?

(b) what are the expressions for the propagation constant $\beta_g = 2\pi/\lambda_g$ (λ_g being the guide wavelength), for the phase and group velocities, v_p and v_g, and for the wave impedance Z_{TE} in terms of β_0, the free space propagation constant, $\beta_c = 2\pi/\lambda_c$ and $c_0 = 1/\sqrt{\varepsilon_0 \mu_0}$?

(c) Calculate λ_c, λ_g, β_g, v_p, v_g and Z_{TE} for the case where $a = 4$ cm, $b = 5$ cm, $\varepsilon' = 2{\cdot}92$ and the frequency of the wave is 10 000 MHz.

2. In a practical case the dielectric will be lossy with a complex permittivity

$$\varepsilon^* = \varepsilon_0(\varepsilon' - j\varepsilon''),$$

where ε' and ε'' are dimensionless and the loss angle is given by $\tan \phi = \varepsilon''/\varepsilon'$.

(a) What is the new propagation constant? Show that it is complex and give the expression for β_g^* in terms of β_0, β_c, ε' and $\tan \phi$.

(b) Find the expression for $\beta_g^* = \beta_g' - j\beta_g''$ with β_g' and β_g'' real, in the case where the losses are small (i.e. $\varepsilon''/\varepsilon' \ll 1$). Hence deduce the attenuation due to the dielectric losses.

(c) Again for $\varepsilon''/\varepsilon' \ll 1$, find the wave impedance Z_{TE} and the phase velocity.

(d) As a numerical example take $\tan \phi = 3 \times 10^{-3}$ and $\varepsilon' = 2{\cdot}92$.

II

A cavity is constructed from the above guide with the aid of two plane conductors P_1 and P_2 separated by a distance d as shown in Fig. 1. The plane P_1 is shown pierced by a coupling hole s, but in this part the cavity is considered independently from such coupling.

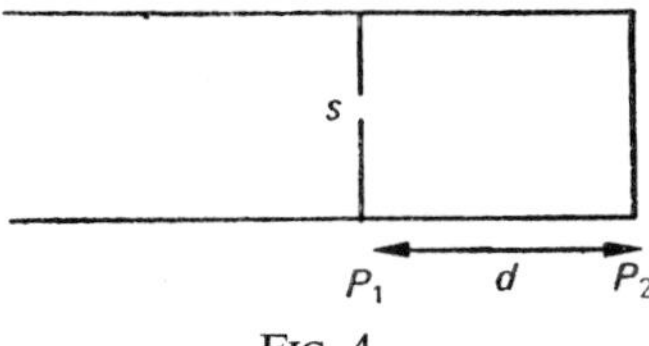

FIG. 4.

1. The cavity is air-filled ($\varepsilon = \varepsilon_0$, $\mu = \mu_0$) and the walls and the planes P_1 and P_2 are no longer considered as perfect conductors. With the conduction losses we associate a quality factor Q_0 which is assumed to be known. We also assume that the boundary conditions imposed by the walls are the same as for perfect conductors. What must be the value of λ_g for there to be resonance in the TE_{22} mode? Hence deduce the resonant frequency f_0 for $d = 8{\cdot}95$ cm.

If W_1 is the energy stored in the cavity and P_1 is the energy dissipated in the walls in unit time, what is the expression for Q_0?

2. What is the new resonant frequency f_0' if the cavity is filled with a dielectric of permittivity $\varepsilon = \varepsilon'\varepsilon_0$? What is the value for $\varepsilon' = 1{\cdot}57$?

The introduction of the dielectric, and the consequent change of resonance frequency, changes the Q-factor from Q_0 to Q_0'. We wish to find the ratio Q_0'/Q_0. If it is supposed that the magnitude of the electric field is not changed by the dielectric, deduce, with the aid of Maxwell's equations, that the magnetic field H_2 in the dielectric is related to the field H_1 of the air-filled cavity by the relation $H_2 = \sqrt{\varepsilon'}H_1$.

Express the energy stored in the dielectric, W_2, in terms of W_1, and the losses in the cavity walls, P_2, in terms of P_1. Hence deduce Q_0' in terms of Q_0.

3. The cavity is now considered to be filled with a lossy dielectric

$$\varepsilon = \varepsilon_0(\varepsilon' - j\varepsilon'').$$

Show that the dielectric losses can be expressed, in effect, by a Q-factor Q_d which is simply related to the loss angle ϕ. Hence give the expression for Q_T, the total Q-factor, in terms of Q_0, ε' and ϕ.

4. Express ε' and ϕ in terms of Q_0, Q_T, f_0 and f_0'. Hence deduce a method for the measurement of a complex permittivity.

III

We now study the equivalent circuit of the cavity.

1. The air-filled cavity can be represented by the equivalent circuit of Fig. 2 with

$$\begin{cases} Q_0 = \omega_0 C/G, & \omega_0 = 2\pi f_0; \\ \omega_0^2 LC = 1. \end{cases}$$

The introduction of the lossy dielectric

$$\varepsilon = \varepsilon_0(\varepsilon' - j\varepsilon'')$$

modifies this circuit to the form shown in Fig. 3.

(a) Calculate C' and $R_p = 1/G_p$, which show the effect of ε on the capacitance C, in terms of C, ε', ε'' and ω.

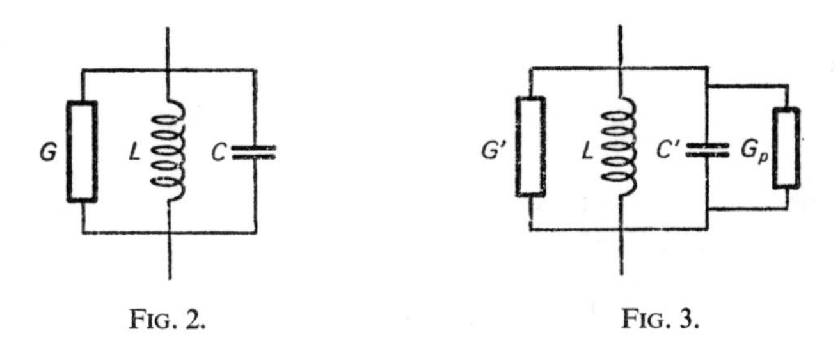

FIG. 2. FIG. 3.

(b) Show that the resonant frequency is changed.

(c) The variation of G is related to the variation of Q_0, which now becomes Q_0'. Taking the results of part II, express G' in terms of G and ε'.

2. If we allow ε'' to become negative, what effect will there be on G_p? What is the physical significance of such an effect? What type of phenomena and what conditions could produce such an effect? (Again ignore the cavity coupling.)

(Partial Examination, Paris, 1965.)

Solution.

I

1. (a) It is known that the cut-off wavelength is given completely by the limiting conditions

$$\beta_c^2 = \left(\frac{2\pi}{\lambda_c}\right)^2 = \left(m\frac{\pi}{a}\right)^2 + \left(n\frac{\pi}{b}\right)^2.$$

For the TE_{22} mode

$$\left(\frac{2\pi}{\lambda_c}\right)^2 = \left(\frac{2\pi}{a}\right)^2 + \left(\frac{2\pi}{b}\right)^2,$$

so that

$$\lambda_c = \frac{ab}{\sqrt{a^2+b^2}}.$$

(b)

$$\beta_g^2 + \beta_c^2 = \omega^2\varepsilon\mu = \omega^2\varepsilon_0\mu_0\varepsilon' = \beta_0^2\varepsilon',$$

from which

$$\beta_g^2 = \beta_0^2\varepsilon' - \beta_0^2,$$

which gives

$$\beta_g = \beta_0\sqrt{\varepsilon' - (\beta_c/\beta_0)^2}.$$

The phase velocity is related to β_g by

$$\beta_g = \omega/v_p,$$

so that

$$v_p = \frac{\omega}{\beta_0\sqrt{\varepsilon' - (\beta_c/\beta_0)^2}} = \frac{c_0}{\sqrt{\varepsilon' - (\beta_c/\beta_0)^2}},$$

where c_0 is the velocity of light in air.

The group velocity is given by

$$v_g = \frac{d\omega}{d\beta_g},$$

with

$$\beta_g^2 + \beta_c^2 = \omega^2 \varepsilon \mu,$$

so that, on differentiating,

$$2\beta_g \, d\beta_g = 2\omega \, d\omega \frac{\varepsilon'}{c_0^2};$$

from which

$$\frac{d\omega}{d\beta_g} = \frac{c_0^2}{\varepsilon'}\frac{\beta_g}{\omega} = \frac{c_0}{\varepsilon'}\sqrt{\varepsilon' - \left(\frac{\beta_c}{\beta_0}\right)^2}.$$

We see that the relation

$$v_g v_p = v^2 = \frac{c_0^2}{\varepsilon'}$$

is verified. Also

$$Z_{TE} = \frac{\omega\mu}{\beta_g} = \frac{\omega\mu_0}{\beta_0\sqrt{\varepsilon' - (\beta_c/\beta_0)^2}} = \frac{\eta_0}{\sqrt{\varepsilon' - (\beta_c/\beta_0)^2}},$$

where

$$\eta_0 = \sqrt{\mu_0/\varepsilon_0} = 377\ \Omega.$$

(c) Numerical example.

$$\lambda_g = \frac{\lambda_0}{\sqrt{\varepsilon' - (\lambda_0/\lambda_c)^2}}$$

$$\left.\begin{array}{l} f_0 = 10\,000 \text{ MHz}, \quad \text{from which} \quad \lambda_0 = 3 \text{ cm} \\ a = 4 \text{ cm}, \quad b = 5 \text{ cm}, \quad \lambda_c = 3{\cdot}12 \text{ cm} \end{array}\right\} \quad \left(\frac{\lambda_0}{\lambda_c}\right)^2 = 0{\cdot}92.$$

Then

$$\varepsilon' - \left(\frac{\lambda_0}{\lambda_c}\right)^2 = 2,$$

and we obtain

$$\lambda_g = \frac{\lambda_0}{\sqrt{2}} = 2{\cdot}12 \text{ cm},$$

$$v_p = \frac{c_0}{\sqrt{2}} = 212\,000 \text{ km/s},$$

$$v_g = \frac{c_0\sqrt{2}}{\varepsilon'} = 145\,000 \text{ km/s}$$

$$Z_{\text{TE}} = \frac{\eta_0}{\sqrt{2}} = 266\ \Omega.$$

2. $\varepsilon^* = \varepsilon_0(\varepsilon' - \mathrm{j}\varepsilon'')$.

(a) We can write the relation

$$\beta_g^{*2} + \beta_c^2 = \omega^2\varepsilon_0\mu_0(\varepsilon' - \mathrm{j}\varepsilon'') = \beta_0^2(\varepsilon' - \mathrm{j}\varepsilon'');$$

so that

$$\beta_g^* = \beta_0\sqrt{\varepsilon'(1 - \mathrm{j}\tan\phi) - (\beta_c/\beta_0)^2}.$$

(b) If $\varepsilon''/\varepsilon' \ll 1$, $\tan\phi \simeq \phi$ and

$$\beta_g^* = \beta_0\sqrt{\varepsilon' - (\beta_c/\beta_0)^2 - \mathrm{j}\varepsilon'\phi},$$

which can be put in the form

$$\beta_g^* = \beta_0\sqrt{\varepsilon' - \left(\frac{\beta_c}{\beta_0}\right)^2}\sqrt{1 - \mathrm{j}\frac{\varepsilon'\phi}{\varepsilon' - (\beta_c/\beta_0)^2}}.$$

If we put β_g as the propagation constant for $\varepsilon'' = 0$, then

$$\beta_g^* = \beta_g\sqrt{1 - \mathrm{j}\frac{\varepsilon'\phi}{\varepsilon' - (\beta_c/\beta_0)^2}}$$

$$\simeq \beta_g\left[1 - \mathrm{j}\frac{\varepsilon'\phi}{2}\,\frac{1}{\varepsilon' - (\beta_c/\beta_0)^2}\right],$$

which, finally, becomes

$$\beta_g^* = \beta_g - \mathrm{j}\frac{\varepsilon'\phi}{2}\,\frac{\beta_0^2}{\beta_g},$$

from which

$$\beta_g' = \beta_g,$$

$$\beta_g'' = \frac{\varepsilon'\phi}{2}\,\frac{\beta_0^2}{\beta_g}.$$

Since the propagation constant is complex, the exponential propagation term is

$$\mathrm{e}^{-\mathrm{j}\beta_g^* z} = \mathrm{e}^{-\mathrm{j}\beta_g' z - \beta_g'' z}, \quad \text{from which} \quad \alpha = \beta_g''.$$

(c) Due to the definition of wave impedance we have

$$Z_{TE}^* = \frac{\omega\mu}{\beta_g^*} = \frac{\omega\mu}{\beta_g\left[1-j\frac{\varepsilon'\phi}{2}\left(\frac{\beta_0}{\beta_g}\right)^2\right]},$$

$$Z_{TE}^* \simeq \eta\left[1+j\frac{\varepsilon'\phi}{2}\left(\frac{\beta_0}{\beta_g}\right)^2\right],$$

and, as $\beta_g' = \beta_g$, the phase velocity is unchanged.

(d) Numerical example. Taking $\tan\phi = 3\times10^{-3}$ we have

$$\beta_g' = \beta_g = 2{\cdot}96 \text{ rad/cm},$$
$$\beta_g'' = 6{\cdot}5\times10^{-3} \text{ Np/cm},$$
$$Z_{TE}^* = 266(1+j\,2{\cdot}2\times10^{-3}) \simeq Z_{TE}.$$

II

1. The resonance condition is $d = p\lambda_g/2$. With $p = 2$

$$d = \lambda_g,$$

from which

$$\beta_g^2 = \left(\frac{2\pi}{d}\right)^2$$

and, with $\beta_0^2 = \beta_g^2+\beta_c^2$, we have

$$\frac{\omega^2}{c_0^2} = \left(\frac{2\pi}{d}\right)^2+\left(\frac{2\pi}{\lambda_c}\right)^2,$$

$$f_0 = c_0\sqrt{\frac{1}{d^2}+\frac{1}{\lambda_c^2}}.$$

$d = 8{\cdot}95$ cm gives $f \simeq 10\,000$ MHz,

$$Q_0 = \omega_0\frac{W_1}{P_1}.$$

2. When the guide is filled with a lossless dielectric, $\varepsilon = \varepsilon'\varepsilon_0$, the equation

$$\beta_g^2+\beta_c^2 = \beta_0^2$$

becomes

$$\beta_g^2+\beta_c^2 = \beta_0'^2\varepsilon'.$$

β_g and β_c are fixed by the geometry of the guide and cavity so that

$$\beta_0' = \frac{\beta_0}{\sqrt{\varepsilon'}}, \quad \text{from which} \quad f_0' = \frac{f_0}{\sqrt{\varepsilon'}} \simeq 8000 \text{ MHz}.$$

There will always be equipartition of energy so that, in air,

$$\varepsilon_0 E_1^2 = \mu_0 H_1^2$$

while, in the dielectric,

$$\varepsilon_0 \varepsilon' E_2^2 = \mu_0 H_2^2.$$

With E unperturbed, $E_1 = E_2$, so that

$$\varepsilon' = \left(\frac{H_2}{H_1}\right)^2, \qquad H_2 = \sqrt{\varepsilon'} H_1.$$

We have then

$$\left.\begin{aligned} W_1 &= \frac{\varepsilon_0}{2} \int_V E_1 E_1^* \, dv \\ W_2 &= \frac{\varepsilon_0 \varepsilon'}{2} \int_V E_2 E_2^* \, dv \end{aligned}\right\} W_2 = \varepsilon' W_1.$$

The losses in the cavity walls are given, in the two cases, by

$$P_1 = \frac{1}{2} \frac{1}{\sigma \delta_1} \int_S |H_{T_1}|^2 \, dS,$$

$$P_2 = \frac{1}{2} \frac{1}{\sigma \delta_2} \int_S |H_{T_2}|^2 \, dS,$$

where σ is the conductivity of the metal, δ is its skin depth and H_T is the tangential magnetic field. Then

$$\frac{P_2}{P_1} = \frac{\delta_1}{\delta_2} \frac{\int_S |H_{T_2}|^2 \, dS}{\int_S |H_{T_1}|^2 \, dS} = \frac{\delta_1}{\delta_2} \frac{\varepsilon' \int_S |H_{T_1}|^2 \, dS}{\int_S |H_{T_1}|^2 \, dS}.$$

The skin depth is inversely proportional to the square root of the frequency and so

$$\frac{P_2}{P_1} = \frac{\delta_1}{\delta_2} \varepsilon' = \sqrt{\frac{f_0'}{f_0}} \varepsilon' = \varepsilon'^{+3/4}.$$

Under these conditions

$$\frac{Q_0'}{Q_0} = \frac{\omega_0' W_2/P_2}{\omega_0 W_1/P_1} = \frac{\omega_0'}{\omega_0} \varepsilon' \varepsilon'^{-3/4} = \varepsilon'^{-1/4},$$

$$Q_0' = Q_0 \varepsilon'^{-1/4}.$$

3.

$$Q_d = \omega_0' \frac{\frac{1}{2} \int_V \varepsilon_0 \varepsilon' E E^* \, dv}{\frac{1}{2} \int_V \sigma E E^* \, dv} = \frac{\omega_0' \varepsilon' \varepsilon_0}{\sigma}.$$

Putting $\varepsilon_0(\varepsilon' - j\varepsilon'')$ in Maxwell's equations requires

$$j\omega_0\varepsilon \rightarrow j\omega_0'\varepsilon_0\varepsilon' + \omega_0'\varepsilon''\varepsilon_0$$

which is equivalent to considering that the lossy dielectric has a conductivity

$$\sigma = \omega_0'\varepsilon''\varepsilon_0,$$

from which

$$Q_d = \frac{\varepsilon'}{\varepsilon''} = \cot\phi$$

$$\frac{1}{Q_T} = \frac{1}{Q_0'} + \frac{1}{Q_d} = \frac{\varepsilon'^{1/4}}{Q_0} + \tan\phi.$$

4.

$$\varepsilon' = \left(\frac{f_0}{f_0'}\right)^2, \qquad \tan\phi = \frac{1}{Q_T} - \frac{1}{Q_0}\sqrt{\frac{f_0}{f_0'}}.$$

The complex permittivity may be measured in the following manner:

(a) f_0 and Q_0 are measured for the empty cavity;

(b) the cavity is filled with the dielectric and f_0' and Q_T are measured. Hence ε' and $\tan\phi$ can be found.

III

The equivalent circuit

1. (a)

$$Q_0 = \frac{\omega_0 C}{G}, \qquad LC\omega_0^2 = 1.$$

If the capacity C is filled with the dielectric $\varepsilon = \varepsilon_0(\varepsilon' - j\varepsilon'')$, its capacitance becomes $C(\varepsilon' - j\varepsilon'')$ and its admittance $j\omega C$ becomes

$$jC\omega\varepsilon' + \omega C\varepsilon'',$$

from which

$$C' = \varepsilon' C,$$
$$G_p = \omega\varepsilon'' C.$$

(b) The resonant frequency is given by

$$LC'\omega_0'^2 = 1, \qquad f_0' = \frac{1}{2\pi\sqrt{LC'}};$$

$$\frac{f_0'}{f_0} = \sqrt{\frac{C}{C'}} = \varepsilon'^{-1/2}.$$

(c)

$$\frac{1}{Q_T} = \frac{G'+G_p}{\omega_0'\varepsilon' C} = \frac{G'}{\omega_0'\varepsilon' C} + \frac{\varepsilon''}{\varepsilon'}.$$

Now

$$\frac{1}{Q_T} = \frac{\varepsilon'^{1/4}}{Q_0} + \frac{\varepsilon''}{\varepsilon'},$$

so that

$$\frac{\varepsilon'^{1/4}G}{\omega_0 C} = \frac{G'}{\omega_0'\varepsilon' C};$$

from which

$$G' = \frac{\omega_0'}{\omega_0}\varepsilon' . \varepsilon'^{1/4}G = G\varepsilon'^{3/4}.$$

2. If $\varepsilon'' < 0$, $G_p = \omega\varepsilon'' C < 0$, i.e. G_p *supplies* energy to the system. In particular, if $G_p = G'$, there can be oscillation at the frequency f_0'. (Such energy can be supplied if the dielectric is "pumped" in some way.)

Problem No. 23

RESONANT CAVITY

We consider a rectangular waveguide with dimensions 1·016 cm×2·286 cm.

1. What frequencies can be transmitted by this guide if there is to be only a single mode excited? What is this mode? Give the complex expressions for the field components for this mode.

2. Given these conditions, state briefly how the properties of a waveguide can be represented with the aid of a line of characteristic impedance Z_0.

3. When the metal of the guide walls is not a perfect conductor, the equivalent line will be lossy and the propagation constant will be $\gamma = \alpha + j\beta_g$ where $\beta_g = 2\pi/\lambda_g$, λ_g being the guide wavelength. If the field components of an electromagnetic wave at 9000 MHz are attenuated at 0·3 dB/m, calculate α and β_g.

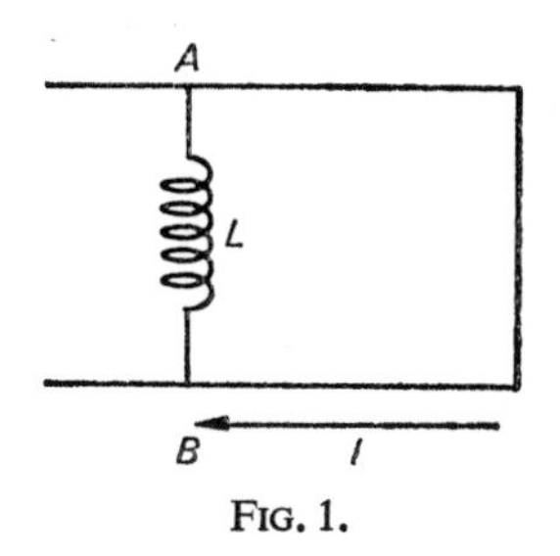

Fig. 1.

4. The waveguide is terminated in a perfect plane conductor (a perfect short-circuit). At a distance $l = 2$ cm from the short-circuit an iris is placed in the form of a perfect plane conductor with a small hole. This iris can be represented by an inductance L (see Fig. 1) placed in parallel with the equivalent line and independent of frequency.

Show that the reduced impedance transferred to the plane AB of the iris can be written as

$$z(l) = \frac{\alpha l + \mathrm{j} \tan \beta_g l}{1 + \mathrm{j}\alpha l \tan \beta_g l}$$

providing that account is taken of the orders of magnitude of the parameters.

5. Show that there exists one frequency and only one in the range 8 200 to 12 400 MHz for which the impedance $z(l)$ is real. Calculate λ_{g0}, λ_0, f_0 and β_{g0}, where λ_0 is the free space wavelength for the frequency f_0 and λ_{g0} the corresponding guide wavelength.

6. Show that, for a frequency close to f_0, the length of line l may be represented, in the plane AB, by a series resonant circuit L_0, C_0, R_0 as in Fig. 2.

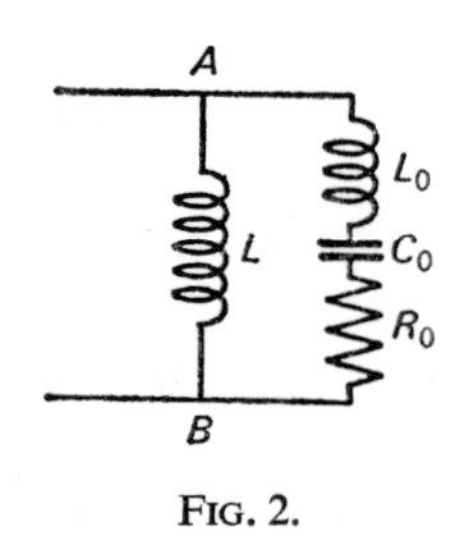

FIG. 2.

Determine the quality factor Q_0 of this circuit, and also L_0, C_0 and R_0, in terms of Z_0, αl, ω_0 and λ_{g0}/λ_0.

Numerical example: Calculate Q_0 given that, for $f = f_0$, $\alpha = 0{\cdot}3$ dB/m.

7. The reactance of L is such that $(\omega L/Z_0)^2 \ll 1$ in the range of frequencies considered. The guide is excited by a perfectly decoupled source. Show that the system comprising the internal impedance of the source, the guide, the iris and the short-circuited guide can be considered, in the plane AB, as a series resonant circuit. Calculate the new components L_s, C_s, R_s and the new resonant frequency f_1.

Compare f_1 with f_0 and show that, taking into account the magnitudes of parameters, the Q-factor can be expressed simply in terms of αl, λ_{g0}/λ_0 and $\omega L/Z_0$. Hence show that $1/Q = A + 1/Q_0$ and state the significance of the term A.

As a numerical example take $\omega L/Z_0 = 2 \times 10^{-2}$.

8. Show that, at the frequency f_1, the reduced admittance of the resonant circuit and of the iris may be written in the form

$$y = \alpha l \left(\frac{Z_0}{\omega L} \right)^2.$$

(To obtain this result, it is necessary to show that $(\alpha l / \tan \beta_g l)^2 \ll 1$ at the frequency f_1 and to deduce the reflection coefficient Γ_A at the plane AB.)

In view of the value of α show that it is permissible to measure this reflection coefficient along the line of the guide.

9. Compare Q and Q_0 if $\Gamma_A = 0$ for $f = f_1$. What conclusion do you draw?

(Paris-Orsay, June 1966)

Solution. 1. For a wave to be transmitted along a guide, the free space wavelength λ must satisfy the condition $\lambda < \lambda_c$ where λ_c, the cut-off wavelength, depends on the waveguide mode.

For a TE_{mn} or TM_{mn} mode

$$\lambda_c = \frac{2}{\sqrt{\left(\frac{m}{a}\right)^2 + \left(\frac{n}{b}\right)^2}},$$

a and b being the guide dimensions.

Thus, for the mode TE_{01}, $\lambda_c = 2b$;
for the mode TE_{10}, $\lambda_c = 2a$;
for the mode TE_{02}, $\lambda_c = b$.

With the values given as $a = 1{\cdot}016$ cm, $b = 2{\cdot}286$ cm, the distribution of cut-off frequencies is as in Fig. 3.

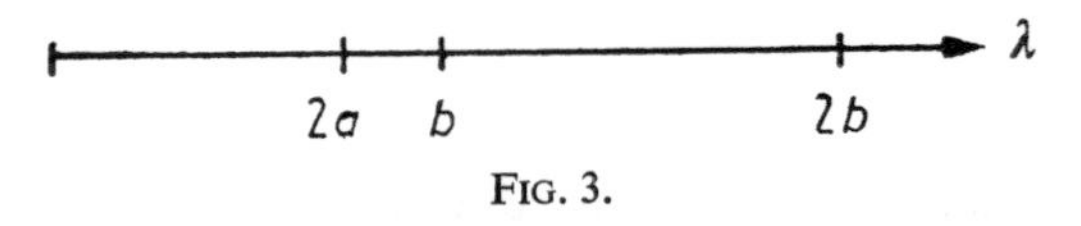

FIG. 3.

In order that, with $b > 2a$, there shall be a single mode it is necessary that $b < \lambda < 2b$. The first TM mode is TM_{11} for which the cut-off wavelength is not in this range. Thus the mode which can be transmitted alone is TE_{01}—the fundamental mode—with

$$2{\cdot}286 < \lambda < 4{\cdot}572 \text{ cm},$$

so that the range of frequency is 6550 MHz $< f <$ 13 100 MHz.

In this mode the wave will propagate in the guide with field components

$$H_z = H_0 \cos \frac{\pi y}{b} e^{-j\beta_g z}, \qquad \beta_g = \frac{2\pi}{\lambda_g};$$

$$H_x = \frac{-j\beta_g}{\beta_c^2} \frac{\partial H_z}{\partial x} = 0, \qquad \beta_c = \frac{2\pi}{\lambda_c};$$

$$H_y = \frac{-j\beta_g}{\beta_c^2} \frac{\partial H_z}{\partial y} = \frac{j\beta_g H_0}{\beta_c} \sin \frac{\pi y}{b} e^{-j\beta_g z};$$

$$E_x = \frac{\omega\mu}{\beta_g} \cdot H_y = \frac{j\omega\mu}{\beta_c} H_0 \sin \frac{\pi y}{b} e^{-j\beta_g z};$$

$$E_y = 0$$

where λ_g is the guide wavelength.

2. An obstacle in a waveguide can be described in terms of a reflection coefficient, as in the case of a transmission line.

In the case where there is only a single mode propagating in the guide it is possible to define an equivalent line. In particular, for the TE_{01} mode, the impedance at a point will be given by the ratio E_x/H_y, where E_x and H_y are the respective fields which result from the superposition of incident and reflected waves in the guide.

In this case we have

$$\frac{E_x}{H_y} = \frac{\frac{j\omega\mu}{\beta_c} H_0 \sin \frac{\pi y}{b} e^{-j\beta_g z} + \frac{j\omega\mu}{\beta_c} H_0' \sin \frac{\pi y}{b} e^{+j\beta_g z}}{\frac{j\beta_g}{\beta_c} H_0 \sin \frac{\pi y}{b} e^{-j\beta_g z} - \frac{j\beta_g}{\beta_c} H_0' \sin \frac{\pi y}{b} e^{+j\beta_g z}}$$

$$\frac{E_x}{H_y} = \frac{\omega\mu}{\beta_g} \frac{1 + \frac{H_0'}{H_0} e^{j2\beta_g z}}{1 - \frac{H_0'}{H_0} e^{j2\beta_g z}}.$$

If there is no propagation in the negative direction, this becomes

$$\frac{E_x}{H_y} = Z_0 = \frac{\omega\mu}{\beta_g}.$$

3. For non-perfectly conducting walls the propagation coefficient is

$$\gamma = \alpha + j\beta_g,$$

$$\alpha \text{ dB/m} = 8{\cdot}686 \; \alpha \text{ Np/m}.$$

With the values given

$$\alpha \text{ Np/m} = \frac{0{\cdot}3}{8{\cdot}686} = 0{\cdot}034 \text{ Np/m}.$$

Also $\beta_g = 2\pi/\lambda_g$ and

$$\frac{1}{\lambda_g^2} = \frac{1}{\lambda^2} - \frac{1}{\lambda_c^2}, \qquad \lambda_c = 2b.$$

We find $\lambda_g = 4{\cdot}86$ cm at 9000 MHz, or

$$\beta_g = \frac{2\pi}{\lambda_g} = 1{\cdot}29 \times 10^2 \text{ rad/m}.$$

The attenuation depends on the frequency and increases at the limits of the waveguide band which, for the dimensions given, is limited to 8200 to 12 400 MHz.

4. The equation for impedance transformation as used in the lines is applicable here. For the reduced impedance

$$z(l) = \frac{z_L + \tanh \gamma l}{1 + z_L \tanh \gamma l}.$$

Here $z_L = 0$, so that

$$z(l) = \tanh \gamma l = \tanh (\alpha + j\beta_g) l$$

$$= \frac{\tanh \alpha l + j \tan \beta_g l}{1 + j \tanh \alpha l \tan \beta_g l}$$

$$\alpha l = 0{\cdot}034 \times 2 \times 10^{-2} = 6{\cdot}8 \times 10^{-4},$$

from which

$$\tanh \alpha l \simeq \alpha l.$$

Thus

$$z(l) = \frac{\alpha l + j \tan \beta_g l}{1 + j\alpha l \tan \beta_g l}.$$

5. For $\beta_g l = p\pi$, with p integral, there will be resonance as $z(l) = \alpha l$.

For $\beta_g l = (2p'+1)\dfrac{\pi}{2}$, $z(l) = \dfrac{1}{\alpha l}$ and there will be antiresonance. These two conditions are

$$\lambda_g = \frac{2l}{p} \begin{cases} p = 1, & \lambda_g = 2l = 4 \text{ cm}; \\ p = 2, & \lambda_g = l = 2 \text{ cm}; \end{cases}$$

$$\lambda_g = \frac{4l}{2p'+1} \begin{cases} p' = 0, & \lambda_g = 4l = 8 \text{ cm}; \\ p' = 1, & \lambda_g = \dfrac{4l}{3} = 2{\cdot}67 \text{ cm}. \end{cases}$$

The bandwidth 8200 to 12 400 MHz corresponds to 2·84 cm $< \lambda_g <$ 6·087 cm. Only $\lambda_{g0} = 4$ cm is included in this range and this corresponds to

$$\lambda_0 = \frac{\lambda_{g0}}{\sqrt{1+\left(\frac{\lambda_{g0}}{\lambda_c}\right)^2}} \simeq 3 \text{ cm}.$$

This gives

$$f_0 = 10\,000 \text{ MHz}$$

and

$$\beta_{g0} = \frac{2\pi}{\lambda_{g0}} = 1{\cdot}57 \times 10^2 \text{ rad/m}.$$

6. For $f = f_0$, $\beta_{g0} l = \pi$.

In the neighbourhood of f_0 we have $\beta_g = \beta_{g0} + \mathrm{d}\beta_g$ and so

$$\beta_g l = \pi + \mathrm{d}(\beta_g l)$$
$$\tan(\beta_g l) = \tan[\pi + \mathrm{d}(\beta_g l)] \simeq \mathrm{d}(\beta_g l).$$

The transferred impedance is here

$$z(l) = \frac{\alpha l + \mathrm{j}\,\mathrm{d}(\beta_g l)}{1 + \mathrm{j}\alpha l\,\mathrm{d}(\beta_g l)} \approx \alpha l + \mathrm{j}\,\mathrm{d}(\beta_g l).$$

Now

$$\beta_g^2 + \beta_c^2 = \omega^2 \varepsilon_0 \mu_0 = \beta_0^2;$$

and, on differentiating,

$$2\beta_g\,\mathrm{d}\beta_g = 2\beta_0\,\mathrm{d}\beta_0,$$

from which

$$\mathrm{d}\beta_g = \frac{\lambda_{g0}}{\lambda_0}\frac{\mathrm{d}\omega}{c_0} = 2\pi\frac{\lambda_{g0}}{\lambda_0^2}\frac{\mathrm{d}\omega}{\omega_0}$$

and

$$\mathrm{d}(\beta_g l) = 2\pi\frac{\lambda_{g0}}{\lambda_0^2}\frac{\mathrm{d}\omega}{\omega_0}\,l \quad \text{with} \quad l = \frac{\lambda_{g0}}{2}.$$

Finally

$$\mathrm{d}(\beta_g l) = \pi\left(\frac{\lambda_{g0}}{\lambda_0}\right)^2\frac{\mathrm{d}\omega}{\omega_0}$$

and

$$z(l) = \alpha l + \mathrm{j}\pi\left(\frac{\lambda_{g0}}{\lambda_0}\right)^2\frac{\mathrm{d}\omega}{\omega_0}.$$

7. If the source is decoupled or isolated then, in the plane of the iris, it will give an equivalent impedance Z_0. The equivalent circuit is shown in Fig. 4 where

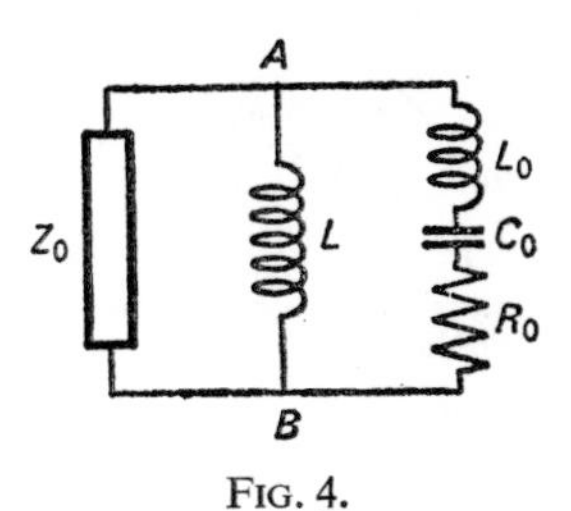

FIG. 4.

we have $(\omega L/Z_0)^2 \ll 1$. Then, the combined impedance of the source and the iris is given by

$$\frac{1}{Z_0}+\frac{1}{j\omega L}=\frac{1}{Z},$$

from which

$$Z=\frac{j\omega L\,Z_0}{Z_0+j\omega L}=j\omega L\cdot\frac{1}{1+\dfrac{j\omega L}{Z_0}},$$

or

$$Z=j\omega L\,\frac{\left(\dfrac{1-j\omega L}{Z_0}\right)}{\dfrac{1+\omega^2L^2}{Z_0}}\simeq\frac{\omega^2L^2}{Z_0}+j\omega L.$$

The total resistance is

$$R_s=R_0+\frac{\omega^2L^2}{Z_0}.$$

In the neighbourhood of resonance the impedance of the series resonance circuit is

$$Z=R_0\left(1+2jQ_0\frac{d\omega}{\omega_0}\right)$$

which must be identical to

$$Z(l)=Z_0\left[\alpha l+j\pi\left(\frac{\lambda_{g0}}{\lambda_0}\right)^2\frac{d\omega}{\omega_0}\right].$$

From this we find

$$R_0=Z_0\alpha l$$

$$Q_0=\pi Z_0\left(\frac{\lambda_{g0}}{\lambda_0}\right)^2\times\frac{1}{2R_0}=\frac{\omega_0L_0}{R_0};$$

and hence

$$L_0 = \frac{\pi}{\omega_0} \frac{Z_0}{2} \left(\frac{\lambda_{g0}}{\lambda_0}\right)^2,$$

$$C_0 = \frac{1}{\omega_0^2 L_0}.$$

Numerical example with $\alpha l = 6{\cdot}8 \times 10^{-4}$:

$$Q_0 = \frac{\pi}{2} \cdot \frac{1}{\alpha l} \left(\frac{\lambda_{g0}}{\lambda_0}\right)^2 = \frac{\pi}{2} \frac{1}{6{\cdot}8 \times 10^{-4}} \left(\frac{4}{3}\right)^2 \simeq 4100.$$

$$L_S = L + L_0,$$

$$C_S = C_0,$$

$$\omega_1 = \frac{1}{\sqrt{(L_0+L)C_0}} = \frac{\omega_0}{\sqrt{1+L/L_0}}.$$

Now

$$\frac{L}{L_0} = \frac{\omega_0 L/Z_0}{\omega_0 L_0/Z_0} \ll 1,$$

since

$$\frac{\omega_0 L_0}{Z_0} = \frac{\pi}{2} \left(\frac{\lambda_0}{\lambda_{g0}}\right)^2 = 2{\cdot}8.$$

From this we see that, near to ω_0,

$$\omega_1 = \omega_0 \left(1 - \frac{L}{2L_0}\right).$$

Further

$$Q = \frac{\omega_0(L_0+L)}{R_0+\omega^2 L^2/Z_0} \simeq \frac{\omega_0 L_0}{R_0+\omega^2 L^2/Z_0} = \frac{\frac{\pi}{2}\left(\frac{\lambda_{g0}}{\lambda_0}\right)^2}{\alpha l + (\omega L/Z_0)^2}.$$

Then

$$\frac{1}{Q} = \frac{\alpha l}{\frac{\pi}{2}\left(\frac{\lambda_{g0}}{\lambda_0}\right)^2} + \frac{(\omega L/Z_0)^2}{\frac{\pi}{2}\left(\frac{\lambda_{g0}}{\lambda_0}\right)^2} = \frac{1}{Q_0} + \frac{1}{Q_E}, \tag{1}$$

Q_E being the external Q-factor arising from the coupling. Numerical example:

$$Q_E = \frac{2{\cdot}8}{4} 10^{+4} = 7000,$$

from which

$$Q = 2580.$$

8.
$$z(l) = \alpha l + \mathrm{j}d(\beta_g l)$$
$$= \alpha l + \mathrm{j}\pi\left(\frac{\lambda_{g0}}{\lambda_0}\right)^2 \frac{\omega_1-\omega_0}{\omega_0},$$

with
$$\frac{\omega_1-\omega_0}{\omega_0} = -\frac{L}{2L_0}.$$

Then
$$z(l) = \alpha l - \mathrm{j}\pi\left(\frac{\lambda_g}{\lambda_0}\right)^2 \frac{L}{2L_0}$$
$$= \alpha l - \frac{\mathrm{j}\omega_0 L}{Z_0},$$

with $\alpha l = 6{\cdot}8\times10^{-4}$, $\omega_0 L/Z_0 = 2\times10^{-2}$.

The admittance of the iris and the resonant circuit is
$$y_A = \frac{1}{\alpha l - \mathrm{j}\omega_0 L/Z_0} - \mathrm{j}\frac{Z_0}{\omega_0 L}$$
$$= \frac{\alpha l}{\alpha^2 l^2 + \omega_0^2 L/Z_0^2} + \mathrm{j}\frac{\omega_0 L/Z_0}{a^2 l^2 + \omega_0^2 L^2/Z_0^2} - \mathrm{j}\frac{Z_0}{\omega_0 L}.$$

As
$$(\omega_0 L/Z_0)^2 \gg a^2 l^2,$$

we obtain, finally,
$$y_A = \alpha l\left(\frac{Z_0}{\omega_0 L}\right)^2$$

and the reflection coefficient is
$$\Gamma_A = \frac{1-y_A}{1+y_A}.$$

Since the attentuation is small, the value of Γ_A will be almost constant along the length of the guide and measurements of Γ_A can be made at some distance from the iris.

9. The condition $\Gamma_A = 0$ implies $y_A = 1$, or
$$\alpha l = (\omega_0 L/Z_0)^2.$$

Reference to equation (1) shows this gives $Q_0 = Q_A$ and so
$$Q = Q_0/2.$$

Problem No. 24

CAVITY EQUIVALENT CIRCUITS

A cavity (C), which resonates at a frequency f_0, is coupled to a guide G as shown in Fig. 1. It is supplied by a matched klystron with a frequency variable between 8200 and 12 400 MHz. A standing wave carriage allows the determination of the voltage standing wave ratio and of the positions of the maxima and minima.

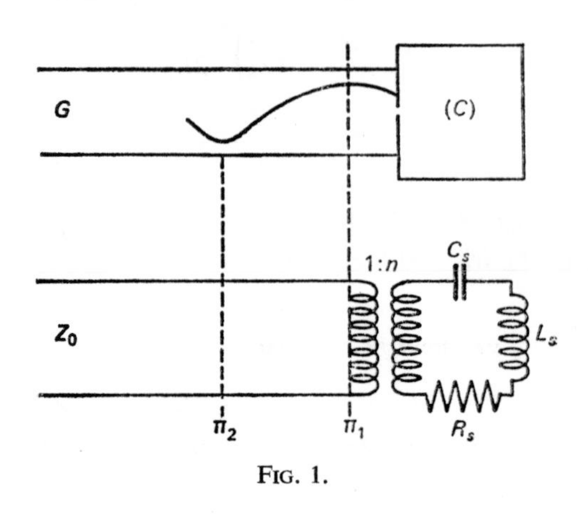

FIG. 1.

I

Series equivalent circuit.

The position of the first maximum of the voltage is found (plane π_1) for a frequency $f \gg f_0$.

1. Show that the equivalent circuit of Fig. 1 correctly represents the system to the right of π_1.

Put $\alpha = n^2 Z_0/R_s$, being the coupling coefficient,
$x = \omega/\omega_0 = f/f_0$, being the reduced frequency,
$Q_0 = \omega_0 L_s/R_s$, being the quality factor of the cavity.

Hence show that:

(a) in the neighbourhood of the resonance the equivalent circuit is a good description of the behaviour of the cavity, and explain the significance of the transformer;

(b) taking note of the magnitude of Q_0, the reduced impedance z_{π_1} transferred to the plane π_1 will result in a voltage maximum for $f \gg f_0$. Take $0 \cdot 1 < \alpha < 10$.

2. Trace on a Smith chart the locus of the reduced impedance in the plane π_1 with the frequency in the following three cases:

$\alpha > 1,$ cavity overcoupled,
$\alpha = 1,$ critical coupling,
$\alpha < 1,$ cavity undercoupled.

What is the value of S_0, the V.S.W.R. for resonance in each case?

3. Experimentally, it is easier to detect a node, or voltage minimum, than a maximum (i.e. for $f \gg f_0$, the plane π_2). What is the locus of z_{π_2}, the reduced impedance transferred to π_2 in the three cases given? What is the physical significance of α?

By studying the reflection coefficient in the plane π_2 for $f = f_0$ can you explain the phenomena which distinguish the cases $\alpha > 1$, $\alpha = 1$ and $\alpha < 1$?

4. With the probe placed at π_2 for $f \gg f_0$, the frequency is varied until $f = f_0$. It is observed that, at resonance, there is a voltage maximum at π_2. Is the cavity over- or under-coupled? Deduce a method of measuring the coupling coefficient.

II

Parallel equivalent circuit

1. Show that the reduced impedance seen in π_2 in question I.2 above is the same as for a parallel resonant circuit.

Show, therefore, that the equivalent circuit of Fig. 2 is a good representation of the system to the right of π_2.

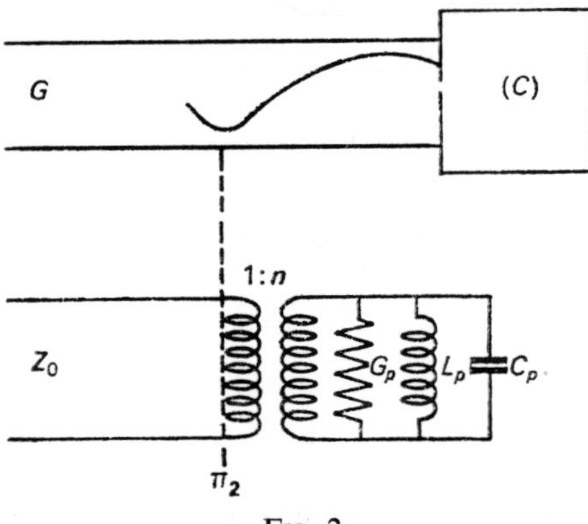

FIG. 2.

Express α and Q_0 in terms of the elements of the parallel circuit.

2. Give the definition of the Q-factor Q_L of the cavity and calculate this quantity for the two equivalent circuits.

III

Measurement of the unloaded Q-factor of the cavity.

1. What is the expression for S, the V.S.W.R. for f near f_0? (Put

$$2Q_0 \frac{\delta f}{f_0} = u, \qquad \delta f = f - f_0.)$$

2. Deduce:
the V.S.W.R. S_0, for $f = f_0$,
the V.S.W.R. S_1, for $f = f_1$
(where f_1 is such that $u = 1+\alpha$).

3. Express S_1 as a function of S_0.

4. To measure Q_0, the variation of $S(f)$ is found with frequency, this variation having the form of Fig. 3.

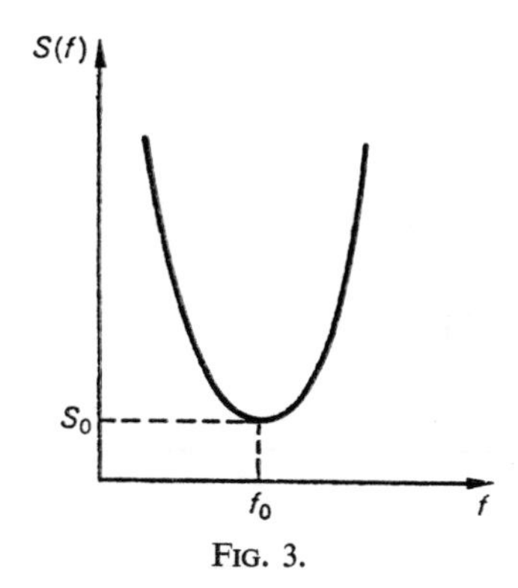

FIG. 3.

Show that S_1 can be found if S_0 is measured and hence Q_0 can be found, independently of the coupling.

Solution

I

1. (a) Since a resonant cavity is known to behave like a low-frequency resonance circuit, it is natural, in the neighbourhood of f_0, to use a representation with an L_s, C_s, R_s circuit and a transformer for the coupling and the guide up to the plane π_1. This is illustrated in Fig. 4. Also, we must have $\omega_0^2 L_s C_s = 1$.

(b) The impedance of the series resonant circuit is

$$Z_s = R_s + \mathrm{j}\left(\omega L_s - \frac{1}{\omega C_s}\right).$$

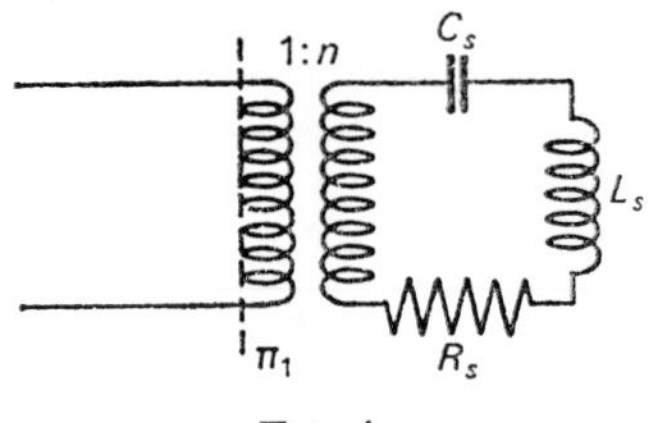

FIG. 4.

At the primary of the transformer this becomes

$$Z_{\pi_1} = \frac{Z_s}{n^2} = \frac{R_s}{n^2} + j\left(\frac{\omega L_s}{n^2} - \frac{1}{n^2 \omega C_s}\right),$$

which is equivalent to

$$Z_{\pi_1} = \frac{R_s}{n^2}\left[1 + jQ_0\left(x - \frac{1}{x}\right)\right].$$

In reduced impedances this is

$$z_{\pi_1} = \frac{R_s}{n^2 Z_0}\left[1 + jQ_0\left(x - \frac{1}{x}\right)\right] = \frac{1 + jQ_0(x - 1/x)}{\alpha}.$$

The Q-factors of resonant cavities are normally large, being in the range 1000 to 10 000. Consider $Q_0 = 1000$.

If the cavity resonates at $f_0 = 10\,000$ MHz and we take a frequency $f = 12\,000$ MHz, then

$$x = 1{\cdot}2, \qquad x - \frac{1}{x} \simeq 0{\cdot}4;$$

and

$$z_{\pi_1} = \frac{1 + j\,400}{\alpha} \simeq \frac{j\,400}{\alpha}.$$

Depending on the strength of the coupling, z_{π_1} will lie between j 40 and j 4000. Referring to the Smith chart it will be seen that such impedances are almost indistinguishable from the point $J(\infty + j\infty)$ and, as a consequence, will correspond to a voltage maximum. The same result is obtained if we calculate the reflection coefficient at π_1, i.e.

$$\Gamma = \frac{z_{\pi_1} - 1}{z_{\pi_1} + 1} = \frac{j\,40 - 1}{j\,40 + 1} = e^{2\psi} \quad \text{where } \tan\psi = 1/40,$$

so that

$$\Gamma \simeq 1.$$

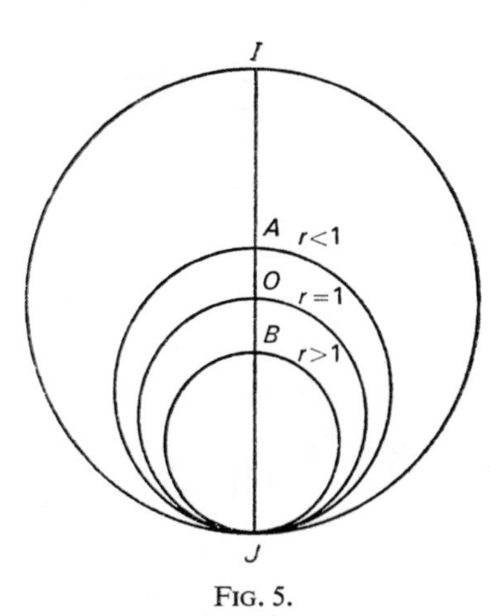

FIG. 5.

2. If the frequency varies, the locus of z_{π_1} is the circle $1/\alpha$ = constant (see the Smith chart in Fig. 5).

For

$$x \gg 1, \quad z_{\pi_1} \to j\frac{Q_0}{\alpha}\left(x-\frac{1}{x}\right) \quad \text{point } J, \quad x-\frac{1}{x} > 0;$$

$$x \ll 1, \quad z_{\pi_1} \to j\frac{Q_0}{\alpha}\left(x-\frac{1}{x}\right) \quad \text{point } J, \quad x-\frac{1}{x} < 0;$$

$$x = 1, \quad z_{\pi_1}(f_0) = \frac{1}{\alpha}.$$

There are three possible cases.

(a) $\alpha > 1$ (over-coupling), $z_{\pi_1}(f_0) = 1/\alpha < 1$, which is represented by the point A.

(b) $\alpha = 1$ (critical coupling), $z_{\pi_1}(f_0) = 1$ (the point 0).

(c) $\alpha < 1$ (under-coupling), $z_{\pi_1}(f_0) = 1/\alpha > 1$ (the point B).

When $\alpha > 1$, the S.W.R. at resonance is S_0 and the point A lies on the real axis in the region $0 < r < l$ and represents a minimum.

At a minimum

$$z_{\pi_1} = \frac{1}{S_0} = \frac{1}{\alpha},$$

from which $\alpha = S_0$.

For $\alpha = 1$, $S_0 = 1$.

For $\alpha < 1$, the point B represents a maximum and $z_{\pi_1} = S_0 = 1/\alpha$ so that $\alpha = 1/S_0$.

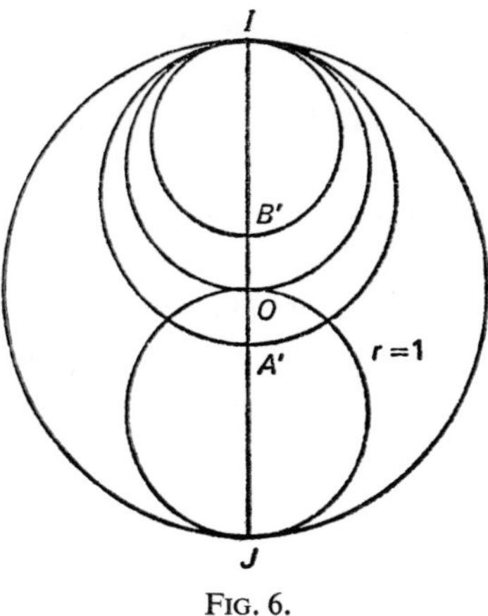

FIG. 6.

3. If we take π_2 as the reference plane with $\pi_1\pi_2 = \lambda g/4$, the loci of impedances are displaced *en bloc* by $\lambda_g/4$ towards the source. This corresponds to a half-circle of the chart as shown in Fig. 6 and

$$z_{\pi_2} = \frac{\alpha}{1+jQ_0(x-1/x)}.$$

For

$$x \gg 1, \quad z_{\pi_2} \to -j\frac{\alpha}{Q_0(x-1/x)} \quad \text{point } I, \quad x-\frac{1}{x} > 0.$$

$$x \ll 1, \quad z_{\pi_2} \to -j\frac{\alpha}{Q_0(x-1/x)} \quad \text{point } I, \quad x-\frac{1}{x} < 0.$$

$$x = 1, \quad z_{\pi_2} = \alpha.$$

For $\alpha > 1$, $z_{\pi_2}(f_0) > 1$ and is represented by the point A',

$\alpha = 1$, $z_{\pi_2}(f_0) = 1$ and is represented by the point 0,

$\alpha < 1$, $z_{\pi_2}(f_0) < 1$ and is represented by the point B'.

α represents the reduced impedance of the cavity transferred to the plane π_2.

$$\Gamma_{\pi_2} = \frac{\alpha-1}{\alpha+1} \quad \text{for} \quad f=f_0.$$

If $\alpha > 1$, $\Gamma = \dfrac{\alpha-1}{\alpha+1}$, the incident and reflected waves are in phase;

$\alpha = 1$, $\Gamma = 0$, we have a matched cavity;

$\alpha < 1$, $\Gamma = -\dfrac{1-\alpha}{1+\alpha}$, the reflected wave has a phase change of π.

4. For $f \gg f_0$ the impedance in the plane π_2 is practically zero (the point I on the chart); there is a voltage node at π_2.

If $f \to f_0$, the impedance is displaced around the circle $1/\alpha$ = constant and, depending on the particular case, will end at A', 0 or B'.

Only the point A' can correspond to a voltage maximum, the impedance at A' being real and greater than unity. In this case the cavity is overcoupled. (If, for $f = f_0$, there is a voltage minimum then the cavity is under-coupled.)

By finding the standing wave pattern at π_2 it is possible to find if the cavity is over- or under-coupled and, by measuring S_0, to find the value of α.

II

1. We have seen that

$$z_{\pi_2} = \frac{\alpha}{1+jQ_0\left(x-\dfrac{1}{x}\right)} = \frac{n^2 Z_0}{R_s+j\left(\omega L_s-\dfrac{1}{\omega C_s}\right)}.$$

This impedance clearly corresponds to an equivalent parallel resonant circuit. From Fig. 2 we get

$$Y_p = G_p+j\left(\omega C_p-\frac{1}{\omega L_p}\right),$$

which gives at the primary of the transformer

$$Y_{\pi_2} = n^2\left\{G_p+j\left(\omega C_p-\frac{1}{\omega L_p}\right)\right\}.$$

The corresponding reduced impedance is

$$z_{\pi_2} = \frac{Y_0/n^2}{G_p+j\left(\omega C_p-\dfrac{1}{\omega L_p}\right)},$$

which is seen to have the same form as z_{π_2} found from the series circuit. At the resonance we must have

$$\omega_0^2 L_s C_s = \omega_0^2 L_p C_p = 1$$

and

$$\frac{Y_0}{n^2 G_p} = \frac{n^2 Z_0}{R_s} = \alpha,$$

while

$$\frac{\omega_0 C_p}{G_p} = \frac{\omega_0 L_s}{R_s} = Q_0.$$

There is thus a dual representation of the cavity with the relations

$$L_s \to C_p, \qquad \alpha = \frac{Z_0}{\dfrac{R_s}{n^2}} \to \alpha = \frac{Y_0}{n^2 G_p},$$

$$C_s \to L_p, \qquad Q_0 = \frac{\omega_0 L_s}{R_s} \to Q_0 = \frac{\omega_0 C_p}{G_p}.$$

2. The loaded Q-factor is the quality factor when the cavity is coupled to the guide and takes account of the losses due to this coupling. In this case the generator gives an impedance Z_0 in the planes π_1 and π_2.

(a) Series circuit:

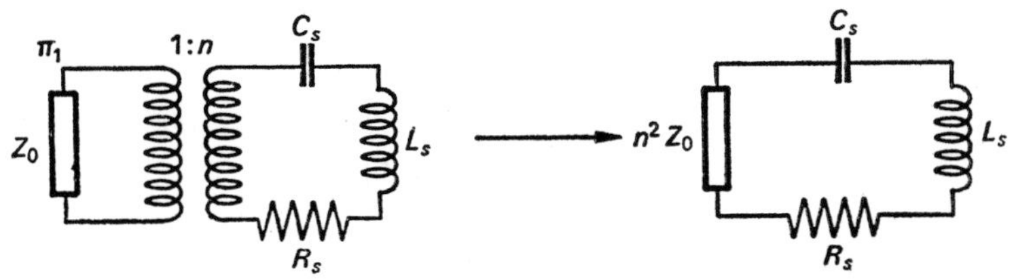

FIG. 7.

We have

$$Q_L = \frac{\omega_0 L_s}{n^2 Z_0 + R_s} = \frac{\omega_0 L_s}{R_s} \cdot \frac{1}{1 + n^2 \dfrac{Z_0}{R_s}} = \frac{Q_0}{1+\alpha}.$$

(b) Parallel circuit:

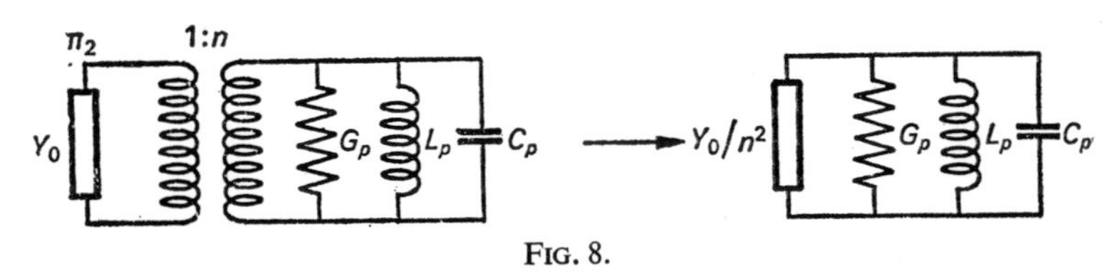

FIG. 8.

Here

$$Q_L = \frac{\omega_0 C_p}{G_p + \dfrac{Y_0}{n^2}} = \frac{\omega_0 C_p}{G_p} \cdot \frac{1}{1 + \dfrac{Y_0}{n^2} G_p} = \frac{Q_0}{1+\alpha}.$$

It is clear that, physically, the two representations are equivalent and give the same result.

III

Measurement of Q_0

1. At π_2, with f close to f_0, we have

$$z_{\pi_2} = \frac{\alpha}{1+jQ_0\left(x-\dfrac{1}{x}\right)} \simeq \frac{\alpha}{1+j2Q_0\dfrac{\delta f}{f_0}} = \frac{\alpha}{1+ju};$$

from which

$$\Gamma = \frac{z_{\pi_2}-1}{z_{\pi_2}+1} = \frac{\alpha-(1+ju)}{\alpha+1+ju},$$

$$|\Gamma| = \sqrt{\frac{(\alpha-1)^2+u^2}{(\alpha+1)^2+u^2}}$$

and

$$S = \frac{1+|\Gamma|}{1-|\Gamma|} = \frac{1+\sqrt{\dfrac{(\alpha-1)^2+u^2}{(\alpha+1)^2+u^2}}}{1-\sqrt{\dfrac{(\alpha-1)^2+u^2}{(\alpha+1)^2+u^2}}};$$

so that

$$S = \frac{\sqrt{(\alpha+1)^2+u^2}+\sqrt{(\alpha-1)^2+u^2}}{\sqrt{(\alpha+1)^2+u^2}-\sqrt{(\alpha+1)^2+u^2}}.$$

2. For $f = f_0$, $u = 0$, $S = S_0$ and

$$S_0 = \frac{\alpha+1+|\alpha-1|}{\alpha+1-|\alpha-1|}.$$

(a) $\alpha > 1$, $|\alpha-1| = \alpha-1$ and

$$S_0 = \frac{\alpha+1+\alpha-1}{\alpha+1-\alpha+1} = \alpha.$$

(b) $\alpha < 1$, $|\alpha-1| = 1-\alpha$ and

$$S_0 = \frac{\alpha+1+1-\alpha}{\alpha+1-1+\alpha} = \frac{1}{\alpha}.$$

The results are the same as those found previously.

For $f = f_1$, $u = 1+\alpha$ and

$$S_1 = \frac{(\alpha+1)\sqrt{2}+\sqrt{(\alpha+1)^2+(\alpha-1)^2}}{(\alpha+1)\sqrt{2}-\sqrt{(\alpha+1)^2+(\alpha-1)^2}} = \frac{1+\alpha+\sqrt{1+\alpha^2}}{1+\alpha-\sqrt{1+\alpha^2}}.$$

3. For an over-coupled cavity $\alpha = S_0$ while for an under-coupled cavity $\alpha = 1/S$.

$$(S_1)_{\substack{\text{over-}\\\text{coupled}}} = \frac{1+S_0+\sqrt{1+S_0^2}}{1+S_0-\sqrt{1+S_0^2}} = \frac{\dfrac{1}{S_0}+1+\sqrt{\dfrac{1}{S_0^2}+1}}{\dfrac{1}{S_0}+1-\sqrt{\dfrac{1}{S_0^2}+1}} = (S_1)_{\substack{\text{under-}\\\text{coupled}}}$$

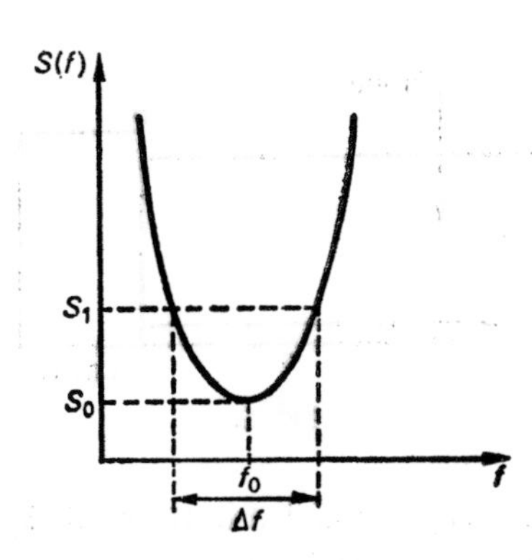

FIG. 9.

4. We trace the curve $S = S(f)$ of Fig. 9. From the value of S_0, S_1 is calculated and the line $S = S_1$ is drawn on the graph. The points at which the curve is cut give

$$\Delta f = 2\delta f \quad \text{such that} \quad u = 1+\alpha.$$

Then

$$Q_0 \frac{\Delta f}{f_0} = 1+\alpha.$$

From this we have

$$\frac{f_0}{\Delta f} = \frac{Q_0}{1+\alpha} = Q_L.$$

To find Q_0 it must be known whether the cavity is under- or over-coupled, i.e. whether $\alpha = S_0$ or $\alpha = 1/S_0$. We must look to find if the minimum which is at π_2 for $f \gg f_0$ moves through $\lambda_g/4$ at the resonance or not. In this way the uncertainty in α is removed.

Problem No. 25

INVESTIGATION OF THE RESONANT CAVITY

A resonant cavity is placed at the end of a microwave bench which works in the range 8200 to 12 400 MHz with waveguide mode TE_{01} and dimensions $a = 1{\cdot}06$ cm, $b = 2{\cdot}286$ cm.

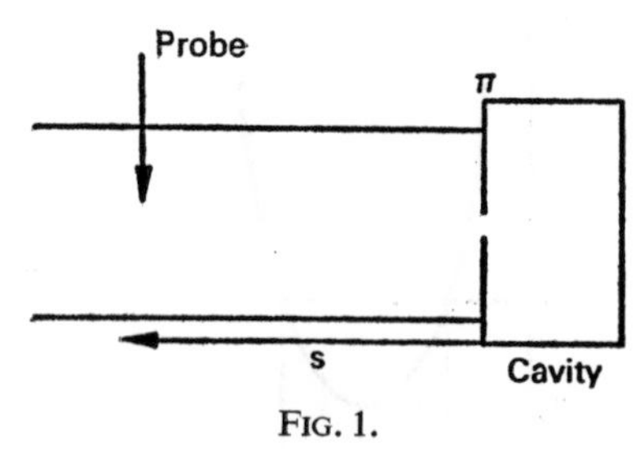

FIG. 1.

As shown in Fig. 1 the cavity is coupled to the guide by an iris and the plane π containing the iris is taken as the origin of co-ordinates.

The following measurements are made:

(a) With the frequency adjusted to the cavity resonance, the iris is replaced by a short-circuit. A voltage minimum is then found at $s_n = 12{\cdot}23$ cm and the adjacent voltage maximum is found at $s_{n+1} = 14{\cdot}34$ cm.

(b) The iris is replaced and, for different frequencies, the S.W.R., S, and the position of the minimum, noted above as s_n, are recorded. The results obtained are shown in the following table, where the frequencies are taken in ascending order:

	S	s (cm)
f_1	∞	13·79
f_2	5·8	13·48
f_3	2·9	13·36
f_4	1·4	13·06
f_5	3·2	12·07
f_6	16	11·85

1. What is the resonance frequency of the cavity?

2. Supposing that, for $f_1 < f < f_6$, the guide wavelength is practically constant, trace the locus of the cavity impedance in the plane π.

3. What length l of waveguide must be added to the assembly of iris and cavity so that the complete system of cavity, iris and length l of guide will behave as a series resonant circuit?

4. What is the coupling coefficient of the cavity?

5. If $f_3 = f_0 - 4$ MHz, what is the Q-factor of the unloaded cavity? What is its loaded Q-factor?

6. If the incident power is 10 mW what is the power absorbed by the cavity on resonance?

7. What is the stored energy in the cavity at resonance?

Solution. 1. With the line short-circuited

$$s_{n+1} - s_n = 14{\cdot}34 - 12{\cdot}23 = 2{\cdot}11 \text{ cm} = \frac{\lambda_g}{2},$$

from which $\lambda_g = 4{\cdot}22$ cm for the resonance frequency. Then

$$\frac{1}{\lambda_g^2} + \frac{1}{\lambda_c^2} = \frac{1}{\lambda_0^2}.$$

Since $\lambda_c = 2b = 4{\cdot}572$ cm, we have

$$f_0 = 9675 \text{ MHz}.$$

2. For each frequency of the table there will be a corresponding impedance in the plane π. To find this impedance we use the classical method. For example, take $f = f_3$ for which the minimum n is displaced by

$$\Delta s = s - s_n = 13{\cdot}36 - 12{\cdot}23 = 1{\cdot}13 \quad \textit{towards the source.}$$

The displacement Δs corresponds, on the Smith chart, to a displacement $\Delta s/\lambda_g = 1{\cdot}13/4{\cdot}22 = 0{\cdot}268$ towards the load. Starting at m_3 on the chart of Fig. 2 we arrive at C with $z \simeq 2{\cdot}6 + j\,0{\cdot}8$.

Repeating this construction for each frequency we obtain the locus of points $A, B, C, \ldots F$ which is the circle (C) centred on K. (Note that, for f_5 and f_6, the displacement on the chart is *towards the load.*) The results are summarized in the following table.

f	S	s (cm)	$\Delta s = s - s_n$	$\dfrac{\Delta s}{\lambda_g}$	Point on chart
f_1	∞	13·79	1·56	0·37	A
f_2	5·8	13·48	1·25	0·296	B
f_3	2·9	13·36	1·13	0·268	C
f_4	1·4	13·06	0·83	0·196	D
f_5	3·2	12·07	−0·16	−0·038	E
f_6	16	11·85	−0·28	−0·09	F

3. If we now move along the guide away from the cavity, the circle (C) turns as a whole towards the source. The reduced impedance of a series resonant circuit is of the form

$$z = r + jx(f),$$

r being independent of the frequency. The locus of such an impedance on the chart is a circle r = constant. The circle (C) will coincide with such a locus if it is turned through (***OK***, ***OJ***) $= 0{\cdot}25 - 0{\cdot}134$. Thus

$$\frac{l}{\lambda_g} = 0{\cdot}116, \quad \text{or} \quad l = 0{\cdot}116\lambda_g = 0{\cdot}49 \text{ cm}.$$

The circle (C) becomes the circle (C_1) which corresponds to $r = 0{\cdot}8$. The points A, B, etc. become the points A_1, B_1, etc.

4. If the cavity is represented by a series equivalent circuit, we know that the reduced impedance presented by the cavity is

$$z = \frac{1 + jQ_0\left(x - \dfrac{1}{x}\right)}{\alpha},$$

where α is the coupling coefficient, Q_0 the unloaded Q-factor and $x = f/f_0$. Here we have

$$1/\alpha = 0{\cdot}8, \quad \text{from which} \quad \alpha = 1{\cdot}25.$$

5. The point C corresponding to $f = f_3$ gives C_1 on the circle (C_1) such that

$$z = 0{\cdot}8 - j.$$

Now

$$f_3 \simeq f_0 = 9675 \text{ MHz}.$$

Therefore we have

$$z = 0{\cdot}8 - j \simeq \frac{1}{\alpha}\left[1 + j2Q_0\frac{f_3 - f_0}{f_0}\right],$$

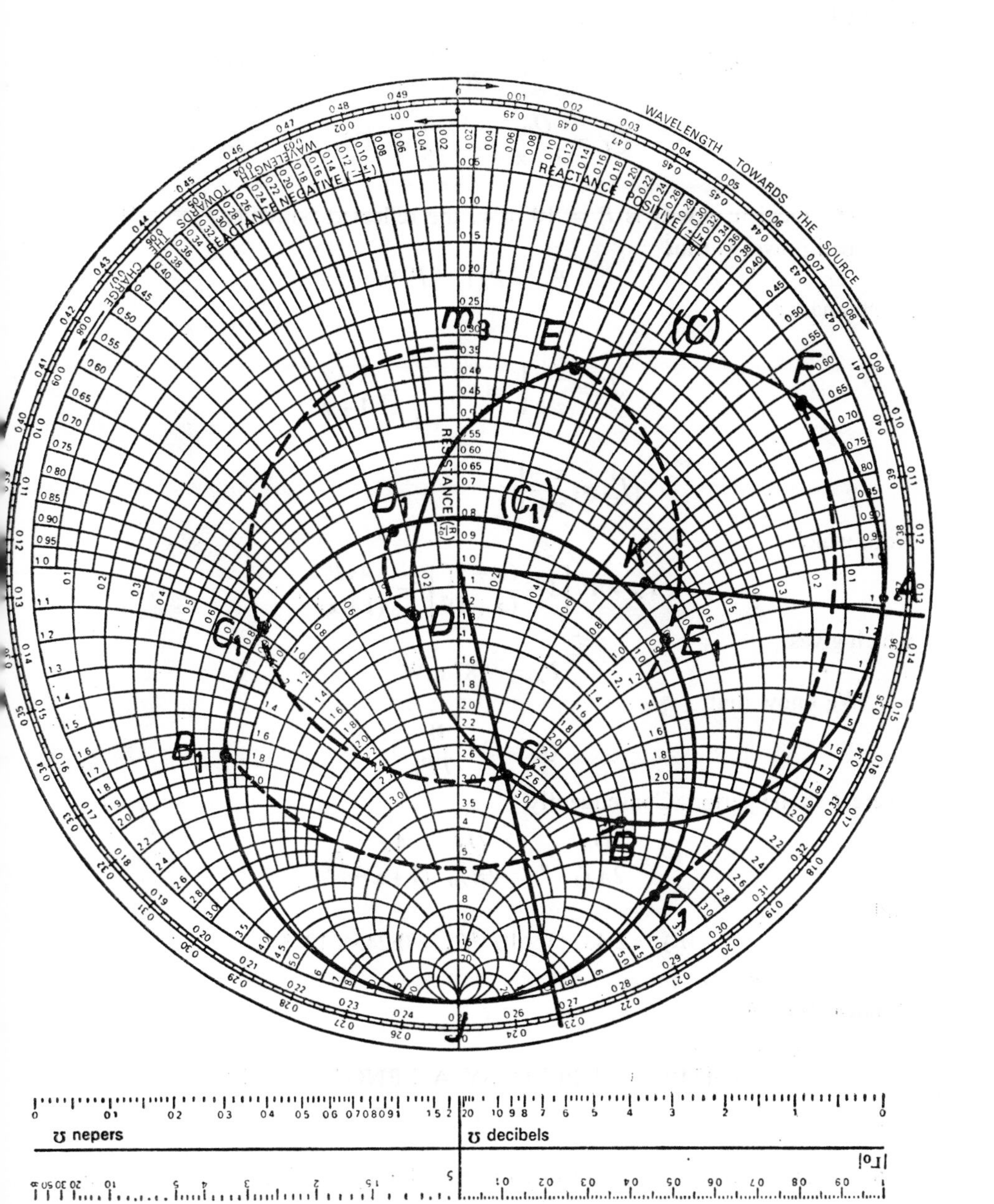
WAVELENGTH TOWARDS THE SOURCE
WAVELENGTH TOWARDS THE CHARGE
REACTANCE NEGATIVE
REACTANCE POSITIVE
RESISTANCE
m3
E
(C)
F
D1
(C1)
K
A
D
C1
E1
B1
C
B
F1
J
ʊ nepers
ʊ decibels

FIG. 2.

from which

$$2Q_0 \frac{f_0-f_3}{\alpha f_0} = 1 \qquad Q_0 = \frac{\alpha f_0}{2(f_0-f_3)} = \frac{\alpha\, 9675}{8} = 1150,$$

$$Q_L = \frac{Q_0}{1+\alpha} = \frac{1510}{2 \cdot 25} \simeq 670.$$

6. The incident power is $P_i = 10$ mW, the reflected power is $P_r = |\Gamma|^2 P_i$ and absorbed power is thus

$$P_a = P_i(1-|\Gamma|^2).$$

At resonance

$$\Gamma = \frac{z-1}{z+1} = \frac{\dfrac{1}{\alpha}-1}{\dfrac{1}{\alpha}+1} = \frac{1-\alpha}{1+\alpha},$$

$$|\Gamma| = \frac{\alpha-1}{\alpha+1};$$

from which

$$P_a = P_i\left[1-\frac{(\alpha-1)^2}{(\alpha+1)^2}\right] = \frac{4\alpha P_i}{(1+\alpha)^2},$$

which gives

$$P_a = 9 \cdot 87 \text{ mW}.$$

7. We know that

$$Q_0 = 2\pi \frac{W}{TP_a},$$

where T is the period. Consequently

$$W = \frac{Q_0}{2\pi f_0} P_a = \frac{Q_0}{2\pi f_0} \frac{4\alpha}{(1+\alpha)^2} P_i$$

and

$$W_J = 2 \cdot 48 \times 1 \times 10^{-8} P_a = 0 \cdot 245 \times 10^{-9} \text{ J}.$$

Problem No. 26

CAVITIES COUPLED BY A LENGTH OF LINE

I

A resonant cavity is coupled to two identical guides which propagate the same electromagnetic mode, for which $Z_0 = 1/G_0$ is the wave impedance.

The equivalent parallel circuit of this cavity is given in Fig. 1: the parallel resonant circuit and the perfect transformers T_1 and T_2 represent, respectively, the cavity and the coupling irises. The waveguides are represented by lines of

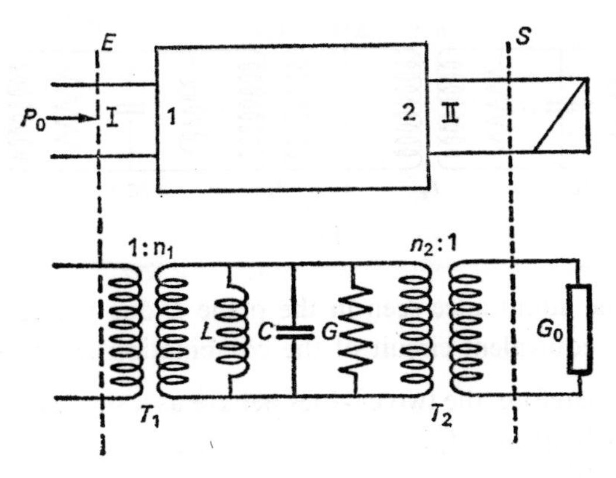

FIG. 1.

characteristic conductance G_0. The microwave source for the system is placed at the end of guide I. It is matched (having an internal conductance G_0) and gives a constant incident power P_0.

The waveguide II is terminated by a matched load of conductance G_0.

(a) Calculate the admittance seen at the input plane E and hence deduce the equivalent circuit of the cavity and load in this plane.

(b) Calculate the loaded Q-factor, Q_L, for the cavity (the unloaded Q-factor being Q_0).

Give expressions for the external Q-factors Q_{E_1} and Q_{E_2} in terms of the parameters of the equivalent circuit.

(c) In the case where the coupling irises are identical ($n_1 = n_2 = n$), give the expression for Q_L in terms of Q_0 and the coupling coefficient $\alpha = G_0/n^2G$. Calculate the power T, which is dissipated in the load G_0 on resonance, in terms of P_0 and α.

(d) In the case where the losses in the cavity can be neglected relative to the coupling losses (i.e. $G = 0$) give the equivalent circuit of the system in the plane E at some frequency away from resonance ($n_1 = n_2 = n, f \neq f_0$). How does Q_L change? Give an expression for Q_L in terms of Q_E.

Calculate the power T_1 dissipated in G_0 in terms of the frequency by writing

$$u = \left(\frac{\omega}{\omega_0} - \frac{\omega_0}{\omega}\right) Q_L.$$

Sketch the curve $T_1(u)/P_0$.

II

Two identical cavities (C) and (C') are placed in series, the output plane of (C) coinciding with the input plane of (C'). The equivalent circuit of this new system is given in Fig. 2, the coupling irises being taken to be identical.

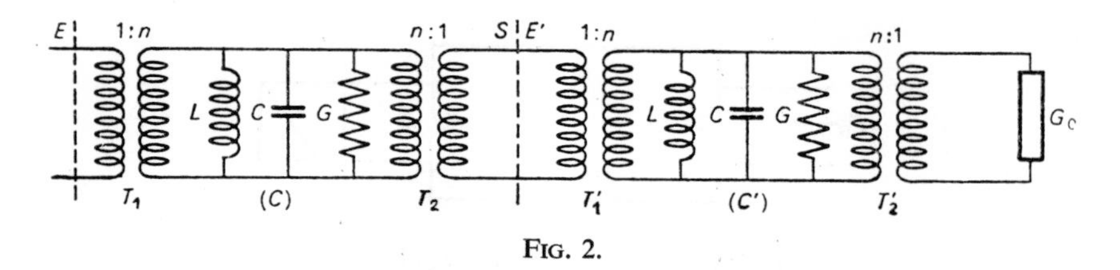

FIG. 2.

(a) Calculate the admittance seen in the plane E due to the load and the two cavities. Give the equivalent circuit of the combination in this plane.

(b) Show that, therefore, the two cavities behave as a single cavity and determine
- the resonant frequency f_0',
- the Q-factor Q_L'.

(c) Calculate the power T', dissipated in the load G_0 at resonance, in terms of P_0 and α. Compare this power with T.

(d) Give the equivalent circuit of the system in the plane E for $f \neq f_0$ and for negligible internal losses in the cavities.

Explain how Q_L' changes and give an expression for Q_L' in terms of Q_0. In this case also, calculate the power $T_2(u)$ dissipated in the load G_0, where u is defined in $I(d)$ above.

Sketch $T_2(u)/P_0$ on the same graph as $T_1(u)/P_0$ and draw a conclusion.

III

The preceding two cavities are placed in series but with the input plane of (C') at a distance $d = \lambda_g/4$ from the output plane of (C). The guide between the two cavities has a characteristic conductance G_0 and the distance d is taken as $\lambda_g/4$ even when the frequency varies. The equivalent circuit of the system is shown in Fig. 3.

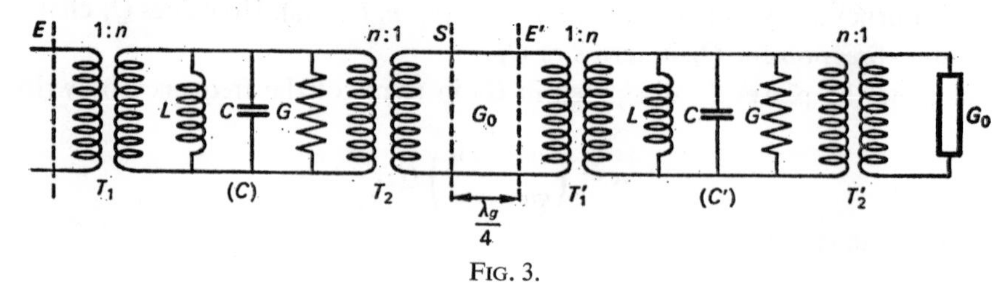

FIG. 3.

(a) Calculate the admittance seen in the input plane E' due to the load and the cavity (C').

(b) Hence deduce the admittance seen in the output plane S of the cavity (C).

(c) Calculate the admittance seen in the plane E.

Show that the equivalent circuit of the system, in the plane E, is composed of a parallel resonant circuit (G_p, C_p, L_p) in parallel with a series circuit (R_s, C_s, L_s) for which the elements are to be calculated (see Fig. 4).

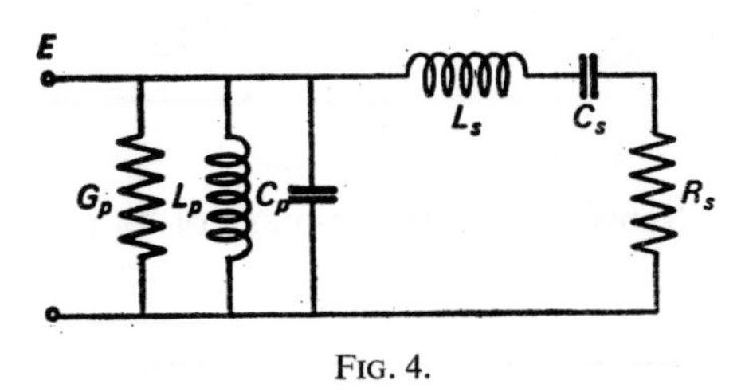

FIG. 4.

(d) Taking the microwave source to be matched to the system, find the Q-factor of the system, Q_L''.

(e) Modify the circuit of Fig. 4 for the case where the internal cavity losses are negligible. How does Q_L'' change? Express Q_L'' in terms of Q_E.

Calculate for this case the power $T_3(u)$ dissipated in the load G_0 for $f \neq f_0$. Sketch $T_3(u)/P_0$ on the graph for $T_2(u)/P_0$ and $T_1(u)/P_0$ and draw a conclusion.

IV

Consider two identical cavities in series, separated by a distance $\lambda_g/4$ and having coupling irises differing from each other as indicated in the equivalent circuit of Fig. 5. The internal losses of the cavities are again negligible.

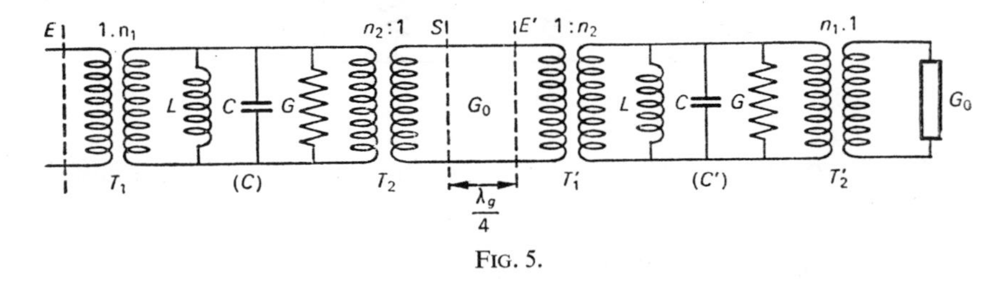

FIG. 5.

(a) Calculate the impedance seen in the plane E due to the complete system and give the equivalent circuit of the system in this plane. Express the elements of this equivalent circuit in terms of n_1, n_2, L, C and G_0.

(b) Give the expression for the reduced admittance in terms of

$$u = \frac{Q_{E_1}}{2}\left(\frac{\omega}{\omega_0} - \frac{\omega_0}{\omega}\right) \quad \text{and} \quad k = \left(\frac{n_1}{n_2}\right)^4.$$

(c) Calculate the power $T_4(u, k)$ dissipated in the load G_0. What is the condition for the system to be matched?

Sketch $T_4(u, k)/P_0$ for $k = 1/4$, 1 and 4 and draw a conclusion.

(Partial examination, Paris, 1966)

Solution

I

(a) The admittance transferred to the primary of T_2 is G_0/n_2^2. The equivalent circuit is thus:

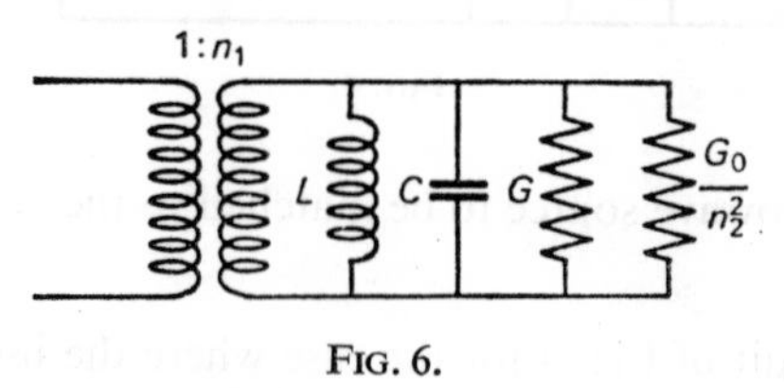

FIG. 6.

from which the admittance in the plane E is

$$Y = n_1^2\left[G + \frac{G_0}{n_2^2} + \mathrm{j}\left(C\omega - \frac{1}{\omega L}\right)\right]$$

which gives the equivalent circuit

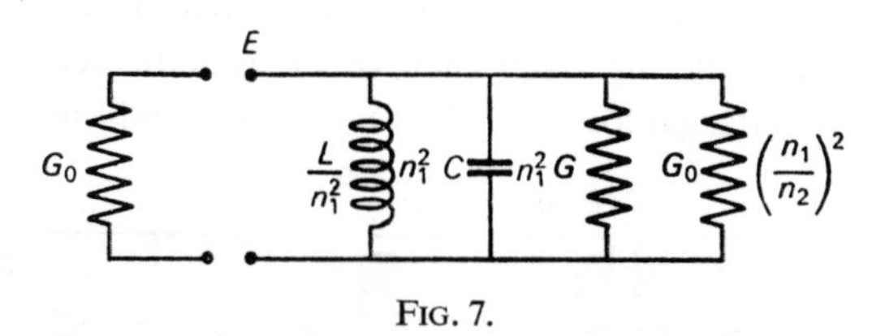

FIG. 7.

(b) The microwave source has an internal conductance G_0 which is transferred to E to give

$$Q_L = \frac{n_1^2\omega_0 C}{\left(\frac{n_1^2}{n_2^2}\right)G_0 + G_0 + n_1^2 G} = \frac{\omega_0 C}{G + \frac{G_0}{n_1^2} + \frac{G_0}{n_2^2}}.$$

Since $Q_0 = \omega_0 C/G$ we have

$$Q_L = \frac{Q_0}{1 + \frac{G_0}{n_1^2 G} + \frac{G_0}{n_2^2 G}}$$

and

$$\frac{1}{Q_L} = \frac{1}{Q_0} + \frac{1}{Q_{E_1}} + \frac{1}{Q_{E_2}} = \frac{G}{\omega_0 C} + \frac{G_0}{n_1^2\omega_0 C} + \frac{G_0}{n_2^2\omega_0 C},$$

from which

$$Q_{E_1} = \frac{n_1^2\omega_0 C}{G_0}, \qquad Q_{E_2} = \frac{n_2^2\omega_0 C}{G_0}.$$

(c) If $n_1 = n_2 = n$ and $\alpha = G_0/n^2G$

$$Q_E = Q_{E_1} = Q_{E_2} = \frac{n^2\omega_0 C}{G_0},$$

$$Q_L = \frac{Q_0}{1+2\alpha}.$$

At resonance $\omega_0^2 LC = 1$ and the equivalent circuit becomes

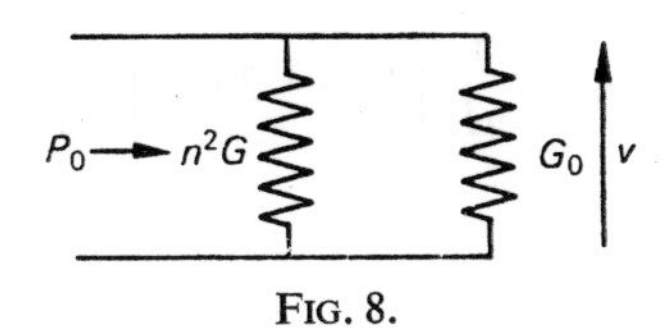

FIG. 8.

The reflection coefficient in the plane E is then

$$\Gamma_E = \frac{G_0 - n^2G - G_0}{G_0 + n^2G + G_0} = -\frac{n^2G}{2G_0 + n^2G} = -\frac{1}{1+2\alpha}.$$

The power actually transmitted to the system is

$$P = P_0[1 - |\Gamma_E|^2] = P_0\left[1 - \frac{1}{(1+2\alpha)^2}\right] = P_0\frac{4\alpha(\alpha+1)}{(1+2\alpha)^2},$$

since the transformers are assumed perfect. The powers dissipated in the various parts of the system can be assumed proportional to the power arriving at the primary of T_1 and to the conductances.

The power dissipated in the load G_0 is

$$T = P\cdot\frac{G_0}{n^2G + G_0},$$

since P and T are given by

$$P = \tfrac{1}{2}(n^2G + G_0)v^2, \qquad T = \tfrac{1}{2}G_0 v^2.$$

Thus

$$T = P_0\frac{4\alpha^2}{(1+2\alpha)^2}.$$

(d) If $G = 0$, the equivalent circuit for $f \neq f_0$ becomes that of Fig. 9.

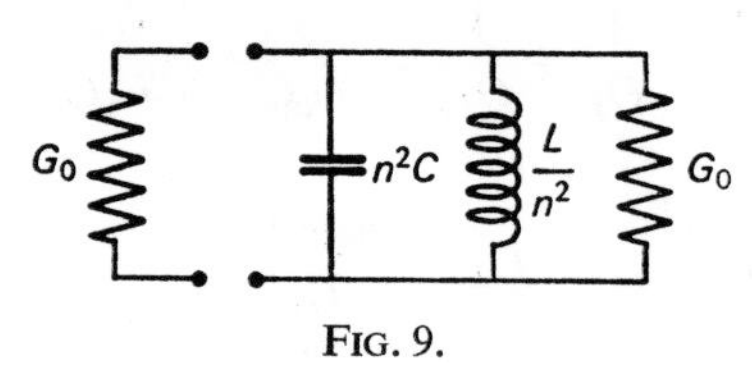

FIG. 9.

$$Q_L = \frac{n^2\omega_0 C}{2G_0} = \frac{Q_E}{2},$$

$$Y_E = G_0 + jn^2\left(\omega C - \frac{1}{\omega L}\right) \quad \text{and} \quad y_E = 1 + j\frac{n^2}{G_0}\left(\omega C - \frac{1}{\omega L}\right)$$

$$T_1 = P_0(1 - |\Gamma_E|^2), \quad \text{with} \quad \Gamma_E = \frac{1 - y_E}{1 + y_E}.$$

Now

$$y_E = 1 + j\frac{n^2}{G_0}\left(\omega C - \frac{1}{\omega L}\right) = 1 + 2jQ_L\left(\frac{\omega}{\omega_0} - \frac{\omega_0}{\omega}\right) = 1 + 2ju,$$

so that

$$\Gamma_E = \frac{-2ju}{2(1+ju)}.$$

We have thus

$$T_1 = \frac{P_0}{1+u^2}$$

and T_1/P_0 has the classical shape for a resonance curve.

II

(a) We have two transformers with ratios $1/n$ and n in series. The equivalent circuit thus simplifies to

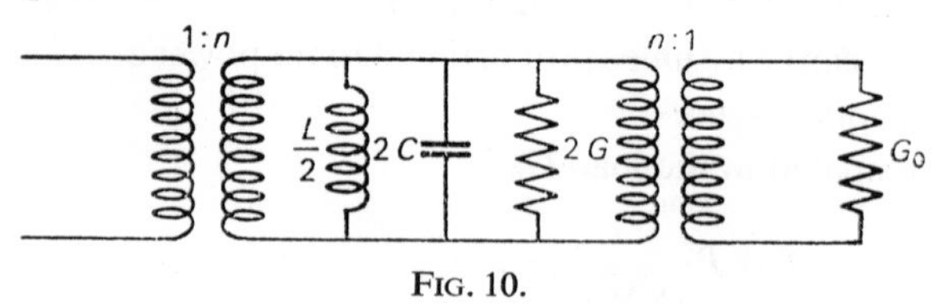

FIG. 10.

We thus have the case of $I(a)$ with $n_1 = n_2 = n$ from which

$$Y_E = n^2\left[2G + \frac{G_0}{n^2} + j2\left(\omega C - \frac{1}{\omega L}\right)\right]$$

corresponding to the equivalent circuit in the plane E.

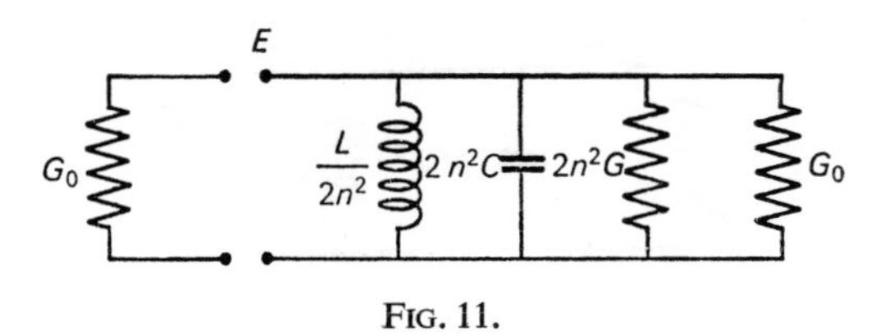

FIG. 11.

(b) The two cavities behave as a single cavity of resonance frequency $f_0' = f_0$ since

$$\omega_0'^2 2C \times \frac{L}{2} \equiv \omega_0^2 LC.$$

The corresponding loaded Q becomes

$$Q_L' = \frac{2n^2\omega_0 C}{2G_0+2n^2G} = \frac{Q_0}{1+\alpha} > Q_L.$$

(c) At resonance the equivalent circuit becomes

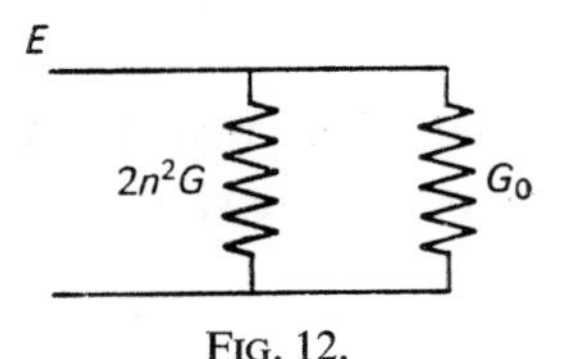

FIG. 12.

and

$$\Gamma_E = \frac{G_0 - 2n^2G - G_0}{2(G_0+n^2G_0)} = -\frac{1}{1+\alpha}.$$

The power transmitted to the system is

$$P = P_0[1-|\Gamma_E|^2] = P_0\frac{\alpha(\alpha+2)}{(1+\alpha)^2}.$$

The power dissipated in G_0 is

$$T' = P\frac{G_0}{2n^2G+G_0} = P\frac{\alpha}{\alpha+2},$$

from which

$$T' = P_0\frac{\alpha^2}{(\alpha+1)^2} < T.$$

(d) If $G = 0$ then, for $f \neq f_0$, the equivalent circuit in E becomes

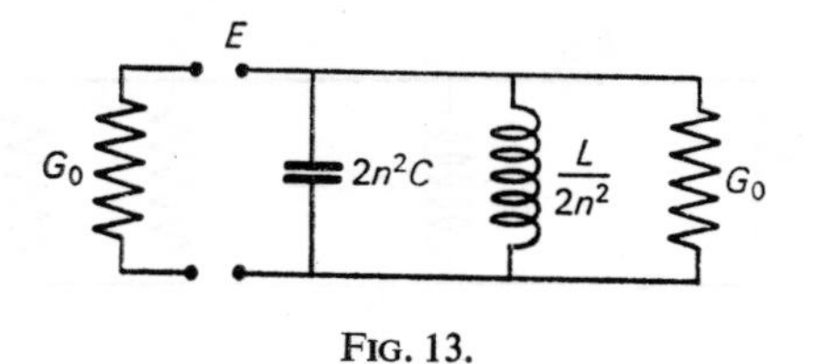

FIG. 13.

from which

$$Q'_L = \frac{2n^2\omega_0 C}{2G_0} = Q_E = 2Q_L,$$

$$y_E = 1+j\frac{2n^2}{G_0}\left(\omega C - \frac{1}{\omega L}\right) = 1+2jQ_E\left(\frac{\omega}{\omega_0}-\frac{\omega_0}{\omega}\right) = 1+4ju,$$

$$\Gamma_E = \frac{-4ju}{2(1+2ju)}, \quad \text{so that} \quad |\Gamma_E|^2 = \frac{4u^2}{1+4u^2}.$$

We obtain

$$\frac{T_2(u)}{P_0} = \frac{1}{1+4u^2}.$$

The curve $T_2(u)/P_0$ is more "selective" than $T_1(u)/P_0$, the quality factor being doubled. The forms of the two curves are shown in Fig. 14.

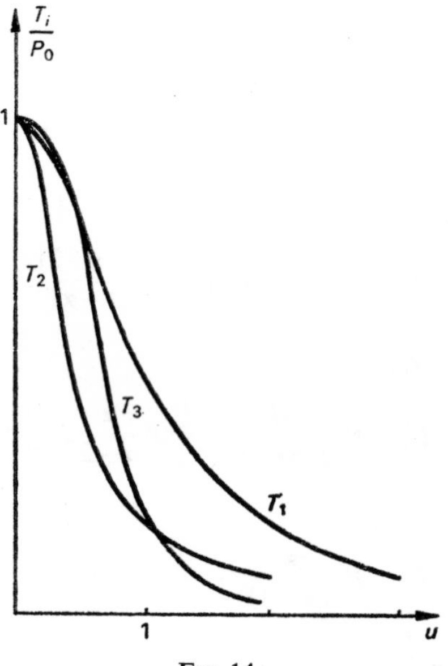

FIG. 14.

III

(a) In every case

$$Y_{E'} = n^2\left[G+\frac{G_0}{n^2}+j\left(\omega C-\frac{1}{\omega L}\right)\right].$$

(b) The quarter-wave line transforms $Y_{E'}$ to Y_S such that $Y_S Y_{E'} = G_0^2$ and so

$$Y_S = \frac{G_0^2}{Y_{E'}}.$$

(c) At the primary of T_2 there is thus an admittance

$$\frac{Y_S}{n^2} = \frac{G_0^2}{n^2 Y_{E'}}.$$

The total admittance at the secondary of T_1 is then

$$Y = G+j\left(\omega C-\frac{1}{\omega L}\right)+\frac{G_0^2}{n^2 Y_{E'}},$$

from which

$$\begin{aligned} Y_E &= n^2 Y \\ &= n^2\left[G+j\left(\omega C-\frac{1}{\omega L}\right)\right]+\frac{G_0^2}{n^2\left[G+\frac{G_0}{n^2}+j\left(\omega C-\frac{1}{\omega L}\right)\right]}, \end{aligned}$$

$$Y_E \equiv G_p+j\left(\omega C_p-\frac{1}{\omega L_p}\right)+\frac{1}{R_S+j\left(\omega L_S-\frac{1}{\omega C_S}\right)},$$

with

$$G_p = n^2 G, \qquad R_S = \frac{n^2 G}{G_0^2}+\frac{1}{G_0},$$

$$C_p = n^2 C, \qquad L_S = \frac{n^2 C}{G_0^2},$$

$$L_p = \frac{L}{n^2}, \qquad C_S = \frac{G_0^2}{n^2 L}.$$

(d) The Q-factor of the system is given by

$$\frac{1}{Q_L''} = \frac{1}{Q_S} + \frac{1}{Q_p},$$

$$Q_p = \frac{\omega_0 C_p}{G_p + G_0} = \frac{n^2 \omega_0 C}{n^2 G + G_0} = \frac{Q_0}{1+\alpha},$$

$$Q_S = \frac{\omega_0 L_S}{R_S} = \frac{n^2 \omega_0 C}{G_0^2 \left(\frac{n^2 G}{G_0^2} + \frac{1}{G_0}\right)} = \frac{Q_0}{1+\alpha},$$

from which

$$Q_L'' = \frac{Q_0}{2(1+\alpha)}.$$

(e) The circuit of Fig. 4 now becomes

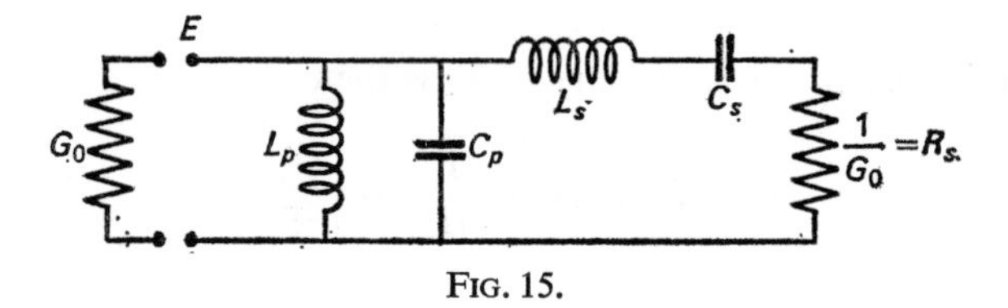

FIG. 15.

from which

$$Q_p = \frac{n^2 \omega_0 C}{G_0} = Q_E,$$

$$Q_s = Q_E;$$

and so

$$Q_L'' = \frac{Q_E}{2} = Q_L,$$

$$y_E = \mathrm{j}\frac{n^2}{G_0}\left(\omega C - \frac{1}{\omega L}\right) + \frac{G_0}{G_0 + \mathrm{j}n^2\left(\omega C - \frac{1}{\omega L}\right)} = 2\mathrm{j}u + \frac{1}{1+2\mathrm{j}u},$$

$$y_E = \frac{2\mathrm{j}u(1+2\mathrm{j}u)+1}{1+2\mathrm{j}u},$$

$$\Gamma_E = \frac{(1+2\mathrm{j}u)(1-2\mathrm{j}u)-1}{(1+2\mathrm{j}u)^2+1} = \frac{4u^2}{2-4u^2+4\mathrm{j}u},$$

$$|\Gamma_E|^2 = \frac{4u^4}{1+4u^4};$$

from which

$$\frac{T_3(u)}{P_0} = \frac{1}{1+4u^4}.$$

The curve obtained for T_3 (see Fig. 14) allows frequencies for which $u \gtrsim 1$ to be strongly attenuated, while frequencies giving $u \lesssim 0{\cdot}5$ are almost unaffected. It thus combines the advantages of both of the curves T_1/P_0 and T_2/P_0.

IV

(a) $$Y_{E'} = n_2^2\left[\frac{G_0}{n_1^2}+\mathrm{j}\left(\omega C-\frac{1}{\omega L}\right)\right] \quad \text{and} \quad Y_{S'} = \frac{G_0^2}{Y_{E'}}.$$

At the primary of T_2 we have

$$Y = \mathrm{j}\left(\omega C-\frac{1}{\omega L}\right)+\frac{G_0^2}{n_2^2 Y_{E'}},$$

from which, in the plane E,

$$Y_E = \mathrm{j}n_1^2\left(\omega C-\frac{1}{\omega L}\right)+\left(\frac{n_1}{n_2}\right)^2\frac{G_0^2}{n_2^2\left[\frac{G_0}{n_1^2}+\mathrm{j}\left(\omega C-\frac{1}{\omega L}\right)\right]}.$$

The equivalent circuit is identical to that of Fig. 15 with

$$L_p = \frac{L}{n_1^2}, \qquad C_p = n_1^2 C,$$

$$L_s = \frac{C}{G_0^2}\left(\frac{n_2}{n_1}\right)^4, \qquad C_s = G_0^2 L\left(\frac{n_1}{n_2}\right)^4,$$

$$R_s = \frac{1}{G_0}\left(\frac{n_2}{n_1}\right)^4.$$

(b) $$y_E = \mathrm{j}\frac{n_1^2}{G_0}\left(\omega C-\frac{1}{\omega L}\right)+\left(\frac{n_1}{n_2}\right)^4\frac{1}{1+\mathrm{j}\frac{n_1^2}{G_0}\left(\omega C-\frac{1}{\omega L}\right)},$$

$$y_E = 2\mathrm{j}u+\frac{k}{1+2\mathrm{j}u} = \frac{2\mathrm{j}u(1+2\mathrm{j}u)+k}{1+2\mathrm{j}u}.$$

(c) $$\Gamma_E = \frac{1-y_E}{1+y_E} = \frac{(1+2\mathrm{j}u)(1-2\mathrm{j}u)-k}{(1+2\mathrm{j}u)^2+k} = \frac{1+4u^2-k}{1-4u^2+k+4\mathrm{j}u},$$

$$\Gamma_E\Gamma_E^* = |\Gamma_E|^2 = \frac{(1+4u^2-k)^2}{(1-4u^2+k)^2+16u^2},$$

$$T_4(u, k) = P_0(1-|\Gamma_E|^2) = P_0 \frac{16u^2+(1-4u^2+k)^2-(1+4u^2-k)^2}{(1-4u^2+k)^2+16u^2}.$$

$$T_4(u, k) = \frac{4kP_0}{16u^4+8u^2(1-k)+(1+k)^2}.$$

There will be matching if

$$4k = 16u^4+8u^2(1-k)+(1+k)^2,$$

or

$$16u^4+8u^2(1-k)+(1-k)^2 = 0.$$

which gives

$$(4u^2+1-k)^2 = 0.$$

Hence

$$u = \pm\tfrac{1}{2}\sqrt{k-1}.$$

1. $k = \frac{1}{4}$; the roots are imaginary for all $k < 1$ and matching is impossible.
2. $k = 1$; gives $u = 0$ for matching. This is, in reality, part III.
3. $k = 4$ (and all $k > 1$); there are two real roots.

The curves for $T_4(u, k)/P_0$ shown in Fig. 16 are identical to those for coupled resonant circuits.

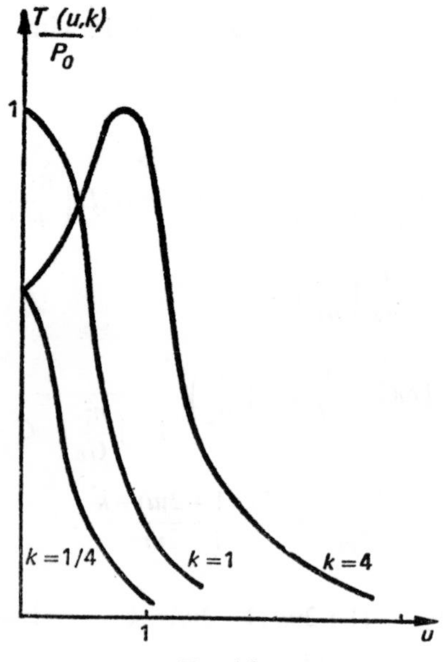

FIG. 16.

Index

Other Titles in the Series in Natural Philosophy

Vol. 1. DAVYDOV—Quantum Mechanics
Vol. 2. FOKKER—Time and Space, Weight and Inertia
Vol. 3. KAPLAN—Interstellar Gas Dynamics
Vol. 4. ABRIKOSOV, GOR'KOV and DZYALOSHINSKII—Quantum Field Theoretical Methods in Statistical Physics
Vol. 5. OKUN'—Weak Interaction of Elementary Particles
Vol. 6. SHKLOVSKII—Physics of the Solar Corona
Vol. 7. AKHIEZER *et al.*—Collective Oscillations in a Plasma
Vol. 8. KIRZHNITS—Field Theoretical Methods in Many-body Systems
Vol. 9. KLIMONTOVICH—The Statistical Theory of Non-equilibrium Processes in a Plasma
Vol. 10. KURTH—Introduction to Stellar Statistics
Vol. 11. CHALMERS—Atmospheric Electricity (2nd Edition)
Vol. 12. RENNER—Current Algebras and their Applications
Vol. 13. FAIN and KHANIN—Quantum Electronics, Volume 1—Basic Theory
Vol. 14. FAIN and KHANIN—Quantum Electronics, Volume 2—Maser Amplifiers and Oscillators
Vol. 15. MARCH—Liquid Metals
Vol. 16. HORI—Spectral Properties of Disordered Chains and Lattices
Vol. 17. SAINT JAMES, THOMAS and SARMA—Type II Superconductivity
Vol. 18. MARGENAU and KESTNER—Theory of Intermolecular Forces (2nd Edition)
Vol. 19. JANCEL—Foundations of Classical and Quantum Statistical Mechanics
Vol. 20. TAKAHASHI—An Introduction to Field Quantization
Vol. 21. YVON—Correlations and Entropy in Classical Statistical Mechanics
Vol. 22. PENROSE—Foundations of Statistical Mechanics
Vol. 23. VISCONTI—Quantum Field Theory, Volume 1
Vol. 24. FURTH—Fundamental Principles of Modern Theoretical Physics
Vol. 25. ZHELEZNYAKOV—Radioemission of the Sun and Planets
Vol. 26. GRINDLAY—An Introduction to the Phenomenological Theory of Ferroelectricity
Vol. 27. UNGER — Introduction to Quantum Electronics
Vol. 28. KOGA—Introduction to Kinetic Theory: Stochastic Processes in Gaseous Systems
Vol. 29. GALASIEWICZ—Superconductivity and Quantum Fluids
Vol. 30. CONSTANTINESCU and MAGYARI—Problems in Quantum Mechanics
Vol. 31. KOTKIN and SERBO—Collection of Problems in Classical Mechanics
Vol. 32. PANCHEV—Random Functions and Turbulence
Vol. 33. TALPE—Theory of Experiments in Paramagnetic Resonance
Vol. 34. TER HAAR—Elements of Hamiltonian Mechanics (2nd Edition)
Vol. 35. CLARKE and GRAINGER—Polarized Light and Optical Measurement
Vol. 36. HAUG—Theoretical Solid State Physics, Volume 1
Vol. 37. JORDAN and BEER—The Expanding Earth
Vol. 38. TODOROV—Analytical Properties of Feynman Diagrams in Quantum Field Theory
Vol. 39. SITENKO—Lectures in Scattering Theory
Vol. 40. SOBEL'MAN—Introduction to the Theory of Atomic Spectra